# 18th Chemnitz Seminar on Materials Engineering



## 18. Werkstofftechnisches Kolloquium

IOP Conference Series: Materials Science and Engineering
Volume 118

Chemnitz, Germany
10 - 11 March 2016

ISBN: 978-1-5108-2233-7
ISSN: 1757-8981

**Printed from e-media with permission by:**

Curran Associates, Inc.
57 Morehouse Lane
Red Hook, NY 12571

**Some format issues inherent in the e-media version may also appear in this print version.**



Printed by Curran Associates, Inc. (2016)



Institute of Physics
Dirac House, Temple Back
Bristol BS1 6BE UK

Phone: 44 1 17 929 7481
Fax: 44 1 17 920 0979

techtracking@iop.org

**Additional copies of this publication are available from:**

Curran Associates, Inc.
57 Morehouse Lane
Red Hook, NY 12571 USA
Phone: 845-758-0400
Fax: 845-758-2633
Email: curran@proceedings.com
Web: www.proceedings.com

# Table of contents

## Volume 118

**18th Chemnitz Seminar on Materials Engineering – 18. Werkstofftechnisches Kolloquium**
**10–11 March 2016, Chemnitz, Germany**

**Accepted papers received: 24 February 2016**
**Published online: 10 March 2016**

### Preface



### Papers

#### Invited Talk

**Joining and Brazing**



**Thermal Coatings**

**Composite Materials Material Compounds**

**Mechanical Testing Fatigue Damage**

**High-Strength Steels**

**Galvanic Deposition Anodization**



**Light Weight Metals Materials Development**

**High-strength Al-based light-weight materials for safety components**

**Merge Technologies for Multifunctional Lightweight Structures**

IOP Conf. Series: Materials Science and Engineering **118** (2016) 011001 doi:10.1088/1757-899X/118/1/011001

# Preface

The Institute of Materials Science and Engineering (Institut für Werkstoffwissenschaft und Werkstofftechnik, IWW) is a well-established part of the Faculty of Mechanical Engineering at Technische Universität Chemnitz, Germany. We are proud to host our traditional scientific meeting, the Chemnitz Seminar on Materials Engineering – Werkstofftechnisches Kolloquium, this year being the 18$^{th}$ Seminar.

The aim of our meeting is to bring together scientists and engineers both from academia and industry, and this is reflected by the different scientific sessions that cover various relevant topics, such as composite materials, microstructural analysis, and surface engineering. Special focus sessions highlight recent developments in the Federal Cluster of Excellence MERGE and in the Collaborative Research Center SFB 692, both established in Chemnitz.

We, as organizers, continuously strive to improve our Seminar, and we therefore decided to provide all contributors with the opportunity to publish their papers, in addition to the printed Proceedings volume, in an international, peer-reviewed journal. Publication in *IOP Conference Series: Materials Science and Engineering* allows for a wide dissemination of the many interesting and relevant results presented in our Seminar. Clearly, this is understood and appreciated by many participants of our meeting: more than 65 % of all authors chose to also publish their Proceedings papers in this special issue, which consequently provides an excellent overview of the main topics of our Seminar.

Chemnitz, March 2016

Prof. Thomas Lampke
Prof. Guntram Wagner
Prof. Martin F.-X. Wagner

IOP Conf. Series: Materials Science and Engineering **118** (2016) 011002 doi:10.1088/1757-899X/118/1/011002

# Characterisation of solidification processes during transient liquid phase bonding with rapidly solidified brazing ribbons from the melt spinning process

**K Bobzin[1], M Öte[1], S Wiesner[1], L Pongratz[1], J Mayer[2], A Aretz[2], R Iskandar[2] and A Schwedt[2]**

[1] RWTH Aachen, Institut für Oberflächentechnik, Kackertstr. 15, 52072 Aachen
[2] RWTH Aachen Gemeinschaftslabor für Elektronenmikroskopie, Ahornstr. 55, 52074 Aachen
pongratz@iot.rwth-aachen.de

**Abstract.** In technical joining processes, such as brazing, locally molten phases are generated in the workpiece, forming a metallurgical bond between filler metal and base material. Heat input, melting and solidification processes can thereby lead to shrinkage, mismatch and warpage. Furthermore, different process parameters, such as heating rate and holding time, influence the formation of the joining zone and especially the development of the microstructure. Due to the limited possibilities to influence these factors directly and by additional mutual interactions of these factors, the precision and reproducibility of the joints is reduced. As a consequence, rejects arise or post-production steps become necessary to remove defects, such as shrinkage, mismatch and warpage. Therefore, the aim of an interdisciplinary research group at the RWTH Aachen University is to achieve high precision in joined components by controlling and manipulating the factors determining the precision during the joining process. To reach this goal, in a first step the factors that influence precision are analysed.

The Transient Liquid Phase (TLP) bonding is a brazing method based on diffusion of braze metal constituents into the base material and of base metal constituents into the braze metal leading to isothermal solidification. The aim of this study is to analyse the timescale of the solidification and the diffusion processes of the constituents. For this purpose rapidly solidified Sn based braze ribbons from the melt spinning process were brazed at high vacuum with aluminium casting alloy for highly dynamically stressed components for automobile and engineering industry. Prior to the brazing experiments the microstructure of the braze ribbons was analysed by electron microscopy. The melting ranges were examined by differential scanning calorimetry (DSC). The braze seam was analysed by scanning electron microscopy with respect to the phase fractions in dependence of the composition of the ribbons.

IOP Conf. Series: Materials Science and Engineering **118** (2016) 011002 doi:10.1088/1757-899X/118/1/011002

# Flux-free brazing in a temperature range of 650 to 850 °C under SiH4-doped nitrogen

**U Holländer, F Weber, K Möhwald and H J Maier**

Institut für Werkstoffkunde, Leibniz Universität Hannover
hollaender@iw.uni-hannover.de

**Abstract.** Results of the public funded project "Development of processes for flux-free protective gas brazing in the temperature range of 650°C to 850°C by means of silane doped process gases" are presented.

In this project the material-based preconditions and the optimal process conditions for brazing relevant parent material combinations made from copper and steel alloys using conventional Cu- and Ag-based braze metals were determined, whereupon the doping of the process gas nitrogen with monosilane as extremely reducing agent enables the brazing without any flux. Brazing processes with silane-doped nitrogen in a conveyor belt furnace as well as in induction brazing under shielding gas were performed and assessed. Especially the latter is a promising alternative to flame brazing processes using flux, which are currently standard processes for industrial brazing in the named temperature range.

IOP Conf. Series: Materials Science and Engineering **118** (2016) 011002 doi:10.1088/1757-899X/118/1/011002

# Brazing of coated thin steel sheets under a metal inert protective gas with the use of Al based filler metal

**S Rentrop[1] and R Winkelmann[2]**

[1] Drahtwerk Elisental W. Erdmann GmbH & Co.
[2] BTU Cottbus - Senftenberg

**Abstract.** On the base of chemical-metallurgical studies and technological development an aluminium-based filler metal was designed with the objective of using it for construction of steel-steel connections. The, adverse to conventional CU-based filler metal, considerably lower joining temperature and the resulting advantages considering construction element geometry and distortion are presented. Additionally, a marked lower mass in the joining zone, whereby the ambitions in lightweight construction is supported, results. The brittle phases in the joining zone which are to be expected due to contact of liquid Al-filler metal with steel, are to be limited by regulated short arc processes in combination with alternating current technology. Metallographic examinations confirm that. Brazing parameters are determined for common sheet thicknesses up to 1,5 mm and common coating variants. Results on bridging and misalignment are presented. Brazing speed is dependent on the desired characteristics of the joint and is in the range of 0,30 to 1,2 m/min. A partially mechanized establishment of a junction is recommended. Investigations of employment of the developed filler metal are completed by construction of Aluminium-Aluminum-Joints and mixed joints Aluminium-Steel. Assessment of the newly developed alloy for the production of Aluminium-Aluminium-welded connections is performed in comparison to conventionally commercial off-the-shelf qualities like AlSi5(A) and AlMg5Cr(A). As assessment criteria, besides process stability, serve mechanically adjustable joint characteristics. Special attention is drawn to abrupt stress on the joints.

Applicability of the developed Al-basis-alloy for production of mixed joints Aluminium-Steel was tested and assessed. Technological parameters for their implementation are presented. Shear tensile strength up to ca. 200 MPa is registered.

IOP Conf. Series: Materials Science and Engineering **118** (2016) 011002 doi:10.1088/1757-899X/118/1/011002

# Brazed joints analyzed by an electrical resistance measurement as an alternative non-destructive testing method

**W Tillmann and N Sievers**

TU Dortmund, Institute of Materials Engineering, Germany
norman.sievers@tu-dortmund.de

**Abstract.** Brazing offers the possibility to connect materials such as cemented carbides and steel with fatigue strength of more than 300 MP. In this regard, this technology is used in a wide range of different applications to produce tools with a high performance for the cutting industry as for example sawblades or hammer drillers. It is necessary to use filler alloys containing a high amount of silver or copper to braze such materials.

If defective brazed joints occur, this can lead to a significant decrease of the quality, which may cause financial damage to the producing companies. A non-destructive quality control with approved methods such as ultrasonic or x-ray testing is very extensive and would increase the costs for the tools. In this regard, many companies do not perform such a precise quality control and rather risk producing failing products.

This paper presents the results of the research project “Qualification of the electrical resistance measurement for a non destructive testing of brazed joints”, which is supported by the German Federation of Industrial Research Associations (AiF, No. 18469 N/1). Electrical resistance measurement is a promising alternative to conventional non-destructive methods to analyze brazed joints. The 4-wire-technique enables a precise measurement of very low electrical resistances. Defects such as cracks and pores improve the resistance of the joint and as intermetallic phases have a lower electrical conductivity, they could be detected using this method. Finally, the limits and possibilities of this innovative method are discussed and some industrial applications presented.

IOP Conf. Series: Materials Science and Engineering **118** (2016) 011002 doi:10.1088/1757-899X/118/1/011002

# Arc brazing with thermo-mechanically pretreated Al-base weld deposits

**C Martens[1], T Klein[2], W Brück[2] and R Winkelmann[1]**

[1] BTU Cottbus - Senftenberg
[2] SKS Welding Systems GmbH

**Abstract.** The process of arc brazing with Al-base weld deposits has a strong presence in the manufacturing industry. Despite all this, the short life of the Al-base weld deposits by oxidation constitutes a strong problem. In addition, the forming oxide layer represents a very good storage for hydrogen, which in turn causes the pores in brazing. Purpose of the carried out development was to reduce the porosity of the compounds by a reduction and equalization of the oxide layer, to increase the lifetime and improve the reproducibility of the compounds.

To reach these objectives a compact wire cleaning and activation unit located next to the welding machine has been developed. The wire cleaning uses a thermo-mechanically transferred plasma jet, using the effect of the discharge of gas and metal ions. The surface of the wire becomes activated and the oxide layer as well as the hydrogen content is reduced. Particles formed meanwhile plasma cleaning on the wires surface are removed using a mechanical component.

The critical parameters of the wire cleaning are the current, the distance between the wire and tungsten electrode, the flow of shielding gas (argon) and the wire velocity. The current and wire velocity take influence on the energy transferred per track segment. The distance of the electrode has an influence on the shape and transmission of the energy of the plasma.

Due to wire cleaning and activation a significant reduction of number and size of pores in the compounds could be observed, resulting in an increased stability. Forming of smoulder by welding of Al-Mg-based wires got suppressed. The clogging of the welding nozzle got reduced, increasing its durability and decreasing the number of process interruptions. All that enables the process reliability.

IOP Conf. Series: Materials Science and Engineering **118** (2016) 011002 doi:10.1088/1757-899X/118/1/011002

# High speed soldering with flat wires – process development

**A Kilian[1], T Klein[2], W Brück[2] and R Winkelmann[1]**

[1] BTU Cottbus – Senftenberg
[2] SKS Welding Systems GmbH

**Abstract.** Flat wires are often used in welding technique to increase the joining velocity. In contrast there is a considerable effort for the haulage of such wires. Goal of the realized development of brazing technology is to earn the benefit of the high velocity and correcting the haulage problems.

Using 2-dimensional high-speed-video-footages and high-resolution U-I-t-measurements significant changes in the arcs structure have been detected meanwhile the shift from round to flat wires. They are interpreted to be casual for the high joining velocity (about 3 m/min) adjustable using flat wires. These data is completed by parameter studies on soldering using Al-flat wires. Checkpoints for the definition of the machines characteristics are represented. The soldered connections have been evaluated metallographical and their mechanical properties have been ascertained. One focus of the investigations described here has been the usability of flat wire for forming flanged seams with preferably small opening angle and big connection area. As an edge condition concave seam geometry has been defined for reduced rework.

Specifications for tolerances and the appearance of spatter are developed. Results for the forming of steel-steel-connections are available.

Process stability has been achieved using round wires which have been transformed into flat wires next to the arc and adapting the transfer of energy - the power contact cone. A new wire feeder unit is presented.

IOP Conf. Series: Materials Science and Engineering **118** (2016) 011002 doi:10.1088/1757-899X/118/1/011002

# Reactive air brazing (RAB) as joining technology for SOFC-applications

**V Wesling, J Hamje, A Lindermeir and L Dörrer**

Institut für Schweißtechnik und Trennende Fertigungsverfahren der TU Clausthal, Agricolastraße 2, D-38678 Clausthal-Zellerfeld

**Abstract.** High efficient energy conversion systems like solid oxide fuel cells (SOFC) have seen a growing demand during the last decades, because of the increasing energy consumption of the world and the shortening of fossil fuels. The centerpiece of a SOFC stack is the ceramic cell (MEA - Membrane Electrode Assembly) with its three functional layers: cathode, electrolyte, and anode. To prevent a direct reaction of the working gases, the gas compartments of the electrodes must be separated gastight from each other. The working temperature of the SOFC is normally 750° – 850°C, since the ceramic cell components show significant ion conductivity only from this temperature on upwards.

A new SOFC concept based on repeating units with two electrically parallel connected cells is presented. Two alternatives were developed: an all ceramic concept with a 3YSZ cell housing and a metallic housing made of Crofer 22 APU. Reactive air brazing (RAB) was used to join the MEAs in the frames. RAB-brazes are typically made of a silver based matrix and a reactive component, usually CuO, which improves the wetting of the ceramic. Reactive air brazing as a relatively novel brazing technique was specifically developed for application in high temperature electrochemical devices. Unlike the previous sealing techniques, based primarily on high-temperature glasses, the RAB joints exhibit excellent thermal cycling performance and long-term stability in both high-temperature oxidizing and reducing atmospheres.

Test units were produced and gas tightness of the joints could be confirmed. Tests with a two cell unit show good overall performance with a fuel utilization above 75 %. The assembly has been operated for 1.200 hours, including full redox and thermal cycles. The results of the RAB-joining confirm the feasibility of the approach with potential benefits concerning durability and production cost.

IOP Conf. Series: Materials Science and Engineering **118** (2016) 011002 doi:10.1088/1757-899X/118/1/011002

# Strength and leak testing of ceramic-steel-bonds (3YSZ-Crofer22APU) produced by means of diffusion brazing

**W Tillmann, C Schaak and N Sievers**

TU Dortmund, Institute of Materials Engineering, Germany
wolfgang.tillmann@udo.edu

**Abstract.** The production of highly efficient ceramic-metal-bonds for high temperature applications is still a challenge. Especially due to the steadily increasing demand for energy, combined with an increasing ecological consciousness, completely new applications are developed, for which ceramic-metal composites are needed. Such a noteworthy application is the Solid Ox-ide Fuel Cell (SOFC). For the mass production of SOFC, a suitable bonding technique and material system is required in order to bond the metallic interconnector (Crofer22APU) to the high temperature electrolyte (YSZ). This challenge is still subject to research. Operating tem-peratures of approximately 800 °C, a corrosive environment, as well as thermal cycling must be considered when selecting suitable joining techniques and materials. Currently, glass sol-ders are widely used for this application. However, glass brazes show a high brittleness; be-cause of this fact the service strength is low. Reactive partial diffusion brazing is an interesting alternative to glass brazes. Especially the ductile character of the metallic brazing zone offers distinct advantages compared to glass brazes. In the present study, 3YSZ was brazed to Crofer22APU by using zirconium and titanium as active elements, combined with nickel inter-layers. The active elements were deposited by means of physical vapor deposition (PVD). Different brazing process parameters were tested to achieve a bond between the materials. After brazing, the morphology was investigated with metallographic methods and ultrasonic testing. The shear strength was measured and correlated with the brazing parameters. Since the leak tightness of the brazing plays a significant role, leak tests were conducted with a vacuum test stand. The results of the experiments showed that the shear strength is clearly dependent on the temperature, the holding time, and the used active element. With zirconium as an active element, an average shear strength of 78 MPa was measured. In addition, it could be proved that the brazing zones are vacuum-tight if the connection between both sam-ples is uniform.

IOP Conf. Series: Materials Science and Engineering **118** (2016) 011002 doi:10.1088/1757-899X/118/1/011002

# Thermal spraying of oxide coatings using solution precursor thermal spraying

**L Pawlowski, R Candidato, I Gutkowski and R Kaczmarek**

SPCTS, UMR CNRS 7315, University of Limoges

**Abstract.** The presentation reviews the techniques of film and coatings deposition using solution as a feedstock. After a short description of the processes of: (i) spray pyrolysis; (ii) solution precursor plasma spray (SPPS); solution precursor flame spray (SPFS); and, (iv) solution precursor high velocity oxy-fuel (SPHVOF) some properties of solution feedstock are discussed. The properties include the chemical composition of solutions used to synthesize typical coatings and some physical ones, such as viscosity, which determine the flow of solution in a pipeline and its injection to plasma or flame. The chemical and physical phenomena occurring at spraying are described including atomization in gas, heating and vaporization of solution, precipitation of solution and formation of solid and heating of this solid. The molten solid is then splashed onto substrate and the phenomena at its formation are shortly discussed. Two examples of the oxide coatings prepared in our laboratory are presented: hydroxyapatite and Indium and tin oxides alloy known as ITO.

IOP Conf. Series: Materials Science and Engineering **118** (2016) 011002 doi:10.1088/1757-899X/118/1/011002

# Investigations of wear caused by solid particle impingement under high temperatures on nickel based hard facings reinforced with carbides

**V Wesling[1] and C Heet[2]**

[1] Technische Universität Clausthal, Clausthal-Zellerfeld
[2] ThyssenKrupp Steel Europe, Duisburg

**Abstract.** The presentation characterizes the tribological behaviour of nickel based alloys reinforced with carbides stressed under solid particle impingement and high temperatures. Influencing variables were the temperature, angle of impact, velocity of the particles and kind of particles. All hard facings were made by PTA-process. The investigations show that the quantity of wear depends on the temperature, the velocity of the particles, the kind of particles, the grain size of the particles and the hardness of the particles. The results of oxidation studies demonstrate that the reinforced nickel alloys already oxidize at low temperatures and therefore they lose the wear resistance. Based on this oxidation the chromium of the nickel based alloys is reduced and the passivation of the materials is lost. Because of the very complex interaction and the understanding of the influence of the parameters it is possible to develop new materials which are able to resist solid particle impingement wear combined with high temperatures under oxidizing atmosphere.

IOP Conf. Series: Materials Science and Engineering **118** (2016) 011002 doi:10.1088/1757-899X/118/1/011002

# Al enriched metal/ceramic interface under cyclic thermal load with dwell time

**I Ali, D Wett, T Grund, and T Lampke**

Materials and Surface Engineering Group, Institute of Materials Science and Engineering, Technische Universität Chemnitz, D – 09125 Chemnitz, Germany
Ibrahim.ali@mb.tu-chemnitz.de

**Abstract.** Ceramic O-stabilized zirconia (O-SZ) top coats (TC), where O most often stands for oxides of Y, Ca, Mg, La, Gd, Dy, and Ce and metallic M-CrAlX bond coats (BC), where M most often stands for a Co, Ni or CoNi base alloy and "X" represents alloying elements like Y, Hf, Si or Re have been investigated as metal/ceramic coating system for high temperature applications. During their service under combined heat and oxygen load, a reaction zone forms at the interface between the TC and the metallic BC. The reaction zone reveals formation of different thermally grown metal oxides (TGO) like (Cr,Al)2O3, (Ni,Co)(Cr,Al)2O4 and NiO. The physical characteristics and volume growth of TGOs lead to system failure. For the prepared M-CrAlY/N-SZ by atmospheric plasma spraying technique, it is proved that enriching the interface with α-Al2O3 is beneficial to an extended operational time by prolonging the steady-state growth stage of the TGO. The corresponding behavior in M-CrAlY/N-SZ systems with heightened Al activity at the BC/TC interface, however, is under study.

In the present study, thin Al layers were deposited by DC-magnetron sputtering on the raw surface of APS sprayed M-CrAlY bond coats. Afterwards, YSZ was sprayed on top. Cyclic load with 30 min dwell time at 1150 °C was conducted. The TGOs formation was investigated by laser scanning microscope, SEM and Raman spectroscopy after 80 cycles. The laser scanning microscopic and SEM analysis were made to choose right place and size of an interesting area of sample for energy dispersive spectroscopy (EDS) and Raman investigations. The designed metal/ceramic systems with Al interlayer (Al-TBC) were compared to reference metal/ceramic systems (R-TBC) concerning their interface characteristics and the oxides formation behavior. The interface enrichment with Al results in less formation of mixed oxides and high content of α-Al2O3 within the formed TGO and hence allows a higher service time of the Al-TBC systems in comparison to the R-TBC systems for the applied thermal load.

IOP Conf. Series: Materials Science and Engineering **118** (2016) 011002 doi:10.1088/1757-899X/118/1/011002

# Development of a new production method for manufacturing of hybrid parts from metal and plastics

**K Bobzin[1], M Öte[1], T F Linke[1], X Liao[1] C Hopmann[2], P Ochotta[2], M Steger[3] and A Gillner[3]**

[1] Institut für Oberflächentechnik (IOT), RWTH Aachen University, liao@iot.rwth-aachen.de
[2] Institut für Kunststoffverarbeitung (IKV), RWTH Aachen University, philipp.Ochotta@ikv.rwth-aachen.de
[3] Lehrstuhl für Lasertechnik (LLT), RWTH Aachen University, Fraunhofer-Institut für Lasertechnik (ILT), michael.steger@ilt.fraunhofer.de

**Abstract.** Increasing requirements for energy and resource efficiency promote the usage of lightweight materials. Moreover, hybrid parts made of different materials can provide a wider range of properties than parts from a single material. One of today's chal-lenges in production technology is the development of efficient manufacturing meth-ods. Within the scope of the Cluster of Excellence "Integrative Production Technolo-gy for High-Wage Countries", the so called In–Mould–Metal–Spraying (IMMS) is be-ing developed for manufacturing of metal/plastics hybrid parts. It is based upon the transfer of thermally sprayed coatings from the mold insert to the plastic coating car-rier during injection molding. A key factor of the IMMS process is the adjustment of the coating bond strength to the mold insert for a residue-free transfer. The surface topography and therefore the surface pretreatment of the mold insert are important for the IMMS process. In the preliminary study, a surface topography fabricated by glass bead blasting has enabled the transfer of wire arc sprayed Zn-coatings. How-ever, further investigations indicated limitations regarding the mold insert material and local coating residuals on the mold insert surface after coating transfer. For the close investigation of this effect, laser structuring is chosen as surface pretreatment method because of its capability of fabricating freely selectable, defined surface to-pographies. Surface structures in micron range overlaid by LIPSS (Laser Induced Periodic Surface Structures) in submicron range can be prepared using ultrashort pulse lasers. Different surface structures produced by regularly drilling on the surface and variation of the drilling diameter, depth and distance, which lead to the variation of the coating bond strength, are investigated. The aim of this investigation is to iden-tify a suitable surface topography and to understand the bonding mechanism regard-ing the IMMS process.

IOP Conf. Series: Materials Science and Engineering **118** (2016) 011002 doi:10.1088/1757-899X/118/1/011002

# Impact of the substrate topography on the tensile bond strength of thermal sprayed coatings

**K Bobzin, M Öte, T F Linke, X Liao and J Sommer**

Institut für Oberflächentechnik, RWTH Aachen Universität, Aachen
sommer@iot.rwth-aachen.de

**Abstract.** The tensile bond strength of a coating is one of the limiting factors of thermally sprayed coatings. To improve the bond strength and for cleaning the surfaces prior to the coating, the substrates are pretreated. Many scientific publications in the field of thermal spraying focus on the spraying process and on developing new coating pa-rameters. The pretreatment process, on the other hand, is not often investigated sci-entificly, rather defined by relied on empirical values. A well-established pretreatment method is grit blasting the surface with corundum. With this method the surface can be cleaned and roughened to support the mechanical clamping between the coating and the substrate. Beside the roughened surface the mechanical clamping is sup-ported by undercuts. Especially the created surface topography depends on the blasting parameters such as grain fraction of the grit, particle velocity and grit blasting time. In this publication the different surface topographies are analysed using confo-cal laser scanning microscopy and described using standardized surface parameters. By varying the above mentioned grit-blasting parameters different surface topogra-phies are produced. After roughening the surfaces the substrates are coated and the bond strength is measured. One goal of this project is finding a surface parameter, which correlates with the bond strength. For an easy illustration a first attempt is to use a linear presentation. The goal of this correlation is to adjust the surface topog-raphy and to minimize the grit blasting time and the usage of grit. It can be shown that the surface parameter Ra shows only minor correlation in terms of coatings bond strength.

IOP Conf. Series: Materials Science and Engineering **118** (2016) 011002 doi:10.1088/1757-899X/118/1/011002

# Wear protection made by welding for components of railroad wagons

**V Wesling[1], T Ryspaev[2] and R Reiter[2]**

[1] Technische Universität Clausthal, Clausthal-Zellerfeld
[2] Kirgisische Staatliche Technische Universität, Bischkek
rolf.reiter@isaf.tu-clausthal.de

**Abstract.** The railroad wagons of the Kirghiz railroad company were made in soviet times in Russia. These wagons are in many cases age-related damaged by wear. A new acquisition of wagons is due to financial reasons not possible. For these reasons important components like the saddle of axis made by cast steel, which are damaged by dry adhesive oscillating wear, should be regenerated by welding. The presentation includes collaborative investigations of the University of Technology Clausthal, the Kirghiz State University of Technology and the new Kirghiz-German Competence Centre of Welding which is supported by the organisation CIM, Germany, for rebuilding those components by welding processes like gas metal arc welding and gas metal arc welding with controlled short arc. These claddings, made by conventional welding wires and flux cored wires, were characterized by metallographical and tribological investigations. The aims of these investigations are reliable and cost-effective repair weldings with improved wear resistance against dry adhesive oscillating wear.

IOP Conf. Series: Materials Science and Engineering **118** (2016) 011002 doi:10.1088/1757-899X/118/1/011002

# Suspension plasma spraying of zirconia coatings using different plasma torches

**P Sokołowski1[1,2], L Łatka1[1], S Kozerski1[1], D Dietrich[3], Z Pala[4,5], T Lampke[3] and L Pawłowski[2]**

[1] Faculty of Mechanical Engineering, Wroclaw University of Technology, Wroclaw, Poland
[2] Laboratory of SPCTS, University of Limoges, UMR 7315, Limoges, France
[3] Materials and Surface Engineering Group, Institute of Materials Science and Engineering, Technische Universität, D – 09125 Chemnitz, Germany
[4] Institute of Plasma Physics, ASCR, Prague, Czech Republic
[5] Faculty of Engineering, University of Nottingham, Nottingham, United Kingdom

**Abstract.** The aim of the present work was to test the commercially available plasma torches for a Suspension Plasma Spraying in view of potential application of SPS technology to the industry i.e. for Thermal Barrier Coatings or Solid Oxide Fuel Cells production. Four different set-ups equipped with plasma torches namely: (i) SG-100, (ii) Triplex, (iii) Axial III and (iv) h-WSP were used. The spray process parameters were optimized for each torch separately. The zirconia coatings were produced using three different precursors based on: (i) YSZ – yttria partially-stabilized zirconia, (ii) Tosoh-YSZ – yttria fully-stabilized zirconia and (iii) YCeSZ – both yttria and ceria stabilized zirconia powders. Moreover three different concentrations of suspensions were tested for each plasma torch in order to observe their influence on the coating morphology and plasma spray process efficiency. The coatings' microstructure was observed using light microscopy, scanning electron microscopy (SEM) as well as electron backscatter diffraction (EBSD). Afterwards the phase composition of coatings was analyzed using X-ray diffraction method. The studies showed that the spray process conditions highly influence coatings microstructure and coatings build-up mechanism.

IOP Conf. Series: Materials Science and Engineering **118** (2016) 011002 doi:10.1088/1757-899X/118/1/011002

# Inner hydrophobizing of thermal sprayed coatings

**R Winkler, T Lindner and T Lampke**

Materials and Surface Engineering Group, Institute of Materials Science and Engineering, Technische Universität Chemnitz, D – 09125 Chemnitz, Germany
ruben.winkler@mb.tu-chemnitz.de

**Abstract.** The applicability of thermal sprayed coatings is strongly limited by corrosion-initiating atmospheres. One central issue is presented with the open, porous structure of the coat, which leads towards functional damage and shortened service life. Currently, thermal sprayed coatings are protected against infiltrating, corrosive substances by an additional sealing. Even with this method being accepted by the industry, the process stability during production as well as the quality of sealed coatings remain subject to research. Especially the strongly heterogeneous build-up of coats poses a fundamental problem in most use cases. A reliable protection requires a homogeneous, deeply effective and complete inclusion of the sealer into the coat, which cannot be sufficiently guaranteed by conventional sealing methods. Subsequent mechanical post-treatments like finishing can already negatively influence the protective effect.

Coming from the known problems of conventional sealing methods, relocating the seal into the coating system offers a reasonable alternative. It is currently impossible to make a reliable statement about the functional depth of a conventional sealing. With the overlaying coating however it is possible to reach an extensive protection of the sealed area. Many sources of defect, which are connected to the insufficient penetration of the sealer, are excluded this way. While conventional methods sometimes require to modify the coating structure to accomplish a deeper penetration, the inner sealing method will get rid of this requirement altogether. Using individually scalable functional layer thickness, it is possible to make life span predictions for the first time, which means an increase in production- and quality safety. Thereby it is possible to strengthen the reliability and market acceptance of these products and furthermore develop new application possibilities.

IOP Conf. Series: Materials Science and Engineering **118** (2016) 011002 doi:10.1088/1757-899X/118/1/011002

# Corrosion investigations by using gel-type electrolytes

**M Babutzka[1] and A Heyn[2]**

[1] Bundesanstalt für Materialforschung und –prüfung, martin.babutzka@bam.de
[2] Korrosionsdiagnostik Dr. Andreas Heyn, heyn@corrdiag.de

**Abstract.** Limitations are encountered regarding the electrolyte when studying atmospheric corrosion reactions on materials that form protective layers on their surface or tend to passivate. For example, the reconstruction of a thin wet film, as well as the interpretation of electrochemical measurements in so-called bulk solutions (a "mass" of electrolyte), often proves to be difficult in view of corrosion behavior under atmospheric conditions. A new approach to this problem in corrosion research includes the use of gel-type electrolytes as an alternative to bulk electrolytes. Gel-type electrolytes form a thin wet film on the surface of the material, whereby the naturally formed protective layer is affected similar to atmospheric conditions and can be examined in an almost non-destructive way. Through electrochemical instrumentation specific values such as polarization resistances and corrosion currents can be determined providing information on kinetics of surface layer formation and stability of surface layers formed under the influence of a thin wet film. Thus, new test methods can be developed that supply a better understanding of corrosion processes in specific atmospheres.

In this article different zinc coatings and aluminium alloys were investigated, their naturally formed protective layers were electrochemically characterized and corrosion relevant values were determined by using a gel pad based on polysaccharide. Corrosion relevant values allowed the differentiation of various coating systems and could describe the current protective effect provided by the coating. It is shown that gel-type electrolytes influence protective layers and coatings considerably less than corresponding bulk electrolytes and that an atmospheric wet film is approached by these test conditions. From the results it is evident that gel-type electrolytes represent a viable and promising field in corrosion research.

IOP Conf. Series: Materials Science and Engineering **118** (2016) 011002 doi:10.1088/1757-899X/118/1/011002

# Experimental and numerical investigation of the strength behavior of corroded aluminum alloy EN AW-2024-T3 under uniaxial loading

**E Schmidl, R Pippig and T Lampke**

Materials and Surface Engineering Group, Institute of Materials Science and Engineering, Technische Universität Chemnitz, D – 09125 Chemnitz, Germany
eric.schmidl@mb.tu-chemnitz.de

**Abstract.** High-strength aluminum alloys are used in a variety of components, such as aeronautical and automotive engineering due to their high strength and good corrosion resistance. However, the corrosion resistance will fail if the passive layer is damaged, e.g. by chloride ions. Local corrosion phenomena (henceforth referred to as defects) and pitting corrosion occur as the damage is proceeding. These defects are the reference of the crack initiation induced by the stress concentration or the functional fatigue, which leads to a strength reduction in the material.

It is well known that the pitting corrosion will decrease the fatigue life time of aluminum alloys under cyclic loading. Additionally, under quasi static uniaxial loading, a decrease of strength and ductility of corroded specimens is noticeable.

This study will present some findings on the influence of pitting corrosion regarding the strength behavior of aluminum alloy EN AW-2024-T3. To generate various defect shapes immersion bath tests using NaCl and "FeC" "l" _"2" solutions as well as galvanostatic tests were carried out. To analyze the pitting corrosion of the specimens a methodology starting from micro-optical 3D pattern surface measurements was developed, which can be easily transferred to other corrosion tests. Subsequently, uniaxial tensile tests of the corroded specimens were carried out. In addition, numerical investigations using Finite Element Method will be presented showing correlations between the shape of the defects and the strength of the aluminum alloy EN AW-2024-T3.

IOP Conf. Series: Materials Science and Engineering **118** (2016) 011002 doi:10.1088/1757-899X/118/1/011002

# Influence of nitrogen on the corrosion resistance of martensitic stainless steels

**P Rosemann[1], C Müller[2], A Meyer[2] and T Halle[2]**

[1] BAM Bundesanstalt für Materialforschung und –prüfung, Berlin
[2] Otto-von-Guericke-Universität Magdeburg, Institut für Werkstoff- und Fügetechnik
paul.rosemann@bam.de; (paul.rosemann@ovgu.de)

**Abstract.** The corrosion resistance of martensitic stainless steels (MSS) depends strongly on the chemical composition and the applied heat treatment. Both determines the distribution of the alloying elements in the microstructure and the resulting material properties. The addition of nitrogen is known to be beneficial for the pitting corrosion resistance of stainless steels. In case of MSS this effect is not only connected to nitrogen itself because nitrogen can be used to substitute carbon which also influences the result of the heat treatment process. This paper shows the effect of nitrogen on the corrosion resistance in relation to the hardening process of MSS. Therefore the effects of austenitization duration, austenitization temperature and cooling rate on microstructure, hardness and corrosion resistance were studied on the MSS X30CrMoN15 1 and X50CrMoV15. The effect of different cooling rates was studied in the range of > 100 K/s down to 1 K/s using the jominy end quench test. The changes in corrosion resistance were detected with electrochemical potentiodynamic reactivation (EPR) and by the determination of critical pitting potentials. Besides this experimental approach thermodynamic calculations with the software thermocalc will be presented and used for the interpretation of the effect of nitrogen on the corrosion resistance of MSS.

IOP Conf. Series: Materials Science and Engineering **118** (2016) 011002 doi:10.1088/1757-899X/118/1/011002

# Methodology for the development of electrolytes for electrodeposition of alloys

**M Müller, I Scharf, D Höhlich, E Schmidl and T Lampke**

Materials and Surface Engineering Group, Institute of Materials Science and Engineering, Technische Universität Chemnitz, D – 09125 Chemnitz, Germany
mueller.markus@mb.tu-chemnitz.de

**Abstract.** Over 500 binary alloys can be deposited from aqueous solution. But only a small amount of them has been used industrially so far. The problem lies mainly in the strong dependence of the alloy composition from electrolyte composition and the working conditions. The empirical approach in the development of new electrolytes depends extremely on the experimenter. The upscaling from laboratory to industrial scale is also very little implemented because of the local optimization of electrolyte properties.

In order to standardize the electrolyte design, new mathematical optimization algorithms are required. They allow a prediction of the alloy composition as a function of essential process parameters.

The presentation defines the bonding between the metal ion and complexing agent with thermodynamic and kinetic parameters. These parameters allow a prediction of the expected type of alloy and the alloy composition. The methodology for the research of basic electrolytes is resulting from it.

IOP Conf. Series: Materials Science and Engineering **118** (2016) 011002 doi:10.1088/1757-899X/118/1/011002

# Determination of the die geometry influence in gradation extrusion using numerical simulation and analytical calculation

**M Bergmann, N Schubert, A Sterzing and D Landgrebe**

Fraunhofer-Institute für Werkzeugmaschinen und Umformtechnik IWU, Chemntiz
markus.bergmann@iwu-fraunhofer.der

**Abstract.** The forming process Gradation Extrusion enables a severe plastic deformation with a large gradient of microstructure. Materials with tailored properties can thus be provided. Controlling the gradation of the microstructure requires knowledge of the interaction of forming process and the special die geometry. Based on an analytical calculation approach different geometry variants are characterized. Additionally, selected variants are studied more comprehensively by numerical simulation. The mechanisms of the interaction and possibilities of influencing the deformation process by die design and process parameter optimization are presented and analysed.

IOP Conf. Series: Materials Science and Engineering **118** (2016) 011002 doi:10.1088/1757-899X/118/1/011002

# Microstructure and mechanical properties of AA6060 after low-temperature extrusion

**N Berndt, P Frint, M Böhme and M F-X Wagner**

Chair of Materials Science, Institute of Materials Science and Engineering, Technische Universität Chemnitz, D – 09125 Chemnitz, Germany
nadja.berndt@mb.tu-chemnitz.de

**Abstract.** The various processes of severe plastic deformation often provide ultrafine-grained materials with high strength and ductility, which are, due to their low thermal stability, commonly processed to near net shape by costly and time-consuming cutting processes. In this study a shaping process operating below recrystallization temperature is investigated. Cast billets of the age hardening aluminum alloy AA6060 were solid solution annealed and extruded at room temperature (RT) with an extrusion ratio (ratio of billet vs. shape area) of 6. Additionally, some billets were processed at the aging temperature (AT, 170 C) of the material. Subsequently, the microstructural features and mechanical properties of the extruded materials were analyzed. EBSD-measurements in the center of the profiles show a coarse-grained microstructure with a pronounced fiber texture for both extrusion temperatures. In the surrounding areas fine-grained, pancake-microstructures were found, whose textures differ from the center. As expected, hardness measurements of the rods show that RT-extrusion leads to a lower hardness in the inner center area (103 HV1) compared to the rim area (108 HV1). This gradient can be compensated by a suitable subsequent heat treatment. Compared to commercially available products the RT-extruded material aged to a maximum of hardness achieves a higher tensile strength of about 350 MPa without significant loss of ductility (elongation to failure: 16 %). Applying the same extrusion parameters (extrusion ratio and speed) for AT-extrusion results in comparable material properties without additional subsequent aging. Besides artificial aging during the process this approach benefits from a lower pressing force and therefore offers a high potential for integration in industrial process chains.

IOP Conf. Series: Materials Science and Engineering **118** (2016) 011002 doi:10.1088/1757-899X/118/1/011002

# Unstable plastic flow during combined extrusion-ECAP-processing – Analysis of local mechanical properties and microstructures

**P Frint, S Pfeiffer and M F-X Wagner**

Chair of Materials Science, Institute of Materials Science and Engineering, Technische Universität Chemnitz, D – 09125 Chemnitz, Germany
philipp.frint@mb.tu-chemnitz.de

**Abstract.** In this study, we report on the observation of heterogeneous microstructures after processing of the aluminum alloy 6060 by extrusion and subsequent equal-channel angular pressing (ECAP) at room temperature. A macroscopic, alternating structure of two types of bands is observed: shear bands carrying large amounts of deformation are located next to matrix bands that contain significantly less deformation. Scanning (SEM) and transmission (TEM) electron microscopy reveal an ultrafine-grained microstructure in the shear bands, while the microstructure of the matrix bands is characterized by a high fraction of low angle grain boundaries and dislocation cells. These characteristic microstructural features affect the local mechanical properties, which are found to be distinctly different for the two types of bands. Micro-hardness measurements and nanoindentation jump-tests are performed to explore the characteristics of the deformation bands. In addition to different hardness values, caused by the different deformation induced strain hardening, the ultrafine-grained shear bands exhibit a considerably increased strain rate sensitivity (by a factor of more than 2) compared to the matrix bands. A comparison of the initial extruded microstructure with the ECAP-processed material reveals that simple shear deformation exclusively takes place within the shear bands while the matrix bands pass through the ECAP-die without getting sheared. This observation contradicts the simple theory of homogeneous simple shear during ECAP-processing; we propose a simple mechanical model describing the evolution of the band-structure, based on an alternating interplay of elastic and plastic deformation inside the shear zone driven by local microstructure-dependent softening of the material.

# Unstable plastic flow during combined extrusion-ECAP-processing – Investigation of engineering factors using FE-simulations

**S Pfeiffer, P Frint and M F-X Wagner**

Chair of Materials Science, Institute of Materials Science and Engineering, Technische Universität Chemnitz, D – 09125 Chemnitz, Germany
steffen.pfeiffer@mb.tu-chemnitz.de

**Abstract.** Equal-channel angular pressing (ECAP) is a method of severe plastic deformation that typically leads to a homogeneous simple shear deformation and that is used to produce ultrafine-grained microstructures. In some special cases, however, single ECAP passes result in the formation of heterogeneous microstructures; this was, for example, observed in a companion study where the formation of alternating types of shear and matrix bands were observed after ECAP of an aluminum alloy 6060. Both microstructural features and mechanical properties (determined by micro- and nanoindentation) demonstrate that the ECAP shear deformation is concentrated in shear bands, whereas the adjacent matrix band regions are hardly deformed. In this contribution, we investigate the influence of different engineering factors on the morphology of the band structure. We use a simple mechanical model to rationalize how a discrete material volume can pass the shear zone of the ECAP die without accumulating significant amounts of plastic shear deformation, and how this process results in the alternating formation of matrix and shear band regions. 2D finite element (FE) simulations with an elasto-plastic material model, contribute to a better understanding how engineering factors directly influence the local deformation behavior. Different engineering factors like inner channel radius, cross sectional area of the ECAP billet, backpressure and pressing speed are investigated. Our results show that the largest equivalent stresses arise at the inner-channel corner, where consequently the formation of shear bands is triggered. Increasing the inner-channel radius or scaling up the cross sectional area of the billet leads to wider matrix bands. The predictions gained from the FE-simulations on band morphologies and dimensions are in good agreement with our experimental observations. These results contribute to a more detailed understanding of the locally heterogeneous deformation of pre-extruded or heavily work hardened materials during ECAP.

IOP Conf. Series: Materials Science and Engineering **118** (2016) 011002 doi:10.1088/1757-899X/118/1/011002

# Tailor-made grain sizes for aluminum alloys by recrystallization after ECAP

**M Härtel, M Lauschner, P Frint, T Kaiser and M F-X Wagner**

Chair of Materials Science, Institute of Materials Science and Engingeering, Technische Universität Chemnitz, D – 09125 Chemnitz, Germany
markus.haertel@mb.tu-chemnitz.de

**Abstract.** Severe plastic deformation is characterized by microstructures with high dislocation density, a large number of vacancies, and the formation of subgrains. These defects can act as nucleation sites for new grains during recrystallization. In this contribution, we study the influence of recrystallization temperature and time on the grain size development after one pass of ECAP of pure aluminum and different aluminum alloys. The materials are deformed in one pass in a 90° ECAP die at room temperature and then subjected to different annealing heat treatments. For all samples, we determine grain size distributions by electro-polishing and employing the line-intercept method. The resulting data-sets (grain size distributions as a function of annealing time and temperature) allow tailoring homogeneous grain sizes in aluminum alloys after ECAP, which can be useful for further studies such as the scientific corrosion test.

IOP Conf. Series: Materials Science and Engineering **118** (2016) 011002 doi:10.1088/1757-899X/118/1/011002

# Investigations in turn milling of SiC particle reinforced aluminium matrix composites using monocrystalline diamond tipped milling cutters

**B Clauß, A Nestler, R Funke and A Schubert**

Technische Universität Chemnitz, Professorship Micromanufacturing Technology
mft@tu-chemnitz.de

**Abstract.** Concerning a steady demand for increased performance and efficiency, in many cases requirements cannot be met using unreinforced, microscopically homogeneous materials. A promising alternative represent metal matrix composites, typically using aluminium alloys for light weight applications and ceramic particles as reinforcement. High potential of these materials is seen for example in aerospace applications, especially requiring a high fatigue strength. Additionally to the material characteristics the surface properties significantly influence the fatigue strength. Reduced surface roughness values and surface imperfections as well as compressive residual stresses are beneficial in this regard. However, in machining of aluminium based composites, processes with geometrically defined cutting edges proved advantageous. Recently turn milling has increasingly become subject to scientific investigations and industrial applications.

Using a fractional factorial design and MCD tipped milling cutters the influence of different machining parameters (e. g. axial feed, tangential feed, cutting speed) and tool geometries (e. g. clearance angle) is investigated, aiming for low roughness values, minimised surface imperfections and highly compressive residual stresses. The specimens required are manufactured from an aluminium matrix composite, consisting of the alloy AA2124 and 25 % volume proportion SiC particles, in the heat treated condition T4.

The investigations provide fundamental results concerning the relationships between the cutting parameters and the properties of the surfaces generated in orthogonal turn milling of aluminium matrix composites. It can be concluded that turn milling is suitable for the generation of surfaces without significant voids. Furthermore, small surface roughness values can be obtained, primarily influenced by the axial feed. As turn milling is capable of non-circular machining the results can be transferred to the machining of rotationally asymmetric workpieces as well.

IOP Conf. Series: Materials Science and Engineering **118** (2016) 011002 doi:10.1088/1757-899X/118/1/011002

# Mechanical performance of short and endless fiber reinforced polypropylene

**J Tröltzsch, J Stiller, F Helbig and L Kroll**

Institut für Strukturleichtbau, Exzellenzcluster MERGE, Technische Universität Chemnitz
juergen.troeltzsch@mb.tu-chemnitz.de

**Abstract.** Glass fiber reinforced polypropylene (PP) is one of the most important thermoplastic, fiber reinforced materials for technical applications. The injection molding process is the thereby the most efficient and flexible technique for producing fiber reinforced thermoplastic structural parts. To meet the required mechanical performance, either highly filled fiber reinforced compounds or textile reinforced thermoplastic prepregs are used. Due to the process technology, the fiber length is limited to a few millimeters in the compound. High shear stress during plastification and injection leads to fiber breakage, which additionally reduces significantly the resulting fiber length in the part. Numerous research and development efforts have therefore dealt with the integration of textile reinforcements in injection molded parts and the provision of appropriate technologies.

In comparative studies, the material performance of short fiber reinforced PP and textile reinforced PP-glass fiber prepregs was compared and considered in relation to the achievable performance in injection molded composites. The achievable strength, stiffness and impact strength run partly contrary to increasing fiber content and fiber length and must be considered in the context of the processing. Increasing fiber contents lead to an increase in the melt viscosity and thus to a higher mechanical load on the fibers during processing, which has a disadvantageous effect on the resulting fiber length. Therefore, the flow properties of the neat polymer have to be matched to the fiber content.

Already at low fiber contents of a view weight-percent a strong change in the mechanical characteristics can be seen. The elongation at break and impact strength decrease sharply, the tensile modulus already increases for 50%. The tensile modulus increases almost linearly to the technological maximum of achievable fiber contents, while tensile strength and impact strength reach the maximum at about 40% fiber content in mass. With increasing fiber length improvements are obtained especially in the strength and toughness. Improving the fiber-matrix adhesion, polymer modifications or electron beam cross-linking lead additionally to an increase in the mechanical properties.

IOP Conf. Series: Materials Science and Engineering **118** (2016) 011002 doi:10.1088/1757-899X/118/1/011002

# Influencing the creep behavior of particle-reinforced aluminium matrix composites (AMCs)

**S Siebeck, D Nestler, B Wielage and G Wagner**

Chair of Composite Materials, Institute of Materials Science and Engineering, Technische Universität Chemnitz, D – 09125 Chemnitz, Germany
steve.siebeck@mb.tu-chemnitz.de

**Abstract.** The presentation focuses on the creep behaviour of an age-hardenable particle-reinforced aluminium alloy. The difference in the creep behaviour compared to unreinforced alloys is documented by means of creep rupture tests. The mechanisms affecting the materials are discussed further. Due to the nature of the powder metallurgical production the AMCs are characterised by a very fine-grained structure, a high concentration of grain boundaries, phase interfaces, as well as a large number of dislocations. All these properties are conducive conditions for diffusion processes to take place, which in turn leads to increased precipitation kinetics of the matrix alloy. Therefore it was examined whether an influence on these effects and thus of the creep behaviour is possible by means of the powder metallurgy. The goal is to lower the possible diffusion of the composite material by adding specific alloying elements. The results showcase the positive impact small quantities of Bor have on the creep behaviour.

IOP Conf. Series: Materials Science and Engineering **118** (2016) 011002 doi:10.1088/1757-899X/118/1/011002

# Identification of Process Parameters for the Anodization of Al-Cu-Alloys Using Design of Experiments

**P Schulze, B Fischer, R Morgenstern, E Schmidl, T Grund and T Lampke**

Materials and Surface Engineering Group, Institute of Materials Science and Engineering, Technische Universität Chemnitz, D – 09125 Chemnitz
pierre.schulze@mb.tu-chemnitz.de

**Abstract.** Precipitation hardenable aluminum-copper alloys exhibit high specific strength levels. Therefore, they are used e.g. in the aircraft industry for safety-related components in lightweight constructions. The application of these alloys is limited by their sensitivity to selective corrosion. In order to effectively protect the components against corrosion, conversion coatings are produced by means of electrolytic anodization. These layers are characterized by very good adhesion, a uniform film thickness distribution and economic producibility. However, the layer quality (thickness, porosity) is negatively affected by the transformation behavior of copper in high-strength aluminum alloys. Thus, an adjustment of the anodizing parameters is required in order to improve the layer properties. As part of the SFB 692 B2, the influence of the process parameters on the anodization of precipitation hardenable copper-containing aluminum alloys is examined on the basis of the alloy EN AW-2024 T4 using methods of the statistical experimental design. This approach minimizes the testing effort and identifies to which extent the interactions of the process parameters have an influence on the thickness and hardness of the conversion layers. Varied parameters are the electrolyte temperature, electric current density, time (at a constant amount of charge) and the electrolyte composition regarding the concentration of sulfuric and oxalic acid. Energetic aspects are considered in order to optimize the anodization process in terms of cost-efficiency. The results indicate that the layer characteristics are particularly affected by the electrolyte temperature and to a lesser extent by the anodic current density and the oxalic acid concentration. The influence of the sulfuric acid concentration is the lowest (within the defined test setup). According to the results, an improvement of the layer thickness and hardness generally demands a much higher amount of electrical energy. Recommendations for industrial anodizing processes are derived and further opportunities for property improvements and reduced energy consumption are discussed. The results of the investigations will be presented in a poster.

IOP Conf. Series: Materials Science and Engineering **118** (2016) 011002 doi:10.1088/1757-899X/118/1/011002

# High-strength aluminum materials with customized damping features for sled in high-performance sports.

**S Schwanitz[1], P Frint[2], M Hill[1], S Odenwald[1] and M F-X Wagner[2]**

[1] Professorship of Sports Equipment & Technology, Technische Universität Chemnitz
[2] Chair of Materials Science, Institute of Materials Science and Engineering, Technische Universität Chemnitz, D-09125 Chemnitz, Germany
stefan.schwanitz@mb.tu-chemnitz.de

**Abstract.** Till now there is little technical application of ultra-fine grained (UFG) aluminum materials. Because of cost-intensive production of these materials just small quantities and dimensions are manufactured. Especially in the field of high-performance sports new high-performance materials could get applied easily because big numbers are not necessary. Smallest batch series with customized performance specification are central in the field of sports equipment and technology. It is the object of this transfer project to apply ultra-fine grained aluminum materials in sports equipment technology for the first time. At the end of the funding period one will have the state of knowledge to build a dynamically loaded component which is operationally reliable. As well as the present light-weight potential will completely be tapped and implemented into a prototype. The focus of the materials science is the investigation of the damping behavior depending on the microstructure of the used UFG materials. Concrete object of investigations in this transfer project are the so-called "bridges" of a sled. The front and back bridges connect as a crossbeam the right and left runner with the seatshell. Normally this connection is an all-steel welded construction. During usage these components get high dynamic loads. Permanent and random vibrations between rail and ice surface bring the whole system sled-rider to vibrate. These vibrations worsen the ride comfort (tractability, acceleration power) of the sled, cause a decreased performance and furthermore affect the reliability of the sled. At the moment there is no systematical and appropriate development of the bridges regarding the requirements on damping, mass reduction and operational stability. In this research program the traditional steel material will be substituted by UFG materials. This will cause a reduction of component mass and an improvement of vibration damping while the operational stability is ensured. Together with the applying partner Thüringer Schlitten- und Bobsportverband e.V. there will get acquired objective methods - beside the measuring of the achieved runtime - to evaluate the functionality of the sports equipment for efficient component development.

IOP Conf. Series: Materials Science and Engineering **118** (2016) 011002 doi:10.1088/1757-899X/118/1/011002

# Electrochemical development of aqueous iridium and iridium alloy electrolyte and characterization of the deposits for electrical contacts

**J Nähter[1], F Köster[1], C Grieger[1], R Freudenberger[2] and C Schöberl[2]**

[1] Hochschule Mittweida – University of Applied Sciences,
naether@hs.mittweida.de
[2] Forschungsinstitut für Edelmetalle und Metallchemie FEM,
r.freudenberger@fem-online.de

**Abstract.** To optimize the tribological behavior of sliding contacts, electrical industry is looking for surface solutions which can decrease abrasive wear by maintaining a low electrical contact. Additionally there is a great interest to replace the expensive gold coatings. Iridium is a conceivably good alternative, because it combines a high hardness with good contact resistance and conductivity. For this project the University of Applied Science Mittweida is concerned with the development of a stable water-based Iridium electrolyte for electrochemical deposits. The investigations are focused on the impact of different additives and parameter changes. The effects on the properties of the Iridium surface and the deposition rates are studied. Another approach is the specific deposition of some alloying elements to the Iridium matrix to create an Iridium alloy with improved properties to be used in electrical applications.

IOP Conf. Series: Materials Science and Engineering **118** (2016) 011002 doi:10.1088/1757-899X/118/1/011002

# Influence of natural weathering and the VDA alternating climate test on the mechanical properties of thermoplastic fibre-metal-laminates

**C Zopp[1], L Kroll[1], M Trautmann[2]and D Nestler[2]**

[1] Institut für Strukturleichtbau, Professur Strukturleichtbau und Kunststoffverarbeitung, Technische Universität Chemnitz, Germany
camilo.zopp@mb.tu-chemnitz.de, lothar.kroll@mb.tu-chemnitz.de
[2] Chair of Composite Materials, Institute of Materials Science and Engineering, Technische Universität Chemnitz, D – 09125 Chemnitz, Germany
maik.trautmann@mb.tu-chemnitz.de, daisy.nestler@mb.tu-chemnitz.de

**Abstract.** Semi-finished products with thermoset and thermoplastic matrix as well as aluminum alloys and titanium alloys are indispensable for the lightweight, because of their specific characteristics. They contribute significantly to the performance of economic, technical and environmentally conscious requirements. Especially the combination of the components for so-called thermoplastic fiber-metal laminates involve a significant potential for lightweight. Thermoplastic hybrid laminates are characterized a comparatively low weight, excellent impact behaviour and very good fatigue crack propagation under cycling loading. Additionally thermoplastics materials exhibit against thermosets essential advantages such as formability, recyclability, cycle time and large-scale process.

The focus of the current paper is based on the investigation of mechanical properties of thermoplastic fibre-metal-laminates, depending on the natural weathering and the VDA alternating climate test. For the laminate manufacturing an aluminum alloy EN AW-6082 is used, which grouted with different fibre-reinforced semi-finished products.

IOP Conf. Series: Materials Science and Engineering **118** (2016) 011002 doi:10.1088/1757-899X/118/1/011002

# New hybrid materials by use of epoxy components in twin polymerization

**M Birkner, M Göring, K Schreiter, K Trommler and S Spange**

Technische Universität Chemnitz
matthias.birkner@chemie.tu-chemnitz.de

**Abstract.** Lightweight constructions play an important role in many fields of mechanical engineering due to weight reduction and energy efficiency. Partial substitution of metal with plastic components combines the advantages of both materials and leads to improved material properties. A well-known method for the construction of metal-plastic-composites is mechanical clamping (riveting, pin welding, forming). In addition, chemical modification of the metal compound through a proper adhesion promoter is used. To obtain a high adhesive strength, it is necessary to adjust the adhesive promoter towards both joining partners.

The twin polymerization is a new approach for the synthesis of suitable adhesive promoters. Based on one monomer, with an organic and inorganic fragment, two polymers are built within one procedure. By reasons of the spatial restriction, nano structured, interpenetrating networks with domain sizes of 2–4 nm were generated. This organic-inorganic hybrid materials are synthesized thermally without a catalyst. The combination of the well-established chemistry of epoxy resin and twin polymerization is crucial for synergistic effects in view of the junction between metal and thermoplastic resin. An amino functionalized twin monomer provides the ability of interactions with both joining partners as well as possible cross linking agent with epoxides.

By use of a selected epoxy resin in the two-component system an additional cross linking in the polymer is observed. The combination of twin monomer and resin enables the variation of monomer structure, ratio, temperature and reaction time to adapt properties towards the fields of application e.g. hot pressing or injection molding. Solid state NMR, REM/EDX as well as nanoindentation were used for investigation of chemical structure and mechanical properties of the obtained hybrid materials.

IOP Conf. Series: Materials Science and Engineering **118** (2016) 011002 doi:10.1088/1757-899X/118/1/011002

# Chemical modification of reinforcing materials to improve adhesion in natural fiber composites

**R John[1], K Trommler[1], K. Schreiter[1], S Spange[1], C Siegel[2] and A Wagenführ[2]**

[1] Institute für Chemie, Technische Universität Chemnitz, rico.john@chemie.tu-chemnitz.de
[2] Institute of Wood and Paper Technology, Technische Universität Dresden, carolin.siegel@tu-dresden.de

**Abstract.** Environmental and economic reasons play an important role in the lightweight construction especially in mechanical engineering and building and construction industry. Natural fiber reinforced thermoplastics have gained great attention, because they have excellent specific properties, which are used inter alia in automobile body construction. The aim in this work is the development of novel biocomposite materials based on thermoplastic biopolymers (bio-polyethylene, bio-polyamide) and renewable reinforcing materials (flax fibers, wood veneer).

The combination of materials with different surface polarities is still a challenging task in the material science. A common approach is the use of adhesion promoters. They can adapt the different surface energies of the single components, which enables an adhesion. A chemical modification or blending of the biopolymers is not allowed due to producer policies. Thus, only the reinforcing materials can be modified, which basically consist of cellulose, hemicellulose and lignin. Therefore, the surface shows a polar character due to the accessible hydroxyl and carbonyl groups. Besides the improved adhesion the scattering of the mechanical properties should be reduced, which is caused by the natural character of the materials.

First attempts to process a veneer-polymer-veneer- (VPV) material with bio-polyethylene (bio-PE) failed without adhesion promoter. An adhesion promoter is necessary to combine the polar veneer with the nonpolar bio-PE. One approach is the hydrophobization of wood veneer surface with polyvinylamine (PVAm, Lupamin®). We present a strategy to hydrophobize wood veneer with PVAm to obtain compatible materials, which can be processed to VPV-materials. The adsorption behavior and the degree of hydrophobization depends on the pH value, the degree of hydrolysis, molar mass and the mass concentration of the PVAm solution. The coated veneers were characterized by contact angle, SEM, EDX and XPS measurements. Furthermore, the tensile properties of the produced VPV-materials were investigated.

IOP Conf. Series: Materials Science and Engineering **118** (2016) 011002 doi:10.1088/1757-899X/118/1/011002

# Wear behavior of high-entropy alloys

**M Löbel, T Lindner, T Uhlig and T Lampke**

Materials Science and Engineering Group, Institute of Materials Science and Engineering, Technische Universität Chemnitz, D – 09125 Chemnitz, Germany
martin.loebel@mb.tu-chemnitz.de

**Abstract.** Due to their outstanding properties, high-entropy alloys have gained much scientific interest recently. These properties include high strength and hardness, comparable to bulk metallic glasses. Nevertheless high-entropy alloys also show distinct ductility. Furthermore wear resistance has been investigated briefly showing promising results. The focus of this contribution is on further investigating the potential of high-entropy alloys for wear resistant applications.

For these investigations three alloy systems have been chosen: AlCrCuFeMnNi, CoCrFeNiTi and AlCoCrFeNiTi. Bulk samples with equimolar composition were produced by inert gas arc melting of elemental raw material. Subsequently Scanning Electron Microscopy (SEM) and Energy Dispersive X-ray Spectroscopy (EDX) were carried out to investigate the microstructure and chemical composition. Furthermore a phase analysis was carried out by XRD (X-ray diffraction). Mechanical properties were investigated by measuring Vickers hardness (HV). The wear resistance against different wear regimes was tested. Therefor ball on disk wear test, oscillating wear test and scratch test were conducted. Results were correlated with microstructure and hardness. To evaluate wear test results a conventional hard chrome reference plating was characterized as well. Comparing wear test results with the reference coating shows that, especially for the AlCoCrFeNiTi alloy, higher wear resistance can be achieved with the new alloying concept of high-entropy alloys.

IOP Conf. Series: Materials Science and Engineering **118** (2016) 011002 doi:10.1088/1757-899X/118/1/011002

# FEM-simulation of hybrid metal/plastic-composites

**E Saborowski, T Lindner, M Scholze and T Lampke**

Materials and Surface Engineering Group, Institute of Materials Science and Engineering, Technische Universität Chemnitz, D-09125 Chemnitz, Germany

**Abstract.** The development of hybrid composite structures for lightweight construction requires an intelligent combination of plastic and metal in variable composition with a defined, process-suitable interface design connecting the joining partners. FEM-simulations are suitable to evaluate the compound in real constructions. To evaluate the interface between the joining partners in terms of adhesive strength and resistance to crack propagation, multiple material tests must be carried out beforehand. The aim of the investigation is to determine the appropriate characteristics, in order to be able to construct the most realistic FEM models possible.

IOP Conf. Series: Materials Science and Engineering **118** (2016) 011002 doi:10.1088/1757-899X/118/1/011002

# Adhesive strength of wood-veneer-biopolymers

**C Siegel, B Buchelt, A Schuberth, D Nickel, A Wagenführ and T Lampke**

[1] Institute of Wood and Paper Technology, Technische Universität Dresden
[2] Materials and Surface Engineering Group, Institute of Materials Science and Engineering, Technische Universität Chemnitz, D – 09125 Chemnitz, Germany
[3] Technische Universität Darmstadt MPA /IfW Darmstadt, Area of Expertise Plastics & Composites, Germany

**Abstract.** Thermoplastic laminates are usually consisting of oriented fibres in thermoplastic matrices. These thin layers are stack in press processes to certain multilayer-composites according to their applications.

In order to produce bio-based laminates wood-veneer is jointed with polymer films in a continuous process. Veneer itself is a naturally grown composite with an obvious anisotropy. The use of veneer as natural fibre reinforcement for polymers requires a strong adhesion between both components.

Different kind of wood-veneers (European Beech, Fagus sylvatica L. and Spruce, Picea abies) and biopolymers were processed to laminates. European Beech as a hardwood and spruce as a softwood differs in their structure, density and mechanical properties. The impregnation of the veneers is influenced by the melt flow index of the polymer films and the cellular structure of the wood. The adhesive properties of produced laminates were evaluated by pull-off test and tensile test.

IOP Conf. Series: Materials Science and Engineering **118** (2016) 011002 doi:10.1088/1757-899X/118/1/011002

# Preliminary study on possibility of application of solution precursor plasma spraying process to obtain indium tin oxide coating

**I Gutkowski, R Kaczmarek, R T Candidato, G Lecomte, L Pawłowski, S Kozerski and W Posadowski**

SPCTS, UMR 7315, 12, rue Atlantis, University of Limoges, 87068 Limoges, France
Wroclaw Technical University, Wroclaw, Poland

**Abstract.** To obtain the indium tin oxide (ITO) coatings in the process of solution precursor plasma spraying (SPPS) tin (II) acetate and indium (III) acetate were used as precursors. They were prepared as a liquid solution in dissolved water in two ratios of In:Sn 90:10 and 90:5 by weight. To allow verification of received ITO solution, the liquid was dried at the temperatures of 120°C, 250°C, 500°C and 750°C. The X-ray diffraction (XRD) results of as dried ITO solution showed the big difference in the received structure. With the increase of the drying temperature the structure changed from amorphous to the crystalline cubic structure. After drying the solution above the temperature of 500°C the received structure matched the structure of stoichiometric indium tin oxide crystalline cubic structure. This confirmed the assumptions of the differential thermal analysis (DTA) that the temperature of crystallization the indium tin oxide structure is around 400°C-600°C. The liquid precursors were injected to plasma torch for single scan spraying process onto stainless steel substrates. In the process of single scan spraying the parameters such as power, spray distance and In:Sn ratio were changed. On the surface of sprayed substrates, observed using scanning electron microscope (SEM), were visible areas with fully melted particles of ITO and characteristic depositions for the solution precursor plasma spraying method called "cracked shells". It was observed that too short distance between the nozzle and the surface can prevent the transformation of solution from droplet to melted solid and its impact on the substrate.

IOP Conf. Series: Materials Science and Engineering **118** (2016) 011002 doi:10.1088/1757-899X/118/1/011002

# Resistance spot welding of aluminum to steel with the use of cold-sprayed interlayers

**M Gumiennik[1], M Korzeniowski[1], M Winnicki[1], M Rutkowska-Gorczyca[1] and P Sokołowski[1,2]**

[1] Faulty of Mechanical Engineering, Wroclaw University of Technology, Wroclaw, Poland
[2] Laboratory of SPCTS, University of Limoges, UMR 7315, Limoges, France

**Abstract.** The role of dissimilar material combinations is more and more popular in many industrial applications. Usually the main attention is paid on the joining of steel to light metals in order to reduce the weight of elements. However the selected joining processes can still cause many difficulties, what arise from the significant differences in their thermophysical properties.

In this work the preliminary study on the Resistance Spot Welding (RSW) of aluminum alloy 5754 to DP600 steel with use of interlayers were performed. Based on equilibrium phase diagrams two types of materials were selected for interlayers: (i) nickel and (ii) zinc. The steel substrates were coated with Ni and Zn with the use of Low-Pressure Cold Spray method prior to RSW. The welding parameters were selected experimentally based on authors own experience. Afterwards the microstructure of coatings was observed by light microscopy and scanning electron microscopy (SEM). Finally the micro hardness and shear strength were investigated also to have basic knowledge about the mechanical properties of RSW welds.

The studies showed that the appropriate selecting and preparation of interlayer allow to join aluminum alloy 5754 to DP600 steel by the means of Resistance Spot Welding with very satisfactory results. Both interlayers allowed to decrease hardness in welding zone and increase shear strength. The greatest increase in mechanical properties was observed when RSW process was conducted using nickel interlayers for which the failure in shear test occurred outside the weld (on the aluminum side).

IOP Conf. Series: Materials Science and Engineering **118** (2016) 011002 doi:10.1088/1757-899X/118/1/011002

# Production and characterisation of a thermoformed complexly curved component from thermoplastic FRP/metal laminate

**M Trautmann, D Nestler, E Schmitteck and G Wagner**

Chair of Composite Materials, Institute of Materials Science and Engineering, Technische Universität Chemnitz, D – 09125 Chemnitz, Germany
maik.trautmann@mb.tu-chemnitz.de

**Abstract.** Hybrid laminates comprised of thin light metal sheets and fibre-reinforced plastics (FRP) combine the advantages of the individual material groups. These laminates offer the great potential to substitute complexly curved and bent thin-walled sheet structures in areas relevant for light weight construction. Compared to thermoset-based fibre-reinforced plastic the utilisation of thermoplastic matrices enables the thermal moulding, reshaping and deformation of the laminates. Part of this contribution is the examination of properties not depending on a direction of hybrid laminates comprised of a multi-directional combination of the aluminium alloy EN AW-6082 T4 and the unidirectional, fibre-reinforced polyamide 6. The orthotropic and mechanical properties of planar laminates with a multi-directional construction are determined by means of tensile and bending tests. The results of these tests are subsequently compared by utilising the classical laminate theory (CLT). A hybrid laminate construction designed for bending and torsion is thermally reshaped to create a complexly curved structural component. Subsequent microstructural examinations showcase the effect the deformation has on the structure of the laminate and on the quality of the consolidation. The hybrid component structure is tested additionally with bending and torsion tests and compared to steel 1.4301. Due to the hybrid laminate the weight was reduced by 40%.

IOP Conf. Series: Materials Science and Engineering **118** (2016) 011003 doi:10.1088/1757-899X/118/1/011003

# Peer review statement

All papers published in this volume of *IOP Conference Series: Materials Science and Engineering* have been peer reviewed through processes administered by the proceedings Editors. Reviews were conducted by expert referees to the professional and scientific standards expected of a proceedings journal published by IOP Publishing.

IOP Conf. Series: Materials Science and Engineering **118** (2016) 012001 doi:10.1088/1757-899X/118/1/012001

# Steel - a Classic Material with a Large Potential for the Future

**B Masek, H Jirkova, D Aisman and S Jenicek**

University of West Bohemia
Univerzitni 22, 306 14 Pilsen

bohuslav.masek@mb.tu-chemnitz.de

**Abstract**. Steel is a traditional material which has been used by mankind for more than five thousand years. We may therefore be tempted to believe that we know practically everything about steel and its forms and variants which offer an extraordinary and broad range of properties. It is this diversity of properties which makes steel such a popular and widely used material. And yet, in recent years new opportunities have emerged for processing steel by unconventional techniques and producing novel, as yet unknown or unusual microstructures. This paper describes several examples of how microstructure evolution can be modified and how new and unconventional processing routes can be developed. These examples present several results of projects carried out in recent years by FORTECH Research Centre of Forming Technology and the University of West Bohemia in collaboration with their research partners.

**Keywords**. AHSS, UHSS, semi-solid, thixoforming, mini-thixoforming, TRIP, Q-P, ODS

## 1. Introduction

The fact that steel is one of the most important technical materials – and the most widely used technical metal – is and will probably remain indisputable in the near future. The key element found in steel is iron. Iron's allotropic transformations are the source of the diverse forms of steel microstructures. The amount of iron in the space is fairly large because iron is the last stable element produced by the nuclear reactions which take place while stars are burning out. This is the reason why Earth is so rich in this raw material. Although a majority of iron is embedded within deep-lying layers of our planet, the Earth's surface always offered enough iron and iron ore for steel to become the essential material that fuelled the mankind's evolution and civilization over the millennia.

Since the conventional structures found in steels – ferrite, pearlite, bainite and martensite – are well known, this article focuses on examples of development of unusual mixed microstructures attainable only by means of unconventional routes or by innovative technology chains.

## 2. Semi-solid treatment and rapid solidification

The first unconventional route involves steel solidifying at a high rate during cooling from the freezing range. This process is known as thixoforming. In high-alloy steels, the resulting structures typically consist of metastable austenite and carbides [1-4]. Most often, the metastable austenite takes the form of polygonal grains embedded in carbide network (Fig. 1). Thanks to being super-saturated with respect to multiple elements, this austenite exhibits relatively high stability and good deformation ability. Nevertheless, the presence of the carbide network means relatively high brittleness and poor toughness. Therefore, efforts were begun to remove these drawbacks. One of the available solutions is alloying with elements which form carbides with high thermal stability. Such carbides do not dissolve during heating to semi-solid processing temperatures and remain present in the melt. After the material cools rapidly, these act as nuclei for crystallization and, finally, become dispersed particles within a ductile austenitic matrix [5-7]. Consequently, the resulting microstructure is inverted with respect to structures that form during conventional thixoforming (Fig. 2).

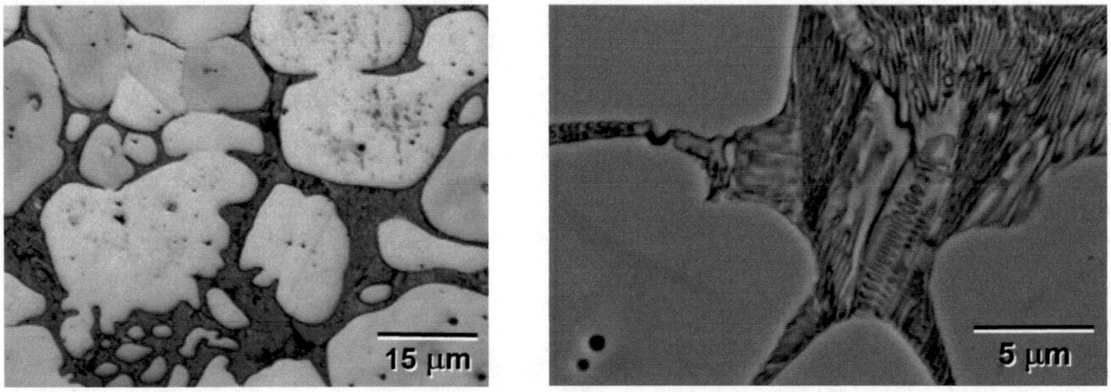


**Figure 1.** Structure of X210Cr12 steel after thixoforming; left: overview of the microstructure under light microscope; right: detail scanning electron micrograph of lamellar network

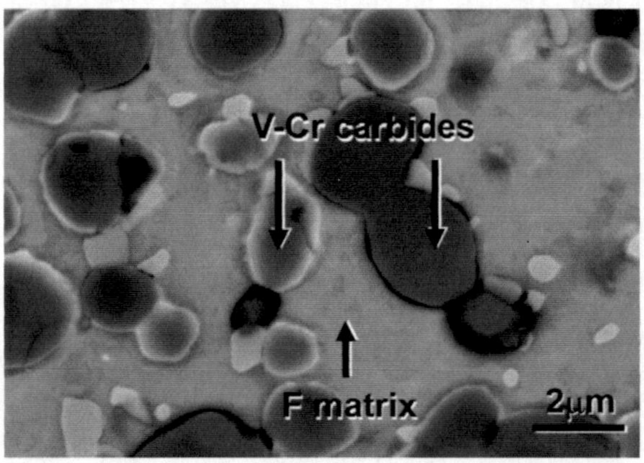


**Figure 2.** Inverted microstructure in CPM15V steel after thixoforming

## 3. New oxide dispersion-strengthened steels

For some applications, namely for operation at high-temperatures, steels must be specially modified to prevent degradation of their mechanical properties. During high-temperature exposure, carbides – which are normally used as the strengthening constituent – become the weak spot. For this reason, ODS steels (Oxide Dispersion-Strengthened Steels) have been developed, in which oxide particles are used to strengthen the matrix, instead of the carbides [8, 9]. As these oxide particles exhibit much better thermal stability than carbides, these steels promise to be usable in high-temperature applications in the future. However, there is a major drawback related to the manufacture of these steels which normally involves embedding the oxide particles in the matrix by mechanical alloying. The procedure is demanding and sometimes fails to produce the desired fine and uniform dispersion of strengthening particles in the matrix. If the required properties are to be guaranteed, this is the major problem that needs to be solved. An alternative procedure was thus proposed, in which oxides precipitate in the matrix which is super-saturated with respect to oxygen. This produces a good distribution of the strengthening oxide particles and reduces their size to several nanometres (Fig. 3). It is also corrosion resistant to temperatures of 1200°C. Thanks to its extraordinary properties, such steel can be expected to find use in high-temperature and other applications.

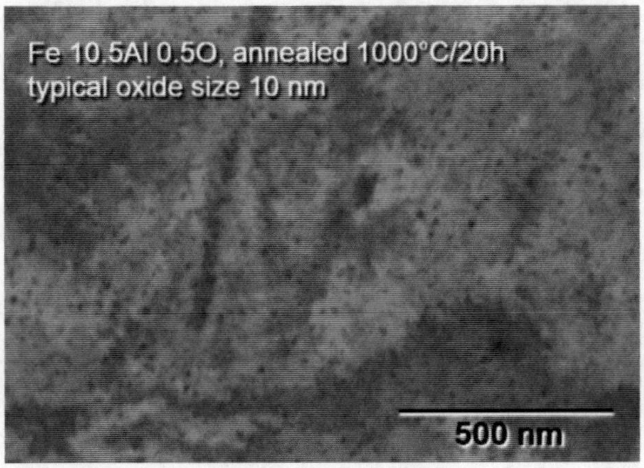


**Figure 3.** The new generation of ODS steels developed in collaboration with Dr. Svoboda from IPM Brno in 2016

## 4. AHSS and UHSS – low-alloy steels for the futureReferences

Intensive development can be seen recently in the fields of AHSS (Advanced High Strength Steel) and UHSS (Ultra High Strength Steel) [10-12]. Both groups comprise a majority of low-alloy steels or steels with cost-effective chemistries whose final properties are obtained by sophisticated treatment. In order to produce the desired structures, and therefore the required properties, the compatibility of the requirements of manufacturing processes and the limits of microstructural evolution should be taken into account for developing the manufacturing processes. Thanks to high-quality monitoring instruments and accurate manufacturing process control tools being available today, the extent of the potential of advanced multiphase steels which can be actually exploited is much greater.

Three main process types are mentioned with respect to producting AHSS-type structures. These include: intercritical processing used for TRIP steels (Transformation Induced Plasticity) [10], Q-P processing (Quenching and Partitioning) [12] and long-time austempering. As the last-named process still involves too long annealing times, it is not suitable for industrial use yet. Hence, only the first two processes were chosen for this demonstration and obtaining AHSS structures in TRIP and Q-P types.

## 5. Warm-formed and cold-formed TRIP steel

The following technology chain [13, 14] (Fig. 4) was developed for manufacturing products of TRIP steels with simple chemistries involving 0.2 % carbon and various amounts of manganese and silicon.

Rolled bar feedstock is heated to a warm-forming temperature and incrementally formed by rotary spin extrusion into a hollow semi-finished product with a deep axial blind hole. This formed part is then annealed in the intercritical range between the $AC_1$ and $AC_3$ temperatures, held at a bainitic transformation temperature, and finally cooled down to ambient temperature. This produces a TRIP microstructure with ferrite, bainite and metastable retained austenite (Fig. 5).

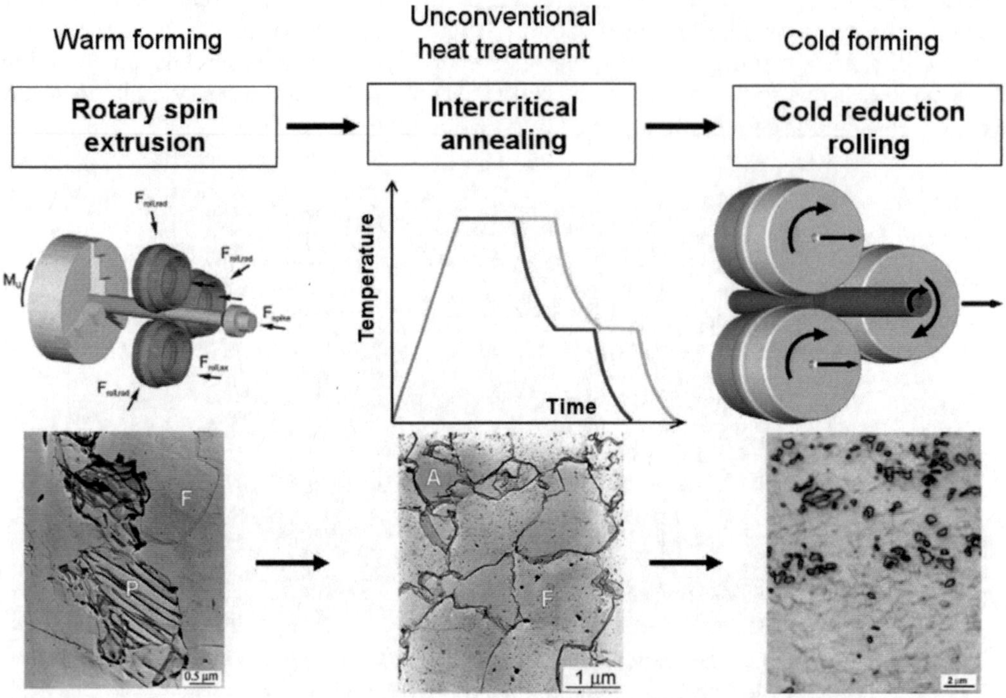


**Figure 4.** Technology chain for making stepped hollow semi-finished products of TRIP steels

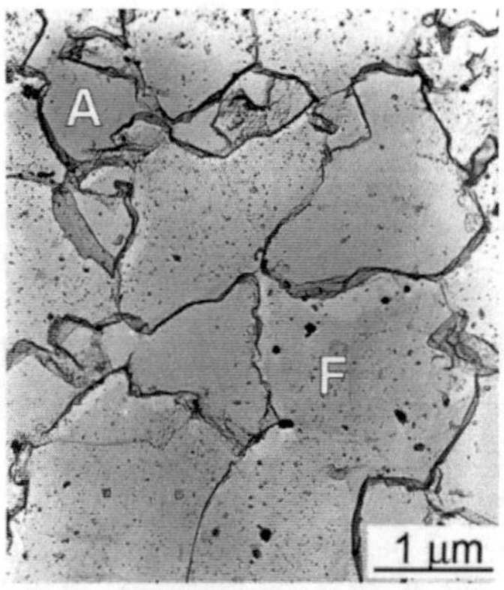


**Figure 5.** Detail of the structure of Mn-Si-Nb steel after intercritical annealing for 30 minutes, extraction carbon replica, TEM

This mixed microstructure generally offers high plasticity characterized by a value of approx. 30 %. The material is therefore very suitable for cold forming. Originally, steels of this kind were developed for the automotive industry for absorbing large amounts of energy in a crash. This ability to undergo plastic deformation can be used for achieving the prescribed deformation in cold forming. This was the case in the present technology chain, where hollow feedstock with a TRIP microstructure was formed by cross rolling to produce multiple steps. This profound deformation causes deformation hardening of the material and transformation of retained austenite to martensite. Furthermore carbides are fragmentated (Fig. 4). The resulting stepped hollow semi-finished product which received no further heat treatment showed a strength of about 1000 MPa and an elongation of approximately 10 %

(Fig. 6). By this process, a hollow, thin-walled and very lightweight product was obtained just by plastic deformation without any machining operation.

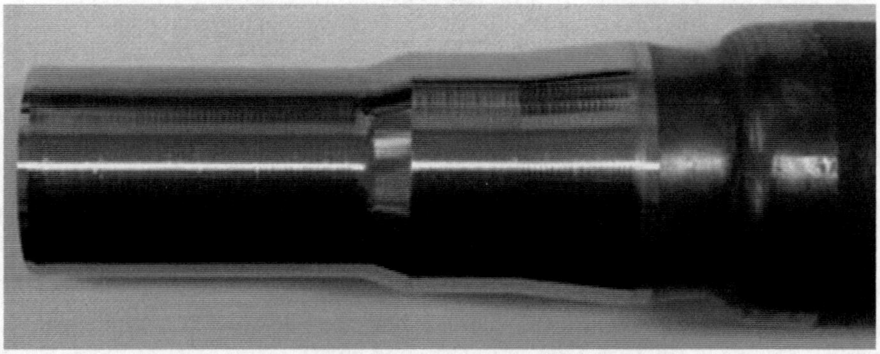

**Figure 6.** Stepped hollow semi-finished product of TRIP steel

## 6. Q-P process integrated into an internal high-pressure forming (IHP) route

Another example of new processes which have been specially developed for AHSS and UHSS steels is the process chain for making ultra-high-strength complex-shaped hollow semi-finished products (Fig. 7). It combines several high-tech aspects, most importantly the Q-P process and IHP forming. The entire process sequence is as follows [15-17].

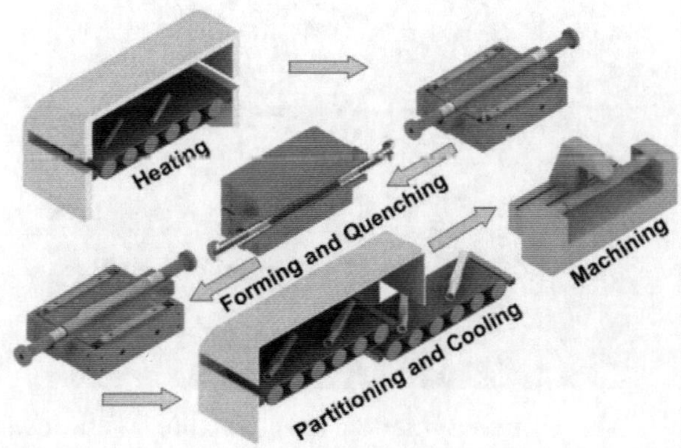


**Figure 7.** Technology chain combining internal high pressure forming, hot stamping and Q-P process

The feedstock material is 0.42 % carbon steel with a cost-effective chemistry containing manganese, silicon and chromium. The feedstock is a thin walled tube which is heated to the forming temperature of 1000 °C. It is then placed into a split die and inflated using nitrogen gas at high pressure until it comes into contact with the die walls (Fig. 8).

**Figure 8.** Experimental feedstock and the resulting shape after processing

This leads to partial quenching as the die is pre-heated to 200 °C. Martensite forms in the microstructure but some austenite remains untransformed due to incomplete quenching. To prevent the retained austenite from transforming during further cooling, the semi-finished product is placed into a furnace at 250 °C for several minutes. At this temperature, carbon diffuses from martensite into retained austenite. The internal stress in the martensite is relieved and the austenite becomes stabilized by this chemical process. The resulting mixed microstructure consists of martensite whose brittleness has been suppressed and retained austenite with good plasticity (Fig. 9). It offers high strength in excess of 2000 MPa and adequate elongation for the service use of approx. 10 %.

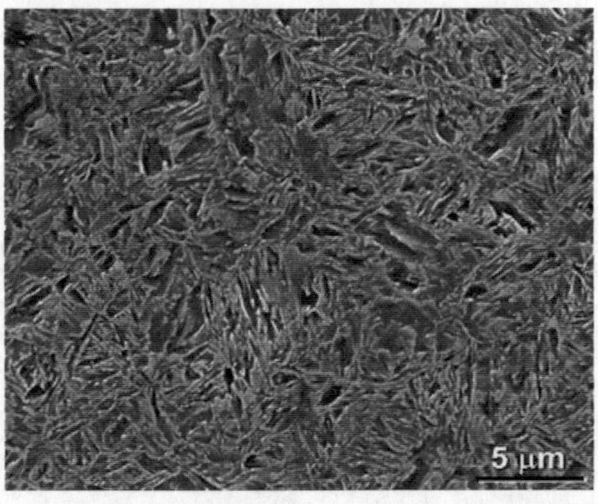


**Figure 9.** UHSS-type mixed martensitic structure with low fractions of bainite, ferrite and retained austenite produced in 42SiCr steel using a manufacturing route with integrated Q-P process

## 7. Summary
Steel – and its treatment – continues to demonstrate its broad innovation potential. Under the pressure from a number of sectors, such as automotive and transport engineering, power generation and manufacture of machines, equipment and tools, research is pursued continuously, yielding new ways of using steels. New types of steels with their sophisticated structures and excellent properties can compete with other alloys, such as non-ferrous and light metals in various applications. In addition, steel is fully recyclable and can be produced using an acceptable amount of energy. Therefore, it will remain a significant material in the future, offering a broad range of options in both traditional and novel applications.

**Acknowledgements**

This paper includes results created within the projects 14-24252S Preparation and Optimization of Creep Resistant Submicron-Structured Composite with Fe-Al Matrix and $Al_2O_3$ Particles subsidised by the Czech Science Foundation, and LO1502 Development of Regional Technological Institute subsidised by the Ministry of Education, Youth and Sports from specific resources of the state budget of the Czech Republic for research and development.

**References**

[1] Masek B, Aisman D, Behulova M and Jirkova H 2010 Structure of Miniature components from steel produced by forming in semi-solid state *Transactions of Nonferrous Metals Society of China* **20** 1037-41

[2] Jirkova H, Aisman D and Masek B 2010 Unconventional Structure of X210Cr12 steel obtained by thixoforming *Journal of Alloys and Compounds* 500-3

[3] Cezard P and Sourmail T 2008 Thixoforming of steel: A state of the art from an industrial point of view *Solid State Phnenomena* **141-143** 25-33

[4] Mikmekova S, Masek B, Jirkova H, Aisman D, Mullerova I and Frank L 2013 Microstructure of X210Cr12 steel after the forming in semi-solid state visualized by very low energy SEM in ultra high vacuum *Applied Surface Science* **275** 403-8

[5] Jirkova H, Aisman D, Sen I, Wagner M F – X, Behulova M, Kusy M and Masek B 2013 Mini-thixoforming of a steel produced by powder metallurgy *Solid State Phenomena* **192-193** 500-5

[6] Sen I, Jirkova H, Masek B, Bohme M and Wagner M F – X 2012 Microstructure and mechanical behavior of a mini-thixoformed tool steel *Metallurgical and Materials Transactions, A* **43** 3034-38

[7] Vancura F, Masek B, Aisman D, Jirkova H, Wagner M F – X and Bohme M 2014 Modification of metastase microstructure of CPM15V steel by heat exposure after treatment in semi-solid state *Journal of Alloys and Compounds* **586** 159-64

[8] Masek B, Khalaj O, Vomela J, Ronesova A, Jirkova H and Svoboda J 2014 *10th International Conference on Composite Science and Technology,* Development of new-generation ODS alloys and ODS composites A.L. Araújo, J.R. Correia, C.M. Mota Soares, et al. Cancun

[9] Masek B, Khalaj O, Jirkova H and Svoboda J 2015 Experimental study on the cyclic and static behavior of new ODS alloys *Materials and Technology* – IN PRESS

[10] Bleck W 2002 *International Conference on TRIP – Aided High Strength Ferrous Alloys* Using the TRIP effect – the dawn of a promising group of cold formable steels Prof. Bruno C. De Cooman Aachen

[11] De Moor E, Lacroix S, Clarke A J, Penning J and Speer J G 2008 Effect of retained austenite stabilized via quench and partitioning on the strain hardening of martensitic steels *Metallurgical and Materials Transactions, A* **39A** 2586-95

[12] Edmonds D V et al 2006 Quenching and partitioning martensite – A novel steel heat treatment *Materials Science and Engineering, A* **438–440** 25–34

[13] Masek B, Stankova H, Malina J, Glass R and Meyer L W 2007 *XXVI. Verfomungskundliches Kolloquim* Integration des interkritischen Glühens in eine Umformprozesskette für die Herstellung von Hohlteilen, Austria

[14] Masek B, Jirkova H, Meyer L, Malina J and Kucerova L 2010 *Proceedings of METAL 2010* Project of unconventional technological process chain for the production of hollow semi-products from low-alloyed TRIP steel Brno

[15] Masek B, Vorel I, Jirkova H and Kurka P 2014 Tváření AHS oceli vnitřním přetlakem *Kovárenství* **50** 5-8

[16] Masek B, Vorel I, Jirkova H and Kurka P 2014 *COMAT 2014* Combination of international high pressure forming and Q-P process for production of hollow products from AHS steel Ostrava

[17] Masek B, Jirkova H, Opatova K, Kurka P and Mahrn U 2015 *4th International Conference on New Forming Technology* Production of High Strength Hollow Shafts Using Hot Stamping and Q-P Process

IOP Conf. Series: Materials Science and Engineering **118** (2016) 012002 doi:10.1088/1757-899X/118/1/012002

# Liquid Metal Embrittlement in Resistance Spot Welding and Hot Tensile Tests of Surface-refined TWIP Steels

**J Barthelmie[1], A Schram[1] and V Wesling[1]**

[1] Institute of Welding and Machining (ISAF), Clausthal University of Technology, Germany

E-mail: office@isaf.tu-clausthal.de

**Abstract.** Automotive industry strives to reduce vehicle weight and therefore fuel consumption and carbon dioxide emissions. Especially in the auto body, material light weight construction is practiced, but the occupant safety must be ensured. These requirements demand high-strength steels with good forming and crash characteristics. Such an approach is the use of high-manganese-content TWIP steels, which achieve strengths of around 1,000 MPa and fracture strains of more than 60 %. Welding surface-refined TWIP steels reduces their elongation at break and produces cracks due to the contact with liquid metal and the subsequent liquid metal embrittlement (LME). The results of resistance spot welds of mixed joints of high-manganese-content steel in combination with micro-alloyed ferritic steel and hot tensile tests are presented. The influence of different welding parameters on the sensitivity to liquid metal embrittlement is investigated by means of spot welding. In a high temperature tensile testing machine, the influence of different parameters is determined regardless of the welding process. Defined strains just below or above the yield point, and at 25 % of elongation at break, show the correlation between the applied strain and liquid metal crack initiation. Due to the possibility to carry out tensile tests on a wide range of temperatures, dependencies of different temperatures of the zinc coating to the steel can be identified. Furthermore, the attack time of the zinc on the base material is investigated by defined heating periods.

## 1. Introduction

The automotive industry seeks for new economic and ecological solutions for new passenger cars to fulfill increasing comfort and safety requirements while complying with the tightening emission standards. The latter demand is mainly met by reducing fuel consumption. The manufacturers can achieve this objective through optimisation of the car engine and/or through reduction of the moving mass. Consequently, the lightweight plays an increasingly important role. Therefore, steel research materials are currently being developed that reach their deformation capacity under mechanical stress not only by dislocation glide, but also due to other deformation mechanisms which make a major contribution to ductility. In this context, a new development of high-manganese austenitic lightweight steels based on the Fe-Mn-Al-Si-C alloy system was introduced for the first time in 1998 by Georg Frommeyer at the Max Planck Institute for Iron Research. Due to a TRIP- and/or TWIP effect, these steels reach fracture strains up to 90 % and strengths up to 1,100 MPa with a simultaneous reduction of density compared to conventional steels, as shown in figure 1 [1, 2, 3, 4].

IOP Conf. Series: Materials Science and Engineering **118** (2016) 012002 doi:10.1088/1757-899X/118/1/012002

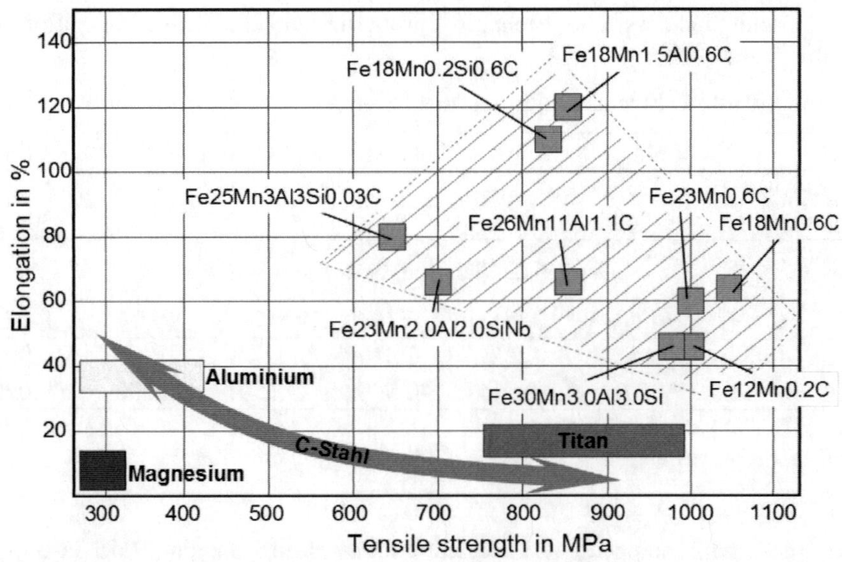


**Figure 1.** Tensile strength and deformation behavior of various metals [5].

When joining blanks in the auto body area, resistance spot welding is used predominantly. Usually unalloyed, but also alloyed steel sheets are processed. In mixed joints of ferritic and austenitic steels problems arise due to the following properties:

- Different composition of steels
- Different material structure
- Significantly different melting points, thermal expansion coefficients and electrical resistances [6]

These typical problems for black and white joints apart, when welding high-manganese-content steels to unalloyed steels, liquid metal embrittlement additionally occurs as follows. The molten zinc of the sheet coating enters locally particularly exposed areas along the grain boundaries in the material, where it leads to cracks. In the worst case, the course of the crack affects the entire thickness of the material, as seen in figure 2.

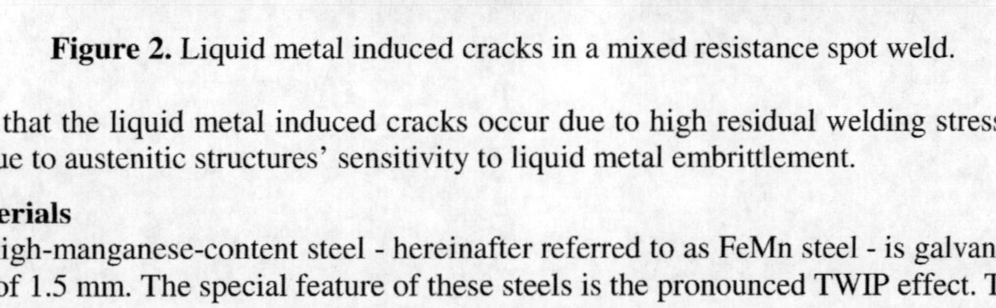


**Figure 2.** Liquid metal induced cracks in a mixed resistance spot weld.

It is proved that the liquid metal induced cracks occur due to high residual welding stresses in mixed joints and due to austenitic structures' sensitivity to liquid metal embrittlement.

## 2. Test materials

The tested high-manganese-content steel - hereinafter referred to as FeMn steel - is galvanized and has a thickness of 1.5 mm. The special feature of these steels is the pronounced TWIP effect. The twinning results in very high strength combined with very high fracture strain. Adding aluminium and silicon reduces the density. The ferritic joint partner is the galvanized microalloyed fine-grained steel

HX340LAD of 1.5 mm thickness. The mechanical properties and chemical compositions of both steels are shown in table 1 and table 2.

**Table 1.** Mechanical properties transverse to rolling direction.

| Steel | Yield Strength (MPa) | Tensile Strength (MPa) | Fracture Strain (%) |
|---|---|---|---|
| FeMn | 592 | 986 | 60 |
| HX340LAD | 405 | 471 | 30 |

**Table 2.** Chemical composition.

| Steel | C (%) | Mn (%) | Al (%) | Si (%) | Ti (%) | Nb (%) | P (%) | S (%) |
|---|---|---|---|---|---|---|---|---|
| FeMn | 0.590 | 15.5 | 2.2 | 2.010 | - | - | 0.002 | 0.001 |
| HX340LAD | 0.11 | 0.5 | 0.015 | 0.500 | 0.150 | 0.090 | 0.025 | 0.025 |

## 3. Test rig and test program

### *3.1. Spot welding machine*

The welds were made with the pneumatic resistance spot welding machine PMS 11-3 of the company Dalex with control system MPS 15043 IQ. The machine has a maximum short-circuit current of 32.2 kA and a maximum electrode force of 6 kN, generated pneumatically. The welding current ranges were determined according to SEP 1220-2 with constant times and pulses as well as variable weld currents, electrode forces and types of electrodes. Only mixed joints were welded. All combinations were also welded with non-galvanized sheets in order to prove that the cracks occur due to LME caused by the zinc only, and not due to the hot cracking which originates from different expansion coefficients.

### *3.2. High temperature tensile testing machine*

The high temperature tensile testing machine, as shown in figure 3, was developed, designed and manufactured at the ISAF. Here, the samples are subject to both mechanical and temperature stresses. The heat input in the specimen is performed by a high frequency alternating current with a frequency of 100 kHz and a maximum power of 20 kW. The test pieces are clamped in a fixed/fixed bearing and extended at a defined speed. Specimens with thicknesses up to 5.0 mm can be tested. The geometry is shown in figure 4. The load cell can measure loads up to 20 kN. The achievable peak temperature depends on the material. The elongation is measured by inductive displacement transducers. The measured values of time, temperature, load and displacement are analyzed with a computer.

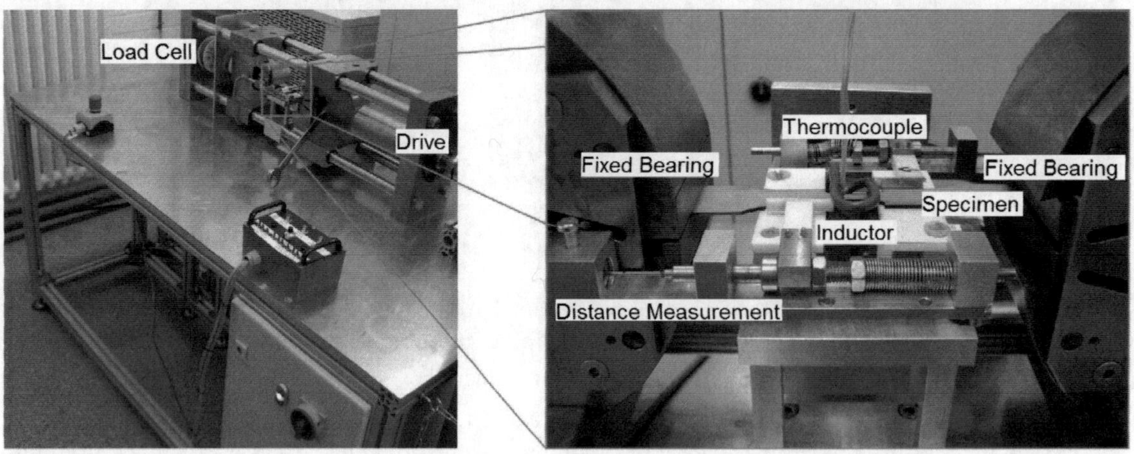


**Figure 3.** High temperature tensile testing machine with specimen chamber.

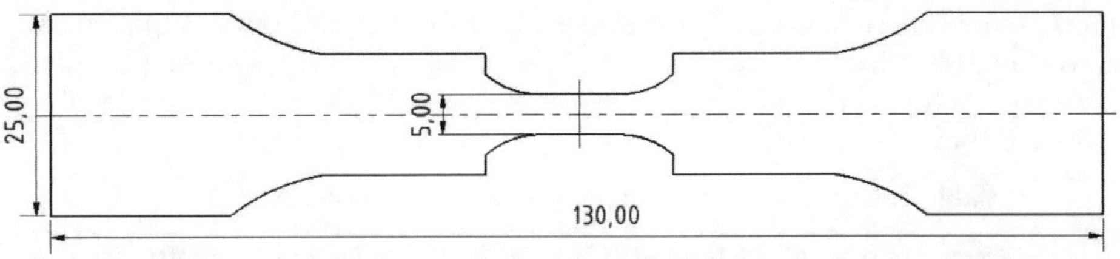


**Figure 4.** Specimen geometry for high temperature tensile tests.

## 4. Results

### *4.1. Influence of the zinc coating on the sensitivity to LME of spot-welded FeMn(+Z)/HX340LAD+Z sheet combinations*

The determination of the relation between the cracks' length and position and the material condition, the current, electrode shape and electrode force was performed based on macro shots of the joints as well as on etched and non-etched metallographic cross section images. In the first test section, resistance spot welds were compared with and without zinc coating in order to study the influence of coating to the cracking. An overview of the influence of the zinc coating to the sensitivity to LME of the FeMn is shown in figure 5.

Welding tests were done with galvanized and non-galvanized FeMn and HX340LAD steel sheets. As a result, all tests with galvanized FeMn had LME cracks. No cracks occurred if both the FeMn steel and the HX340LAD steel were non-galvanized. Mixed joints in which only the HX340LAD steel was coated showed no cracks on the surface of the FeMn sheet. However, some few welds showed cracks in the cross section between the sheets. The zinc of the HX340LAD sheet could get into the non-galvanized FeMn and induced the LME crack.

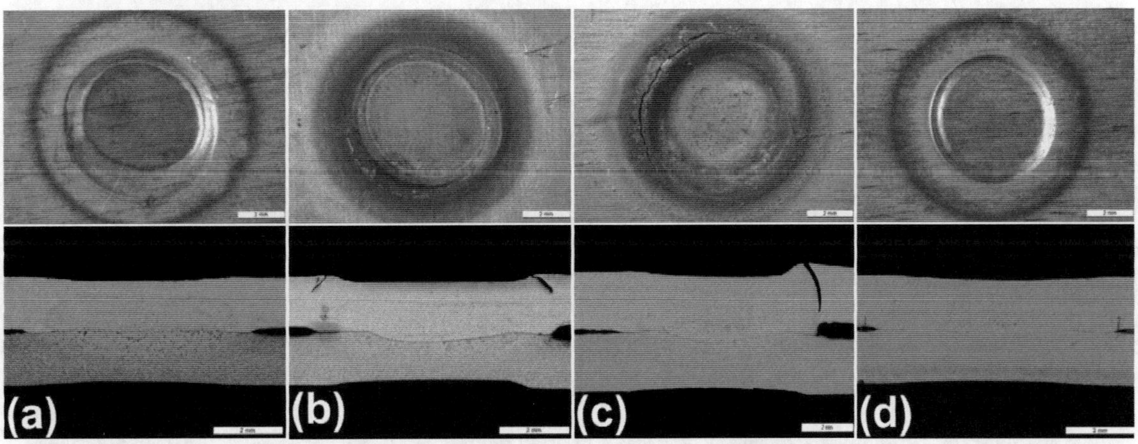


**Figure 5.** Macro shots as well as metallographic cross sections of spot welded mixed joints. (a) FeMn/HX340LAD; (b) FeMn+Z/HX340LAD; (c) FeMn+Z/HX340LAD+Z; (d) FeMn/HX340LAD+Z.

### *4.2. Influence of the energy input on the sensitivity to LME of spot-welded FeMn(+Z)/HX340LAD(+Z) sheet combinations*

After examining the zinc influence on the sensitivity to LME, the energy input was analyzed. In this case, the tests were carried out exclusively on galvanized sheets. The minimum, medium and maximum currents of the detected welding current ranges were investigated with different electrode diameters and forces. It was discovered that the LME cracks increase in length as the current rises. The specimens showed no surface cracks when splatter occurred. Obviously shrinkage stresses were reduced, so that a residual stress field necessary for cracking was missing. However, the splatter created large cavities in connection with interdendritic extending cracks inside of such spot welds.

This was due to large volume shrinkage by the displaced material. Also the sheet failed at the joint line outside the weld nugget, which was caused by the bending of the sheets as a result of splashes. Table 3 shows the average crack length in mixed joints of FeMn+Z/HX340LAD+Z, welded with different electrode caps and forces.

**Table 3.** Crack lengths in mixed joints of FeMn+Z/HX340LAD+Z

| Electrode cap (mm) | Electrode force (kN) | Average crack length (mm) |
|---|---|---|
| F16x20 5.5 | 4.5 | 12.6 |
| F16x20 5.5 | 5.6 | 11.5 |
| F16x20 8.0 | 4.5 | 8.1 |
| F16x20 8.0 | 5.6 | 6.7 |

It was found that the cracks decrease in length with larger cap diameters and larger electrode forces, so they can be reduced down to 50 % compared to the smaller electrode caps and forces.

### *4.3. Influence of the electrode cap geometry on the sensitivity to LME of spot-welded FeMn(+Z)/HX340LAD(+Z) sheet combinations*

The initiation of the LME cracks mostly takes place in the edge area of the electrode indentation on the sheet due to particularly high stresses in this area. For this reason, different types of electrode caps were tested. The idea was to reduce the tensions at the most stressed areas by larger electrode cap radii, as seem in figure 6.

**Figure 6.** Influence of different radii of electrode caps (schematic).

For comparison purposes, the influence of the electrode cap geometry was tested with constant parameters: electrode force 5.6 kN, welding current 6.7 kA. As shown in figure 7, three different diameters and geometries of electrode caps were used in addition to the standard types (4 and 5).

**Figure 7.** Different types of electrode caps.

Type 3 has an even surface and a diameter of 14 mm. It did not provide satisfactory results, since the current of 6.7 kA is not sufficient for an adequate heat development in the weld area. Increasing the current to 9.0 kA led to excessive spatter formation. When welding with rounded cap type 1, invariably sparks and splashes occurred due to the small contact area, so that these specimens were also not satisfying. The combinations of the flat and slightly crowned type 2 initially proved to be useful. The surfaces of the weld spot did not show any cracks on the edge. Metallographic cross sections of the joints uncovered, however, that the cracks moved inwards. In conclusion, the electrode cap geometry could minimize the size of cracking to a small extend, but could not completely prevent the cracks.

### *4.4. High temperature tensile tests*

In addition to the influence of temperature, all test materials were mechanically tested for the present stresses. Knowing the influences of temperature, time and stress, the existent range of LME cracks was determined. Of major importance was the question whether the zinc had to be in molten state, above 420 °C, or whether it could penetrate even at lower temperatures along the grain boundaries into the base material by diffusion. High temperature tensile tests were carried out in 50 °C steps between 300 °C and 600 °C.

In the galvanized FeMn specimens, macro shots show LME cracks on the surface starting from 450 °C. This was also perceptible in the mechanical and technological characteristics. For example the values of tensile strength in comparison to galvanized and non-galvanized specimens were on one level up to a test temperature of 400 °C. From 450 °C on, and finally at the maximum test temperature for galvanized specimens of 600 °C, the values varied already by 50 MPa, as shown in figure 8.

To illustrate the superficial LME cracks, the fractured surfaces of the specimens of galvanized and non-galvanized FeMn sheets at test temperatures of 300 °C up to 600 °C were recorded. The specimens at 300 °C and 400 °C showed no cracks and an equally reduction of area in both conditions. From 450 °C on and most clearly at 600 °C, widening cracks could be observed. A reduction of area as happens in non-galvanized sheets did not occur. The reduction in area in both states at test temperatures up to the melting point of zinc is at the same level. With further increase in temperature, however, the curves move away from each other, so that the elongation at break of non-galvanized FeMn steel is at 48 % and the galvanized at 22 %.

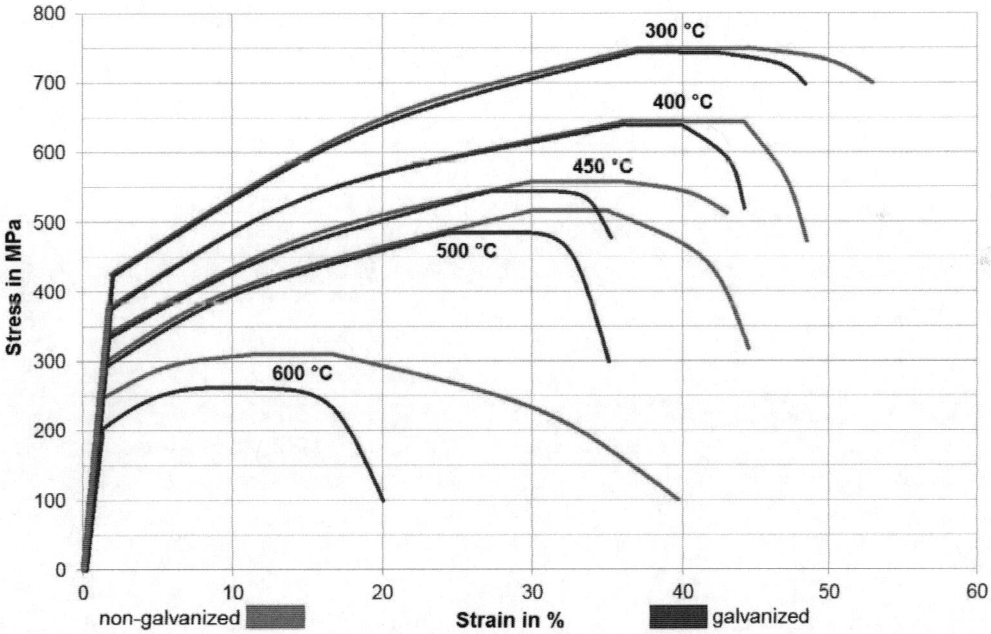


**Figure 8.** Stress-strain-diagram of high temperature tensile tests with galvanized and non-galvanized FeMn steel.

In a further step, the influence of strain on the sensitivity to LME was examined by tensile tests on galvanized and non-galvanized FeMn specimens at elevated temperatures of 450 °C and 500 °C.

For the respective temperature levels, three different yield strengths were considered:

- just below $R_{p0,2}$
- just above $R_{p0,2}$
- 25 % total elongation

The non-galvanized specimens had no cracks regardless of temperature or yield stress levels. Also in galvanized specimens, no cracks were determined below the yield point. However, at 450 °C and just above the yield point, the galvanized specimens had cracks of an average length of 2 µm. At 25 % total elongation, the cracks extended up to 5 µm. Galvanized samples at 500 °C and just above the yield stress evinced cracks with an average length of 10 µm. At 25 % total elongation, significantly more cracks occurred with an average length of 8 µm. In addition, there were sporadic cracks with a length of 100 µm. The increased mechanical load on the material in combination with the high temperature led to a sudden rise in crack length.

Further investigations were carried out with various pre-elongations, dwell temperatures, dwell stresses and cooling media. The tests showed no dependence on the dwell time and no cracks.

The pre-elongation of the specimens leads to cold working of the material. The tensile strength could be partially increased from the original value of 1,000 MPa up to 1,300 MPa. This cold working did not affect the sensitivity of the FeMn steel to LME.

In further tests, the specimens were heated up to 500 °C and then elongated at this temperature level until a stress of 450 MPa was achieved. At this temperature and stress level the specimens were maintained again for different dwell times (0.5 minutes, 1 minute, 1.5 minutes, 2 minutes…5 minutes). Thereafter, the sample was released from load and heat and cooled down in ambient atmosphere. Finally, a tensile test was carried out at room temperature. Here, a significant influence on the yield strength and reduction of area was found and is shown in figure 9.

The tests have shown that the thermal treatment of the galvanized material significantly affects the yield strength and the reduction of area. Depending on the dwell time, the yield strength drops by 100 MPa down from 455 MPa at 0.5 min to 355 MPa at 5 min. The reduction of area decreases from 48 % at 0.5 min down to 33 % at 5 min. The tensile strength is not affected.

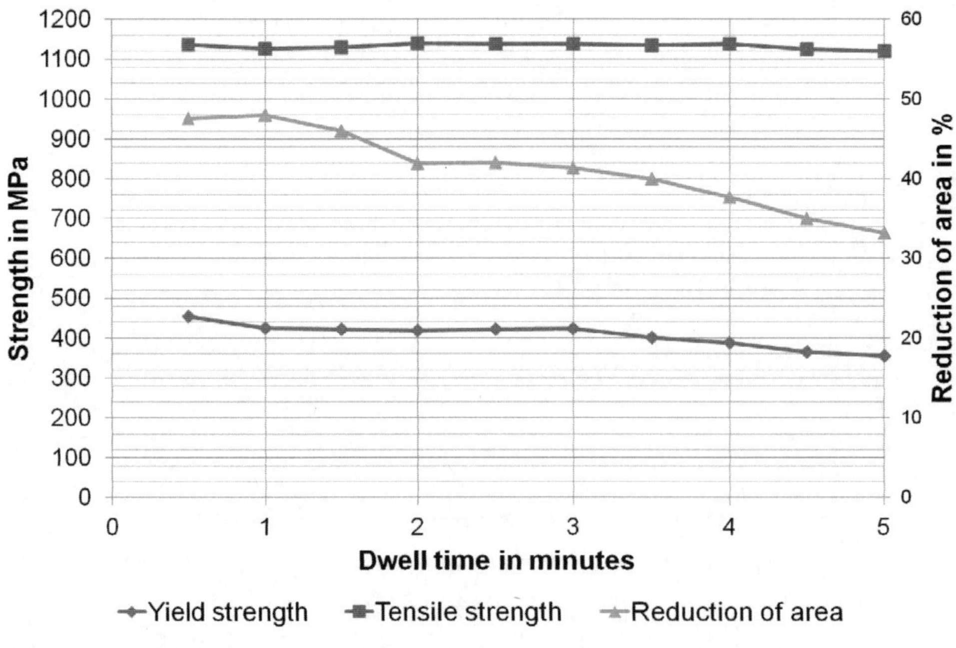


**Figure 9.** Strength and reduction of area of the galvanized FeMn specimens elongated to 450 MPa and heated to 500 °C depending on the dwell time.

## 5. Summary and conclusion

Liquid metal induced cracks occur during spot welding of high-manganese steel and microalloyed fine-grained steel HX340LAD with 1.5 mm thickness, whenever at least one of the sheets is galvanized. No cracks occur, when both steels are non-galvanized. Using larger electrode caps and forces reduced the crack lengths down to 50 %. Also, the use of different electrode cap geometries could not completely prevent the cracks.

The construction of the high temperature tensile testing machine has laid the foundation for systematic determination of the LME cracks' existence range regardless of the welding process.

At 600 °C, the tensile strength and the elongation at break of galvanized FeMn steels are already lower by 20 % and 50 % than the values of the non-galvanized material.

At a temperature of 500 °C and stress of 450 MPa, the increased mechanical load on the material leads to a sudden rise in crack length.

Experiments with different dwell times have shown that these temperatures and stresses result in a strong decrease of the yield strength and reduction of area.

Further work is necessary and will be performed, in order to achieve ideal joints completely without liquid metal induced cracks or whose length is at least reduced to a tolerable minimum.

## References

[1] Frommeyer G and Gräßel O 1998 Light constructional steel and the use thereof *Patent specification WO001999001585A1*

[2] Frommeyer G and Gräßel O 1998 High strength TRIP/TWIP and superplastic steels: development, properties, application *La Revue de Metallurgie-CIT* **10** 1299–1310

[3] Gräßel O and Frommeyer G 1998 High strength TRIP/TWIP steels for crash resistance of structural car body parts *ISATA, International Symposium on Automotive Technology & Automation* B **31** H31

[4] Gräßel O and Frommeyer G 1998 Hochfeste FeMn-(Al,Si)-TRIP/TWIP-Stähle für crash-stabile Fahrzeugstrukturen *Werkstoffwoche* B **2**

[5] Frommeyer G and Rablbauer R 2008 Entwicklung hochfester und supraduktiler Leichtbaustähle für die Verkehrstechnik *technical report Max-Planck-Institut für Eisenforschung GmbH*

[6] Pohle C 1999 Schweißen von Werkstoffkombinationen, Metallkundliche und fertigungstechnische Grundlagen sowie Ausführungsbeispiele *Verlag für Schweißen und Verwandte Verfahren DVS Verlag*

IOP Conf. Series: Materials Science and Engineering **118** (2016) 012003 doi:10.1088/1757-899X/118/1/012003

# Mechanical and microstructural behavior of brazed aluminum / stainless steel mixed joints

**V. Fedorov[1], S. Weis[1] and G. Wagner[1]**

[1] Chair of Composite Materials, Institute of Materials Science and Engineering, Technische Universitaet Chemnitz, 09125 Chemnitz, Germany

vasilii.fedorov@mb.tu-chemnitz.de

**Abstract.** There is a requirement to combine different materials such as aluminum and stainless steel in industrial applications like automotive heat exchangers. Brazing offers the possibility to reduce the joining temperature in comparison to welding due to the lower liquidus temperature of the fillers. In the present work, the mechanical and microstructural behavior of aluminum / stainless steel mixed joints is investigated. The specimens are produced by induction brazing using an AlSi10filler and a non-corrosive flux. To evaluate the mechanical properties of the joints, tensile tests at elevated temperatures are carried out. Additionally, long-term thermal exposure experiments are done in order to investigate the changes in the microstructure.

## 1. Introduction

In the last years, mixed joints became more and more interesting in the automotive industry due to the potential in weight and cost reduction. Especially joints of aluminum and stainless steel are required. However, joining of these materials is complicated due to significant differences in the melting ranges and the formation of brittle AlFe intermetallic compounds (IMC) [1]. The phase formation is caused by the diffusion and reaction of Fe and Al, these effects are highly dependent on the temperature and holding time of the joining process. A higher thermal impact leads to thicker IMC layers at the interface to the stainless steel. These layers are generally considered to be the main cause for the mechanical degradation of the joints. In order to minimize the layer thickness, it is necessary to reduce the heat input and therefore the intensity of chemical reactions and diffusion. In this case, brazing offers a possibility to reduce the joining temperature in comparison to welding because of the lower liquidus temperature of the used fillers. Due to the short brazing time and the local heat input into the joint, induction-brazing leads to the formation of IMC layers less than 10 μm in thickness, resulting in a sufficient mechanical strength [2]. In the present work, the microstructural evolution of the joints during long-term thermal exposure experiments is investigated. The mechanical properties of the joints are evaluated by tensile tests at elevated temperatures.

## 2. Experimental procedures

Aluminum alloy 3003 (AA 3003) and AISI 304 austenitic stainless steel plates with dimensions of 40 mm × 20 mm × 1,5 mm were used as base materials. The AlSi10 filler (composition corresponds to aluminum alloy 4045) was applied as a paste. Before brazing, the surfaces were cleaned by abrasive paper and ethanol. A non-corrosive flux was applied and dried. The aluminum / stainless steel mixed joints were produced by induction brazing in an argon atmosphere at 600 °C according to the eutectic temperature of the AlSi10 filler (575 °C) and liquidus temperature of 3003 aluminum alloy (630 °C). The temperature measurement was carried out using a twin-channel pyrometer. The brazing process including cooling time takes about 2 min. The thickness of the produced brazed joint was adjusted at 50 µm. After brazing, the specimens were prepared metallographically. Long-term thermal exposure experiments were done in a muffle furnace at 200 °C for 6 h, 48 h and 120 h to observe the growth kinetics the IMC layer. The exposure temperature corresponds to the maximum application temperature of automotive heat exchangers. The microstructure of the mixed joints was characterized by optical microscopy (OM) and scanning electron microscopy (SEM). The chemical composition was analyzed by energy-dispersive X-ray spectroscopy (EDX). The mechanical properties were determined by tensile tests at 20, 100, 150 and 200 °C. Furthermore, the fracture behavior was observed and discussed.

## 3. Results and discussion

### *3.1. Microstructure*

Figure 1 shows the typical microstructure of the produced aluminum / stainless steel mixed joints. A primary Al solid solution and in between an Al-Si eutectic are formed. At the interface to the stainless steel, the expected IMC layer is formed due to diffusion and reaction of Fe, Al and Si. This layer is shown in more detail in the SEM micrographs, figure 2. Without a holding time at brazing temperature, a thickness of about 0.5 µm can be observed, it increases with extended holding time.

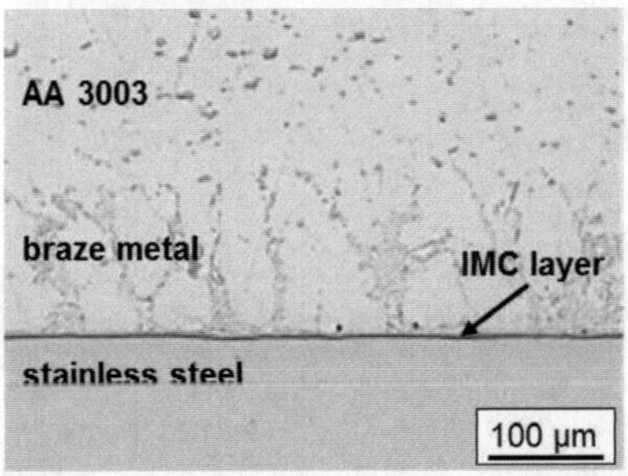


**Figure 1**. Microstructure of the brazed aluminum / stainless steel joint (OM)

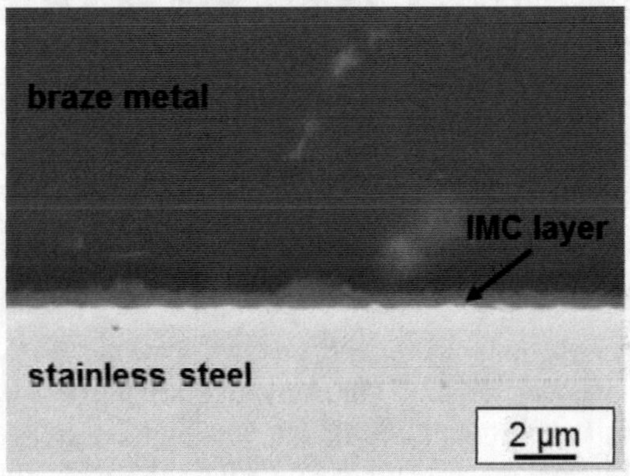


**Figure 2**. Interface to stainless steel (SEM)

Due to its low thickness, the chemical composition of the IMC layer is difficult to determine quantitatively by EDX analysis. A maximum thickness of about 10 µm was reached at holding times of 10 min. These specimens were used to measure the chemical composition of the IMC layer by EDX: 71–74 at.% Al, 17–19 at.% Fe and 7–10 at.% Si. Additionally, traces of Cr and Ni were detected. The results indicate that the IMC layer conforms the $Al_7Fe_2Si$ phase. Song et al. reported that an $Al_{7.2}Fe_{1.8}Si$ layer is formed at the interface of TIG welded joints of aluminum alloy und stainless steel [3]. Literature data about the Al–Fe–Si system also confirm this conclusion [4].

### *3.2. Long-term thermal exposure*

In order to investigate the growth of IMC layers depending on of the application temperature, long-term thermal exposure experiments were conducted at 200 °C for 6 h, 24 h and 120 h, figures 3, 4, 5 and 6 show the results:

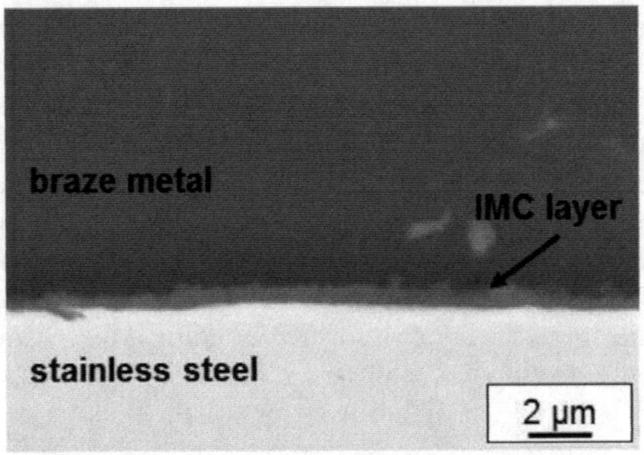


**Figure 3**. Micrograph of long-term thermal exposure experiments, initial state

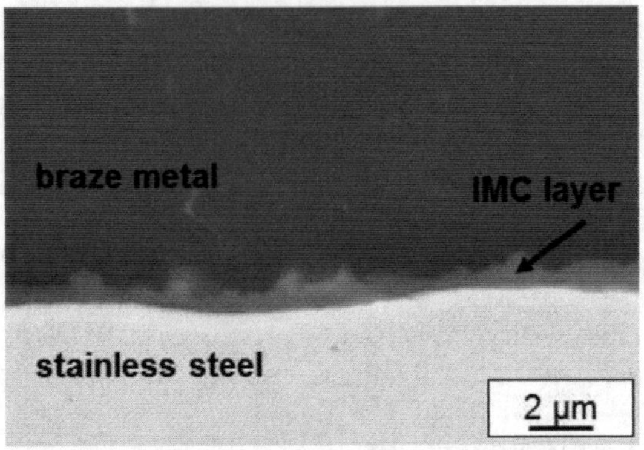


**Figure 4**. Micrograph of long-term thermal exposure experiments at 200 °C for 6 h

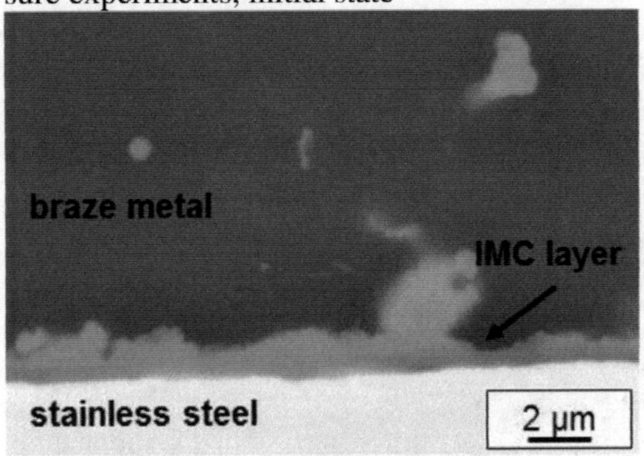


**Figure 5**. Micrograph of long-term thermal exposure experiments at 200 °C for 24 h

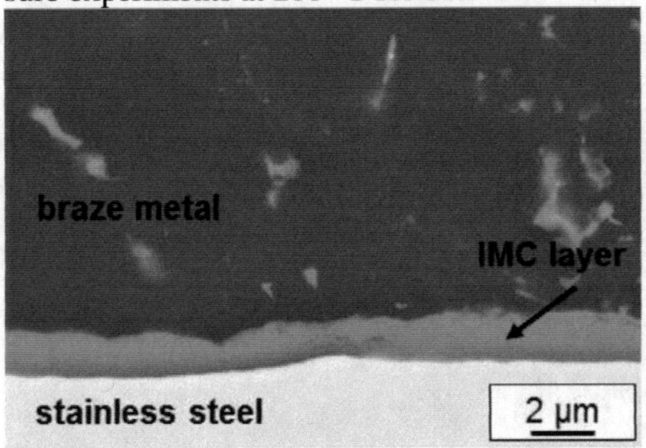


**Figure 6**. Micrograph of long-term thermal exposure experiments at 200 °C for 120 h

The IMC layer in the as brazed condition is about 0,5 µm thick, figure 3. It slowly begins to grow during thermal exposure at 200 °C for 6 h, figure 4. The diffusion of Fe, Al and Si is reactivated. In addition, the time has a great influence on the diffusion process. With further exposure time for 24 h and 120 h, the thickness is increasing to approx. 2 µm, figures 5 and 6. Summarized, the long-term thermal exposure has a significant effect on the growth of the IMC layer. Nevertheless, the thickness is lower than 10 µm, which is known to be uncritically from literature data [2]. Therefore, good mechanical strength of the brazed joints is expected.

### *3.3. Tensile tests at elevated temperatures*

To evaluate the mechanical properties of the joints, tensile tests were carried out at 20, 100, 150 and 200 °C in a material testing machine Zwick Allround-Line 20 kN. Three samples per testing temperature were produced with an overlap of 10 mm. The fractured samples are shown in figure 7. In addition, the front and side projections of the brazed samples are plotted schematically. The brazed area is tagged with the dashed box. The fracture of all samples occurs in the Al base material. The brazed joints did not break. Consequently, the results of the tensile tests correspond to the strength of the Al base material, figure 8. At room temperature (20 °C), a tensile strength of 145 MPa is determined. Thus, the brazing process did not significantly affect the strength of the Al base material: strain hardened, 120–160 MPa [5]. With an increase of testing temperature, a strength decrease is registered. With regard of these results to the brazed area, minimal joining strengths of 22 MPa at 20 °C, 20 MPa at 100 °C, 14 MPa at 150 °C and 10 MPa at 200 °C can be ensured. This is matching with results of Lugscheider et al. (15–30 MPa) [6].

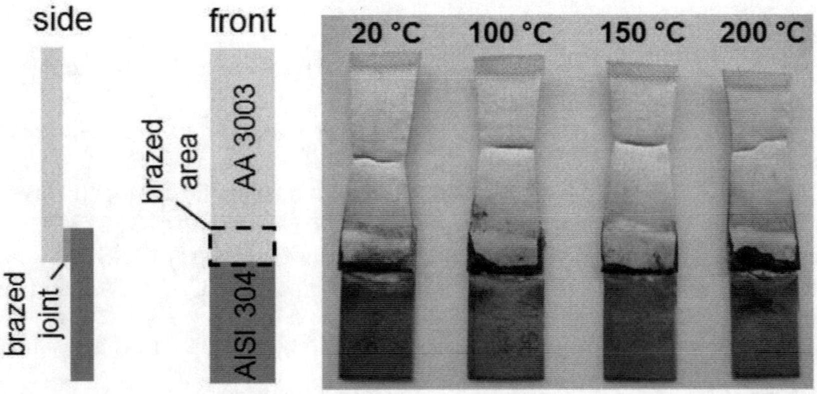


**Figure 7.** Fracture image of brazed aluminum / stainless steel samples

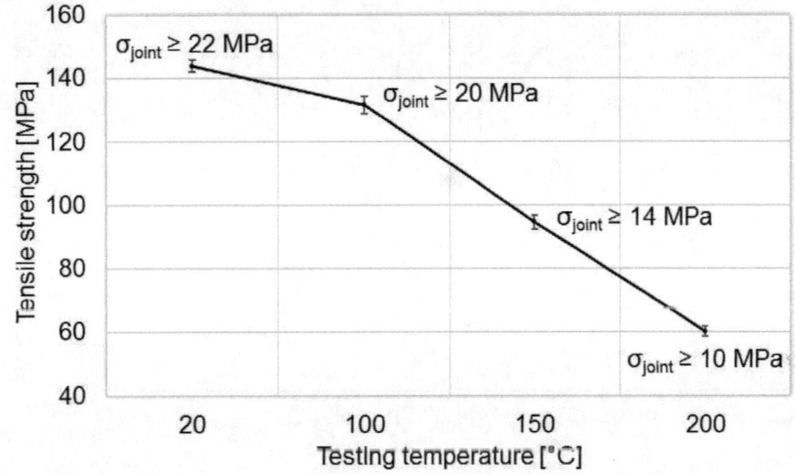


**Figure 8.** Tensile strength of the base material AA 3003 after the brazing process as a function of the testing resp. application temperature

## 4. Conclusions

Aluminum / stainless steel mixed joints, produced by induction brazing using an AlSi10filler, show thin IMC layers at the stainless steel interface. EDX analyses indicate a chemical composition of $Al_7Fe_2Si$. It was figured out that the layers grow depending on exposure time in long-term thermal exposure experiments. However, the thickness remains under 10 µm, expecting good mechanical properties of the brazed joints. The results of the tensile tests show that the fracture of all brazed samples occurs in the Al base material. The brazed joints did not break. With regard to the brazed area, minimal joining strengths of 22 MPa at 20 °C, 20 MPa at 100 °C, 14 MPa at 150 °C and 10 MPa at 200 °C can be ensured. In further investigations, the overlap will be reduced in order to cause the fracture within brazed joint.

**Acknowledgments**
The authors thank the Federal Ministry for Economic Affairs and Energy for the financial support of the presented work and the colleagues of MAHLE Behr GmbH & Co. KG for the cooperation.

IOP Conf. Series: Materials Science and Engineering **118** (2016) 012003 doi:10.1088/1757-899X/118/1/012003

## 5. References

[1] Dong H and Yang L 2012 Improving arc joining of Al to steel and Al to stainless steel *Materials Science and Engineering* A **534** 424–35

[2] Kreimeyer M and Sepold G 2002 *Proc. Int. Conf. on Applications of Lasers and Electro-Optics* (Jacksonville: Laser Institute of America)

[3] Song J L and Lin S B 2009 Effects of Si additions on intermetallic compound layer of aluminum–steel TIG welding–brazing joint *Journal of Alloys and Compounds* **488** 217–22

[4] Du Y and Schuster J C 2008 A thermodynamic description of the Al-Fe-Si system over the whole composition and temperature ranges via a hybrid approach of CALPHAD and key experiments *Intermetallics* **16** 554–70

[5] Standard DIN EN 485-2 2013 *Aluminium and aluminium alloys – Sheet, strip and plate – Part 2: Mechanical properties* (Berlin)

[6] Lugscheider E and Schlimbach K 1998 Fluxfree brazing of aluminium and Cr-Ni-steel. *Proc. Conf. Brazing High Temperature Brazing and Diffusion Welding* (Aachen) pp. 191–95

IOP Conf. Series: Materials Science and Engineering **118** (2016) 012004 doi:10.1088/1757-899X/118/1/012004

# New hermetic sealing material for vacuum brazing of stainless steels

**S Hildebrandt[1], G Wiehl[1] and F Silze[2]**

[1]Umicore AG & Co.KG, Rodenbacher Chaussee 4, 63457 Hanau, Germany
[2]Institut für Komplexe Materialien, Helmholtzstr. 20,01069 Dresden, Germany

**Abstract.** For vacuum brazing applications such as in vacuum interrupter industry *Hermetic Sealing Materials* (HSM) with low partial pressure are widely used. AgCu28 dominates the hermetic sealing market, as it has a very good wetting behavior on copper and metallized ceramics. Within recent decades wetting on stainless steel has become more and more important. However, today the silver content of HSMs is more in focus than in the past decades, because it has the biggest impact on the material prices. Umicore Technical Materials has developed a new copper based HSM, CuAg40Ga10. The wettability on stainless steel is significantly improved compared to AgCu28 and the total silver content is reduced by almost 44 %. In this article the physical properties of the alloy and its brazed joints will be presented compared to AgCu28.

## 1 State of the art

Brazing is a technology for joining, in general, two metal or metallized parts without introduction of major microstructural changes into the base materials. This is in contrast with technologies such as welding [1].

A prerequisite for brazing is the sufficient wetting of the brazing filler metal of both base materials as well as spreading on the parts subsequently. A characteristic for the wetting is the wetting angle $\theta$. The wetting angle is influenced by the surface energies of the brazing filler metal and the base material as well as the brazing atmosphere.

[2]. Therefore, the absence of scale forming oxygen on the surface is required to allow a proper wetting process. Among others, this can be achieved by the application of reactive fluxes, brazing in a protective gas atmosphere, or in a vacuum [3]. The vacuum brazing process of metal parts results in high quality brazing joints with low porosity, because fluxes and other gas forming impurities that could lead to gas pores inside the brazing joints are avoided. In addition, vacuum brazing allows the sealing of high purity vacuum inside of e.g. vacuum interrupter tubes, if high purity and non-evaporating brazing filler metals are utilized [4].

The purity of the vacuum as well as the surface condition of the parts to be joined, can have a significant effect on the wetting and therefore joining behavior of the HSM [5]–

[7]. A prominent example of a typical HSM is the eutectic alloy AgCu28 [4]. AgCu28 features excellent wetting on copper base materials while the wetting on steels could be improved by additions such as Ge, Pd, Ni, and Co [4]. The aim of the present study is to investigate a new alloy CuAg40Ga10 that exhibits independently improved wetting behavior on stainless steels compared to AgCu28. The improved wetting behavior is an advantage since the chemical concentration and therefore the quality of the stainless steel parts can vary within the limits of the steel grade standard. Furthermore, significantly less silver is required while still maintaining a similar brazing temperature to AgCu28 [8]. In order to demonstrate the performance of the new HSM its characteristics and properties of its joints will be discussed.

## 2 Methods

### *2.1 Material's selection*

The HSM's evaluated in this study were manufactured according to the tolerances for volatile elements and impurities of standard DIN EN ISO 17672 (special vacuum requests: class 1) and with the tolerances of the alloying components given by table 1.

**Table1**. Chemical concentration ranges of investigated HSM's

| Alloy name | guaranteed chemical concentration | | | identical to standardized alloy of DIN EN ISO 17672 |
|---|---|---|---|---|
| | Ag | Cu | Ga | |
| AgCu28 | 71-73 % | 27-29 % | - | Ag272 |
| CuAg40Ga10 | 39-41 % | bal. | 9,5-10,5 % | not applicable |

Hence wetting and brazing on copper and nickel is due to the high solubility of copper and nickel in Ag-Cu brazing alloys not challenging and the interactions of the brazing material and the substrate are well investigated[9, [10]. The present study focusses on application tests of the HSM's on two typical stainless steel grades well-known in the vacuum industry: 1.4307 and 1.4404. These steel grades are specified according to DIN EN 10088-3 and also known under the following material names X2CrNi18-9 and X2CrNiMo17-12-2 respectively. In figure 1 ring-shaped preforms can be seen.

**Figure 1**. Ring-shaped brazing preforms produced from new CuAg40Ga10 HSM

*2.2 Melting range, density and coefficient of thermal expansion (CTE)*

Essentially for the technical function of HSM's is their brazing temperature influenced by their melting range. Therefore, liquidus and solidus temperatures were determined by thermal analysis using a TGA/DSC 1 machine from Mettler-Toledo. Samples of 150-200 mg weight were heated in aluminum oxide crucibles. The heating rate was set to 10 K/min and flowing argon atmosphere was used. Liquidus and solidus temperatures were derived from the measured data as described by *Schnee et al.* [11]. The density of the brazing alloys was evaluated with a hydrostatic balance according to the Archimedes principle [12].

The CTE was measured by dilatometry using a Netzsch 402S. Samples with a length of 30 mm of HSM were applied while the heating rate was constant at 2 K/min. An argon atmosphere was injected to avoid oxidation of test samples during measurement. The CTE was determined for the temperature range from room temperature up to 600 °C.

*2.3 Wetting behavior and spreading factor*

The formation of the brazing joint and therefore the later strength and vacuum tightness of the joint depends on the wetting and spreading of the HSM on the base materials. By in-situ measurement of the wetting angle between liquid HSM and base material the wetting process was quantified. For determining the wetting angle the sessile drop method was used [13]. The wetting behavior of AgCu28 and CuAg40Ga10 on two standardized stainless steels (1.4307 and 1.4404) was evaluated. The HSM was positioned on a horizontal sheet of stainless steel base material, 15 x 15 x 1 $mm^3$ in size. The wetting tests were performed in a vacuum chamber with gas pressures below 5·10-5 mbar. A molybdenum wire spiral was used to heat the sample while the local temperature was measured with a thermocouple positioned in a distance of only 2 mm above the sample. The wetting angle measurement was conducted in a temperature-dependent mode by applying a constant heating rate of 5 K/min throughout the experiment. Left and right wetting angles of the HSM droplet on the base material were examined optically and averaged afterwards. The spreading factor S is a crucial property for the technical performance of the HSM combining wetting behavior and the kinetics of flowing of the liquid at a given temperature. It was determined, by measuring the wetted area A1 of the HSM on base material after a temperature treatment of 820 °C for 5 min. S is calculated with the equation 1, where *A1* and *A0* are the areas of the applied alloy before and after temperature treatment, figure 2.

IOP Conf. Series: Materials Science and Engineering **118** (2016) 012004 doi:10.1088/1757-899X/118/1/012004

For cleaning the stainless steel base materials, they were mechanical grinded (No chemical cleaning or degreasing). Initial surface A0 of the alloy discs is given by their diameter of 10 mm². The thicknesses of the discs were 0.15mm.

$$S = A1/A0 \tag{1}$$

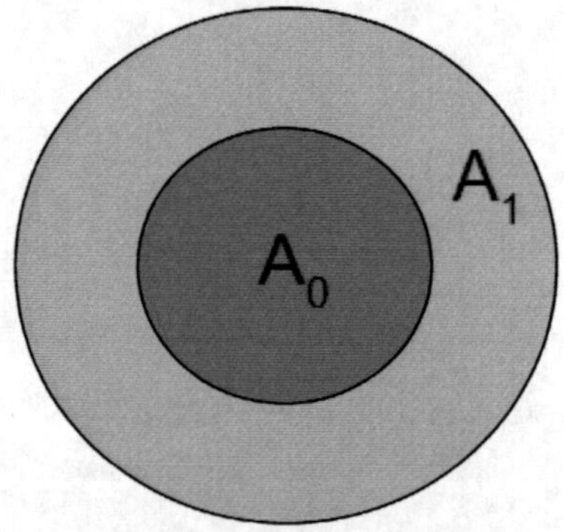


**Figure 2.** Determination of the spreading factor S from brazed HSM discs

*2.4 Micrographs and ultimate tensile strength of brazing joints*

Typically for brazing joints there is a strong structure-property relationship between features in the microstructure and their mechanical behavior such as the responsiveness to tensile loads. Therefore, in this study micrographs were taken from brazed samples by mechanical metallographic preparation and optical microscopy. During optical microscopy, bright field imaging was applied to render the material contrasts.

The typical load case during the application of a vacuum interrupter during its lifetime is tensile load in the brazing joints. The ultimate tensile strength of the brazing joints can provide a quantitative degree for the maximum resistance of the brazing joint against tensile loads. The ultimate tensile strength was tested according to DIN EN12797:2000. The geometry of the tensile specimen is shown in figure 3.

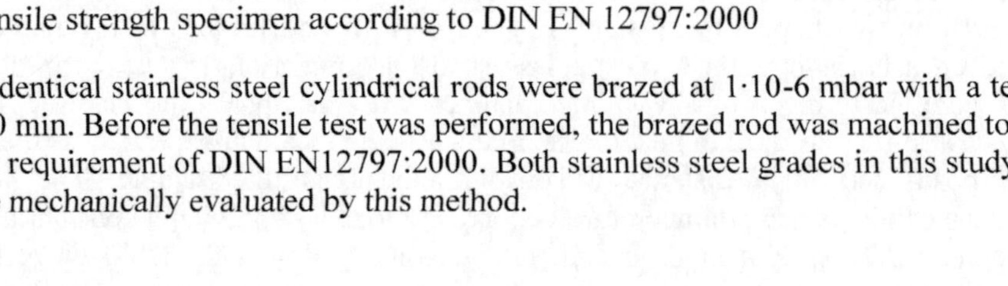


**Figure 3.** Tensile strength specimen according to DIN EN 12797:2000

The two identical stainless steel cylindrical rods were brazed at 1·10-6 mbar with a temperature of 830 °C for 10 min. Before the tensile test was performed, the brazed rod was machined to comply with the geometry requirement of DIN EN12797:2000. Both stainless steel grades in this study (1.4307 and 1.4404) were mechanically evaluated by this method.

**3 Results**

*3.1 Physical properties of HSM*

The melting range of a HSM is one of the core properties of the material, because it is influencing other technological parameters of the material such as the brazing temperature. Compared to AgCu28, which is a eutectic alloy with a melting point at 779 °C, the CuAg40Ga10 has a melting range. The

IOP Conf. Series: Materials Science and Engineering **118** (2016) 012004 doi:10.1088/1757-899X/118/1/012004

melting range is about 100 K from 726 °C to 831 °C, table 2. In consequence liquid of CuAg40Ga10 is already available at lower temperatures than 779 °C ($T_S$ of AgCu28).

**Table 2.** Chemical concentration ranges of investigated HSM's

| Material | TSolidus [°C] | TLiquidus [°C] | Density [g/cm³] | CTE 0°C-600°C [10-6/K] |
|---|---|---|---|---|
| AgCu28 | 779 | 779 | 9.97 | 23.60 |
| CuAg40Ga10 | 726 | 831 | 9.33 | 21.56 |

From figure 4 it can be seen that the main melting event is at 756 °C. Under the assumption that latent heat consumed for melting at 780 °C (~70 % of total heat according to measurement) correlates with the molten volume fraction it can be concluded that the major volume fraction of the material is melted below 780 °C.

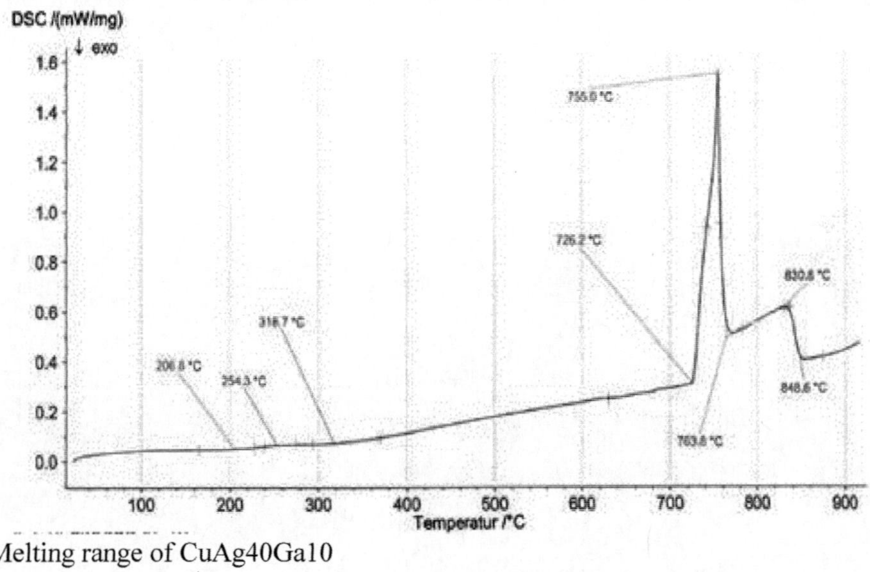


**Figure 4:** Melting range of CuAg40Ga10

The density of CuAg40Ga10 is slightly lower than that of AgCu28, which is beneficial hence the required volume and not the required weight for filling a brazing joint is constant, table 2. For CuAgGa10 less weight and therefore again less silver per brazing joint is required by the applicant. The thermal expansion coefficient of CuAg40Ga10 is very similar to AgCu28, table 2. Therefore, additional thermal residual stresses between the base materials caused by use of CuAg40Ga10 instead of AgCu28 are not expected if the base materials are not exchanged.

*3.2 Brazing joint properties*

*3.2.1Wetting and spreading behavior during brazing*

In the literature, it is reasoned that partial wetting occurs for wetting angles $\theta < 90°$ [14]. From an application perspective the working point of the HSM is achieved at the temperature when the wetting angle falls below 90°. Figure 4 and figure 5 show the measured development of the wetting angle over temperature. The critical temperatures for $\theta < 90°$ are summarized in table 3.

**Table 3:** Temperature T [θ < 90°] for which wetting is initiated

| T[θ < 90°] in °C | | HSM | |
|---|---|---|---|
| | | AgCu28 | CuAg40Ga10 |
| Steel Grades | 1.4307 | 827 | 829 |
| | 1.4404 | 861 | 821 |

The wetting angle measurements show that especially for the case of the 1.4404 stainless steel CuAg40Ga10 leads to significantly lower wetting angles than AgCu28. For 1.4307 the difference is smaller, but still measureable. In addition, for the wetting on 1.4404 steel the temperature for θ < 90° is significantly lower than for AgCu28. Therefore, it can be concluded that the wetting behavior on the two tested steels grades is improved for CuAg40Ga10 compared to AgCu28.

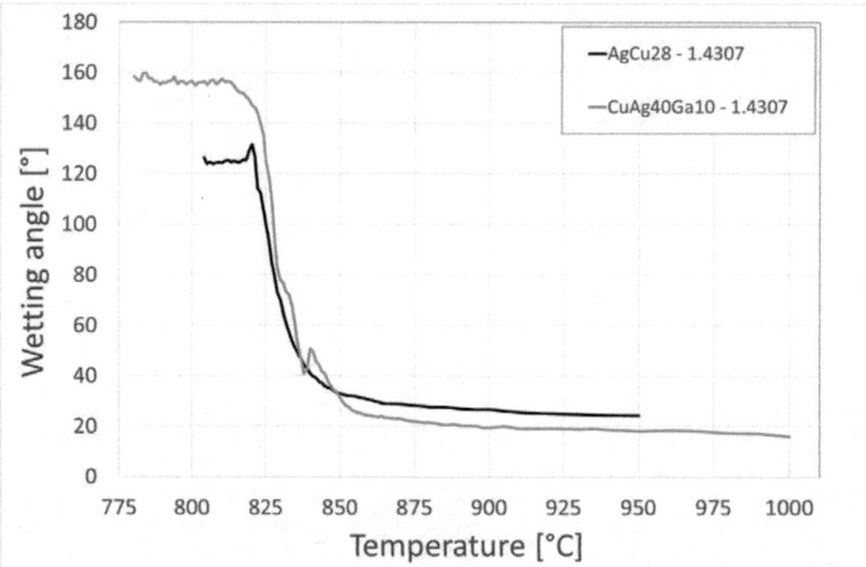


**Figure 5**. Temperature-dependent wetting of AgCu28 and CuAg40Ga10 on 1.4307

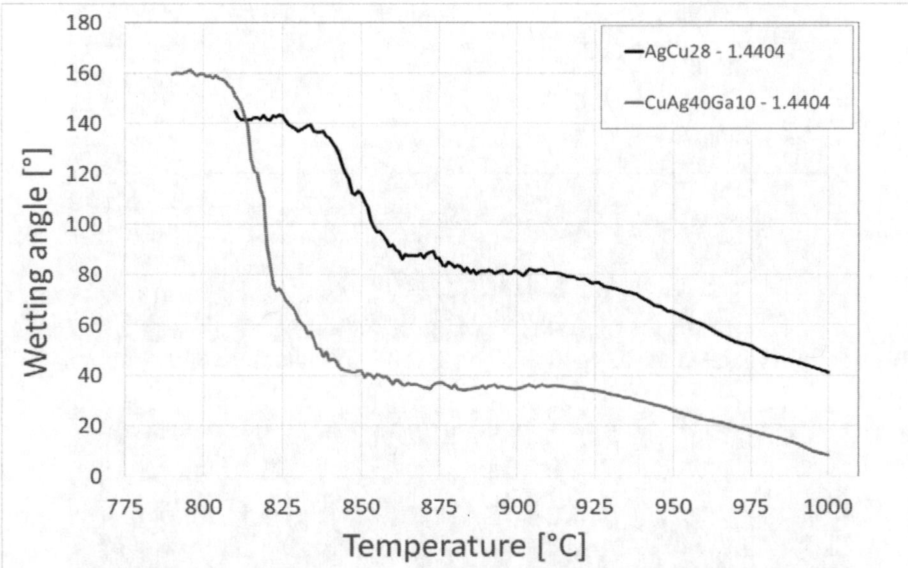


**Figure 6**. Temperature-dependent wetting of AgCu28 and CuAg40Ga10 on 1.4307

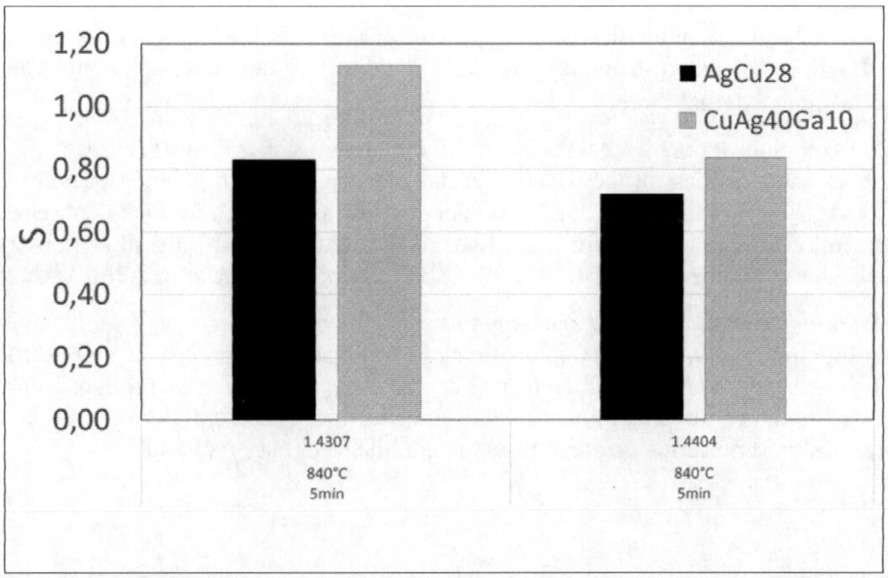


**Figure 7.** Spreading factor S of AgCu28 (black) and CuAg40Ga10 (grey) on 1.4307 and1.4404 after a temperature treatment of 840 °C for 5 minutes

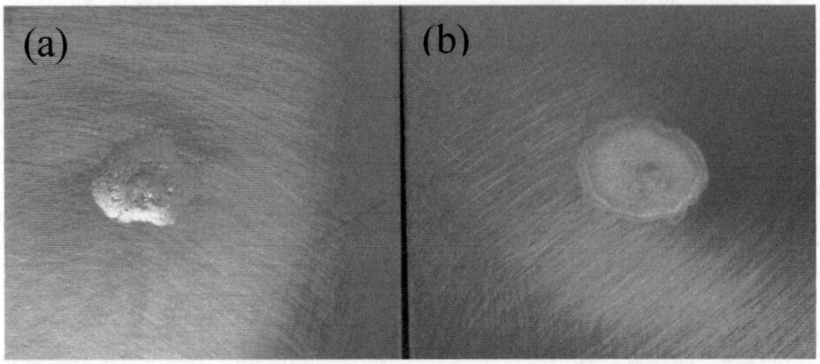


**Figure 8**. Wetting of AgCu28 and CuAg40Ga10 on 1.4307, Temperature 840 °C, 5 min

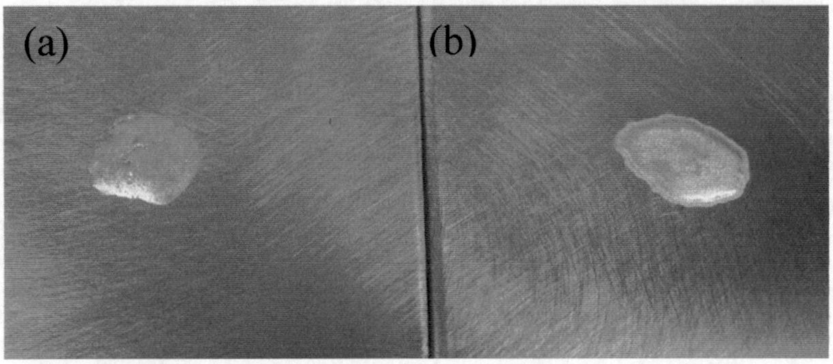


**Figure 9.** Wetting of AgCu28 and CuAg40Ga10 on 1.4404, Temperature 840 °C, 5 min

Next to the wetting, behavior the other important property is the tendency to spread on a base material. It describes the extent to which the HSM increases or decreases the contact area with the base material without external force.

For AgCu28 on plain stainless steel surfaces without any specialized surface treatment such as e.g. electro-polishing this test can be considered as an extreme test setup. So especially under these conditions for AgCu28 instead of spreading a reduction of the contact interface is observed. This can be seen in the following figure 6, figure 7 and figure 8. For CuAg40Ga10 it could be observed on both stainless steels, that the wetted area after spreading test is significant larger than for AgCu28.

*3.2.2 Metallographic cross sections of brazing joints*

The microstructure of CuAg40Ga10 is dominated by copper-rich primary crystals. At the interface of brazed steel parts a pronounced diffusion zone can be observed, figure9a and figure10a. In the cross sections shown in figure 9b and figure10b the eutectic microstructure of AgCu28 is visible. For AgCu28 there is almost no diffusion zone observable in the steel base material

IOP Conf. Series: Materials Science and Engineering **118** (2016) 012004 doi:10.1088/1757-899X/118/1/012004

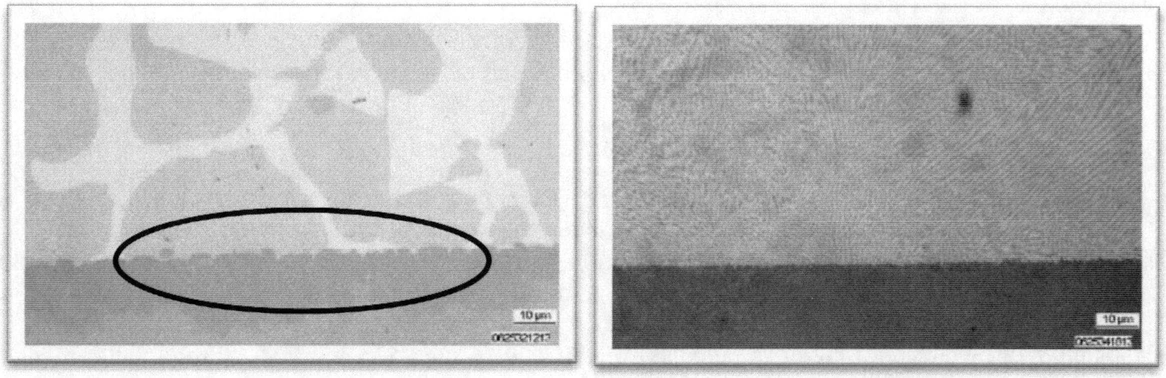

(a) (b)

**Figure 10**. Cross section of the 1.4307 joint with CuAg40Ga10 (a) and with AgCu28 (b); the oval indicates a diffusion zone. The scale is 10 µm.

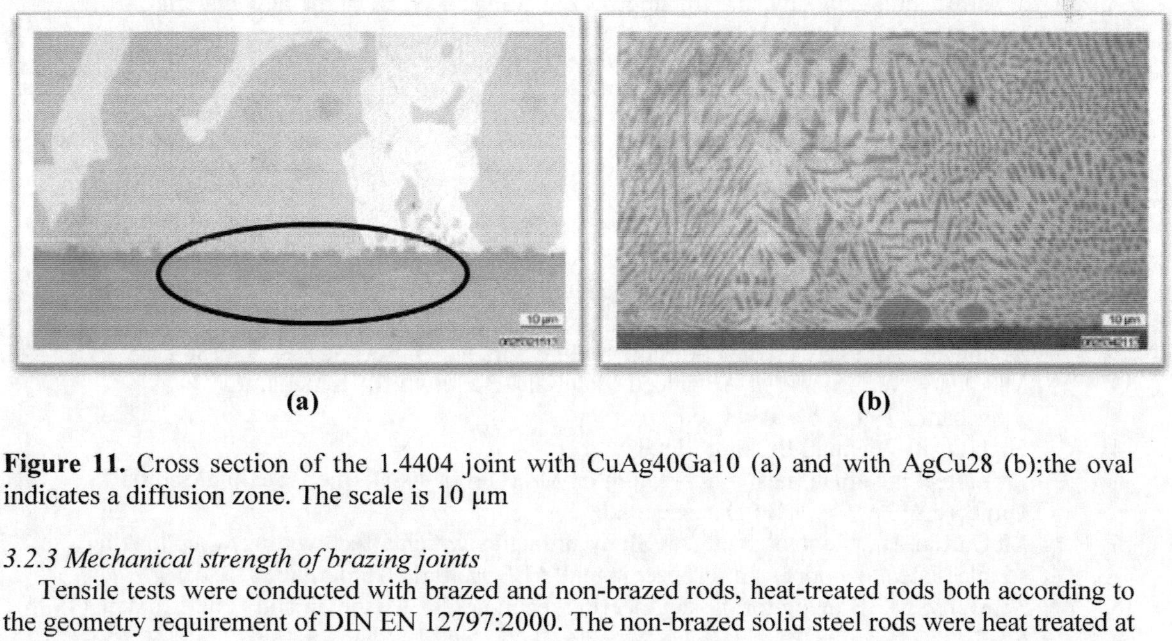

(a) (b)

**Figure 11.** Cross section of the 1.4404 joint with CuAg40Ga10 (a) and with AgCu28 (b);the oval indicates a diffusion zone. The scale is 10 µm

*3.2.3 Mechanical strength of brazing joints*

Tensile tests were conducted with brazed and non-brazed rods, heat-treated rods both according to the geometry requirement of DIN EN 12797:2000. The non-brazed solid steel rods were heat treated at 790 °C and 890 °C to simulate the microstructural changes and therefore changes in ultimate tensile strength of the base materials. This means that Rm values of the brazing joint (830 °C) can be at maximum as high as a value between these two values. The results are provided in table 4.

**Table. 4.** Ultimate tensile strength Rm in MPa.

| | stainless steel 1.4307 | | stainless steel 1.4404 | |
|---|---|---|---|---|
| after brazing with AgCu28 at 830 °C | 480 | | 468 | |
| after brazing with CuAg40Ga10 at 830 °C | 535 | | 519 | |
| 790 °C annealed steel | | 679 | | 625 |
| 890 °C annealed steel | | 625 | | 583 |

It is obvious that with CuAg40Ga10 the strength of the brazing joint can be increased on both stainless steels. Therefore, with the new HSM, joints can be produced that exhibit a higher tolerance against catastrophic failure of the brazed parts

## 4 Conclusion

CuAg40Ga10 is a silver-reduced specialized vacuum grade brazing alloy developed to show an improved technical profile for applications involving the joining of stainless steels.

As shown with the technical characterization of CuAg40Ga10 the new HSM exhibits superior ultimate tensile strength and wettability on the tested steel grades compared to AgCu28. Explanations for the favorable characteristics of CuAg40Ga10 as the improved cohesion within the joint can be found in the abundance of a diffusion zone formed between the HSM and steel. Secondly, the effect of mixed crystal hardening of the third alloying component gallium in the CuAg40Ga10 alloy might have its contribution to the higher strength of its brazing joints.

Hence, the melting range of the CuAg40Ga10 is enclosing the melting point of AgCu28, users do not have to change the parameters of the brazing process in many applications. Future investigations will have to show if additional surface pre-treatment steps of stainless steel surfaces such as electro-polishing are still required when using the improved wetting potential of the new material.

In addition, the new alloy provides two ways to reduce the silver consumption for the vacuum brazing processes: The absolute silver content could be reduced by 32 wt-% and the density was reduced by 6.4 %: resulting in lower alloy mass per identical brazing preform consumed by the applicant

Examples taken from published papers:

## 5 References

[1] DIN ISO 857-2; Schweißen und verwandte Prozesse. DIN Deutsches Institut für Normung e. V., 2007.

[2] Young, T.: An essay on the cohesion of fluids. In: M. D. For. Sec. R. S. (1804.)

[3] Messler, R.W.: Joining of Advanced Materials. Butterworth-Heinemann, Stoneham, (1993), S. 303 et seq.

[4] High Purity Vacuum Brazing Metals (http://technicalmaterials.umicore.com/CPM/en/Products/HermeticSealingMaterial/), Umicore AG & Co. KG : Datasheet, 2013.

[5] McGurran, B.; Nicholas, M.G.: A Study of factors which affect wetting when brazing stainless steel to copper. In: Proceedings, BABS Autumn Conference (1984.)

[6] Schwartz, M.: Brazing for the engineering technologist. Chapman and Hall, London, (1995), S. 12.

IOP Conf. Series: Materials Science and Engineering **118** (2016) 012004 doi:10.1088/1757-899X/118/1/012004

[7] Asthana, R.; Kumar, A.; Dahotre, N.: Kapitel 2: Materials Science in Manufacturing. Elsevier, Burlington, 2006, S. 280–281.
[8] Technical Data Sheet AgCu28, Umicore AG & Co. KG, 2009
[9] Nicholas, M.G.: Joining Processes. Kluwer Academic Publishers, Dordrecht, (1998), S. 57 et seq.
[10] Jacobson, D.M. ; Humpston, G.: Principles of Brazing. ASM International, Ohio, (2005),S. 52.
[11] Schnee, D.; Wiehl, G.; Starck, S.: Development of Ag-Cu-Zn-Sn brazing filler metals with a 10 weight-% reduction of silver and same liquidus temperature.2014 International conference on Brazing, Soldering and special Joining Technologies, Beijing, China, June 2014, pp. 53-58.
[12] N.N.; Seilnacht Verlag (http://www.seilnacht.com/versuche/dichteb.html), 2016.
[13] Eustathopoulos, N.; Sobczak, N.; Passerone, A.; Nogi, K.: Measurement of contact angle and work of adhesion at high temperature. In: J. Mater. Sci. 40 (2005), S. 2271–2280.
[14] Starov, V.M.; Velarde, M.G.; Radke, C.J.: surfactant science series. Bd. 138: Wetting and Spreading Dynamics. CRC Press, London, (2007), S. 2–3.

IOP Conf. Series: Materials Science and Engineering **118** (2016) 012005 doi:10.1088/1757-899X/118/1/012005

# Development of a brazing process for the production of water-cooled bipolar plates made of chromium-coated metal foils for PEM fuel cells

**M Mueller[1], D Hoehlich[1], I Scharf[1], T Lampke[1], U Hollaender[2], H J Maier[2]**

[1] Materials and Surface Engineering Group, Institute of Materials Science and Engineering, Technische Universität Chemnitz, 09125 Chemnitz, Germany
[2] Institute of Materials Science, Field joining and surface engineering (Fortis), Leibniz Universität Hannover, 58453 Witten, Germany

**Abstract.** Beside lithium batteries, PEM fuel cells are the most promising strategy as a power source to achieve the targets for introducing and increasing the usage of electric vehicles. Due to limited space and weight problems, water cooled, metallic bipolar plates in a fuel cell metal stack are preferred in motor vehicles. These plates are stamped metal sheets with a complex structure, interconnected media-tight. To meet the multiple tasks and requirements in use, complex and expensive combinations of materials are currently in use (carbon fiber composites, graphite, gold-plated nickel, stainless and acid resistant steel). The production of such plates is expensive as it is connected with considerable effort or the usage of precious metals. As an alternative, metalloid nitrides (CrN, VN, W2N, etc.) show a high chemical resistance, hardness and a good conductivity. So this material category meets the basic requirements of a top layer. However, the standard methods for their production (PVD, CVD) are expensive and have a slow deposition rate and a lower layer thicknesses. Because of these limitations, a full functionality over the life cycle of a bipolar plate is not guaranteed. The contribution shows the development and quantification of an alternative production process for bipolar plates. The expectation is to get significant advantages from the combination of chromium electrodeposition and thermochemical treatment to form chromium nitrides. Both processes are well researched and suitable for series production. The thermochemical treatment of the chromium layer also enables a process-integrated brazing.

## 1. State of the Art

The solution to the energy and exhaust problems of our time, e.g. the exhaustion of greenhouse gases or the energy consumption from fossil sources, could be the fuel cell. It provides a development to a zero emission vehicle during the life time. Recent developments favor the *Polymer Electrolyte Fuel Cell (PEFC)*. The principle of this fuel cell is to produce a cell voltage by a redox-reaction between hydrogen and oxygen to water (see figure 1). Figure 2 demonstrates a full PEFC stack for the use in vehicles by Daimler AG. Advantages of the PEFC are: easiness to control, short circuit protection and quick response to changing loads. The electrolyte is less corrosive, and the cells are relatively easy to manufacture. Typical challenges of the PEFC technology are the moderate peak performance, limited service life of the membrane, electrode materials and the production costs of the full stack. To reduce production costs, a starting point are the bipolar plates (see figure 2-5). The bipolar plates are the main elements of a stack. They essentially determine cost, weight and function. These plates have high requirements to ensure a long life of the PEFC (see table 1). Therefor the manufacturing of the bipolar plates requires high effort and costs. With regard of the very high numbers of bipolar plates needed in case of a launch of fuel cell vehicles, there is a need for developing new materials and production techniques for mass production.

IOP Conf. Series: Materials Science and Engineering **118** (2016) 012005 doi:10.1088/1757-899X/118/1/012005

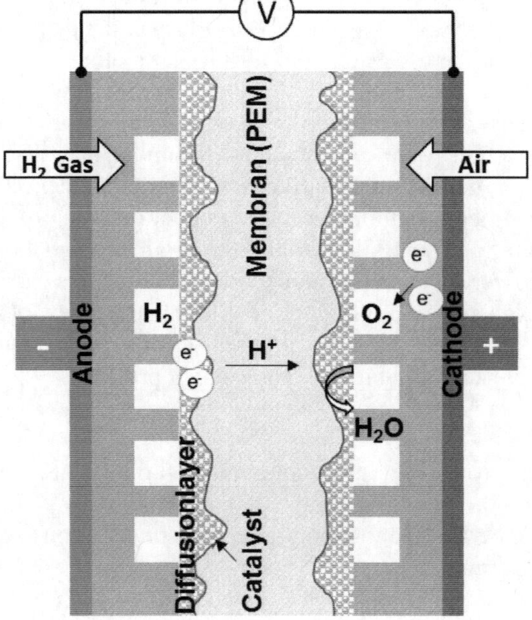


**Figure 1.** Basic design and schematic functionality of PEFC: feeding of pure hydrogen leads to anodic oxidation of hydrogen; diffusion through the polyelectrolyte membrane (PEM); feeding of air to cathode leads to reduction of oxygen; reaction between hydrogen and oxygen ions leads to water [own illustration]

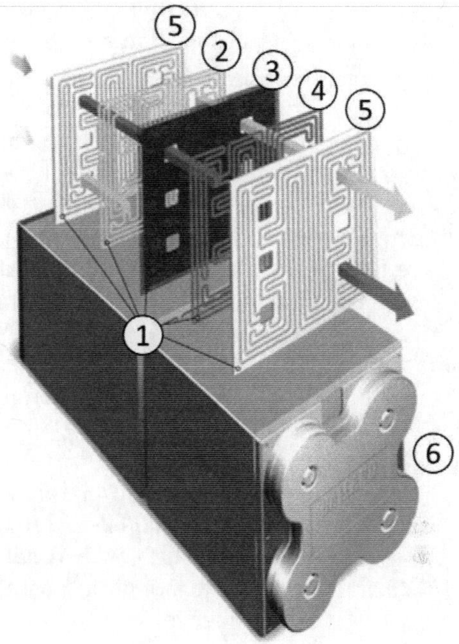


**Figure 2.** 1: PEFC stack; 2: H2 flow field; 3: membrane-electrode assembly; 4: air flow field; 5: bipolar plate; 6: endplate [1]

**Table 1.** Tasks and requirements of bipolar plates in a PEFC and the current materials in use [1]

| Tasks | Requirements | Materials |
|---|---|---|
| electrical contact | electric conductivity | graphite |
| gas supply | gas-tightness | carbon composite |
| water supply | corrosion resistance | titanium, zirconium |
| cooling | tensile and transverse strength | gold plated nickel |
| | | electrically conductive ceramic layer |

## 2. Approach

Materials selection: Most stack manufacturers use graphitic bipolar plates made of different materials and different production methods. However, the production technologies for mass production are not sufficiently developed. Therefor metallic bipolar plates, made of stainless steel, offer an advantage. The progress of deep drawing enables the production of 0.1 mm plates with fine channels. Though the passivation of stainless steel prevents a long-term use. The electric conductivity decreases noticeably.

Previous coatings (e.g. gold) are too expensive for mass production. Some patents introduce chromium nitride (CrN) as topcoat [e.g. 2–4]. It combines high corrosion resistance with electric conductivity. Resulting from this, CrN meets the high requirements for the use as topcoat for bipolar plate.

New production route: The production technology focuses on vapor deposition only. The process times are long and the layers are very thin. The new production route recommends chromium plated steel plates or foils as the starting point. They are a general alternative for stainless steel. The electrodeposition of chromium from chromic acid is a process suitable for mass production. The process times are low, thickness of layers are adjustable (from 5 µm to 1000 µm) and the production costs are also low, in comparison to PVD/CVD. To produce the CrN layer, a thermochemical treatment is carried out in a continuous furnance under $H_2/N_2$ atmosphere. The brazing of the plates takes place simultaneously. This ensures an economical process, suitable for mass production of bipolar plates. Furthermore, the choice of the basic material is variable according to the requirement profile.

Procedure for development:

a) Thermochemical treatment (Conveyor belt-protective gas-continuous furnace; Kohnle, Birkenfeld HT 1200-200/80-1500)
   - Parametric study with variation of temperature, annealing time, gas composition

b) Brazing (temperature monitoring with trailing thermocouples)
   - Braze filler selection
   - Wetting behavior
   - Brazing experiments at different temperatures

c) Characterization
   - XRD (x-ray diffraction) – phase analysis for detection of $Cr_xN_y$
   - SEM (scanning electron microscope) – optical evaluation of the coating and joining connection
   - Cross section and optical microscope – evaluation of the connection to the substrate

## 3. Thermochemical treatment

A parametric study takes place to identify the optimal parameters for the thermochemical treatment of chromium-plated samples. The parameters temperature, annealing time and the composition of the atmosphere in the furnace serve to secure the production of a chromium nitride layer. Preliminary tests carry out the restriction of the parameters (see Table 2). The full factorial design of experiments (32 experiments) ensures to find an optimal manufacturing route. The temperatures are based on typical brazing temperatures using standard braze filler material.

**Table 2.** Variables for the full factorial design of thermochemical experiments

| Temperature | Annealing time | Atmosphere |
|---|---|---|
| 900 °C | 15 min | 80 : 20 - $N_2 : H_2$ |
| 950 °C | 30 min | 50 : 50 - $N_2 : H_2$ |
| 1000 °C | | 20 : 80 - $N_2 : H_2$ |
| 1120 C | | $N_2$ + 15 ppm $SiH_4$ |

Steel (S235) acts as substrate, coated with a hard chromium layer (10 µm–20 µm) from a commercial chromic acid electrolyte. Figure 3 and 4 display two different thermo-chemical treatments. The conversion, from Cr to $Cr_xN_y$, is evident by the dark coloration of the coating. Chromium nitride layers

IOP Conf. Series: Materials Science and Engineering **118** (2016) 012005 doi:10.1088/1757-899X/118/1/012005

do not obtain in the atmospheres with lower nitrogen content. In comparison, the pure nitrogen atmosphere is the best, be-cause of a closed and partly continuous nitride layer at lower temperatures. The conversion of Cr to $Cr_xN_y$ closes residual cracks of the chromium coating due the increase of the volume.

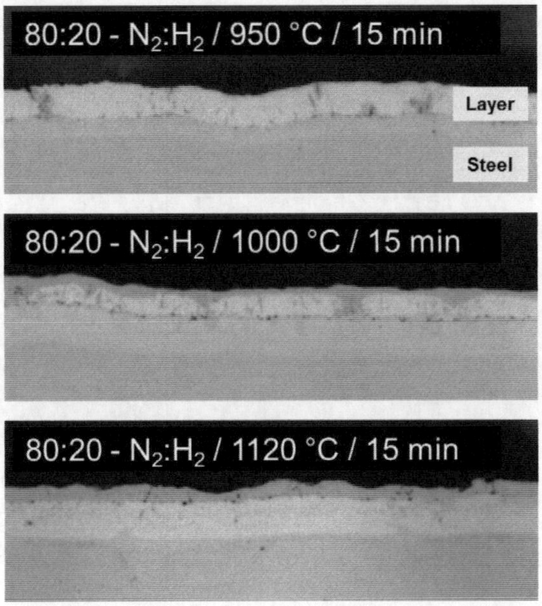


**Figure 3.** Cross section of chromium plated steel after thermochemical treatment at different temperatures in N2 / H2 atmosphere; nitride layer (dark) only few µm thick not continuously

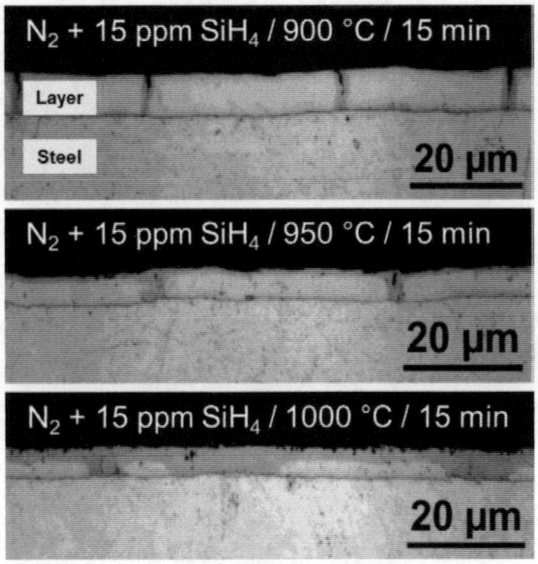


**Figure 4.** Cross section of chromium plated steel after thermochemical treatment at different temperatures in N2 atmosphere doped with SiH4; at 950 °C cracks of chromium coating closed; at 1000 °C about 5 µm thick nitride layer partly continuous

The detection of the composition of the $Cr_xN_y$ is carried out with XRD at low diffraction angles (2° / 6°). This method permits the detection of the phase composition at the surface of the samples. The penetration depth is maximum 10 µm. Figure 5A illustrates SEM pictures of the surface of electrodeposited chromium as reference and samples after thermochemical treatment at different temperatures. The atmosphere ($N_2$ with doped $SiH_4$) and the annealing time are constant. These are examples of the different parameters of table 2. XRD phase analysis results in the detection of polycrystalline $Cr_2N$ as the main phase. CrN phase does not appear under these conditions and production route. This means an advantage for the use as topcoat for bipolar plates. The transition resistance of $Cr_2N$ is up to 30 % lower compared to CrN, which means a much better electric conductivity. The use of nitrogen gas as atmosphere prevents also the oxidation of the chromium layer. Starting from a minimum temperature of 950 °C, appears a closed $Cr_2N$ layer with a minimum thickness of 4 µm.

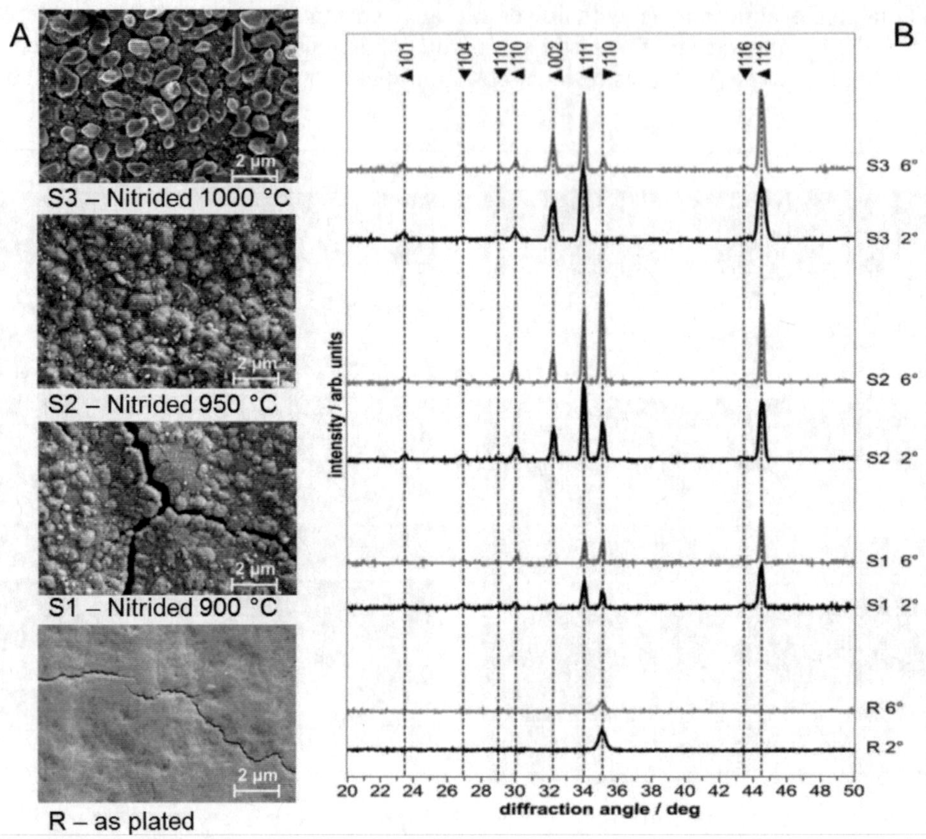


**Figure 5.** A – SEM surface pictures of different thermochemical treated samples in $N_2$ atmosphere doped with 15 ppm $SiH_4$ / annealing time 15 min | B – X-ray diffraction patterns of the samples of Fig. 5A under different diffraction angles (2° / 6°); marked crystal orientations Cr (►), $Cr_2N$(◄) $Cr_2O_3$ (▼); no CrN detected and very low intensity of $Cr_2O_3$ under these conditions; instead $Cr_2N$ and metallic Cr as main phases

## 4. Brazing Tests

The annealing temperature range between 900 °C and 1120 °C and holding time for the thermochemical treatment allows a simultaneous brazing. The process integrated joining reduces the process cost and saves a production step regarding to the PEFC stack. The typical joining of metallic bipolar plates is made by laser welding. The selected braze filler materials have a melting range similar to the temperatures of thermochemical treatment (see table 3).

**Table 3.** Selection of the used braze filler as a paste

| Label | Alloy composition [wt.%] | Melting range | Minimum brazing temperature |
|---|---|---|---|
| Ni 710 | Ni-14Cr-10P | 890 °C | 900 °C |
| B-Ni60CrPSi-980/1020 | Ni-30Cr-6P-4Si | 980 °C – 1020 °C | 1120 °C |

Chromium plated steel with different braze filler undergoes a wetting test at the same conditions as the thermochemical treatment (see figure 6). The experiments show the braze filler Ni 710 needs a minimum temperature of 1000 °C to have a good adhesion to the chromium plated substrate. At this temperature, a wetting angle is barely visible. The second braze filler needs 1120 °C as a minimum temperature. Under these conditions, the connection to the substrate is good; the diffusion zone is homogenously developed. The wetting angle is nearly 0°. Both braze filler are suitable for a simultaneous thermochemical treatment and brazing. The definition of the parameters for thermochemical treatment and brazing leads to the production of test panels. Therefore, the simulation under production conditions starts with easy structured steel sheets (see figure 7).

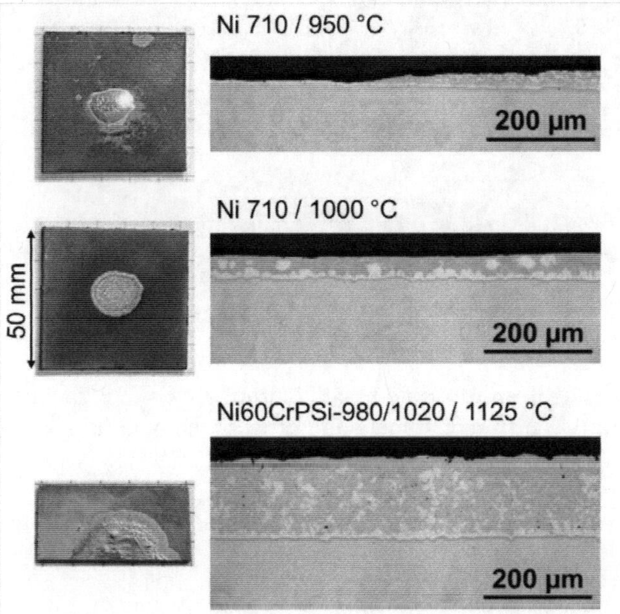


**Figure 6.** Wetting test with chromium plated steel and braze filler of Table 3 after annealing for 15 min and different temperatures

## 5. Conclusions and Outlook

The state of the art shows the need of a mass production process for bipolar plates. The combination of electroplating, thermochemical treatment and brazing could solve this problem, not only for the plates but also for the production of the stack. The thermochemical treatment and brazing of chromium coated steel produces closed $Cr_2N$ coatings joined together adherent. To perform these processes simultaneously, the material selection and the use of equal conditions is critical. The article clarifies both points. The next tasks are the corrosion tests, in the medium of the fuel cell, and the bond strength testing. Both test should clarify if the material selection is suitable for the application in PEFC. Because of the REACH regulation, the use of chromic acid, as an electrolyte for chromium electrodeposition, is restricted at 2017. As an alternative plating technology, chromium coatings obtain from Chromium-(III)-salt electrolytes. At this time, no commercial thick coatings are available on the market. The base of the process is a chromium coated substrate. Switching to a new chromium layer requires further experiments.

The whole development ends with the thermochemical treatment and koining of a real produced bipolar plate. A performance test will show, if the proposed production route and material selection suits the requirements of a life cycle of bipolar plates.

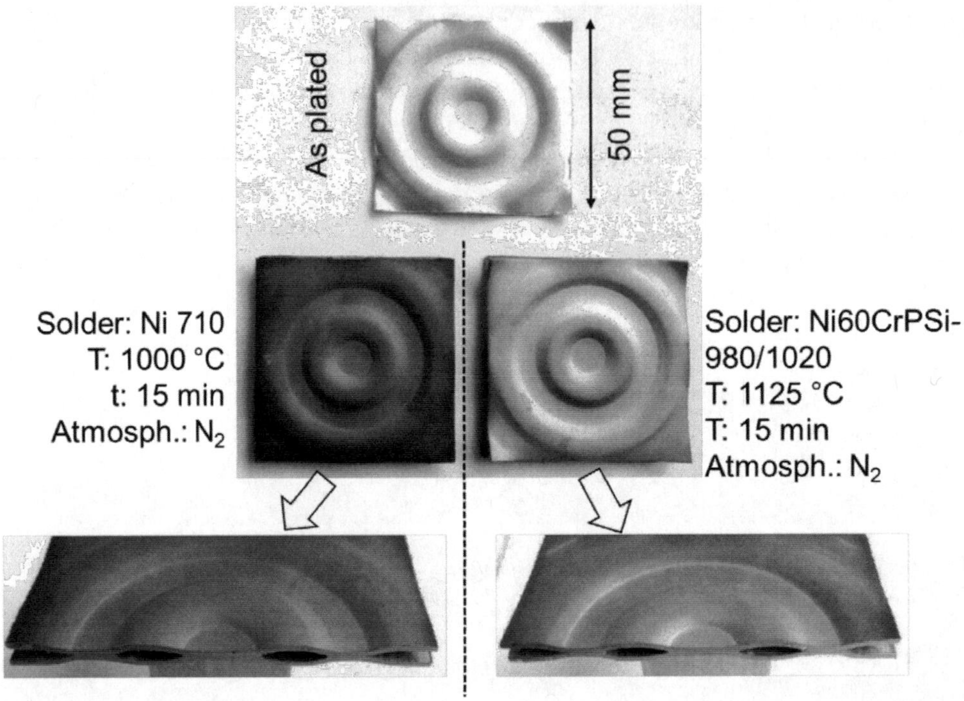


**Figure 7.** Production route of a stamped steel sheet coated with chromium; simultaneous thermochemical treatment and brazing with two different solders and process temperatures in a continuous furnance; annealing time 15 min; atmosphere $N_2$ doped with $SiH_4$

**Acknowledgement**

The results are a product of the IGF-project No. 18422 BG/DVS-No. 07.077 *"Entwicklung eines Lötverfahrens für die Fertigung von wassergekühlten Bipolplatten aus chrombeschichteten Metallfolien für PEM-Brennstoffzellen"*. The AiF provides the funding, as part of the program for cooperative industrial research. The authors are grateful for financial support.

**Literature**

[1] Kurzweil, P: Brennstoffzellentechnik. 2. Auflage. Wiesbaden: Springer, 2013

[2] EP1273060B1: METALLISCHE BIPOLARE PLATTE / Ballard Power Systems Inc. REPENNING, Detlev, 23569 Lübeck (DE); SPÄH, Richard 88662 Überlingen (DE); KAISER, Wolfram 79848 Bonndorf (DE); WIND, Jörg 88090 Immenstaad (DE) (19.01.2005)

[3] DE10139930A1: Verfahren zur Herstellung von elektrisch leitfähigen Nitridschichten auf chromhaltigen Metalllegierungen / DaimlerChrysler AG. Steinwandel, Jürgen, Dr., 88690 Uhldingen-Mühlhofen, DE ; Späh, Richard, Dr., 88662 Überlingen, DE ; LaCroix, Annette, Dr., 88048 Friedrichshafen, DE (30.04.2003)

[4] DE102008044340A1: Metallische Bipolarplatte für eine Brennstoffzelle und Verfahren zum Bilden einer Oberflächenschicht für diese / Hyundai Motor Company. Lyo, In Woong, Suwon, Kyonggi, KR; An, Jeong Uk, Seoul, KR; Ahn, Seung Gyun, Suwon, Kyonggi, KR; Nam, Young Min, Siheung, Kyonggi, KR; Yang, Yoo Chang, Gunpo, Kyonggi, KR; Kirn, Sae Hoon, Yongin, Kyonggi, KR; Chin, Sang Mun, Seoul, KR (21.01.2010)

IOP Conf. Series: Materials Science and Engineering **118** (2016) 012006 doi:10.1088/1757-899X/118/1/012006

# Influencing the arc and the mechanical properties of the weld metal in GMA-welding processes by additive elements on the wire electrode surface

**V Wesling, A Schram, T Müller and K Treutler**

Institute for Welding and Machining, TU Clausthal, 38678 Clausthal-Zellerfeld, Germany

**Abstract.** Under the premise of an increasing scarcity of raw materials and increasing demands on construction materials, the mechanical properties of steels and its joints are gaining highly important. In particular high- and highest-strength steels are getting in the focus of the research and the manufacturing industry. To the same extent, the requirements for filler metals are increasing as well. At present, these low-alloy materials are protected by a copper coating (<1 µm) against corrosion. In addition, the coating realizes a good ohmic contact and good sliding properties between the welding machine and the wire during the welding process. By exchanging the copper with other elements it should be possible to change the mechanical properties of the weld metal and the arc stability during gas metal arc welding processes and keep the basic functions of the coating nearly untouched. On a laboratory scale solid wire electrodes with coatings of various elements and compounds such as titanium oxide were made and processed with a Gas Metal Arc Welding process. During the processing a different process behavior between the wire electrodes, coated and original, could be observed. The influences ranges from greater/shorter arc-length over increasing/decreasing droplets to larger/smaller arc foot point. Furthermore, the weld metal of the coated electrodes has significantly different mechanical and technological characteristics as the weld metal from the copper coated ground wire. The yield strength and tensile strength can be increased by up to 50 %. In addition, the chemical composition of the weld metal was influenced by the application of coatings with layer thicknesses to 15 microns in the lower percentage range (up to about 3 %). Another effect of the coating is a modified penetration. The normally occurring "argon finger" can be suppressed or enhanced by the choice of the coating. With the help of the presented studies it will be shown that Gas Metal Arc Welding processes are significantly affected by thin film coatings on solid wire electrodes for Gas Metal Arc welding. The influences are regarding the stability of the arc, the properties of the weld metal in terms of geometric expression, chemical composition and mechanical properties, the composition of the arc-plasma and the dynamics of the molten metal.

## 1. Introduction

In times of decreasing resources, lightweight high strength construction materials are getting important. Thus the mechanical properties of low alloyed high strength steels are more and more interesting. In particular the water-quenched and thermomechanically rolled fine-grained get propagated in many different applications in the manufacturing industry. To the same extend as the use of high strength steel increases, the demands on the weldability of these steels and the mechanical properties of the used filler materials are increasing. At present, these additional materials for GMA-welding are mostely protected by a copper coating or in rare cases a coating of titanium, brass or bronze against corrosion. In addition, the coating provides a good ohmic contact and good sliding

properties. These three characteristics form the three core requirements of current solid wire electrode coatings [1]. In this context arises the question whether other coating elements meet these requirements but also merge into the metal vapor of the arc plasma and into the melt pool and affect the process characteristic, the melt pool and the mechanical properties of the weld metal. This hypothesis is supported by the familiar effects of various alloying elements in steel melts and their segregates on the surface of the melt pool, also called Marangoni effect, [2]. The surface tension is a primary factor of influence in the formation of the melt pool dynamics and is changed by a variety of alloying elements of iron, [3]. In addition to the Marangoni effect, other influences on the formation of the penetration and the melt pool dynamics in GMA welding could be found. According to Takasu et al. [4], four forces are involved: a stirring force by the moment of the falling drop, a radial force by the normal percentage of the drop motion, a buoyant force induced by density differences between the droplets and the molten pool and forces due to the Marangoni effect. Furthermore, according to Mills [2] aerodynamic drag forces, and electromagnetic Lorentz forces resulting from the flow of current, have to be added.
A possible influence of the arc can also be based by known facts. The plasma, in which the arc passes, consists of proportions of the used welding gas and vaporized metal. The metal vapor has depending on its composition different ionization energies and this influences the shaping of the arc and the current transition, [5].
Apart from these effects, it is expected that a change of the mechanical properties and the solidification behavior of the weld metal by alloying the melt pool via the coating, occurs. By introducing finely divided nitrides or carbides, which act as solidification nuclei, a finer micro structure can be achieved. Due to the alloying elements and the micro structural changes, among other things, an increase in strength of the weld metal is expected [6], [7].

## 2. Preparation and characterization of test materials

For the preparation of coated wires, the physical vapor deposition has been used, as it allows a wide range of coating compositions. The base wire for the coatings was a copper-coated steel wire (DIN EN 14341-A G50 7 4Mo M21) designed for temperature steels. **Figure 1** shows the coating chamber used. During the coating process the copper-coating of the wire electrode was almost completely removed by a plasma etching process and had no influence on the welding experiments with the coated solid wire electrodes.

**Figure 1.** PVD system for coating of the wire electrodes

**Figure 2** shows an example of a coating produced in the micrograph. In the lower region the etched structure of the steel wire can be seen. The individual layer thicknesses of each investigated coating system are listed in **Table 1**.

IOP Conf. Series: Materials Science and Engineering **118** (2016) 012006 doi:10.1088/1757-899X/118/1/012006

**Table 1.** Layer thicknesses

| Coating system | TiN | Cr + C | Al + TiN | Al + Cr | NiCr + SiCr | Cr | Ti |
|---|---|---|---|---|---|---|---|
| Layer thickness in µm | 4.5 | 2 | 4.5 | 16 | 6 | 8 | 15 |

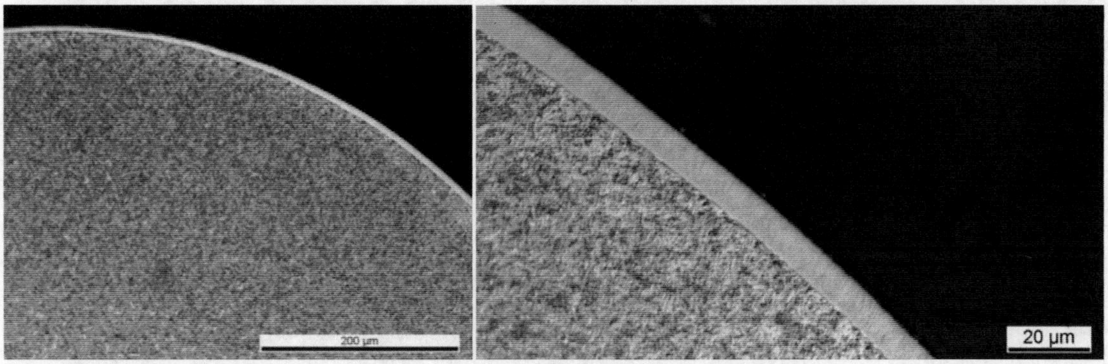


**Figure 2.** Ti-coated solid wire electrode

The welding processing of the coated wire electrodes with a GMA process was carried out at comparable process conditions. First dummy seams are carried out to assess the penetration behavior, followed by the creation of three-layer deposit-welding for taking spectroscopy and tensile specimens. The welding tests were carried out on the fine grained steel S700MC.

## 3. Influence of arcing, penetration behaviour and mechanical properties

To detect the influences on the welding-arc, high-speed imaging of the arc and the droplet transition were performed. The imaging clearly shows that influencing the arc and the droplet transfer can be done via the coatings. As an example, the pictures of the welding arc in argon and in 82 % + Ar 18 % $CO_2$ atmosphere under the use of the copper coated base wire and chromium- respectively titanium-coated welding wire are shown in **Figure 3**.

Pictures are taken during welding with comparable power and voltage values. The different arc-length for the different coatings can be seen clearly in Fig. 3. Using Ti-coatings the arc has been significantly extended. It is even visible that the arc begins to form the typical shape of a rotating arc.

Using a mixed gas (82% Ar, 18% $CO_2$) a similar behavior can be observed, **Figure 3**. It differs only in the absence of the rotating arc using the Ti-coating. Furthermore, the recordings show that the chromium coating resulting in a huge drop which performs a transition without short-circuits.

In summary, titanium content in the coatings leads to an extension of the arc-length and chromium content is accompanied by a shortening of the arc.

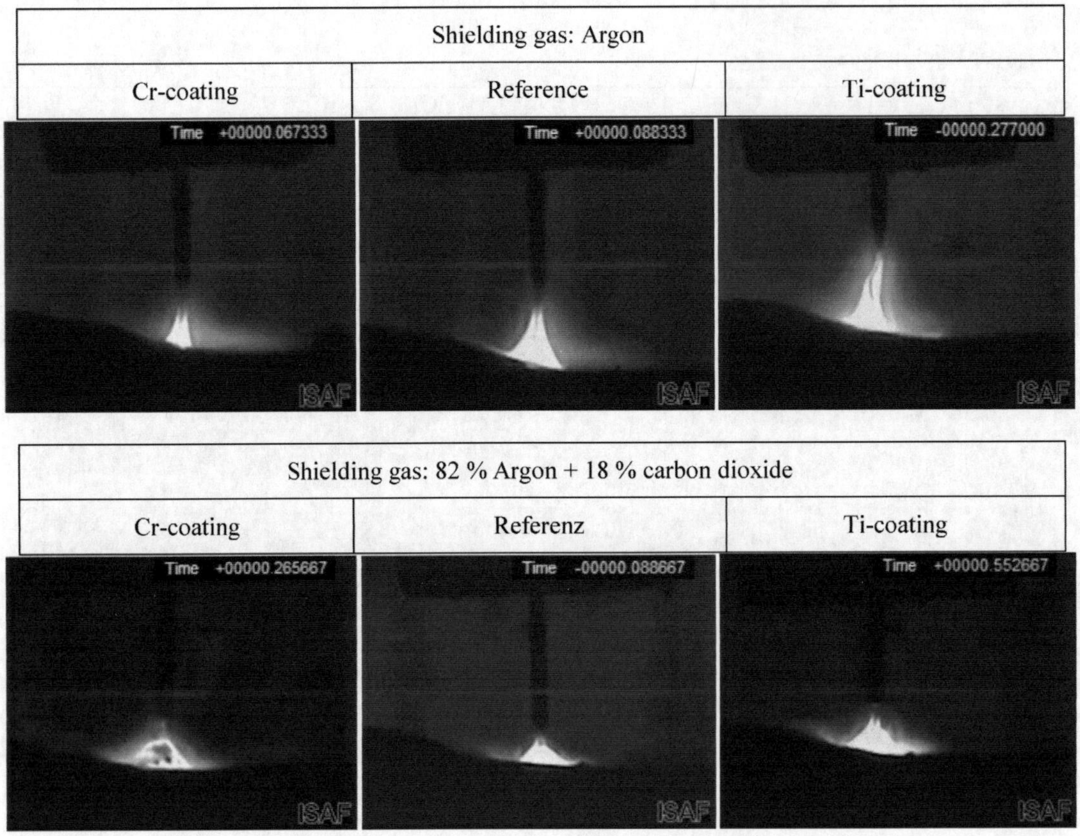


**Figure 3.** Comparison of the arc length, under argon, and mixed gas (82 % Ar + 18 % $CO_2$)

In addition to an altered arc behavior, the impact of the wire electrode coating on the penetration behavior was found and investigated. In **Figure 4** the penetration behavior for the various coatings is shown. A change in the shape of the weld metal penetration can clearly be seen. Some coatings such as TiN or Al + Cr suppress the normally occurring argon finger. In contrast, the Cr + C coating increased the argon finger. The most visible change could be achieved by the coating of pure titanium. The penetration was wider and the weld metal shows a significantly flatter transition between weld and base metal. Further, the argon finger is less pronounced than in reference weld metal formed by the coppered wire electrode.

In addition to the shape of the weld penetration also the chemical composition of the weld metal was changed by the coating on the solid wire electrode. To quantify the chemical composition of the weld metal it was determined by optical emission spectroscopy for the three-layer deposit weldings. In order to prevent a possible influence on the measurement results by contamination on the sample surface, the samples were abraded before the tests.

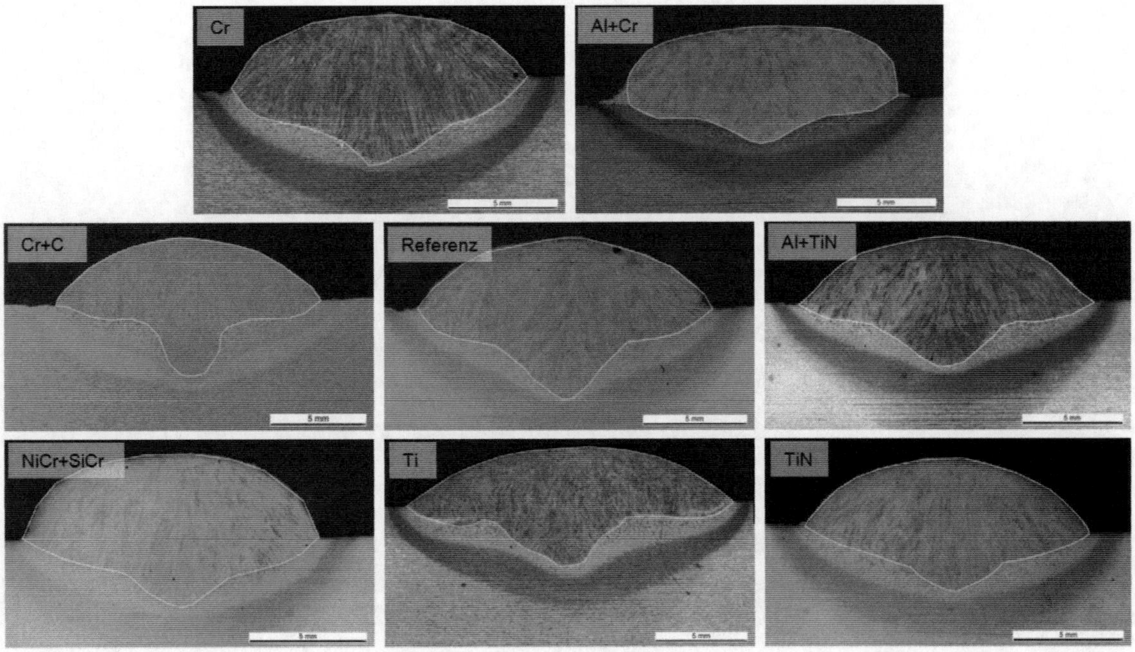


**Figure 4.** Penetration behavior

After the abrading, the measurement point is located approximately 2 mm below the original sample surface. It turns out that all elements of the coating pass into the weld metal. An increase in the corresponding mass contents can be clearly seen, *Tab. 2.*

**Table 2.** Mass content of the weld metal for various elements (selection)

| Layer | Mass contents in % | | | | | |
|---|---|---|---|---|---|---|
| | Al | C | Cr | Ni | Si | Ti |
| 4Mo (Referenz) | - | 0.07 | 0.10 | 0.07 | 0.59 | - |
| TiN | - | 0.07 | 0.21 | 0.07 | 0.63 | 0.54 |
| Cr + C | 0.01 | 0.12 | 0.56 | 0.08 | 0.46 | - |
| Al + TiN | 0.29 | 0.08 | 0.1 | 0.07 | 0.66 | 0.25 |
| Al + Cr | 0.37 | 0.09 | 2.39 | 0.07 | 0.65 | - |
| NiCr + SiCr | - | 0.07 | 0.77 | 1.79 | 0.76 | - |
| Cr | - | 0.07 | 3.24 | 0.06 | 0.44 | - |
| Ti | - | 0.07 | 0.11 | 0.06 | 0.58 | 1.11 |

Of particular note is the increase in the chromium content in the weld metal to 3.24 %. This amount corresponds to the percentage of the coating in the total volume of the coated solid wire electrode.
The change in chemical composition is accompanied by a change in solidification behavior and consequently of the weld microstructure. These structural changes and the effect of the coating elements as an alloying element in the weld lead to significantly altered mechanical properties.

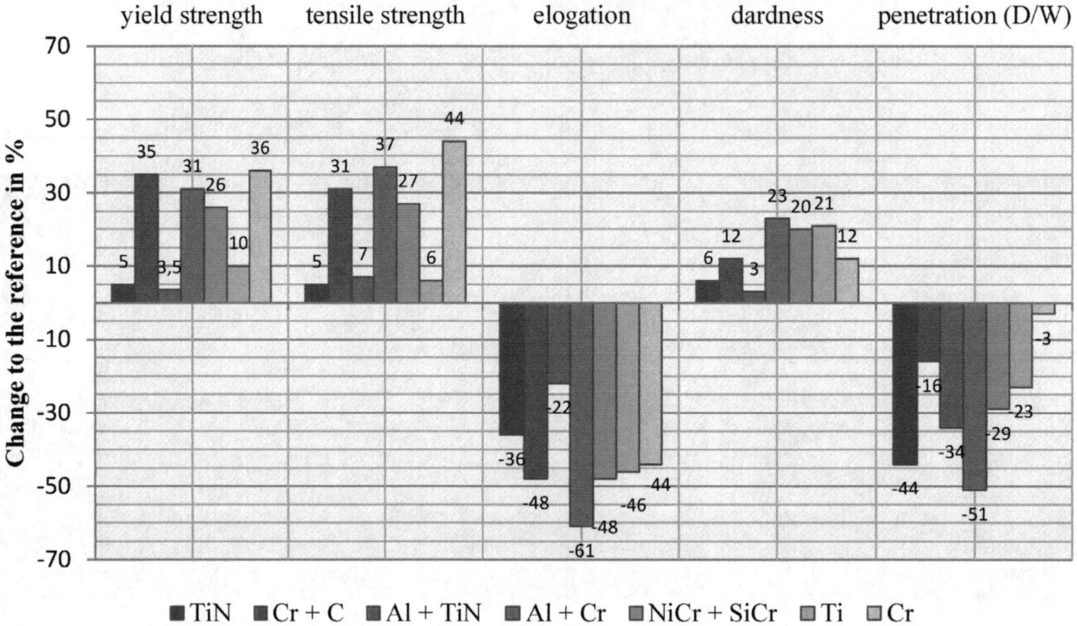


**Figure 5.** Mechanical properties of the weld metal and fusion penetration compared to the reference wire

**Figure 5** shows the change of selected mechanical properties of the weld from the coated wire electrodes compared to the reference. It can be distinguished according to their characteristics between the titanium-containing coatings and chromium coatings.

Using titanium-containing coatings no significant increase in yield or tensile strength could be observed. However, the elongation at break decreases significantly and the hardness. This suggests the formation of brittle phases. Furthermore, by the participation of titanium, the weld penetration is considerably changed. This is evident by the decrease of the ratio D / W in the coating with TiN by up to 44 % with pure Ti and up to 23 %. Where D is penetration width and W is the penetration depth.

Under the use of chromium-containing coatings on the solid wire electrodes a significant increase in the yield and tensile strength of the weld metal was found. This increase can reach 44 % using a layer thickness of 15 microns. The elongation at break decreases significantly, but remains in areas which comparable high strength steels show. According to this fact chromium coated welding consumables could possibly used for high strength steels. Since the penetration behavior has not been changed compared to the reference wire using the chromium-coating it can be assumed that the change in the penetration behavior, shown in **Figure 4** is attributable to the use of Al, Ni, Si, C in addition to chromium for the coatings. It follows that the carbon leads to a narrow and deep penetration, aluminum to a flattening of the weld penetration. However, these assumptions need to be verified in further studies.

Besides the strength of materials under static loading, in many cases the strength und cyclic loading is of interest and was investigated. For this purpose single staged cycling loading test were carried out on different load horizons. A method of consecutive load steps was used and offers the advantage that a lack of knowledge of the approximate location of the S-N curve in prior to the tests is possible and therefore lends itself to the characterization of the cyclical strength of the weld metal of coated solid wire electrodes. Since, as already described, the coatings can be separated into two major groups in accordance with the participation of chromium or titanium, the cyclic tests were performed for coatings without any additional elements beside pure chromium or titanium. For determining the cyclical strength tensile specimens were taken from the three-layer deposit welding. The tension ratio for the tests was $R = 0.1$ and as failure criterion the complete specimen rupture was defined. The

IOP Conf. Series: Materials Science and Engineering **118** (2016) 012006 doi:10.1088/1757-899X/118/1/012006

results for a survival probability of 95 % confidence interval of 75 % are shown in **Figure 6.** It shows that the fatigue strength can be significantly raised by the addition of titanium or chromium via the coating. The reference shows a value of 275 MPa stress range at $2x10^6$ load cycles. In comparison, the weld metal from the chromium or titanium coated welding wire can survive a value of 510 MPa stress range for $2x10^6$ load cycles. Furthermore, it is conspicuous that the low cycle fatigue strength is by the addition of titanium only at 600 Mpa stress range and thus results in a very shallow S-N curve with an slope exponent of k = 54. In contrast, the weld metals with added chromium behave similar to the reference. The slopes of the curves are nearly identical. The cyclical strength will be significantly raised by chrome in all the area, so that in the low cycle fatigue strength is approximately 820 Mpa in stress range.

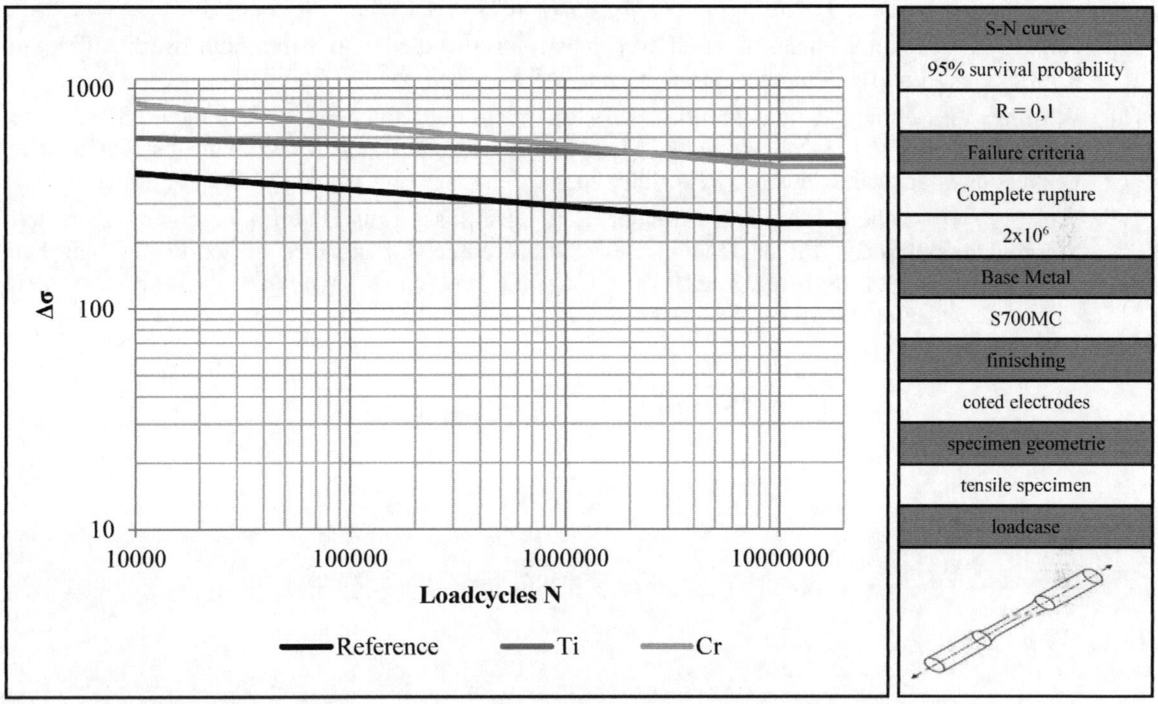


**Figure 6.** S-N curves of weld metals of different coated solid wire electrodes

## 4. Conclusion

Like postulated, a significant influence on the welding process and the mechanical properties of the weld metal can be applied by the deposition of thin film coatings on the welding wire surfaces.

Chromium-containing coatings on solid wire electrodes reduced the arc length significantly. Furthermore they increased the yield and tensile strength up to 44 %. The cyclical strength of the weld metal was increased significantly as well. The elongation at break decreased.

Titanium-containing coatings reduced the penetration ratio to 56 %. A pronounced increase of the yield strength and the static tensile strength could not be seen. In contrast, however, a significant increase in strength und cyclic loading was observed.

Finally, it should be noted that these studies contribute to the knowledge of the targeted modulation of the mechanical properties of weld filler by functional thin film coatings. Further studies on the influence of these thin film coatings on the arc plasma, the micro structural changes and the underlying effects as well as the characterization of these influences by further coating elements should follow.

## 5. References

[1] Schwarz, B.; „Neues innovatives Drahtoberflächenkonzept für niedriglegierte MAG-Drahtelektroden für warmfeste hochfeste Stähle," in *Schweißen und Schneiden, Große Schweißtechnische Tagung*, Aachen, 2006.

[2] Mills, K.; Keene, B.; Brooks, R. and Shirali, A.; „Marangoni effects in welding," *Phil. Trans. r. Soc. Lond. A 356,* pp. 911-925, 1998.

[3] Keene, B.; „Review of data for the surface tension of iron and its binary alloys," *International Materials Reviews 33,* pp. 1-37, 1988.

[4] Takasu, T. and Toguri, J.; „Pyrometallurgical Significance of Marangoni Flow: Mechanism and Contributions to Processing," *Philosophical Transactions: Mathematical, Physical and Engeneering Sciences Vol 356. No. 1739,* pp. 967-980, 15 4 1998.

[5] Hülsen, R.; „Vollmechanisches WIG-Gleichstromschweißen von Aluminium-werkstoffen im Rohrleitungs- und Behälterbau" *Schweissen und Schneiden,* pp. 219-223, 2001.

[6] Wesling, V.; Schram, A. and Treutler, K.; „Potentiale neuartiger Schweiß-drahtbeschichtungen in der Prozessgestaltung" Fakultät für Mathematik/Informatik und Maschinenbaus, Technische Universität Clausthal, Clausthal-Zellerfeld, 2015.

[7] Wesling, V.; Schram, A. and Treutler, K.; „Einfluss von Dünnfilmbeschichtungen auf Massivdrahtelektroden für das Metallschutzgasschweißen auf die mechanisch-technologischen Eigenschaften des Schweißgutes" in *1. Niedersächsisches Symposium Materialtechnik*, Clausthal-Zellerfeld, 2015.

IOP Conf. Series: Materials Science and Engineering **118** (2016) 012007 doi:10.1088/1757-899X/118/1/012007

# High-Temperature Corrosion and Radiation Characteristics of Thermal Sprayed Molybdenum Disilicide-Based Coatings

**S Weis[1], T Uhlig[1], G Wagner[1], T Lampke[2], W Bauer[3] and A Moldenhauer[3]**

[1] Chair of Composite Materials, Institute of Materials Science and Engineering, Technische Universitaet Chemnitz, 09125 Chemnitz, Germany
[2] Chair of Materials and Surface Engineering, Institute of Materials Science and Engineering, Technische Universitaet Chemnitz, 09125 Chemnitz, Germany
[3] University Duisburg-Essen, Germany

E-mail: sebastian.weis@mb.tu-chemnitz.de, thomas.lampke@mb.tu-chemnitz.de, wolfgang.bauer@uni-due.de

Due to their poor oxidation resistance, most metallic components of industrial furnaces can be used only under high-temperature service conditions in inert atmospheres or vacuum. The operation in oxidizing atmospheres is normally limited to temperatures up to 1200 °C. However, it is known that molybdenum disilicide shows a good oxidation resistance in the temperature range up to 1650 °C. Therefore, the aim of this work is the application of molybdenum disilicide coatings for high-temperature oxidation protection of molybdenum-based substrates. Thermal spraying of agglomerated and sintered powder was carried out by atmospheric plasma spraying (APS). Feedstock and coatings were characterized by means of optical microscopy. The high-temperature oxidation behavior was investigated by exposure tests in air at 500 °C and 1200 °C. All coatings didn't show any degradation after this exposure tests. Additionally the formation of the $SiO_2$ protection layer on the surface of the $MoSi_2$-based coatings was detected by in-situ-measurements of spectral emissivities by Fourier transform infrared spectroscopy (FTIR).

## 1. Introduction

A lot of furnace linings consist of refractories and high-temperature ceramics like $Al_2O_3$ or SiC with additional metallic radiation shields, subdivisions or muffles. These parts are widely used in furnaces that are operating in inert atmospheres or vacuum. Especially because of the better thermal dynamics, metallic components have advantages in comparison to the conventional high-temperature ceramics. Nevertheless, these parts cannot be utilized when an oxidizing atmosphere and high temperatures occur at the same time. Molybdenum disilicide ($MoSi_2$), which was discovered in 1907, has a unique profile of physical properties. Because of its high melting point, its low electrical resistance up to high temperatures and its excellent oxidation resistance it is predestinated for applications in oxidizing atmospheres in the temperature range of 600…1650 °C. The oxidation resistance of $MoSi_2$ is based on the so called self-protection. This effect can be explained by the partial oxidation of Si on the surface resulting in a glass like $SiO_2$-layer that protects the material from further oxidation [1, 2]. Furthermore the material has a self-healing effect so that the parts whose surface is damaged by mechanical or thermal stresses are nevertheless protected. The application range at lower temperatures is limited by the so-called pest oxidation that occurs in the temperature range from 400 °C up to 600 °C. In this temperature range there is no formation of a dense $SiO_2$-layer so that the components, especially the Mo rich parts of them, oxidize very fast. The maximum service temperature is 1650 °C because the protecting $SiO_2$-layer melts at 1723 °C [3, 4]. The temperature dependent properties of $MoSi_2$ are summarized in figure 1.

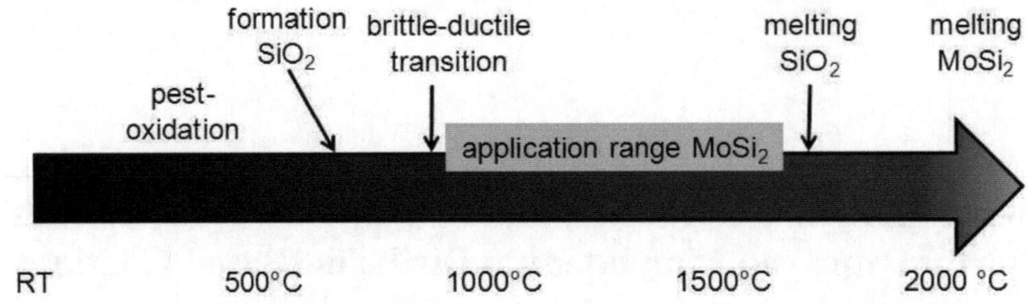


**Figure 1.** Temperature dependent properties of $MoSi_2$

Apart from the utilization as heating conductor the excellent properties of $MoSi_2$ can be used in the production of glow plugs, susceptors and gas turbines [5, 6]. Because of the limited possibility of geometrical shaping of pure $MoSi_2$, the aim of this work is the application of molybdenum disilicide coatings as high-temperature oxidation protection of molybdenum-based substrates that are easier machinable. Coating of metallic and ceramic components with $MoSi_2$ is possible by using different processes like PVD/CVD or thermal spraying [7, 8]. Thermal spraying was selected due to the high efficiency in producing well adhering coatings. The applicable processes are vacuum plasma spraying (VPS), atmospheric plasma spraying (APS) and high velocity oxygen fuel (HVOF) spraying, therefor powdery raw materials are needed [9, 10].

## 2. Experimental Procedure

The powder production line contains the steps production of suspension, spray drying and sintering. Within this investigations the used powders are $MoSi_2$ and $MoSi_2$+$ZrB_2$ in fractions of 5 µm < d < 20 µm and 20 µm < d < 45 µm, respectively. The powders were characterized in the final sintered state by laser diffraction, X-ray diffraction and SEM. The addition of $ZrB_2$ was carried out in order to lower the formation temperature of $SiO_2$. The powders are processed by means of APS in a GTV F6 torch, using different argon/hydrogen mixes. The parameter sets are optimized for the two particle fractions of the powders because the surface to volume ratio differs. The current is kept constant at 600 A. The coating process is carried out at a surface velocity of 0.5 m/s on grit blasted substrates in form of rods and metal sheets. The spray distance is adjusted at 125 mm. The heat-resistant steel 1.4762 and Mo are used as substrate materials. For application of $MoSi_2$-based coatings in industrial furnaces it is necessary to know their radiation properties, especially the emissivity which depends on wavelength and temperature. An experimental determination is the only way to get these data for thermotechnical calculations [11]. In the literature the emissivity of $MoSi_2$ is not specified, also spectral and temperature dependent data are missing. It can be assumed that the radiation properties show ceramic and metal characteristics. Due to the formation of the protective $SiO_2$-layer, the emissivity changes during the usage of the coated parts. The temperature dependent spectral emissivities are measured in-situ during the formation of this layer. Therefor Fourier transform infrared spectroscopy (FTIR) and a special heating and measuring device are used [12, 13, 14]. The results are correlated with the different parameter sets of the coating process. Additionally, exposure tests with Mo-Substrates are done. The coated samples are cut into cylinders by a metallographical cutter. This strategy allowes the production of all samples in one setting. The emerging uncoated Mo-surfaces are protected by ceramic slurry. The samples are exposed in air in the range of the pest oxidation at 500 °C and in the application range at 1200 °C for one hour. After exposure, the cylinders are cut into slices, polished metallographically and characterized by light microscopy.

## 3. Results

The results of the X-ray diffraction using Cu-Kα-radiation (figure 2) show that the predominantly present phase in the $MoSi_2$-powder without $ZrB_2$ is $MoSi_2$, $Mo_5Si_3$ is detected additionally. This second phase ($Mo_5Si_3$) doesn't occur in the $MoSi_2$+ 15 % $ZrB_2$-powder, figure 3. Induced by the process heat of the APS followed by a rapid cooling, phase transformations take place in the coating, so that α-$MoSi_2$ (hexagonal) und β-$MoSi_2$ (tetragonal) co-occur.

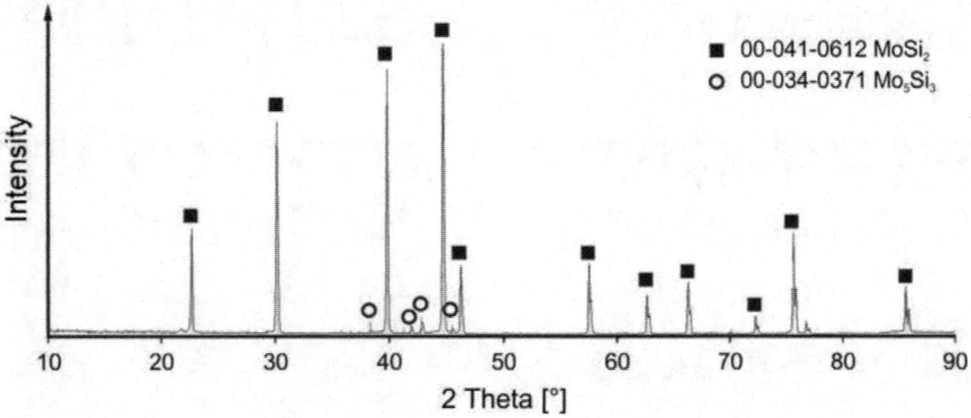


**Figure 2.** Diffractogram of the $MoSi_2$ powder: main phase $MoSi_2$, additionally $Mo_5Si_3$

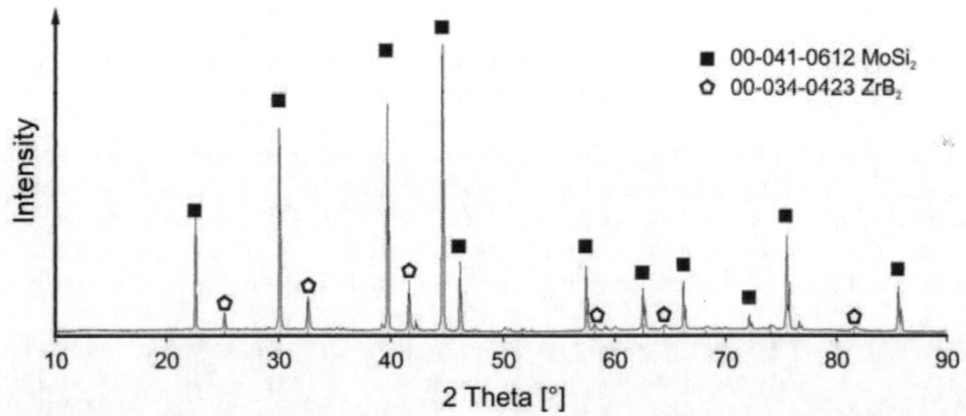


**Figure 3.** Diffractogram of the $MoSi_2$+ 15 % $ZrB_2$ powder: main phase $MoSi_2$, additionally $ZrB_2$

After the application by means of APS all coatings are investigated by SEM, Figure 4. A good adhesion to the substrates and no delamination can be observed. Depending on the used powder fraction (5 µm < d < 20 µm or 20 µm < d < 45 µm) different pore and layer structures are reached. A graded structure can be obtained by the combination of the two powder fractions.

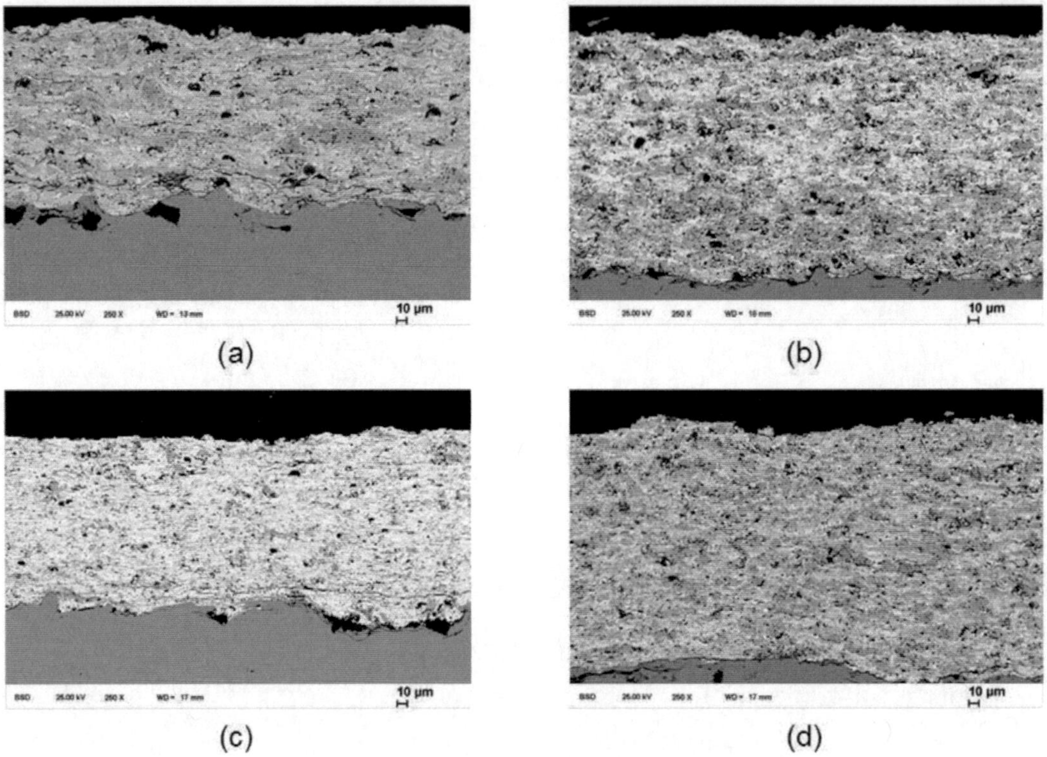


**Figure 4.** SEM-Images of $MoSi_2$-based coatings, as sprayed, a) $MoSi_2$ -45+20 µm; b) $MoSi_2$ + 15% $ZrB_2$ -45+20 µm c) $MoSi_2$ -20+5 µm; d) $MoSi_2$ + 15% $ZrB_2$ -20+5 µm

To check the behavior of the coatings at high temperatures, exposure tests are done, Figure 5. After the exposure at 500 °C a strong difference in oxidation behavior occurs. Whilst pure $MoSi_2$-coatings show a higher level of oxidation, coatings containing $MoSi_2$ and $ZrB_2$ are more resistant against the pest oxidation. After high-temperature oxidation at 1200 °C there is no significant difference between the two powder compositions. All coatings didn't show any degradation after the exposure tests.

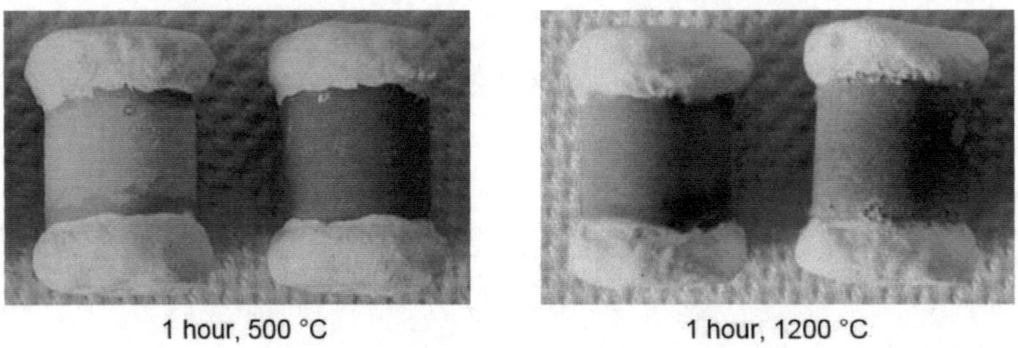


**Figure 5.** Photographs of the coatings after exposure in air

Because of the differences in oxidation behavior, that can already be seen optically, all heat treated samples are cut into slices and prepared metallographically. As shown in Figure 6, a smoothing effect occurs on the surface of pure $MoSi_2$-coatings at 500 °C. The $ZrB_2$-containing coatings don't show this effect, they can be characterized by a much rougher surface. After exposure at 1200 °C the self-protection layer can be observed at the surfaces of the coatings.

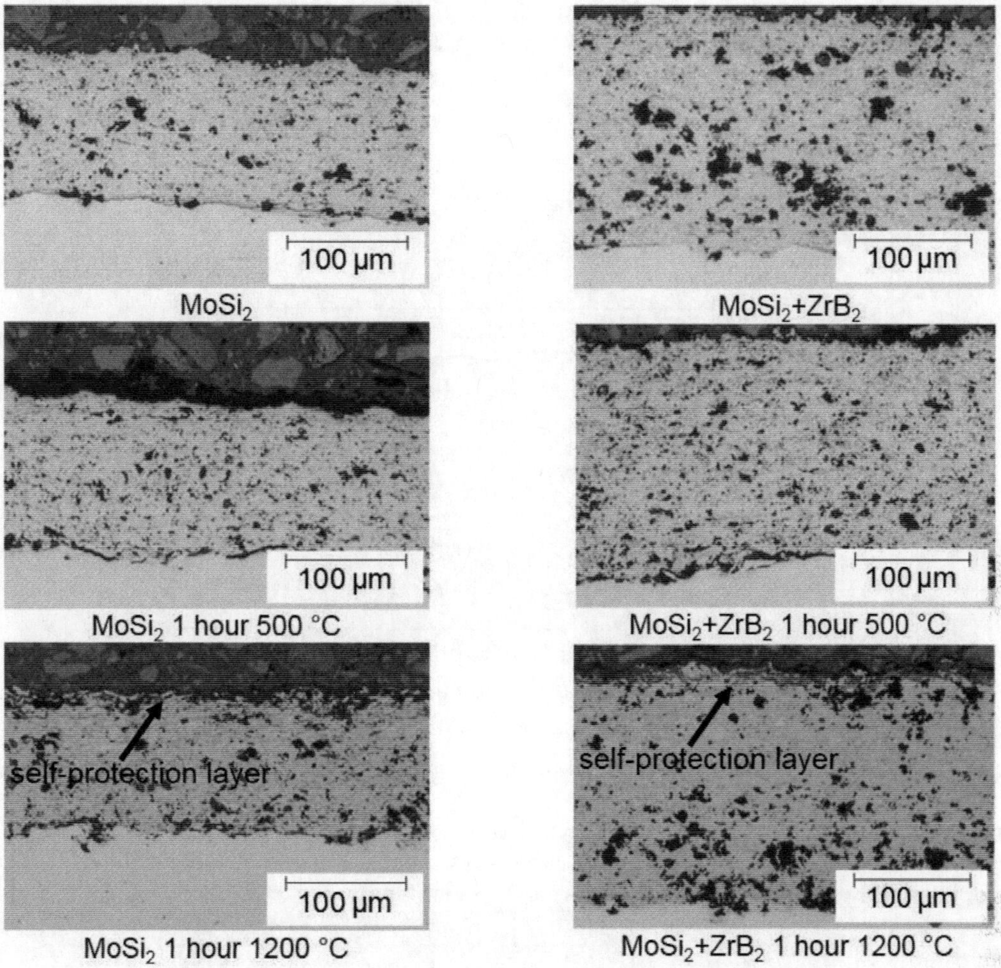


**Figure 6.** Optical micrographs of the coatings after exposure in air

The spectral emissivities of the $MoSi_2$ and $MoSi_2$+ 15 % $ZrB_2$ coatings as function of wavelength and temperature are investigated by FTIR (figure 7). The measurements are done in air to enable the formation of the $SiO_2$ protection layer. The results show that the emissivities of both types of coatings increase significantly for all wavelengths with increased temperatures. In the temperature range of 800 °C up to 1100 °C the values become almost stable. That behavior indicates the formation of the $SiO_2$ protection layer. The emissivities of the $MoSi_2$+ 15 % $ZrB_2$ coatings are higher than those without $ZrB_2$. Remarkably is there a local minimum at 9 µm in the 1100 °C curve of the MoSi2+ 15 % $ZrB_2$ coating, which is typical for $SiO_2$-based refractories too [15].

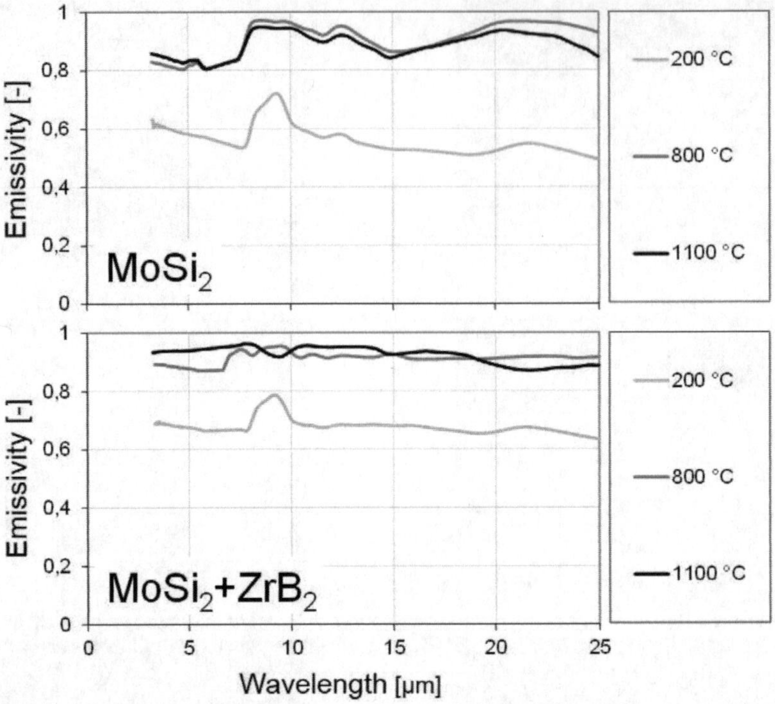


**Figure 7.** Emissivity of the $MoSi_2$ and $MoSi_2$+ 15 % $ZrB_2$ coating as a function of the wavelength and the temperature, measured in air

Summarized the in-situ measurements of spectral emissivities by FTIR is an applicable tool to detect the formation of the $SiO_2$ protection layer of $MoSi_2$-based coatings. Furthermore different varieties of this layer induced by different heat treatments can be identified.

## 4. Conclusions

Oxidation resistant coatings were applied by APS using agglomerated and sintered powders of two different compositions ($MoSi_2$ and $MoSi_2$+$ZrB_2$). Processing of the powders was carried out by GTV F6 torch. The coatings were applied on Mo-substrates with optimized spraying parameters. After exposure in air at 500 °C a strong difference in the oxidation behavior was observed. It can be concluded that the addition of $ZrB_2$ to the powder feedstock strongly decreases the tendency to pest oxidation at lower temperatures. At high temperatures there is no significant difference in oxidation. Furthermore the temperature dependent spectral emissivities of the coatings were measured in-situ by FTIR using a special heating and measuring device. Thereby the formation of the $SiO_2$ protection layer on the surface of the $MoSi_2$-based coatings was detected. Further investigations will deal with the chemical characterization of the coatings, especially the oxygen content and the distribution of Mo and Si, before and after the exposure tests.

**Acknowledgement**

Part of the research was carried out within the scope of the project # 16858 BG/1 funded by the Bundesministerium für Wirtschaft und Energie (BMWi) via the Arbeitsgemeinschaft industrieller Forschungsvereinigungen "Otto von Guericke" e.V. (AiF). The authors thank the BMWi for financing this project and the Fraunhofer IKTS for supplying the powders.

## References


[1] Chang Y A, 1969. *Oxidation of Molybdenum Disilicide. J. Mater. Sci., 641-643.*

[2] Schulle W, 1991. *Feuerfeste Werkstoffe. Deutscher Verlag für Grundstoffindustrie.*

[3] Jeng Y L, Lavernia E J, 1994. *Processing of Molybdenum Disilicide. J. Mater. Sci., 2557-2571.*

[4] Bundschuh K, Schütze M, 2001. *Materials for Temperatures above 1500 °C in Oxidizing Atmospheres – Part I: Basic Considerations on Materials Selection. Mater. Corr., 204-212.*

[5] Stoloff N S, Liu C T, 2000. *Emerging Applications of Intermetallics. Intermetallics, 1313-1320.*

[6] Yao Z J, Stiglich T S, 1999. *Molybdenum Silicide Based Materials and Their Properties. J. Mater. Eng. Perf., 291–304.*

[7] Yoon J K, Byun J Y, 2002. *Growth Kinetics of Three Mo-Silicide Layers Formed by Chem. Vapor Deposition of Si on Mo substrate. Surf. Coat. Techn., 85-98.*

[8] Govindarajan S, Moore J J, 1994. *Physical Vapor Deposition of Molybdenum and Silicon Thin Films. Surf. Coat. Techn., 45-50.*

[9] Fukumoto M, Ueda T, 1994. *Gas Permeability of Plasma Sprayed $MoSi_2$ Coating. J. Mater. Sci. Let., 656-657.*

[10] Motojima S, Fujimoto A, 1982. *Molybdenum Disilicide Coating on Steel and its Oxidation Resistance. J. Mater. Sci. Let., 19-22.*

[11] Bernhard F, 2004. *Technische Temperaturmessung. Springer Verlag Berlin Heidelberg.*

[12] Oertel H, Bauer W, 1998. *Facility for the measurement of spectral emissivities of bright metals in the temperature range 200 – 1200 °C. High Temperatures-High Pressures, 531-536.*

[13] Bauer W, Moldenhauer A, 2004. *Device for spectral emissivity measurements using a FT-IR-spectrometer. Proc. Spie Interferometry XII: Applications*

[14] Bauer W, Moldenhauer A, Gräfen W, 2009. *Emissionsgrade von NE-Metallen. Gaswärme International, 139-145.*

[15] Bauer W, Moldenhauer A, 2003. *Messungen spektraler Emissionsgrade von feuerfesten Baustoffen mit einem FT-IR-Spektrometer. Temperatur 2003: VDI-Berichte, 169-177.*

IOP Conf. Series: Materials Science and Engineering **118** (2016) 012008 doi:10.1088/1757-899X/118/1/012008

# Surface modification of austenitic thermal-spray coatings by low-temperature nitrocarburizing

**T. Lindner**[1], **T. Mehner**[1], **T. Lampke**[1]
[1] Materials and Surface Engineering Group, Institute of Materials Science and Engineering, Technische Universität Chemnitz, D-09125 Chemnitz, Germany

th.lindner@mb.tu-chemnitz.de

**Abstract.** Thermal-spray coatings of austenitic materials are mainly used under corrosive conditions. The relatively poor wear resistance strongly limits their use. In comparative studies between nitrocarburized and untreated thermal-spray coatings, the influence of the nitrogen and carbon enrichment on the properties of the coatings and the microstructure was investigated. The cross-section micrograph of the nitrocarburized coating shows the S-phase formation in the surface layer region. The depth profile of the nitrogen and carbon concentration was determined by glow discharge optical emission spectroscopy (GDOS) analysis. A selective enrichment of the surface layer region with nitrogen and carbon by means of thermochemical heat treatment increases the wear resistance. The interstitially dissolved nitrogen and carbon causes the formation of strong compressive residual stresses and high surface hardness. Increases in the service life of existing applications or new material combinations with face-centred cubic friction partners are possible. In the absence of dimensional change, uniform as well as partial nitrogen enrichment of the thermal spray coating is possible. Nitrocarburized coatings demonstrate a significant improvement in adhesive wear resistance and extremely high surface hardness.

## 1. Introduction

The properties of thermal spray coatings are mainly determined by the coating material and the spraying process. An influence on the properties of the coatings by variation of the process parameters is only possible within narrow limits. To deliver a combined wear and corrosion resistance, it is usually necessary to apply expensive coating materials. In recent years, there have been increasing efforts to develop low-cost alternatives, mainly materials based on iron. Successes could be achieved by further development of recent alloy compositions [1, 2]. In addition, improvements by mechanical finishing of manganese steel coatings are described [3, 4].

Iron-based bulk parts can be treated by means of thermal, thermochemical or mechanical treatment. This will change the surface properties to achieve the desired wear resistance. Stainless and acid-resistant iron-based materials with face-centred cubic lattice structure are mainly used under corrosive conditions. Due to the wear behavior, pairings and tight fits are so far limited. Particularly critical are metal combinations with similar face-centred cubic lattice structure, where adhesion wear behavior restricts a further practical use. The low-temperature thermochemical treatment of face-centred cubic materials provides the opportunity to extend the applicability by a nitrogen enrichment of the surface layer region. It is necessary to ensure the sufficiently low operation temperature to prevent the precipitation of chromium phases that will lower the corrosion resistance. Promising results for bulk material parts show a significant raise in hardness combined with improved adhesion-wear resistance [5, 6]. Due to the improvement of the properties, a successful treatment provides a high innovation potential with certain new applications.

This article describes exemplarily how a low-temperature thermochemical treatment influences and changes the microstructure as well as the functional properties of thermal-spray coatings with face-centred cubic structure.

## 2. Experimental

Due to promising results for bulk parts [7, 8], the alloy AISI316L was investigated. The thermal-spray coatings were produced with the HVOF system GTV-K2 with the parameters shown in *Tab. 1*.

**Table 1.** Process parameters HVOF GTV-K2

| | |
|---|---|
| kerosine | 24 l/h |
| oxygen | 900 l/min |
| $\lambda$ | 1.09 |
| combustion chamber pressure | 7.1 bar |
| spraying distance | 350 mm |
| nozzle | 150/14 mm |
| transverse speed | 1 m/s |
| offset | 5 mm |
| powder feed rate | 70 g/min |
| gas feed | 2 x 8 l/min (argon) |
| wiper | NL |

The alloy powder was used in fraction of -53+20 µm. The coating layer was polished followed by nitrocarburation with the industrial process called HARD-INOX®-S by Gerster.

The chemical, mechanical, microstructural, and tribological properties were investigated for the untreated and carburized coatings. The nitrogen and carbon concentration as a function of the surface distance was measured by GDOS (SPECTRUMA GDA750) with a diameter of 4 mm. Furthermore, the microhardness was measured on the surface and on the cross-section using a FISCHERSCOPE HM2000 XYm. XRD measurements (diffractometer D8 DISCOVER) on the surface of the coating were used to identify phases and to determine the lattice parameters. The microstructure was characterized on the cross-section micrograph.

The tribological properties of the coating were investigated with the reciprocating ball-on-plane test leant on ASTM G 133 as dry couple and the ball-on-disc test leant on ASTM G 99 as a dry sliding system, both with a spherical counterface.

**Table 2.** Testing parameters of wear test

| reciprocating ball-on-plane test | | | ball-on-disc test | | |
|---|---|---|---|---|---|
| force | 26 | N | force | 10 & 20 | N |
| frequency | 40 | Hz | radius | 5 | mm |
| time | 900 | s | speed | 96 | U/min |
| amplitude | 0.5 | mm | cycles | 15916 | |
| Ø $Al_2O_3$ | 10 | mm | Ø $Al_2O_3$ | 6 | mm |

The test parameters as well as the counterbody material and diameter are shown in *Tab. 2*. The wear death was profilometrically determined with the 3D profilometer MikroCAD. The benchmarks for all testings were the determined values for the untreated samples.

## 3. Results and Discussion

The aim of the nitrocarburization was the saturation and interstitial deposition of nitrogen and carbon in the surface layer region of the thermal spray coating. This was supposed to create high compressive residual stresses and to increase the wear performance.

IOP Conf. Series: Materials Science and Engineering **118** (2016) 012008 doi:10.1088/1757-899X/118/1/012008

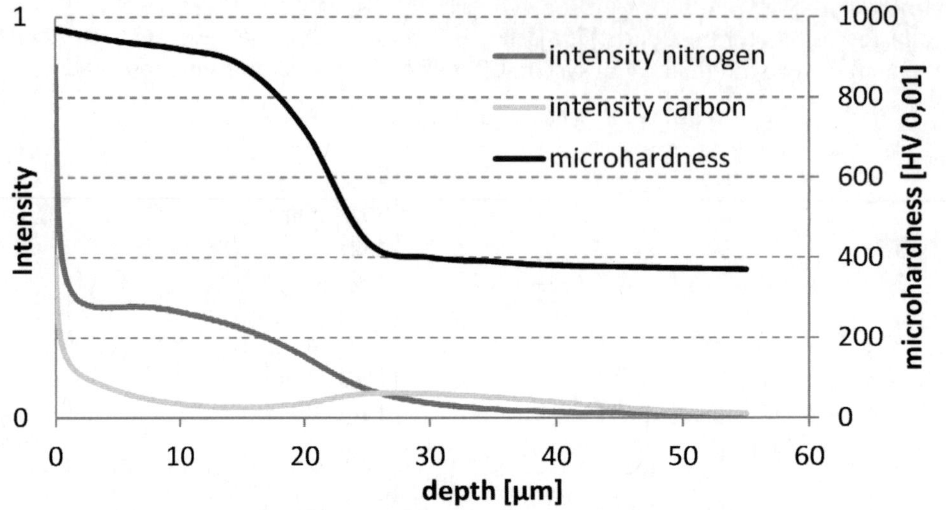


**Figure 1.** Nitrogen and carbon GDOS intensities as well as microhardness of nitrocarburized AISI316L HVOF coatings as a function of the surface distance

The determined nitrogen and carbon intensities by GDOS analysis show a decrease with increasing surface distance, *Fig. 1*. The microhardness varies deeply in dependence of local saturation depth. The shown values are the minimum microhardness values. In the surface-layer region the microhardness increases by about factor of 2 compared with the untreated zone.

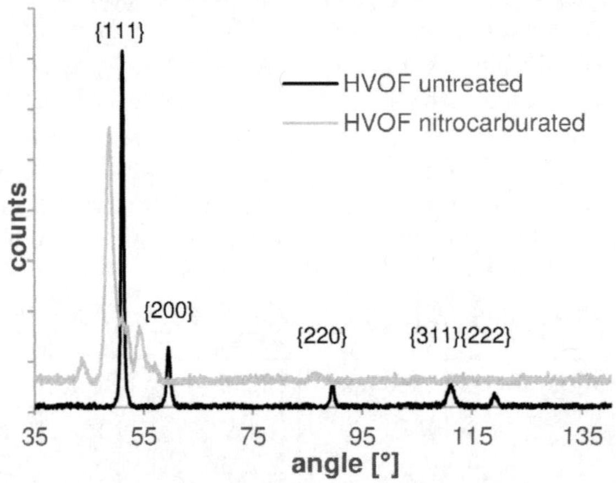


**Figure 2.** XRD diffraction diagrams of the nitrocarburized and untreated AISI316L HVOF coating with the Miller indices of the austenite peaks

The XRD studies show that the lattice parameter of the measured austenite phase strongly expands after nitrocarburization, *Fig. 2*. In comparison with the untreated sample, a clear shift of the characteristic peaks of austenite to lower angles was determined, which can be explained by the increase of lattice spacing due to the interstitial deposition of nitrogen and carbon atoms. In addition to the main austenitic phase, a small amount of iron chromium carbid phase was determined with a good accordance in the nitrocarburated sample.

The measurement of the residual stress state was not possible because of the broadening of the peaks due to microstrains. With assumption of a constant stress state the qualitative shift of the austenite peaks can be explained by an interstitial solvation of nitrogen or carbon, *Tab. 3*.

**Table 3.** Lattice parameters of the austenite phase for the untreated and nitrocarburized AISI316L HVOF coatings

| lattice plane | Young's modulus $E$ [9] | lattice parameter $a$ | |
|---|---|---|---|
| | | untreated | nitrocarburized |
| {111} | 250 GPa | 3.59 Å | 3.77 Å |
| {200} | 140 GPa | 3.59 Å | 3.92 Å |

*Table 3* reveals the lattice spacing of the samples determined by XRD. The nitrocarburized sample shows a great expansion in lattice parameter that follows the anisotropic elastic behavior of austenitic steel. Due to the different Young's moduli, the expansion of the lattice planes {111} and {200} is expected as a consequence of the residual stress. This can be proven by the following relation:

$$\frac{E_{\{200\}}}{E_{\{111\}}} \approx \frac{a_{\{111\}\text{nitrocarb.}} - a_{\text{untreated}}}{a_{\{200\}\text{nitrocarb.}} - a_{\text{untreated}}}$$

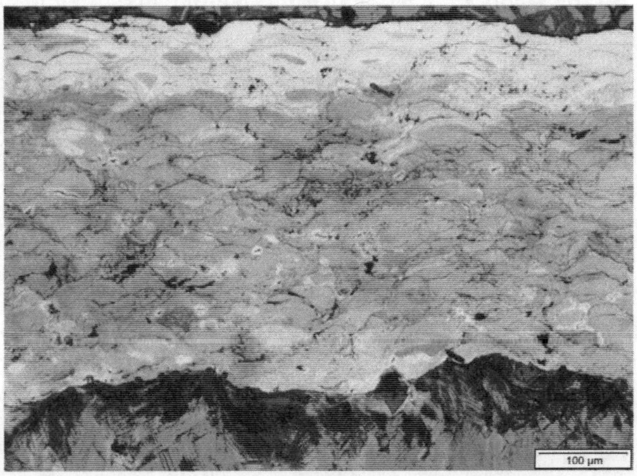


**Figure 3.** Cross-section of the nitrocarburized AISI316L HVOF coating (etched)

After the etching of the cross-section, two regions are identifiable. The topmost layer (white area in Fig. 3) surface layer region is the so-called expanded austenite phase or S-phase that shows a strong resistance to the etchant. The depth of the S-phase varies between 20 µm and 100 µm.

The marked increase of hardness and compressive stress causes a significant increase in wear resistance. This is particularly applicable for the studied sliding and reciprocating testing methods. For the untreated and nitrocarburized samples, the values of wear are shown in Tab. 4.

**Table 4.** Wear testing of nitrocarburized and untreated AISI316L HVOF coatings

| | | untreated | nitrocarburized |
|---|---|---|---|
| reciprocating ball-on-Plane | wear volume [mm³] | 0.176 | 0.056 |
| | wear depth [µm] | 92 | 51 |
| ball-on-disc 10 N | wear volume [mm³] | 4.975 | < 0.005 |
| | wear depth [µm] | 101 | 3 |
| ball-on-disc 20 N | wear volume [mm³] | 12.329 | < 0.05 |
| | wear depth [µm] | 175 | 7 |

After the wear testing, the wear depth was profilometrically determined and the volume loss was calculated. The wear volume is determined in the whole trace of wear. For depth of wear, the deepest point in the trace of wear was measured. The comparison of untreated and nitrocarburized samples shows a reduction of the wear volume in reciprocating wear of about 68 %. For the nitrocarburized samples in ball-on-disc test, almost no wear could be measured.

IOP Conf. Series: Materials Science and Engineering **118** (2016) 012008 doi:10.1088/1757-899X/118/1/012008

## 4. Conclusion

The great potential of low-temperature thermochemical treatment of thermal-spray coatings has been demonstrated in this paper. Exemplarily for AISI316L HVOF coatings, it is possible to build up a carbon rich surface layer with improved tribological and wear properties without forming chromium carbides. With the enhanced residual stress state, significant improvements in fatigue strength and abrasive wear resistance are possible.

## 5. Literature

[1] B. Wielage, S. Schuberth: Entwicklung eisenbasierter Verschleißschutzlegierungen. Tagungsband zum 13. Werkstofftechnischen Kolloquium (2010), S. 188-193.

[2] K. Müller: Werkstoffkundliche Qualifizierung des Randschichthärtens mit Laserstrahlung. Dissertation, Lehrstuhl Metallische Werkstoffe, Universität Bayreuth, 1999.

[3] J. Kopac: Hardening phenomena of Mn-austenite steels in the cutting process. Journal of Material Processing Technology (2001) 109, S. 96-104.

[4] H. Schumann: Verformungsinduzierte Martensitbildung in metastabilen austenitischen Stählen. Kristall und Technik (1975) 10, S. 401-411.

[5] R. H. Van Der Jagt: Kolsterising - surface hardening of austenitic and duplex stainless steels without loss of corrosion resistance. Heat Treatment of Metals (2000) 27, S. 62-65.

[6] G. M. Michal, et al : Carbon Supersaturation Due to Paraequilibrium Carburization: Stainless Steels With Greatly Improved Mechanical Properties. Acta Materialia (2006) 54, S. 1597-1606.

[7] P. Blau, et al.: Tribological Properties of Stainless Steels Treated by Colossal Carbon Supersaturation. Wear (2007) 263, S. 719-726.

[8] W. Hubner. Phase transformation in austenitic stainless steels during low temperature stressing. Tribology International (2001) 34, S. 231-236.

[9] B. Eigenmann, et al.: Röntgenographische Untersuchung von Spannungszuständen in Werkstoffen. Teil III. Materialwissenschaft und Werkstofftechnik (1996) 27, S. 426-437.

# Analytical methods to characterize heterogeneous raw material for thermal spray process: cored wire Inconel 625

**T. Lindner[1], S. Bonebeau[2], R. Drehmann[1], T. Grund[1], L. Pawlowski[2], T. Lampke[1]**
[1] Materials and Surface Engineering Group, Institute of Materials Science and Engineering, Technische Universität Chemnitz, 09125 Chemnitz, Germany
th.lindner@mb.tu-chemnitz.de

[2] Faculty of Science and Technology, University of Limoges, France
simon.bonebeau@etu.unilim.fr

**Abstract.** In wire arc spraying, the raw material needs to exhibit sufficient formability and ductility in order to be processed. By using an electrically conductive, metallic sheath, it is also possible to handle non-conductive and/or brittle materials such as ceramics. In comparison to massive wire, a cored wire has a heterogeneous material distribution. Due to this fact and the complex thermodynamic processes during wire arc spraying, it is very difficult to predict the resulting chemical composition in the coating with sufficient accuracy. An Inconel 625 cored wire was used to investigate this issue. In a comparative study, the analytical results of the raw material were compared to arc sprayed coatings and droplets, which were remelted in an arc furnace under argon atmosphere. Energy-dispersive X-ray spectroscopy (EDX) and X-ray fluorescence (XRF) analysis were used to determine the chemical composition. The phase determination was performed by X-ray diffraction (XRD). The results were related to the manufacturer specifications and evaluated in respect to differences in the chemical composition. The comparison between the feedstock powder, the remelted droplets and the thermally sprayed coatings allows to evaluate the influence of the processing methods on the resulting chemical and phase composition.

Keywords: Inconel 625; Chemical analyzes; Arc spray; Coating; Cored wire

## 1. Introduction

In thermal spraying wire-shaped feedstock material is mainly used for arc spray techniques. It is available as massive material or as hollow sheath with pre-processed powder inside. This so-called cored wire material is a special wire type that is predominantly used for processing hard materials as core material in a ductile and electrically conductive sheath to form pseudo-alloys. Generally cored wire material is more expensive than massive wire. However, this applies only for big production quantities, since the production of smaller amounts of massive wire is often unprofitable. Thus for small quantities, cored wires are usually cheaper than massive wires or even the only wire form, which is commercially available. This can be attributed to the fact that most wire manufacturers have a wide selection of powders and thin metal sheets in stock, which allows the on-demand production of cored wires in the customer-requested quantity. Cored wires offer the opportunity to adjust almost any desired material mixture by filling the appropriate powder in the hollow sheath. This allows the production of coatings, that cannot be achieved by other coating processes. However, very often it is difficult to predict the final chemical composition in the coating based on the used sheath and filler material. The reason for that is the complexity of chemical reactions that take place during the processing of the wires. This includes reactions in the sheath and the powder material as well as reactions with each other and with the surrounding atmosphere. By the example of Inconel 625 - a widely used corrosion-resistant Ni-based material - it was intended to investigate the changes in the chemical composition by a comparison of the feedstock cored wire material, arc-sprayed coatings and droplets, which were remelted in an arc furnace under argon atmosphere.

## 2. Experimental

### *2.1. Coating*

As stated previously, this study focuses on the Ni-base material Inconel 625 (IN625). Therefore, a cored wire DURMAT® AS-755 provided by DURUM Verschleißschutz GmbH (Willich, Germany) was used. The wire has an outer diameter of 1.6 mm. The nominal composition in weight percentage as stated by the manufacturer is shown in Table 1.

**Table 1.** Chemical composition of IN625 cored wire as stated by the manufacturer

| Element | Ni | Cr | Mo | Nb | Fe | C |
|---|---|---|---|---|---|---|
| wt.-% | balance | 20–23 | 7–10 | 3–4 | < 0.1 | < 2 |

A wire arc spray system VisuArc 350 (Oerlikon Metco, Switzerland) was used to produce the coatings. The spray parameters were adjusted as recommended by the gun manufacturer (Table 2).

**Table 2.** Spray parameters for wire arc-sprayed Inconel 625

| | |
|---|---|
| Substrate | S235 |
| $\varnothing_{substrate}$ (mm) | 25 |
| Gas pressure (bar) | 3.5 |
| Voltage (V) | 30 |
| Current (A) | 120 |
| Spray distance (mm) | 130 |
| Spray velocity ($m \cdot s^{-1}$) | 1 |
| Shift z (mm) | 5 |
| No. of passages | 7 |
| Mean coating thickness | 280 |

The used substrate was mild steel containing about 99 % of iron (S235). Prior to the coating process, the substrate surface was roughened by grit blasting.

### *2.2. Sample preparation and characterization techniques*

In order to produce solid Inconel 625 droplets, the cored wire was molten in a self-constructed electric arc furnace. The furnace was flooded with argon when vacuum reaches ($2 \times 10^{-4}$ mbar). By using argon as inert gas, it was intended to avoid uncontrolled oxidation. After the ignition of the arc with a tungsten electrode, the cored wire was molten in a water-cooled copper crucible to droplets with a diameter of about 20 mm. The parameters are shown in Table 3.

**Table 3.** Arc furnace parameters

| Mean quantity (g) | Droplet size (mm) | Vacuum (mbar) | Argon pressure (bar) | Gas purity (%) | Current (A) |
|---|---|---|---|---|---|
| 15 | 20 | $2\times10^{-4}$ | 1.1 | 99.9999 | 250 |

The feedstock cored wire, the arc-sprayed coating as well as the droplet were firstly subjected to XRF analysis in a Fischerscope X-ray XAN device (year 2005) [1–3]. This device does not need any sample preparation and provides direct qualitative measurement in less than one minute. The number of imposed strokes was roughly 3000 $stk \cdot min^{-1}$. An acceleration voltage of 30 kV was applied and the current passing through the anode was 950 μA.

The samples were then analyzed by a scanning electron microscope (SEM) LEO 1455VP (Zeiss, Oberkochen, Germany) equipped with an EDAX Genesis EDX spectroscope. According to the device and the specimen-holder prerequisite, samples were subjected to a special specimen preparation. Metallographic cross-sections of the cored wire, the IN625 coating and the droplet were prepared according to standard metallographic procedures. The resulting samples were mounted into conductive resin, followed by grinding and polishing. Subsequently, the morphology, microstructure and chemical composition were investigated by means of secondary and backscattered electron imaging in conjunction with EDX measurements [4].

Crystallographic study was performed on all the previous samples by XRD using a D8 DISCOVER device from Bruker Corporation (Billerica, USA). The used X-radiation was a monochromatic wavelength Cu-$K_\alpha$. A voltage of 40 kV, a current of 40 mA, a scanning rate of 24° $h^{-1}$ and diffraction angles of 20° to 130° were utilized. The used X-rays sensor was a LynxEye-XE detector, the aperture size after the beam gun was 1 mm and a polycap was utilized in order to obtain a better parallel beam emission. The evaluation of the data was done by using Rietveld refinement, requiring knowledge of the theoretical structure which has to be close to the real structure [5].

## 3. Results and discussion

### *3.1. Characterization of Inconel 625 cored wire*

The analyses of the heterogeneous cored wire were performed on metallographic cross-sections (Fig. 1). Therefore, the wire was mounted in epoxy resin to avoid a trickling out of the powder.

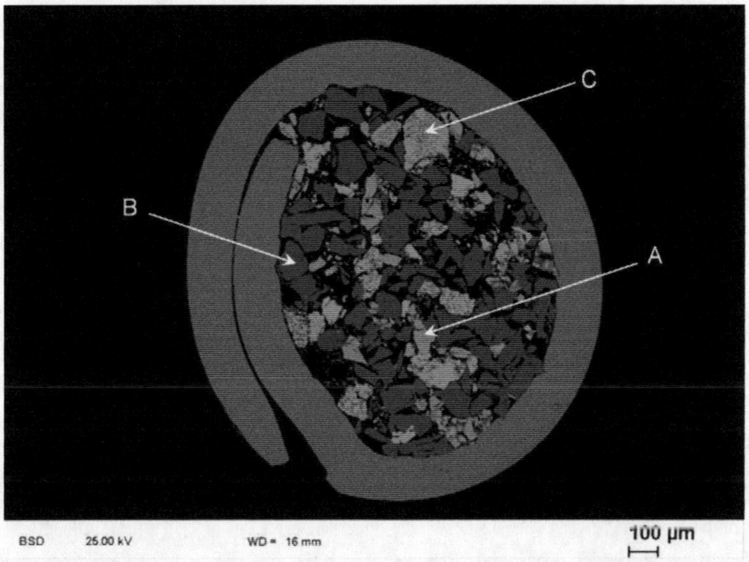


**Figure 1.** SEM image (BSE mode) of the cross-section of IN625 cored wire

In a first step, global XRF and EDX analyses were carried out. Therefore, the investigated area comprised of the whole cross-section of the cored wire. The back scattered electron image (Fig. 1) shows the atomic number contrast that illustrates clearly the heterogeneous character of the powder and the whole wire, respectively.

Unlike XRF, EDX enables the measurement of the local chemical composition. By this means, it was proven that the sheath consists of pure nickel. The results for the marked grains in Fig. 1 are shown in Table 4.

**Table 4.** Local concentration in IN625 cored wire as determined by EDX

| | EDX (relative wt-%) | | | |
|---|---|---|---|---|
| | Sheath | Grain A | Grain B | Grain C |
| Ni | 100 | - | - | - |
| Cr | - | - | 100 | - |
| Mo | - | 100 | - | - |
| Nb | - | - | - | 68.6 |
| Fe | - | - | - | 28.2 |
| Si | - | - | - | 3.2 |

The heterogeneity of the cored wire allows the detection of the different phases of the contained components because - in contrast to the remelted droplet and, to a certain extent, the arc-sprayed coating (see the results below) - the elements are not solved in the nickel matrix. Consequently, the nickel phase of the sheath shows the highest intensity level. Also the characteristic peaks of pure molybdenum and pure chromium can be found in the XRD pattern (Fig. 2). The niobium-rich phase that was previously detected by EDX (Tab. 4) was not stored in the database. However, an undefined phase was found, which most probably can be attributed to the mentioned niobium-rich component.

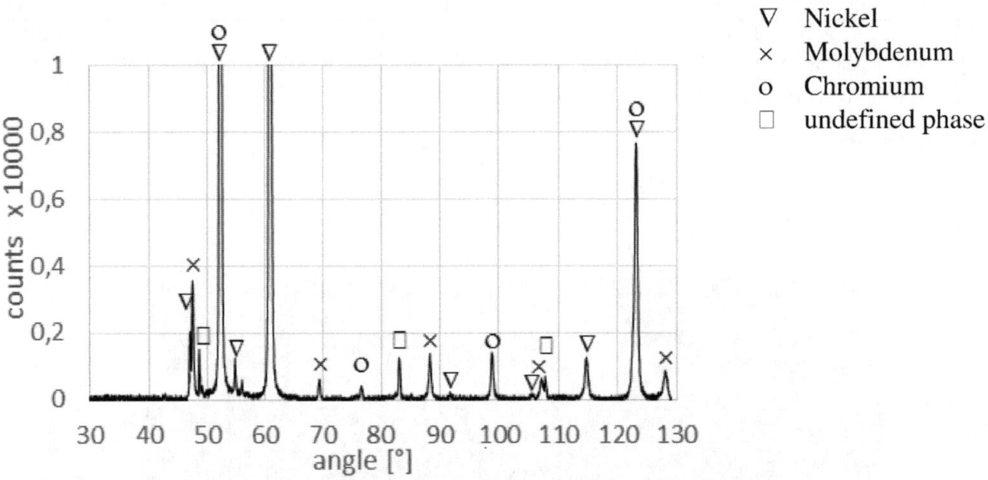


**Figure 2.** XRD of IN625 cored wire

*3.2. Characterization of arc-sprayed Inconel 625 coating*

The metallographic cross-section of the Inconel 625 coating shows the typical lamellar structure of arc-sprayed coatings (Fig. 3). The different brightness values in the back scattered electron SEM image indicate an incomplete blending of the different alloying components (Fig. 3). However, since no unmolten powder particles are found in the coating, it can be assumed that the feedstock material was fully molten during the arc spray process. As can be seen in Fig. 3b, an oxide layer may have formed on a few of the molten droplets, preventing them from being flattened upon impact.

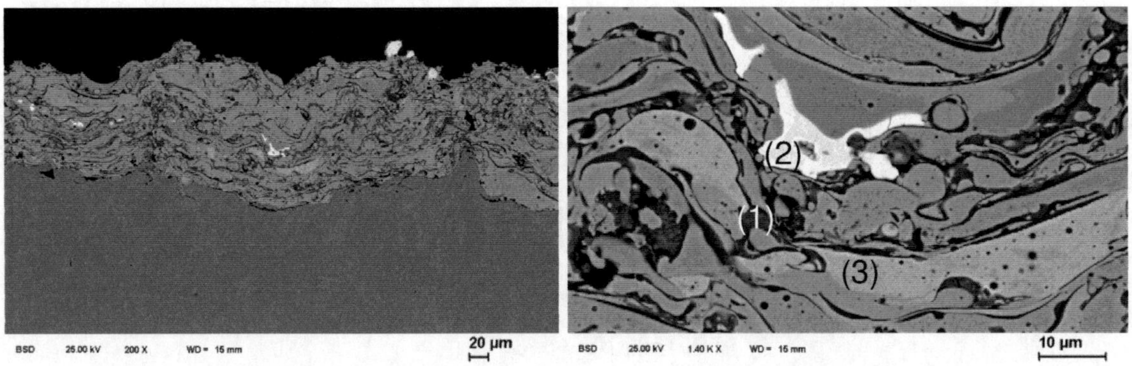


*a) Overview* *b) Detail*

**Figure 3.** SEM image (BSE mode) of arc-sprayed IN625 coating

It is presumed that the time window, in which the material is in a liquid state, is too short to obtain homogeneous material alloying. This is also connected to the differing solubility time of the materials.

In comparison to the cored wire, the phase composition has changed due to the thermal spray process. The EDX measurements (Table 5) show that the main phase (3), which appears grey in Fig. 3, is nickel-rich. Since no precipitations were observed, it is assumed that all remaining alloy elements are solved in this phase. Furthermore, a chromium-rich phase (1, displayed dark grey in Fig. 3) was detected. Presumably, these are stable chromium oxides, which were formed in the oxidative process atmosphere. The white phase (2) is nearly pure molybdenum, which is known for its slow solubility. A pre-alloyed phase - comprising e.g. iron - could improve the solubility.

IOP Conf. Series: Materials Science and Engineering **118** (2016) 012009 doi:10.1088/1757-899X/118/1/012009

**Table 5.** Local concentration in IN625 coating as determined by EDX

| | EDX (relative wt-%) | | |
|---|---|---|---|
| | Dark grey phase (1) | White phase (2) | Grey phase (3) |
| Ni | 16.1 | 9.4 | 49.9 |
| Cr | 50.1 | - | 14.4 |
| Mo | 8.7 | 90.6 | 28.6 |
| Nb | 20.4 | - | 4.6 |
| Fe | 3.6 | - | 2.3 |
| Si | 1.1 | - | 0.2 |

Even though the mainly pure elements from the feedstock cored wire were mixed for the most part due to the thermal spray process, the previously described (Table 1) composition of IN625 was not reached homogeneously across the whole coating. The coating microstructure contains different phases in a heterogeneous distribution. This has to be considered as disadvantageous in terms of corrosion resistance since a heterogeneous microstructure bears the danger of local corrosion damage.

The EDX results were confirmed by XRD measurements by which several phases were detected (Fig. 4). The assumed process-induced oxidation of chromium was proven by the presence of a chromium oxide peak in the XRD pattern.

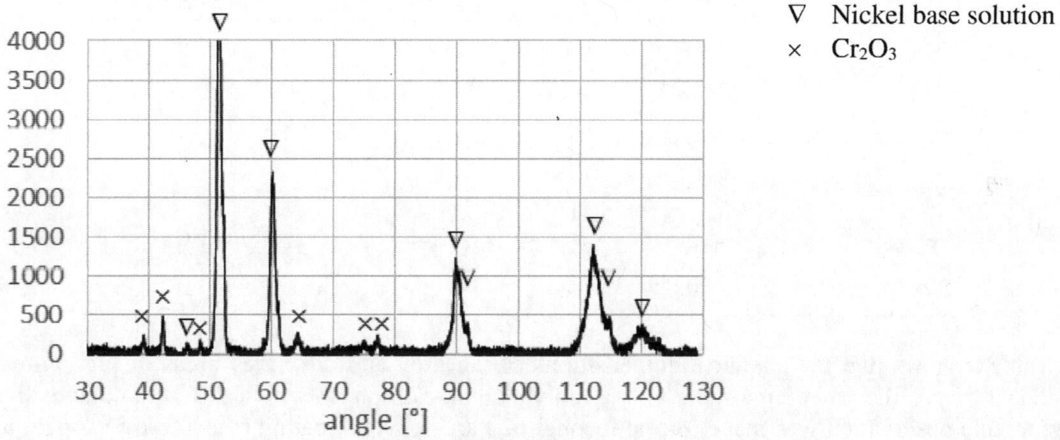


**Figure 4.** XRD of IN625 coating

*3.3. Characterization of Inconel 625 droplet*

In order to reach a homogeneous element distribution, the cored wire was remelted to droplets in an electric arc furnace. In contrast to the heterogeneous cored wire, the droplets microstructure exhibit a more homogeneous element distribution, by alloying in process.

A cross-section of the droplet was prepared in order to analyze the microstructure. As can be seen in Fig. 5, a homogeneous microstructure was not achieved completely, even though the droplet was held in molten state for several seconds.

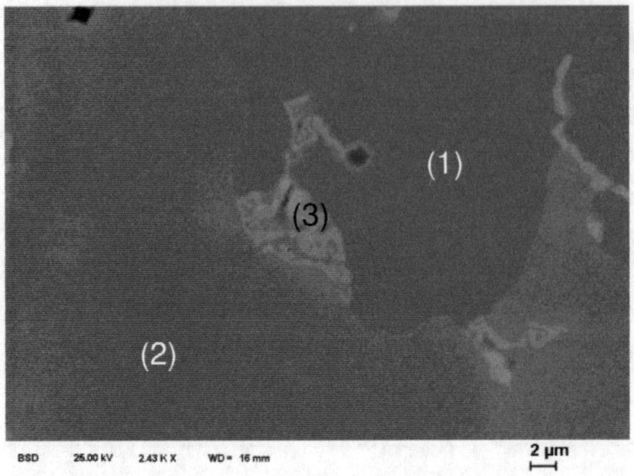


**Figure 5.** SEM image (BSE mode) of IN625 droplet

Due to the very coarse-grained microstructure (Fig. 5), a quantitative evaluation by means of XRD was not possible. The investigated area covered too few grains, leading to insufficient statistical data. Therefore, no XRD measurements are presented for the droplet.

**Table 6.** Local concentration in IN625 droplet as determined by EDX

| | EDX (relative wt-%) | | |
|---|---|---|---|
| | Dark grey zone (1) | Grey zone (2) | White zone (3) |
| Ni | 62.0 | 61.8 | 45.9 |
| Cr | 22.2 | 22.0 | 15.3 |
| Mo | 9.0 | 9.5 | 16.8 |
| Nb | 3.8 | 3.7 | 17.4 |
| Fe | 2.2 | 2.4 | 1.6 |
| Si | 0.8 | 0.6 | 3.0 |

Table 6 shows that the chemical compositions of the grey and dark grey areas in Fig. 5 are very similar, whereas the white area represents a molybdenum-rich phase with a different composition. A comparable observation was made for the Inconel 625 arc-sprayed coating (see Table 5). Again it can be assumed that the heterogeneity is attributed to the slow solubility of molybdenum.

### *3.4. Comparison of XRF and EDX measurements*

Table 7 shows the comparison of the results of the XRF and EDX measurements of the feedstock cored wire, the droplet and the arc-sprayed coating.

A special precalibration for the investigated alloy system was not performed. However, this does not have a negative influence on the validity of the measurements, since the comparability of the results of wire, droplet and coating with each other is not affected. Lightweight elements such as oxygen and nitrogen, that are basically detectable with EDX, were not analyzed because of the high error probability. The measured values in Table 7 are relative values summed to 100 %.

IOP Conf. Series: Materials Science and Engineering **118** (2016) 012009 doi:10.1088/1757-899X/118/1/012009

**Table 7.** EDX and XRF analyses presenting the average element concentrations

| | XRF (relative wt-%) | | | EDX (relative wt-%) | | |
|---|---|---|---|---|---|---|
| | W* | D† | C‡ | W | D | C |
| Ni | 58.0 | 63.2 | 63.7 | 52.3 | 60.8 | 60.8 |
| Cr | 21.0 | 21.6 | 20.2 | 24.6 | 21.2 | 20.7 |
| Mo | 11.5 | 9.5 | 9.7 | 12.2 | 10.0 | 10.3 |
| Nb | 7.5 | 3.7 | 4.3 | 7.5 | 4.9 | 4.8 |
| Fe | 2.0 | 2.0 | 2.1 | 2.3 | 2.3 | 3.0 |
| Si | 0 | 0 | 0 | 1.1 | 0.8 | 0.4 |

* Cored wire † Droplet ‡Coating

For both analysis methods, a significant decrease of the element content of several alloy components could be observed when comparing the feedstock cored wire on the one hand with the droplet and the coating on the other. This led to a shift in the chemical composition, which is similar for droplet and coating. The strongest decrease in concentration shows niobium followed by molybdenum. The fact that nickel is the only element that increases in amount leads to the conclusion that the iron and chromium contents also decrease, but the relative amount does not change. This suggests that alloy systems for arc spraying in form of cored wire, which include certain amounts of refractory elements and elements that tend to vanish during the spraying process, need to be composed in consideration of the change in chemical composition by processing.

Silicon was only detected by EDX, but not by XRF. However, the low Si content hardly affects the comparability, so this is not considered to be the main reason for the differing measurement results of XRF and EDX. The highest deviation between the two measurement techniques can be found in the results of the cored wire. Especially the nickel and chromium contents differ significantly (Table 8).

**Table 8.** Relative differences between the values determined by EDX and XRF (wt-%)

| | ΔW | ΔD | ΔC |
|---|---|---|---|
| Ni | 5.7 | 2.4 | 2.9 |
| Cr | -3.6 | 0.4 | -0.5 |
| Mo | -0.7 | -0.5 | -0.6 |
| Nb | 0 | -1.2 | -0.5 |
| Fe | -0.3 | -0.3 | -0.9 |
| Si | -1.1 | -0.8 | -0.4 |

Also for the droplet and the coating, the XRF analysis revealed a higher Ni content in comparison to EDX, even though the difference is smaller than for the cored wire. Apart from Ni, the XRF and EDX results showed only small deviations for the elemental concentration in the droplet and the coating. This suggests that the comparability of the results of these two analysis methods increases with an increasing homogeneity of the investigated samples.

## 4. Conclusion

This paper shows the potential and the limitations of different chemical analysis methods by the example of an Inconel 625 cored wire. Contrary to hard particle-filled cored wires, the main application for Inconel 625 coatings is not wear protection, but corrosion protection, which implicates the requirement for a homogeneous microstructure after the arc spray process. The investigated samples, however, show a heterogeneous phase distribution in the coating and, to a lesser extent, even in the remelted droplet. An approach to solve this problem might be the use of pre-alloyed components that have to be determined in a further study.

IOP Conf. Series: Materials Science and Engineering **118** (2016) 012009 doi:10.1088/1757-899X/118/1/012009

**References**

[1] L. Van de Voorde, B. Vekemans, E. Verhaeven, P. Tack, R. De Wolf, J. Garrevoet, P. Vandenabeele, and L. Vincze, "Analytical characterization of a new mobile X-ray fluorescence and X-ray diffraction instrument combined with a pigment identification case study," *Spectrochim. Acta Part B At. Spectrosc.*, vol. 110, pp. 14–19, Aug. 2015.

[2] G. Artioli, "Science for the cultural heritage: the contribution of X-ray diffraction," *Rendiconti Lincei*, vol. 24, no. S1, pp. 55–62, Feb. 2013.

[3] J. H. Burton and T. D. Price, "The ratio of barium to strontium as a paleodietary indicator of consumption of marine resources" *J. Archaeol. Sci.*, vol. 17, no. 5, pp. 547–557, Sep. 1990.

[4] S. Canlı, T. Raşit, and B. A. Kurç, "Thickness analysis of thin films by energy dispersive x-ray spectroscopy," METU, Ankara, 2010.

[5] A. Hooda, S. Sanghi, A. Agarwal, and R. Dahiya, "Crystal structure refinement, dielectric and magnetic properties of Ca/Pb substituted SrFe12O19 hexaferrites" *J. Magn. Magn. Mater.*, vol. 387, pp. 46–52, Aug. 2015.

IOP Conf. Series: Materials Science and Engineering **118** (2016) 012010 doi:10.1088/1757-899X/118/1/012010

# Binary WC– and $Cr_3C_2$–containing hardmetal compositions for thermally sprayed coatings

**L-M Berger**

Fraunhofer Institute for Ceramic Technologies and Systems (IKTS), Dresden
E-Mail: Lutz-Michael.Berger@ikts.fraunhofer.de

**Abstract.** Compositions of thermally sprayed hardmetal coatings for wear protection are based on the hard materials WC and $Cr_3C_2$ with Co and Ni as the most important binders, which are often alloyed with Cr. There are a few commercial compositions containing WC and $Cr_3C_2$ together, which have a high potential for the improvement of coating properties, in particular for service in corrosive environments and high temperature applications. However, the combined application of WC and $Cr_3C_2$ in the coating compositions leads to very complex reactions between these components both during feedstock powder preparation and the spray process. This contribution summarizes the knowledge about the interactions of WC and $Cr_3C_2$ for the most important commercially available compositions: WC–10Co–4Cr, WC–20‘CrC’–7Ni, 45$Cr_3C_2$–37WC–18NiCo. These three compositions show remarkable differences of the interaction between WC and $Cr_3C_2$ and the coating properties.

## 1 Introduction

Commercially available feedstock powder compositions for thermally sprayed hardmetal coating solutions are based on the hard materials WC and $Cr_3C_2$. The most important binder metals are Co and Ni, which are often alloyed with Cr. WC–12Co, WC–17Co, WC–10Co–4Cr, WC–6Co–8Cr, WC–20‘CrC’–7Ni, 45$Cr_3C_2$–37WC–18NiCo and $Cr_3C_2$–(20–25) NiCr (all compositions in weight percent unless otherwise indicated) are widely used commercially available compositions. Thus, in the case of an alloyed binder, chromium is distributed between the hard and the binder phase. According to the state of the art, these coatings are prepared by high velocity oxy-fuel (HVOF) and high velocity air-fuel (HVAF) spraying and are applied for wear protection including service at temperatures up to 900 °C and in aggressive media [1–3].

Since in thermal spraying, the coating is formed from solid feedstock, the feedstock quality is essential to the coating properties. In the past (1960s – 1980s), feedstock powders with irregular particle shape were widely used. Sintering and crushing (s&c) is a standard technology to produce such powders. The flowability, and consequently processability, can be significantly improved by use of feedstock powders with a spherical particle shape. Currently, such powders are produced by agglomeration (spray drying) and sintering (a&s) and are predominantly used in industrial practice [1–3]. Due to lower sintering temperatures, the preparation of feedstock powders with finer carbide grains is possible.

Coatings obtained from the complex feedstock powder compositions WC–10Co–4Cr, WC–6Co–8Cr, WC–20‘CrC’–7Ni and $45Cr_3C_2$–37WC–18NiCo show advantages regarding their properties over plain compositions WC–Co or $Cr_3C_2$–NiCr [2–5]. Due to the harsh conditions of the spray process, changes of the chemical (e.g. carbon loss, oxygen uptake) and phase composition of the coating material occur. As-sprayed coatings are in a strong chemical non-equilibrium state [1–3]. Thus, these designations are rather trading names than a true description of their chemical and phase composition. The ratio of WC and $Cr_3C_2$ (assumed as the initial material for feedstock powder preparation also for WC–10Co–4Cr and WC–6Co–8Cr) is shown in Figure 1.

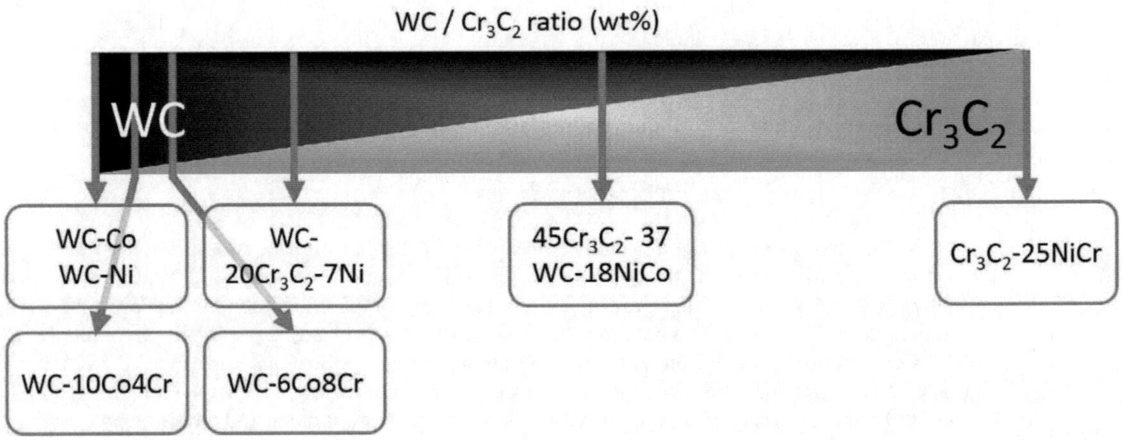


Figure 1: Ratio of WC and $Cr_3C_2$ (in wt%) for commercial composition. It is assumed that for WC-10Co–4Cr and WC-6Co–8Cr also $Cr_3C_2$ is used as initial component for powder preparation.

In this contribution, the phase composition, microstructures as well as some properties of the feedstock powders and of three of these compositions, namely WC–10Co–4Cr, WC–$20Cr_3C_2$–7Ni and $45Cr_3C_2$–37WC–18NiCo, are discussed. These compositions are characterized by very different ratios of WC and $Cr_3C_2$. The discussion focusses on the distribution of chromium between the carbide phases and the binder alloy, as well as the interaction of WC and $Cr_3C_2$ with one another and with the binder metals during feedstock powder preparation and spraying.

## 2 Main Binary Hardmetal Compositions

### *2.1 Initial Powders for Feedstock Preparation*

For feedstock powder preparation by spray drying and sintering, finely dispersed carbide and metal powders are required. Due to its outstanding importance in tool materials, a wide range of WC powders, including nm- and submicrometer grades are available. In comparison, the preparation of fine chromium carbide grades meets difficulties due to other synthesis reactions, but chromium carbide has also a much lower importance for tools and wear parts prepared by sintering. As demonstrated below, the use of fine WC and coarse $Cr_3C_2$ powders is reflected in the microstructures of the feedstock powders and the coatings. Fine metallic powders of cobalt and nickel are also easily available, while it is rather difficult to prepare binder alloys in grain size suitable for spray drying.

### *2.2 WC–10Co–4Cr*

Nowadays, the composition WC–10Co–4Cr is most commonly applied for the preparation of hardmetal coatings by HVOF having simultaneously high wear and corrosion resistance [2, 3]. The increasing importance of this composition is closely connected with the fast development of HVOF for coating deposition in the 1990s. Still in the middle of the 1990s, only very few references dealing with powders and coatings of this composition were identified [6]. However, the origin of this composition is not clear. The nominal ratio of cobalt to chromium in this composition is 2:1 by volume [7].

For this composition, it is expected and also suggested by the designation "WC-10Co4Cr" that fine WC, Co and Cr (or a CoCr binder alloy) initial powders are used for feedstock powder preparation and that chromium is fully contained in the binder. However, the analysis of this composition [7] has shown that WC, the $\eta$-phase $(Co,Cr)_3W_3C$, and a (Co,Cr,W) alloy binder are expected to be in thermodynamic equilibrium. The $\eta$-phase will also appear in the case of carbon deficiency. In the case of higher carbon content, in equilibrium conditions, a $(Co,Cr,W)_7C_3$ phase will appear instead of the $\eta$-phase [7, 8]. In all cases, chromium is contained both in the metallic binder and in the hard phase [7–9]. Another important feature of chromium addition is a significant melting point reduction compared to plain WC-Co [8].

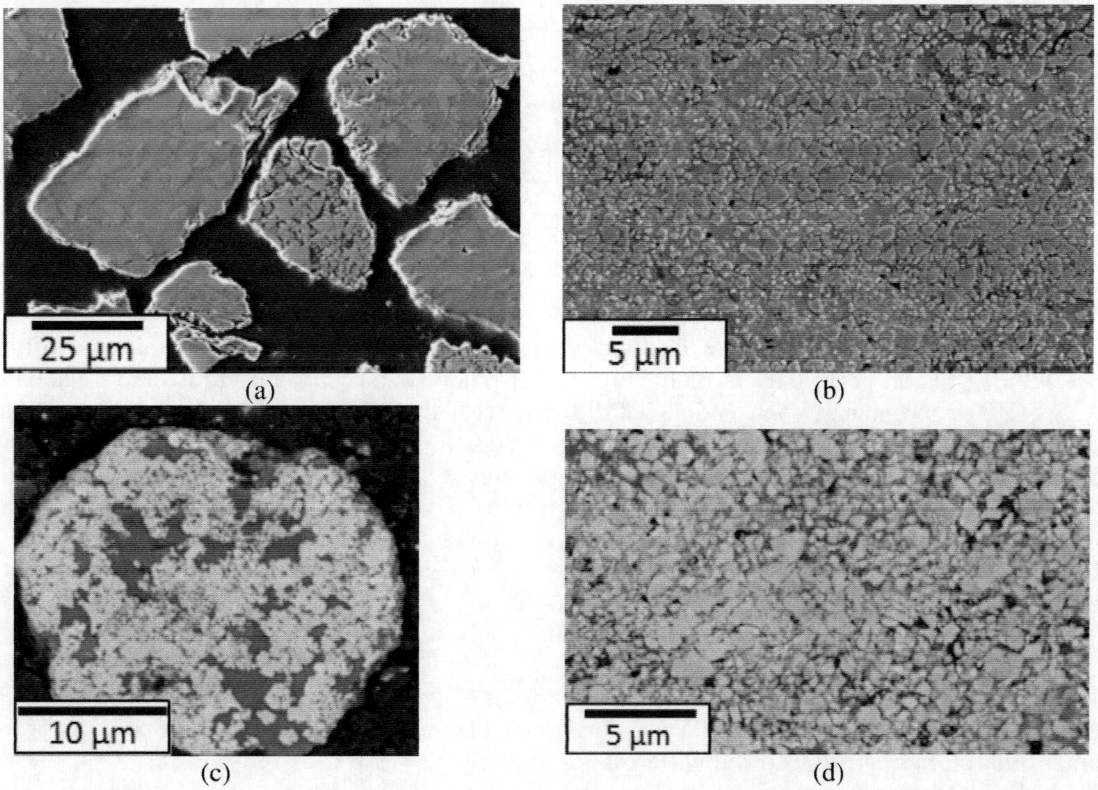


Figure 2: Microstructures of WC–10Co–4Cr feedstock powders and coatings: a) cross section of a sintered and crushed powder with low carbon content [7], b) typical microstructure of a HVOF (JP-5000) coating sprayed from an agglomerated and sintered powder, c) cross section of a particle of an a&s powder with high carbon content [10], d) microstructure of an HVAF (M3) coating sprayed from the powder shown in Figure 3c [10].

An analysis of feedstock powders prepared by sintering and crushing available in the 1990s in the market [7] has shown that those were characterized by a low carbon content. The cross section of such a powder is shown in Figure 2a. It is characterized by an inhomogeneous distribution of the elements, leading to individual particles of different structure. In most of the powder particles, bright WC grains are embedded in large areas of the $\eta$-phase $(Co,Cr)_3W_3C$, appearing darkish gray [7]. For such powders, low deposition efficiencies were reported [7]. Figure 2b shows the typical microstructure of a coating sprayed by a liquid-fueled (JP 5000) HVOF process from an a&s powder, where fine WC grains appear in the binder matrix of varying grayscale. In a recent study [10], large dark gray areas in powder particles as shown in Figure 2c were identified as $(Co,Cr,W)_7C_3$. This leads to the assumption that $Cr_3C_2$ is used as starting component for feedstock powder preparation instead of metallic chromium, causing an excess of carbon in the composition as proved by chemical analysis [10]. Currently, it is not known, if such feedstock powder structures are typical for all commercial products in the market. However, similar large dark gray areas were observed in other studies as well [11, 12], but were not identified as $(Co,Cr,W)_7C_3$. Due to the carbon loss during spraying, these areas are not observed in the coatings, even in the case of an HVAF spray process [10].

It was stated that WC–10Co–4Cr seems to be a good empirically derived and balanced formulation for the preparation of simultaneous wear- and corrosion-resistant coatings [7]. Besides alloying of Co by Cr, the formation of the $\eta$-phase decreases the amount of the binder, which is more sensitive to corrosion. Compared to plain WC-Co there is no increase in the oxidation resistance [1-3, 10, 11]. Thus, application of these coatings in atmospheric conditions is limited to a service temperature of about 600 °C.

### 2.3 *WC–20"Cr₃C₂"–7Ni*

Another WC-based commercial feedstock powder composition is WC–20"$Cr_3C_2$"–7Ni, which has its origin in a patent originally filed in 1958 [13]. Coatings prepared by detonation gun spraying (DGS) of such a powder are traded under the designation UCAR LW-5 [14]. Currently, there are variations in contents of the components in the range of 1–4 % from different manufacturers [2, 3]. The composition contains a lower binder metal content than other WC-based compositions but can be sprayed with a high deposition efficiency [15, 16].

The metallurgical processes during feedstock powder preparation and spraying for this composition have been studied in detail only recently [11, 15, 16]. The most specific feature is the formation of $(W,Cr)_2C$. The appearance, composition and distribution of the $(W,Cr)_2C$ phase depends on the heat treatment temperature during feedstock powder preparation. There are also commercial grades not containing this phase. However, it can be formed (when not present in the powder) or can change its W/Cr ratio during the spray process. Recently, a hypothesis for the formation of this phase has been proposed for the reaction between WC and $Cr_3C_2$ [11]. It was postulated that, after the initial formation of $Cr_7C_3$ due to the interaction of $Cr_3C_2$ with nickel, $(W,Cr)_2C$ can be formed with a high chromium content according to the following formal equation:

$$WC + Cr_7C_3 \rightarrow 4\,(W_{0.125}Cr_{0.875})_2C$$

This explained earlier findings, where Cr-rich $(W,Cr)_2C$ was found in several powders [16]. The occasional absence or occurrence of $(W,Cr)_2C$ might be one reason why the feedstock powders of this composition are currently traded with different designations such as WC–'CrC'–Ni, WC–'$Cr_3C_2$'–Ni or WC–NiCr [2, 15, 16].

For illustration, Figure 3 shows the powder cross sections of a s&c (Figure 3a) and an a&s (Figure 3c) feedstock powder, as well as the corresponding HVOF (JP-5000) sprayed coatings (Figure 3b and Figure 3d, respectively). In both powders, $(W,Cr)_2C$ was already formed. Due to the higher sintering temperature, the content of this phase, appearing as large gray areas in the micrographs, is higher in the s&c powder. At the same time, fine WC is embedded in Ni in other areas [15]. These structures are basically transferred from the powder to the coatings. Unlike WC–Co and $Cr_3C_2$–NiCr, WC–$(W,Cr)_2C$–Ni is not a simple binary hard phase-binder metal composite. The $(W,Cr)_2C$ as the second hard phase is inhomogeneously distributed in the feedstock powders and coatings [2, 15, 16]. Although the structure of $(W,Cr)_2C$ is described in the literature, e.g. [17, 18], there is very limited data on its physical and mechanical properties and their dependence on the chromium content available. Recently, for a composition $(W_{0.2}Cr_{0.8})_2C$ a hardness of 20.0 GPa and a Young's modulus of about 370 GPa were measured for grains appearing in the coating [11].

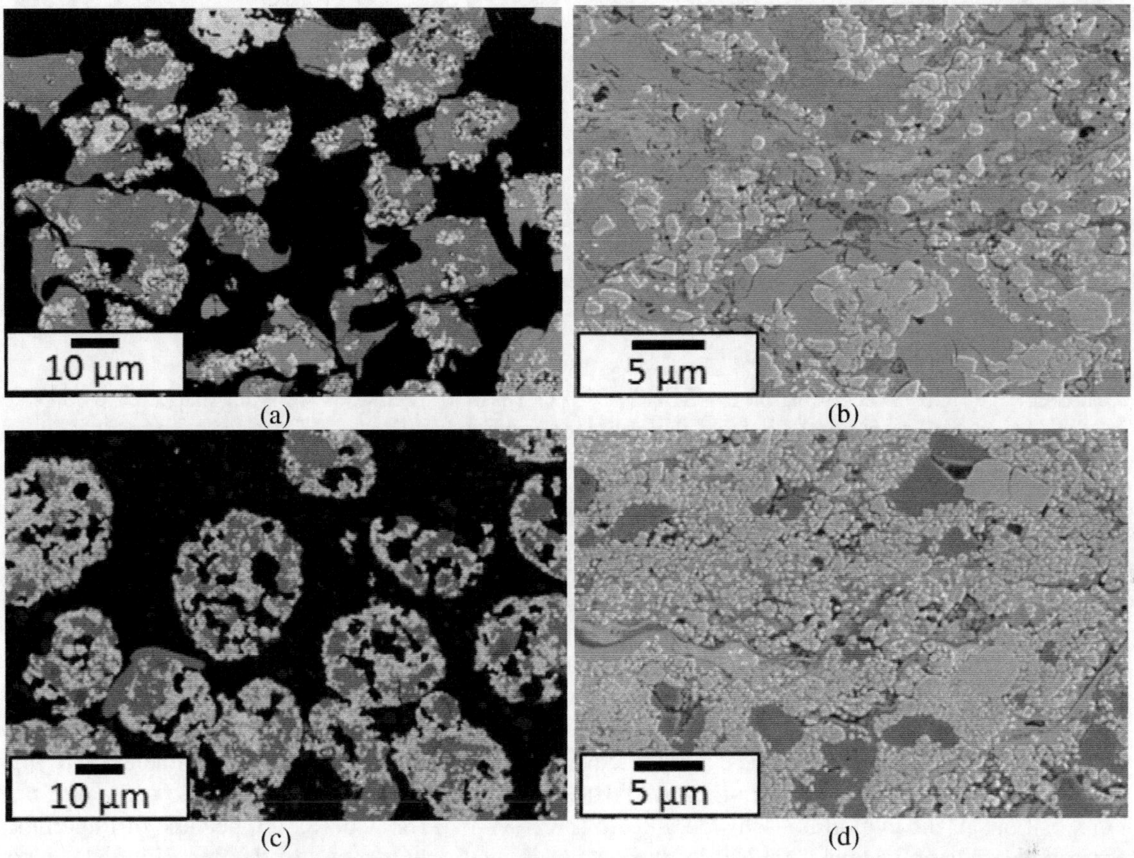


Figure 3: Cross sections of WC–$(W,Cr)_2C$–Ni feedstock powders and coatings [15]: a) s&c powder, b) HVOF sprayed coating from the powder shown in a); c) a&s powder, d) HVOF sprayed coating from the powder shown in c).

Some studies have shown that WC–$(W,Cr)_2C$–Ni coatings are characterized by a significantly higher oxidation resistance than other WC-based compositions including WC–10Co–4Cr, and can be applied for high-temperature sliding wear applications, e.g. [11, 15].

### *2.4 45$Cr_3C_2$–37WC–18NiCo*

First information on this coating composition was published in 2005 [19]. The powder of this composition is available from only one manufacturer. Recently, a powder of a similar composition of another manufacturer entered the market [22]. But so far, feedstock powders and coatings were only rarely topic of investigations [4, 5, 20, 21]. Figure 4a shows the cross section of a 45$Cr_3C_2$–37WC–18NiCo powder particle and Figure 4b the cross section of a coating, sprayed with a liquid-fueled HVOF process (K2) [4]. In both the powder and the coating, fine WC and large $Cr_3C_2$ grains are coexisting. The fast formation of $(W,Cr)_2C$ was not observed for this composition. However, there is some beginning reaction between both carbides, as some large $Cr_3C_2$ grains show a core-rim structure, as supposed by the appearance of tungsten in the brighter rims.

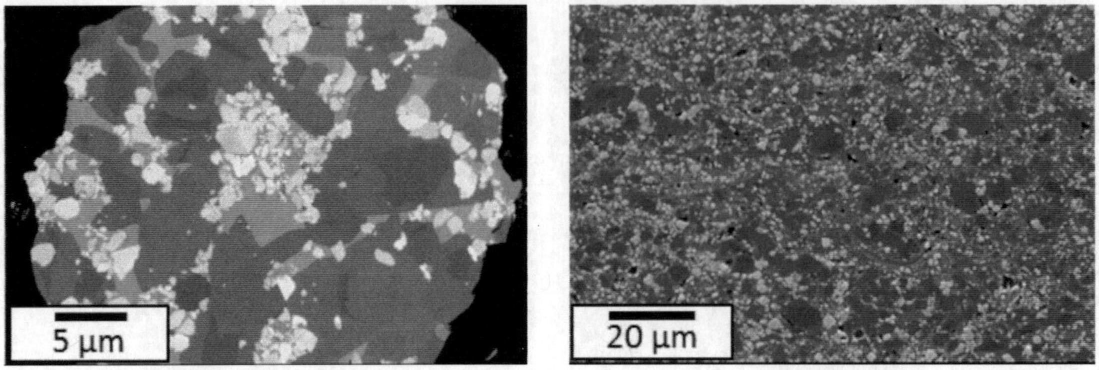


(a) (b)

Figure 4: a) Cross sections of a $45Cr_3C_2$-37WC-18NiCo feedstock powder particle, b) HVOF sprayed coating (K2) from the powder shown in a) [4].

The comparison with plain $Cr_3C_2$-NiCr showed an increase of the deposition efficiency and a better abrasion wear resistance as well [4, 5]. At the same time, the oxidation resistance is similar [5]. Thus, the combined application of both carbides on the $Cr_3C_2$–rich side of compositions was found to be a promising approach to increase the coating performance as well as the cost-effectiveness of coating preparation.

## 3 Summary and Outlook

The combined application of WC and $Cr_3C_2$ in coating compositions has a high potential for the improvement of coating properties, in particular for service in corrosive environments and high temperature applications. In addition to WC–10Co–4Cr and WC–20"$Cr_3C_2$"–7Ni, now also the new compositions of the chromium-rich side, like $45Cr_3C_2$–37WC–18NiCo seem to become of importance for commercial applications [22, 23]. In the case of the latter an increase of the deposition efficiency and a better abrasion wear resistance in comparison with plain $Cr_3C_2$–NiCr was observed [4, 5]. An initial $Cr_3C_2$–content of 20 % and higher increases the oxidation resistance. However, the reactions between these components, both during feedstock powder preparation and the spray process, are very complex, and are by far not understood. For the three powders investigated here, there are sudden changes in microstructures and phase compositions, in particular the formation $(W,Cr)_2C$, in the feedstock powders and the coatings depending on the W/Cr ratio. So far, due to availability, the investigation of these compositions was focused on commercially available powder grades. It requires a much better understanding of the interaction of WC and $Cr_3C_2$ in order to explore the application potential of these compositions.

## References

[1] Berger L-M 2012 Metallurgische Vorgänge beim thermischen Spritzen von Hartmetallen und deren Auswirkungen auf die Schichteigenschaften *Tagungsband zum 15. Werkstofftechnischen Kolloquium in Chemnitz, 20.-21. September 2012, Schriftenreihe "Werkstoffe und werkstofftechnische Anwendungen", Band 47,* B. Wielage (ed.), Chemnitz, TU Chemnitz, p. 16–28 – ISBN 978-3-00-039358-7, ISSN 1439-1597.

[2] Berger L-M 2014 Coatings by Thermal Spray, in: V. Sarin (ed-in-chief), D. Mari, L. Llanes (vol. eds.), *Comprehensive Hard Materials – Vol. 1: Hardmetals*, Elsevier, Amsterdam, The Netherlands 471-506

[3] Berger L-M 2015 Application of Hardmetals as Thermal Spray Coatings *Int. J. Ref. Met. Hard Mater.* **49** 350-64

[4] Berger L-M, Trache R, Toma, F-L, Thiele S, Norpoth J and Janka L 2015 Development of Cost-Effective Hardmetal Coating Solutions for High-Temperature Applications, Part one: Feedstock powders, cost-effectiveness and coating properties *Therm. Spray Bull.*, **8** (2) 126-36

[5] Berger L-M, Trache R, Toma F-L, Thiele S, Norpoth J and Janka, L 2016 Development of Cost-Effective Hardmetal Coating Solutions for High-Temperature Applications, Part two: Influence of Heat Treatments and Tribolog-ical Investigations *Therm. Spray Bull.* **9** accepted manuscript.

[6] Berger L-M, Vuoristo P, Mäntylä T, Kunert W, Lengauer W and Ettmayer P 1996 Microstructure and Properties of WC-Co-Cr Coatings *Thermal Spray: Practical Solutions for Engineering Problems, Proc. 9th National Thermal Spray Conference*, C.C. Berndt (ed.), 7–11 October 1996, Cincinnati, Ohio, USA, Materials Park/Ohio, ASM International 97-106

[7] Berger L-M, Ettmayer P, Vuoristo P, Mäntylä T and Kunert W 2001 Microstructure and Properties of WC-10%Co-4%Cr Spray Powders and Coatings. I. Powder Characterization *J. Therm. Spray Technol.*, **10** (2) 311-25

[8] Frisk K and Markström A 2008 Effect of Cr and V on Phase Equilibria in Co-WC Based Hardmetals *Int. J. Mater. Res.* **99** (3) 287-93

[9] Gries B and Zimmermann S 2012 Corrosion Behaviour of HVOF Sprayed Carbide Coatings *Proc. 9th HVOF Kolloquium,* Gemeinschaft Thermisches Spritzen eV, Unterschleißheim, Germany, 71-8

[10] Bolelli G, Berger L-M, Börner T, Koivuluoto H, Lusvarghi L, Lyphout C, Markocsan N, Matikainen V, Nylén P, Sassatelli P, Trache R and Vuoristo P 2015 Tribology of HVOF- and HVAF-Sprayed WC-10Co4Cr Hardmetal Coatings: A Comparative Assessment *Surf. Coat. Technol.* **265** 125-44

[11] Bolelli G, Berger L-M, Bonetti M and Lusvarghi L 2014 Comparative Study of the Dry Sliding Wear Behaviour of HVOF-Sprayed WC-$(W,Cr)_2C$-Ni and WC-CoCr Hardmetal Coatings *Wear* **309** (1-2) 96-111

[12] Varis T, Suhonen T, Ghabchi A, Valarezo A, Sampath S, Liu X and Hannula S-P 2014 Formation Mechanisms, Structure, and Properties of HVOF-Sprayed WC-CoCr Coatings: An Approach Toward Process Maps *J. Therm. Spray Technol.* **23** (6) 1009-18

[13] Pelton J F and Koffskey Jr J M 1963 Process of Flame Spraying a Tungsten Carbide-Chromium Carbide-Nickel Coating, and Article Produced Thereby <u>*US 3,071,489*</u>, filed: 28 May 1958, granted 1 January (also *GB 874,463*)

[14] Ashary A A and Tucker Jr R C 1991 Corrosion Characteristics of Several Thermal Spray Cermet-Coating/Alloy Systems *Surf. Coat. Technol.* **49** (1–3) 78-82

[15] Berger L-M, Saaro S, Naumann T, Kašparova M and Zahálka F 2008 Microstructure and Properties of HVOF-Sprayed WC-$(W,Cr)_2C$-Ni Coatings *J. Therm. Spray Technol.* **17** (3) 395-403

[16] Berger L-M, Saaro S, Naumann T, Kašparova M and Zahálka F 2010 Influence of Feedstock Powder Characteristics and Spray Processes on Microstructure and Properties of WC-$(W,Cr)_2C$-Ni Hardmetal Coatings *Surf. Coat. Technol.* **205** (4) 1080-7

[17] Stecher P, Benesovsky F and Nowotny H 1964 Untersuchungen im System Chrom-Wolfram-Kohlenstoff *Planseeber Pulvermetall* **12** 89-95

[18] Gladyshevskii E I, Telegus V S, Fedorov T F and Kuz'ma Y B 1967 Tungsten-Chromium-Carbon Ternary System *Izv. Akad. Nauk. SSSR, Metally* **1** 190-3

[19] Kirsten A, Oechsle M and Moll R.F 2005 Carbide Containing Materials for Hard Chromium Replacement by HVOF-Spraying *Proc. Int. Thermal Spray, Conf. ITSC* 2005, 2–4 May 2005, Basel, Switzerland. Düsseldorf, DVS-Verlag, CD, 6 p.

[20] Hulka I, Şerban V A, Secosan I, Vuoristo P, Niemi K 2012 Wear Properties of CrC-37WC-18M Coatings Deposited by HVOF and HVAF Spraying Processes *Surf. Coat. Technol.* **210** 15-20

[21] Hulka I., Utu D, Şerban V A, Vuoristo P, Niemi K, Koivuluoto H 2014 Properties and Microstructure of HVOF Thermally Sprayed Carbide Based Coatings *Solid State Phen* **216** 261-6

[22] Gries B 2015 Neuartige Karbidwerkstoffe für HVOF und HVAF *Proc. 10th HVOF Kolloquium, Gemeinschaft Thermisches Spritzen eV, Unterschleißheim, Germany* 97-102

[23] Reisel G and Zimmermann S 2015 Chromkarbid-basierte Pulver für das HVOF-Spritzen: Pulvereigenschaften und ihr Einfluss auf das Schichtverhalten *Proc. 10th HVOF Kolloquium, Gemeinschaft Thermisches Spritzen eV, Unterschleißheim, Germany*, 105-12

IOP Conf. Series: Materials Science and Engineering **118** (2016) 012011 doi:10.1088/1757-899X/118/1/012011

# Highly carbide filled composite materials for the mining and drilling industry

**Ch Gerk[1, 3], L Yan-Gerk[1], V Wesling[2] and R Reiter[2]**

[1] Barenberg Special Materials GmbH, R&D Center in the Clausthal Center of Material Technology, D-38678 Clausthal-Zellerfeld, christian.gerk@barenberg.solutions
[2] ISAF, Technische Universität Clausthal, D-38678 Clausthal-Zellerfeld
[3] Author to whom correspondence should be addressed

The mining and drilling industry has a need for highly wear resistant large diameter drill heads. The manufacturing of such drill heads takes place by near net shape infiltration of a powder matrix using tungsten carbides with copper based infiltration alloys. The drill heads are exposed to different geologic formations which triggers a desire for very flexible design parameters resulting in a complex need for both high wear resistance, but also high bending strength as well as good repair-ability. Such contradictory needs have to be solved using adjusted carbide compositions. The paper covers an area where only little information has been published so far. It is written from an Industry point of view and presents the industrial production routine for the drill heads as well as the current state of the art material solutions. The influences of different types of carbides, particle sizes and compositions on the resulting properties will be discussed. Additionally, the paper will show recent challenges of the drilling industry resulting from the shale gas boom.

## 1 Technical challenges for drill heads used in oil and gas exploration

The drilling service industry offers a large variety of drill heads in order to optimize the drilling performance in the field. Roller Cones and Matrix bits as well as Steel Body bits equipped with PDCs (Poly Crystalline Diamond Cutters) are the most important examples. The drilling performance is often expressed as Rate of Penetration (ROP), which is a term for the drilled distance per time unit and strongly depends on the selection of the drill head as well as the drilling parameters, e.g. the rotation speed. Optimizing the ROP usually results out of the right balance between drilling speed and durability of the drill head. Therefore, different drill head designs have been developed. For PDC bits, the number and height of the blades are the most important design factors next to the number, angle and location of the diamond cutters. Such geometrical designs require the adjustment and optimization of material related properties such as TRS (Transverse Rupture Strength, ASTM B406) and abrasion/erosion resistance (e.g. according to ASTM B611, ASTM G65, ASTM G75) in order to avoid early failures in the field. Generally speaking, the advantage of matrix body bits is the outstanding wear resistance whereas steel body bits have a lower wear resistance requiring an additional coating but superior TRS. The focus of this paper is the material design of matrix body bits representing a highly heterogeneous composite material based on a metallic binder and carbide hard phases. Therefore, the measurement and in depth understanding of the related mechanical properties of such composite materials is a major precondition for the development of optimized drill head compositions.

## 2 Industrial drill head manufacturing process

The industrial manufacturing of matrix PDS drill heads can be divided into three major process steps as shown in figure 1. In the first step the desired geometry of the drill head is machined as a negative form out of a graphite block. Subsequently the carbide powders will be poured out into the graphite mold and infiltrated with a binder to form the solid bit body. The infiltration binder is a Cu/Mn/Ni/Zn alloy with a melting point above 900 °C. The target of step one is to produce a crack-free solid carbide body. This will be tested in step two using dye penetration of the body and pressure testing of the nozzle areas. Step three requires careful preheating of the bit body as well as the cutter pockets to prepare the manual diamond cutter brazing process. This is a critical step with respect to the occurrence of cracks which may result in scrapping of the drill head.

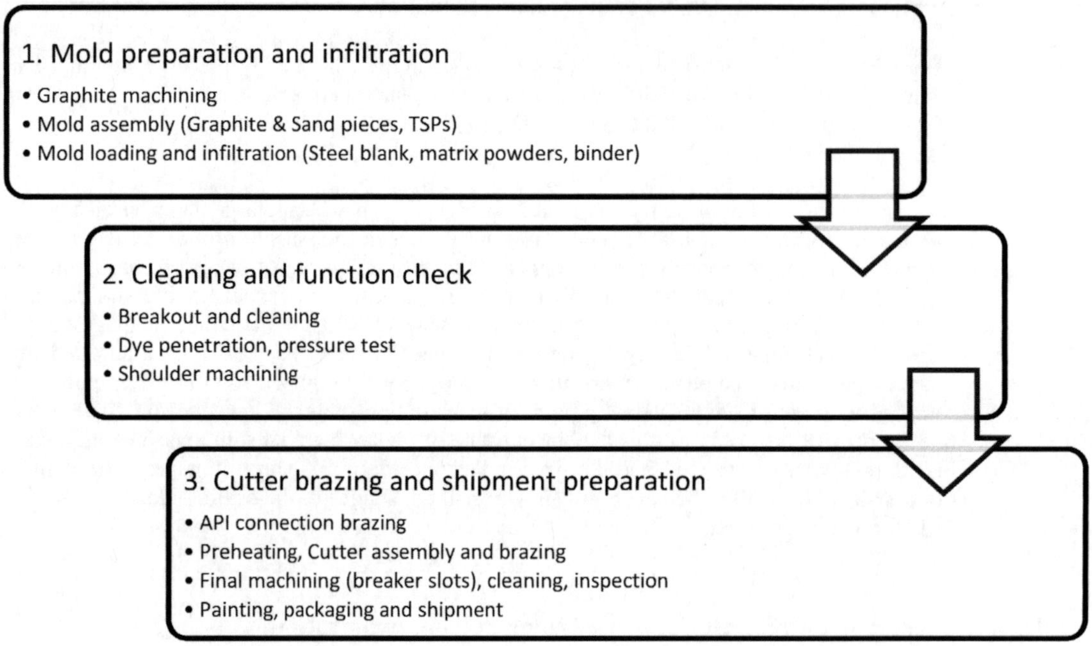


**Figure 1:** Three step manufacturing process for matrix body PDC drill heads

Upon usage of the drill head in the field two typical types of wear may occur. Firstly, the diamond cutters can get lost or wear down. Secondly, the matrix body gets eroded away and loses its original shape. Both wear losses can be repaired several times in repair centers to elongate the drill head life as much as possible. The life time of a drill head usually ends due to severe erosion or cracking in the blades after 2-10 repair cycles. The analysis such failure cases as well as the design requirements translate into the desired material properties for the matrix body powder. The reverse engineering of failure mechanisms as well as the development of new matrix powders is a complex matter bearing a high risk of costly issues in the field. A close cooperation and knowledge exchange of the drill bit manufacturers with powder producers and R&D centers has proven to be a successful concept for this challenge. Although some of the established matrix powders did not change a lot over decades and still represent an industry standard, new powder formulations for special applications are continuously under development in order to save cost and improve the most important KPI for drilling, the ROP. As a rule of thumb, 50-60 m/h is a good figure for the ROP.

## 3 State of the art material solutions

The established state of the art materials for matrix PDC drill head manufacturing are exclusively recruited from the family of tungsten carbides. This includes macrocrystalline WC, sintered WC, cast tungsten carbide (CTC) as well as reclaimed powders, crushed hard metals (WC-Co) and sintered WC-Co granules. The wear resistance and bending strength of the final infiltrated drill head strongly depends on the carbide origin as well as the grain size or particle size distribution respectively. Some basic rules for the selection of the best application fits are summarized in table 1.

**Table 1:** basic rules for the selection of tungsten carbides for drill head manufacturing

| Selection criteria | Variable | Impact on intrinsic carbide properties | Impact on drill head properties |
|---|---|---|---|
| Type of carbide | Sintered, macrocrystalline, crushed, cast | hardness and fracture toughness depend on manufacturing process | Wear resistance increases with higher fracture toughness, e.g. CTC is superior to WC due to the fiber structure. But CTC also dissolves more easily in the binder and forms brittle eta phase. |
| Grain size and particle size distribution | Roughly 25-1000µm | A wide PSD reduces the porosity of the powder in the mold and therefore the mean free path of the binder | Coarse grains lower abrasion loss but also decrease TRS. A wide PSD improves soft erosion. |
| Chemistry | Raw material selection, manufacturing process | Impurity level, e.g. O, S, free carbon, trace metals | Usually low TRS because certain impurities disturb the infiltration process resulting in porosity and therefore lower TRS. W2C (e.g. out of CTC) dissolves more easily in the binder and forms brittle eta phase. |

Figure 2 documents the visual fracture toughness differences of three types of carbides after G65 (rubber wheel test) wear test treatment. Figure 2 (a) shows a macrocrystalline WC which tends to lose larger chips during the relatively hard erosive G65 test. In figure 2 (b), a polycrystalline WC is shown. In this case the particle had a hollow center and therefore tends to break into larger pieces either. In contrast to this, figure 2 (c) represents a typical CTC matrix powder with high intrinsic fracture toughness. The CTC particles therefore only lose small chips in the G65 test and tend to be more stable, i.e. they usually perform superior in the G65 (also B611) test.

Figure 3 shows the decrease of B611 wear losses but also the decrease of TRS with increasing average grain size of the hard phases (different materials contribute to the average figures). The mean free path (distance between particles which is filled with binder) is by nature larger for coarse particles. However, this does not explain the decrease of TRS.

Figure 4 illustrates the formation of brittle $Ni_2W_4C$ eta phase in a CTC/binder system.

(a) (b)

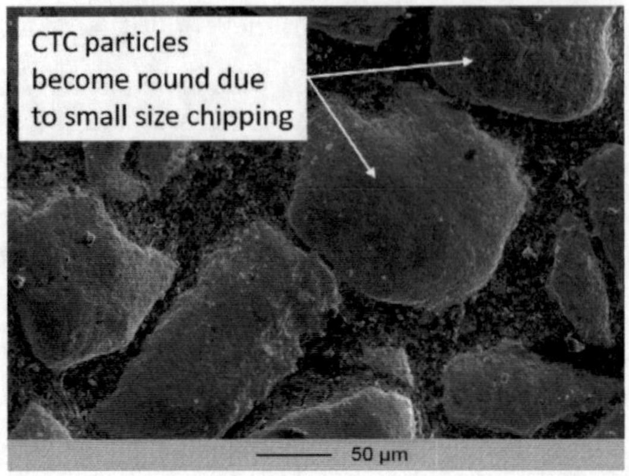


(c)

**Figure 2:** (a) Macrocrystalline WC, (b) Sintered WC, (c) CTC, all after G65 treatment

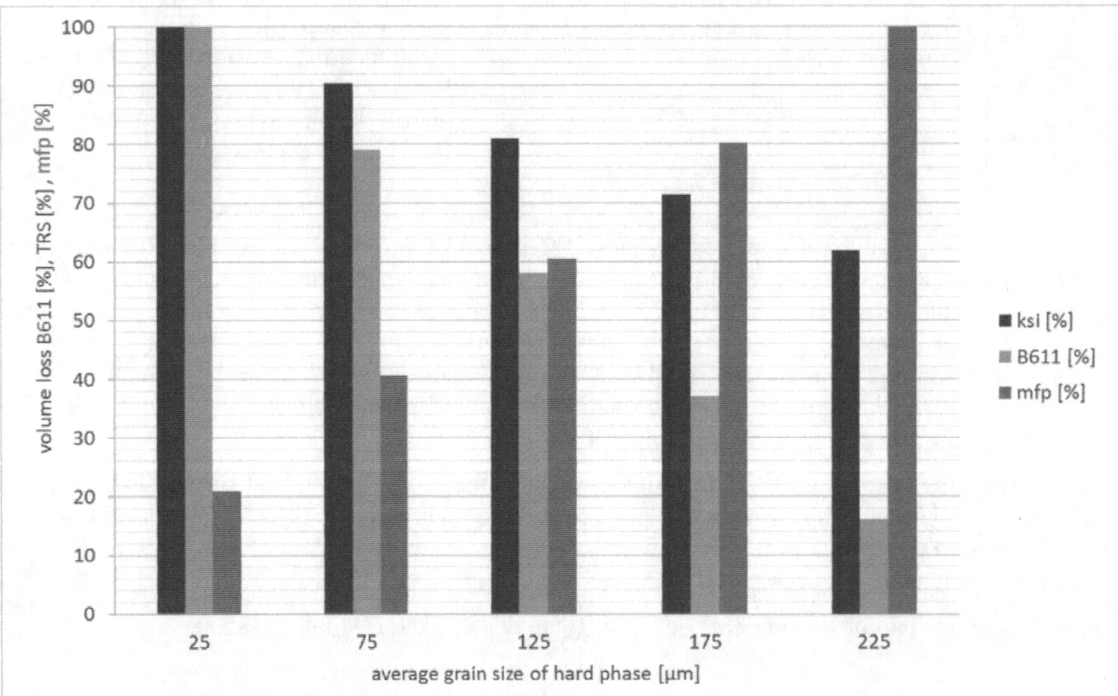


**Figure 3:** Average wear losses according to B611 and average TRS of hard phases with different average grain sizes relative to the figures at 25µm and mean free path (binder) relative to the figures at 225µm

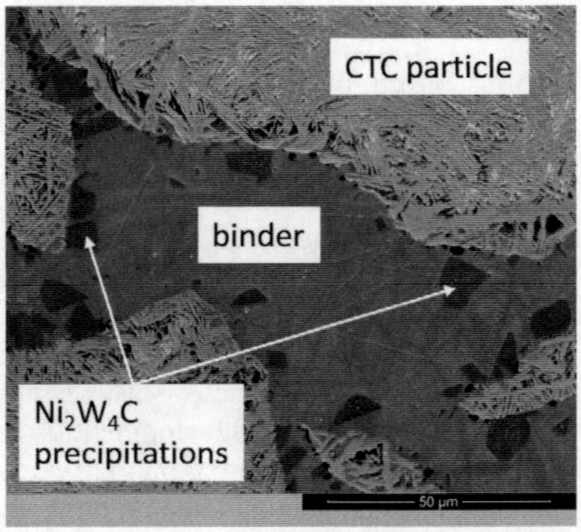


**Figure 4:** $Ni_2W_4C$ eta phase precipitated in the infiltration binder near the CTC particles

The fact that there is still a huge improvement potential for drill head performance but only a few systematic rules are known for material design has led to an increased R&D activity recently.

However, the historic development as shown in figure 5 proves that some key performance figures like the bending strength (TRS) and especially the wear resistance according to ASTM B611 have improved significantly in the recent years (figures relative to year 2000). The current development trends are shown in figure 5 as well. Trend (A) represents the need to continue the historic development of improving the wear resistance without giving up on TRS. Trend (B) reflects a philosophy change into the direction of ultrahigh TRS which has been influenced from the shale gas exploration.

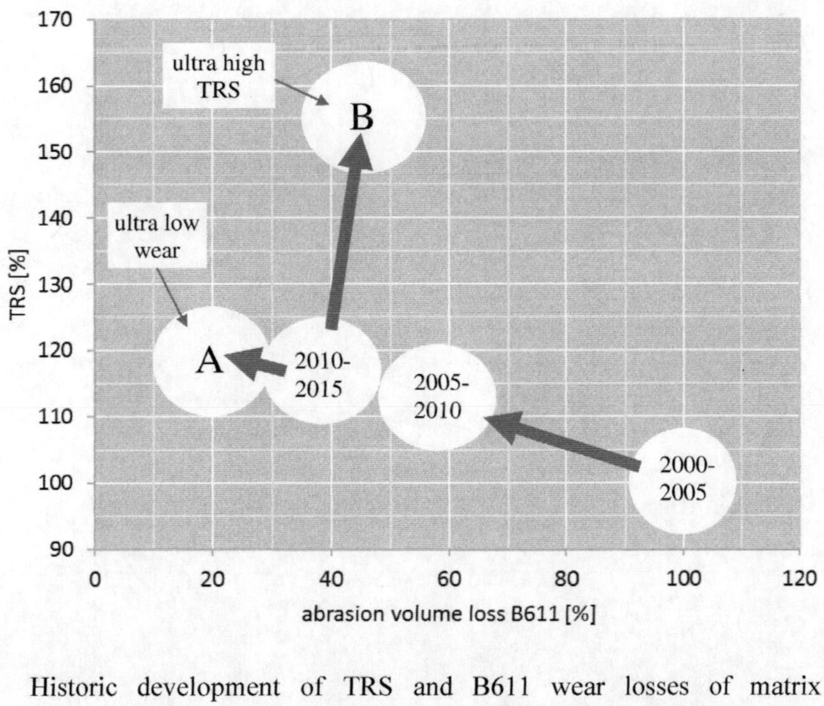


**Figure 5:** Historic development of TRS and B611 wear losses of matrix powders and current development trends A (ultrahigh TRS) and B (ultralow wear)

## 4 Shale Gas exploration – Conclusions and consequences for the design of new materials

Shale deposits are known as a "non-conventional" resource for oil and especially natural gas. The exploration of shale gas has become very common in the US in the recent years but in contrast to conventional resources it usually needs both fracking and horizontal drilling of the relatively soft to medium hard formations. This gives priority to the usage of PDC bits, especially those who can be used with a high rotation speed. In relatively soft formations the blades of the drill head should be taller than in hard formations in order to improve the transportation of the crushed stones in the mud. Both, the high rotation speed as well as the tall blades triggers a need for a higher TRS of the matrix body. Therefore, steel body bits are a favorite choice in such environments. The disadvantage of steel body bits is the low wear resistance, which is about 10-20 times lower compared to matrix body bits, e.g. according to soft erosion (G75) testing. Therefore, the development direction (B) as indicated in figure 5 has gained importance recently. In case (B) the matrix bits are supposed to achieve a significant improvement of TRS whereas the wear resistance can be compromised. Materials fulfilling the requirements of trend (B) according to figure 5 are currently under development and/or field testing. Further progress will be reported elsewhere.

**Acknowledgement**

This research was partially supported by Clausthal Center of Material Technology, a member of Clausthal University of Technology. We thank our colleagues from CZM who provided insight and expertise that greatly assisted the research.

IOP Conf. Series: Materials Science and Engineering **118** (2016) 012012 doi:10.1088/1757-899X/118/1/012012

# Experimental study of the mechanical behaviour of pin reinforced foam core sandwich materials under shear load

**M A Dimassi, C Brauner and A S Herrmann**

Faserinstitut Bremen e.V., Am Biologischen Garten 2, 28359 Bremen, Germany

Email: dimassi@faserinstitut.de

**Abstract.** Sandwich structures with a lightweight closed cell hard foam core have the potential to be used in primary structures of commercial aircrafts. Compared to honeycomb core sandwich, the closed cell foam core sandwich overcomes the issue of moisture take up and makes the manufacturing of low priced and highly integrated structures possible. However, lightweight foam core sandwich materials are prone to failure by localised external loads like low velocity impacts. Invisible cracks could grow in the foam core and threaten the integrity of the structure. In order to enhance the out-of-plane properties of foam core sandwich structures and to improve the damage tolerance (DT) dry fibre bundles are inserted in the foam core. The pins are infused with resin and co-cured with the dry fabric face sheets in an out-of-autoclave process. This study presents the results obtained from shear tests following DIN 53294-standard, on flat sandwich panels. All panels were manufactured with pin-reinforcement manufactured with the Tied Foam Core Technology (TFC) developed by Airbus. The effects of pin material (CFRP and GFRP) and pin volume fraction on the shear properties of the sandwich structure and the crack propagation were investigated and compared to a not pinned reference. It has been concluded that the pin volume fraction has a remarkable effect on the shear properties and damage tolerance of the observed structure. Increasing the pin volume fraction makes the effect of crack redirection more obvious and conserves the integrity of the structure after crack occurrence.

## 1. Introduction

Compared to stiffened monolithic shell structures sandwich panels offer a high specific bending stiffness with superior buckling stability, which could bring new design principles and make the assembly process more efficient [1]. Closed cell foam core materials combined with dry fibre preforming enable to manufacture integral sandwich structures with complex geometries using resin infusion technologies [2]. Nevertheless, foam core sandwich structures are vulnerable to concentrated impact loads, which could result in serious structure damages [3]. Core shear cracks with large damage sizes could occur and degrade the load carrying capacity of the structure [3]. With monolithic profiles reinforced foam cores or Z-pinned sandwich structure like the X-Cor™ and K-Cor™ sandwich [4] the out-of-plane properties are improved. The pins used in this study are cut with a pre-defined extensions on both pin ends that lie in the interface between the face sheets and the foam core, thus the Mode I fracture toughness of the interface and the damage tolerance are improved [5]. During the manufacturing process thermal stresses could arise in the interface between pins and foam material, which could lead in critical cases to early failure of the core material [6].

Using the TFC pinning technology reduces the production costs and time, and improves the damage tolerance as well as the fatigue behaviour of the sandwich structure [7]. The aim of this work is to study the shear crack stopping behaviour and the effect of the pins on the mechanical properties of the structure under shear loading.

## 2. Materials and manufacturing

The Modified Vacuum Infusion (MVI) system as described in [8] was used for the manufacturing of the studied sandwich panels. The sandwich structure consists of a pin-reinforced ROHACELL® 71 HERO foam core and two CFRP face sheets made of Toho Tenax HTS carbon fibre non-crimp fabrics (NCF) impregnated by Hexcel RTM6 epoxy resin. The mechanical properties of the used materials are depicted in the manufacturers' data sheets [9–11]. The core has a thickness of about 25.7 mm and the face sheets consist of four triaxial NCF-layers, with the layup (45/0/-45), which makes a face skin thickness of about 1.5 mm.

Four different pin configurations with different pin distance, materials and same pin angle of 50° as well as a reference configuration without pin reinforcement were tested. The pins have a diameter of about one millimetre. T800H carbon fibre from TORAYCA [12] and 1383 Yarn glass fibre from PPG Fiber Glass [13] were used for the pins. Figure 1 shows the used pin pattern and a 3D-view of the unit cell.

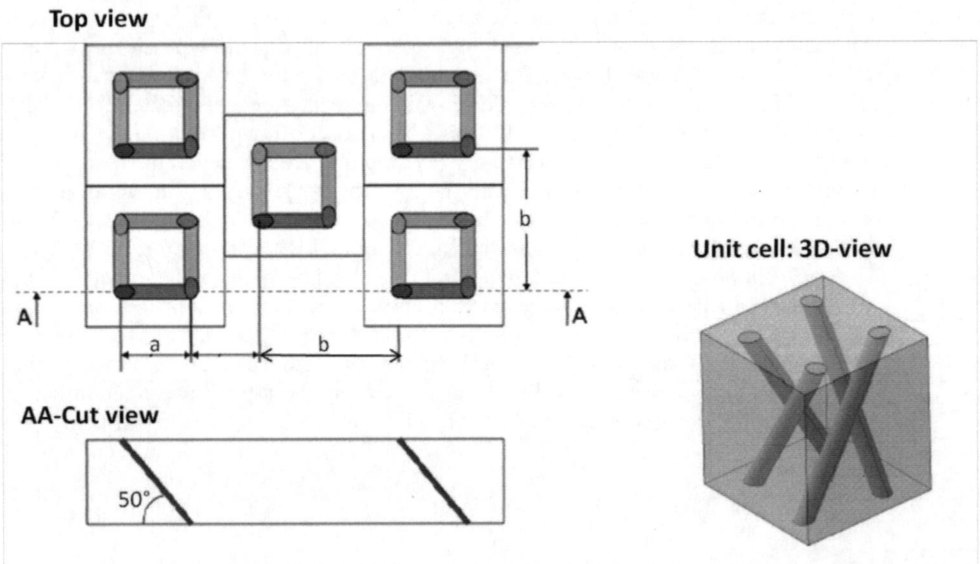


**Figure 1.** Different views of the unit cell and pinning pattern [14].

The nomenclature of the tested configurations is depicted in table 1.

**Table 1.** Parameters of the tested configurations.

| Configuration | Pin material | Unit cell width (a) [mm] | Distance between two unit cells (b) [mm] | Pin Volume fraction [%] |
|---|---|---|---|---|
| CFRP.10 | CFRP | 10 | 20 | 0.75 |
| CFRP.20 | CFRP | 20 | 40 | 0.2 |
| GFRP.10 | GFRP | 10 | 20 | 0.75 |
| GFRP.20 | GFRP | 20 | 40 | 0.2 |
| H.71 | Reference configuration with unreinforced foam core | | | 0 |

The pin volume fraction was calculated based on the test specimen geometry and the number of pins inserted in a test specimen. Only pins with fully constrained ends were considered.

## 3. Out-of-plane shear test

*3.1 Test description and specimen preparation*

To determine the shear moduli and strengths for the tested configurations, shear tests in accordance to DIN 53294-standard [15] were conducted. From every test configuration, four test specimens were tested. A test specimen has a length of 390 mm and a width of 100 mm. The test specimens were bonded to steel rigid plates by means of the adhesive epoxy system Scotch-Weld® 9323 B/A [16]. The tests were conducted with sandwich skins bonded to the rigid plates to enable a homogeneous load introduction to the pin-reinforced core. Figure 2 shows the test setup with all parts labelled.

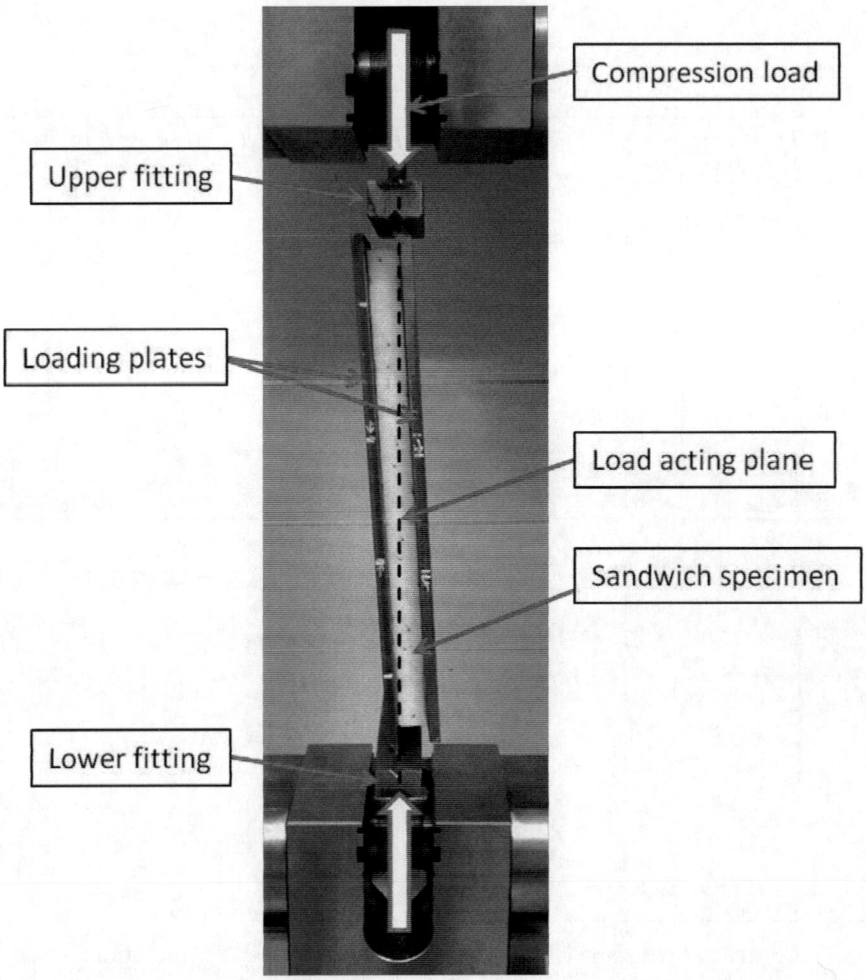


**Figure 2.** Test specimen under shear load.

The load was applied to the ends of the rigid plates in compression through a knife-edge bearing at a constant movement rate of the testing machine cross head of 1 mm/min. The load-shear displacement curves were automatically recorded and used to calculate the shear modulus and strength of the core material. Only specimens with shear failure mode were considered for the evaluation. The shear strength was calculated as follow [15]:

$$\tau_S = \frac{F_B}{A_0} = \frac{F_B}{l_1 \cdot b} \qquad (1)$$

where $F_B$ is the load at shear failure in N, $l_1$ the length of the specimen in mm and b is the specimen width in mm.

The shear modulus is determined as follow [15]:

$$G_S = \frac{\Delta\tau}{\Delta\gamma} = \frac{h}{l_1 \cdot b} \cdot \frac{\Delta F}{\Delta\vartheta} \qquad (2)$$

where h is the thickness of the core in mm and $\frac{\Delta F}{\Delta\vartheta}$ is the slope of initial linear portion of load-shear displacement curve in N/mm. Since the used standard doesn´t contain an explicit definition of $\frac{\Delta F}{\Delta\vartheta}$ , it was calculated by using a best fit to the linear region of the load-shear displacement curves.

### *3.2. Test results*

The stress-shear strain curves of the tested configurations are depicted in figure 3.

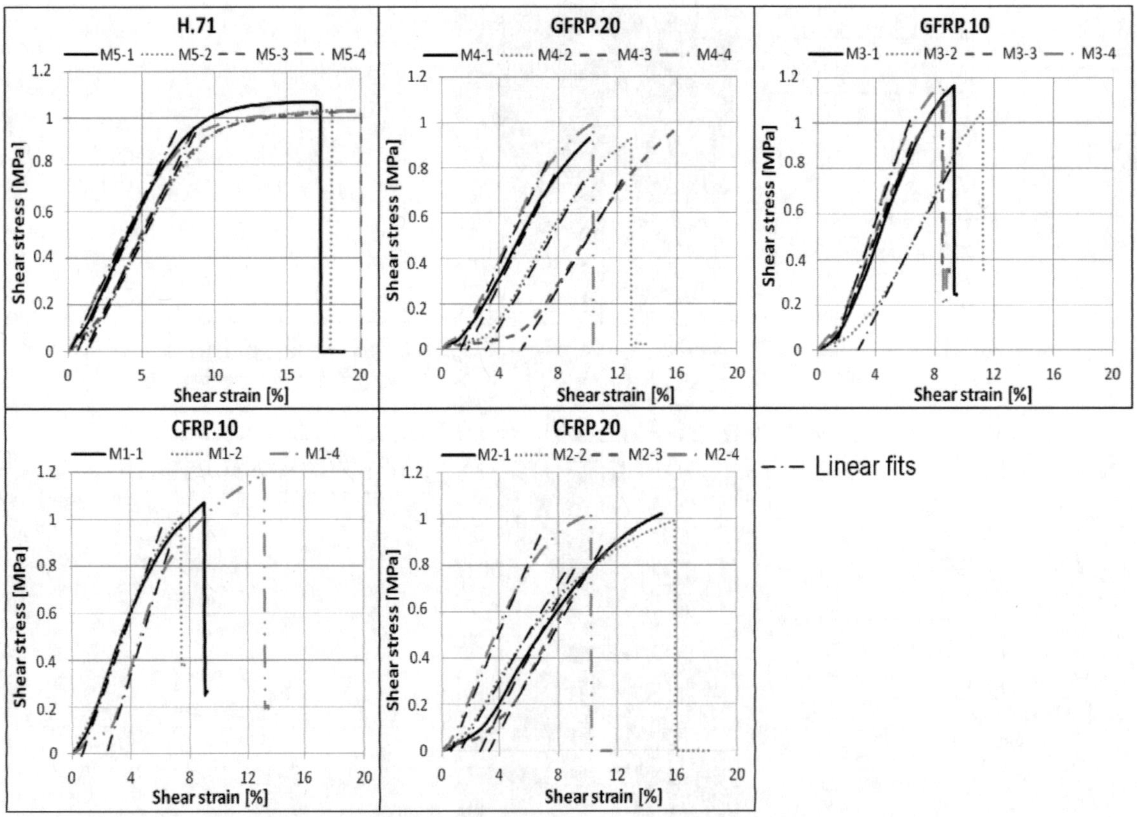


**Figure 3.** Stress-shear strain response of the tested configurations.

Mx-y is the identifier of the test specimen where x is the configuration number and y is the number of the test specimen in the tested configuration. In every configuration test results of four test samples were considered except the configuration CFRP.10 where the specimen M1-3 failed due to debonding of the adhesive layer and was not considered in the interpretation.

Comparing the different stress-strain curves shows that using pin-reinforcement increases the non-linearity at the initial portion of the curve. This makes the determination of the shear-modulus of pin-reinforced test specimens more complicated. Moreover, the response of pin-reinforced specimens shows higher results scatter and a reduced shear strain at failure compared to the reference configuration.

The obtained values for shear strength and shear modulus for the tested configurations are summarised in bar graph-form in figure 4. First observations of the shear properties show that an improvement of shear behaviour is only reached for the configurations CFRP.10 and GFRP.10, which have about four times higher pin volume fraction than the configurations with 20 mm unit cell width. While the configurations CFRP.10 and GFRP.10 led to an increase of shear modulus of about 35% and shear strength of about 15%, the configurations CFRP.20 and GFRP.20 decrease the shear properties. Considering the pin material, CFRP and GFRP-pins led nearly to the same results when maintaining the pin-pattern unchanged. The latter finding leads to the conclusion that an improvement of the mechanical properties could also be reached when using low-cost pins made of glass fibres.

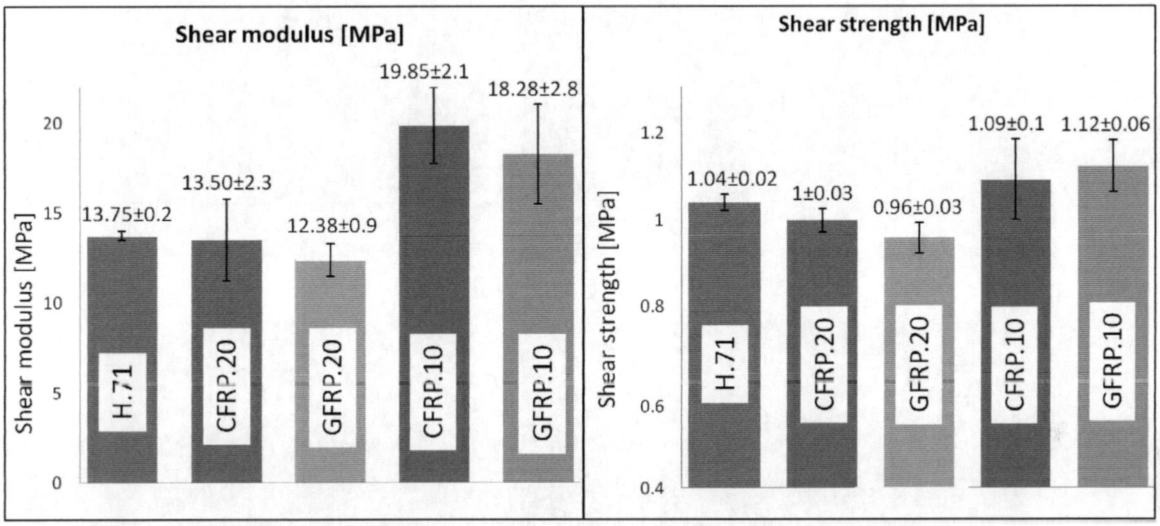


**Figure 4.** Average values and standard deviations of the calculated shear properties of the tested configurations.

*3.3. Crack propagation*

When the shear strength of a specimen is reached, a shear crack with an angle between 35° and 45° arises and propagates in upper and lower interfaces between the core material and the skins thus the test specimen is divided in two parts. Figure 5 shows an example of a shear crack.

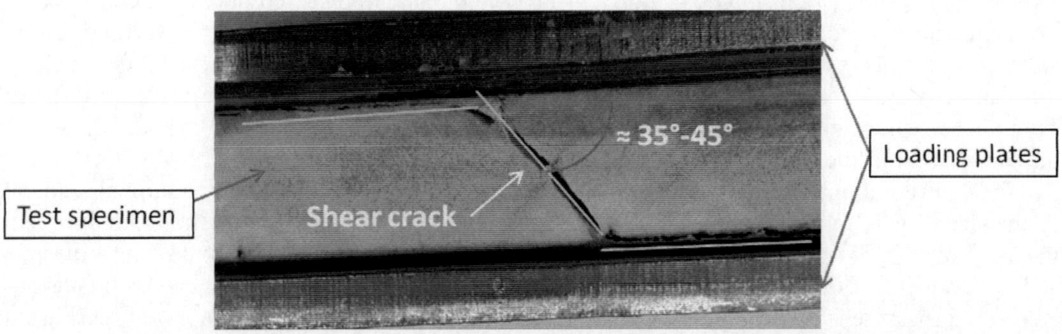


**Figure 5.** Under shear load failed specimen-close-up view.

Regarding the crack redirection and maintaining the integrity of the structure, observations from test execution and failed specimens investigations showed that only specimens with 10 mm unit cell width prevented the abrupt failure and the collapse of the tested specimen. With high pin volume fraction the expected effects of crack redirection and alternative load path are reached. After crack initiation, the crack propagation is slowed down and the load is mainly carried by the pins. Figure 6 shows the crack redirection effect of two different pin volume fractions. Test specimens with unit cell width of 20 mm showed only a minor effect of the pins on the crack propagation and they had nearly no residual load carrying ability, while the specimens with 10 mm unit cell width showed a change of the direction of the crack propagation and the pins at the fracture surface assured the integrity of the test specimens.

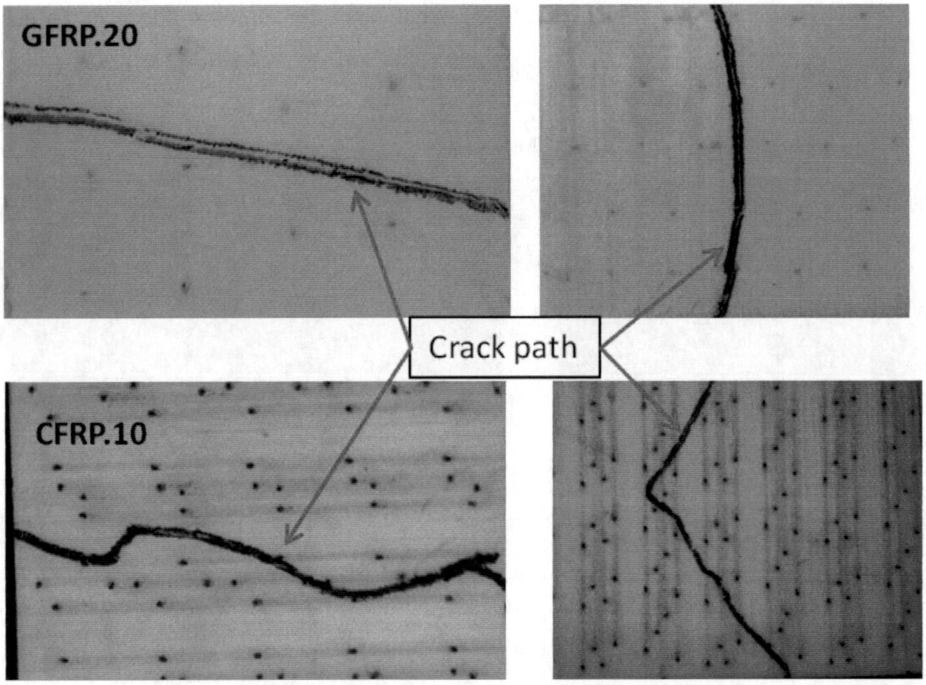


**Figure 6.** Top view of tested specimens after removing the skins.

## 4. Conclusions

This study is a baseline about the influence of pin-material and pin volume fraction on the shear properties and crack propagation of TFC-sandwich. From the results presented above an enhancement of the shear properties and crack redirection are only possible when a sufficient pin volume fraction is existent. It was expected that the low pin volume fraction of the configurations CFRP.20 and GFRP.20 would lead to a minor improvement of the shear properties compared to the reference configuration without pin reinforcement, but the test results showed that a pin volume fraction of about 0.2% led to a slight degradation of the shear properties of the foam core. It is possible that the low pin volume fraction of the configurations CFRP.20 and GFRP.20 was not sufficient to create an additional load path to carry the shear loads and the inserted pins created local stress concentrations in the foam core that led to an early collapse of test samples. Considering the pin-material, GFRP-pins showed, as expected, the same improvement of the shear properties like the CFRP-pins, which would reduce the material cost. To get a full characterisation of TFC-sandwich, out-of-plane tension and compression tests are to be performed in further investigations.

**Acknowledgements**
This work was funded by the German Federal Ministry for Economic Affairs and Energy through the Federal Aeronautical Research Program LuFo V-1 within the project Sandwich². The authors wish to acknowledge this support.

**References**

[1] Zuardy M I, Block, T B and Herrmann A S 2011 Structural analysis of centre box of a vertical tail plane with a side panel from composite foam sandwich *Deutscher Luft- und Raumfahrtkongress* pp 1377–89
[2] Herrmann A S, P Zahlen and Zuardy M I 2005 Sandwich structures technology in commercial aviation *Proc. 7th Int. Conf. on Sandwich Structures (Aalborg, Denmark)* pp 13–26
[3] Olsson R and Block T B 2015 Criteria for skin rupture and core shear cracking induced by im pact on sandwich panels *Compos. Struct.* **125** pp 81–87
[4] Marasco A I 2005 Analysis and evaluation of mechanical performance of reinforced sandwich structures: X-Cor and K-Cor *PhD Thesis* Cranfield University, UK
[5] John M, Geyer A, Schäuble R and Schlimper R 2015 Comparing unreinforced and pin-reinforced CFRP/PMI foam core sandwich structures regarding their damage tolerance behav-iour 20th *Int. Conf. on Composite Materials* (Copenhagen, Denmark)
[6] Brauner C, Block T B, Herrmann A S 2012 Meso level manufacturing simulation of sandwich structures to analyse viscoelastic dependent residual stresses *J. Compos. Mater.* **46** pp. 783-99
[7] Endres G 2014 Tied Foam Core Technology and the achieved impact performance of sandwich constructions based on reinforced foam materials *CFK-Valley Stade Convention* (Stade, Ger many)
[8] Zahlen P C, Rinker M and Heim C 2008 Advanced manufacturing of large, complex foam core sandwich panels *8th Int. Conf. on Sandwich Structures* (Porto, Portugal)
[9] HexFlow RTM6 data sheet December 2014 Document ITA 065f Hexcel Corporation Stamford, USA
[10] Product information ROHACELL HERO February 2014 edition Evonik Industries AG Darmstadt, Germany
[11] Tenax HTS Filamentgarn data sheet Bd. Version 5, 30-09-2014 Toho Tenax Europe GmbH Wuppertal, Germany
[12] Torayaca T800H data sheet Doc-Nr. CFA-007 Toray Carbon Fibers America INC Santa Ana, CA, USA
[13] 1383 Yarn data sheet Product Bulletin_04_2015 PPG Fiber Glass Cheswick, PA, USA
[14] Eckhardt J 2010 Schadensbeurteilung von verstärkten Kernverbundstrukturen nach Schlagstoß belastung *diploma thesis* University of a applied sciences Aachen, Germany
[15] Standard DIN 53294 Prüfung von Kernverbunden - Schubversuch Edition **2.1982** (Berlin: Beuth Verlag)
[16] Scotch-Weld 9323 B/A data sheet June 2002 edition 3m Deutschland GmbH Neus, Germany

# Simulations and measurements of annealed pyrolytic graphite–metal composite baseplates

**F Streb[1,*], G Ruhl[1], A Schubert[2], H Zeidler[2], M Penzel[2], S Flemmig[2], I Todaro[3], R Squatrito[3] and T Lampke[4]**

[1] *Infineon Technologies AG, Germany*
[2] *Professorship Micromanufacturing Technology, Technische Universitaet Chemnitz, Germany*
[3] *CIRI - Advanced Applications in Mechanical Engineering & Materials Technology, University of Bologna, Italy*
[4] *Materials and Surface Engineering Group, Institute of Materials Science and Engineering, Technische Universität Chemnitz, 09125 Chemnitz, Germany*
[*] *fabian.streb@infineon.com*

**Abstract.** We investigated the usability of anisotropic materials as inserts in aluminum-matrix-composite baseplates for typical high performance power semiconductor modules using finite-element simulations and transient plane source measurements. For simulations, several physical modules can be used, which are suitable for different thermal boundary conditions. By comparing different modules and options of heat transfer we found non-isothermal simulations to be closest to reality for temperature distribution at the surface of the heat sink. We optimized the geometry of the graphite inserts for best heat dissipation and based on these results evaluated the thermal resistance of a typical power module using calculation time optimized steady-state simulations. Here we investigated the influence of thermal contact conductance (TCC) between metal matrix and inserts on the heat dissipation. We found improved heat dissipation compared to the plain metal baseplate for a TCC of 200 kW/m²/K and above.To verify the simulations we evaluated cast composite baseplates with two different insert geometries and measured their averaged lateral thermal conductivity using a transient plane source (HotDisk) technique at room temperature. For the composite baseplate we achieved local improvements in heat dissipation compared to the plain metal baseplate.

## 1. Introduction

A common method to improve heat dissipation in electronic devices is the usage of anisotropic materials [1–4]. These materials, like annealed pyrolytic graphite, have a very high thermal conductivity in two spatial directions and a low conductivity in the third direction [3]. Figure 1 shows an illustration of a typical power module. Most real life power modules, like figure (b), consist of several IGBTs (insulated-gate bipolar transistor) and FWDs (freewheeling diode) pairs connected to each other via wire bonds.

IOP Conf. Series: Materials Science and Engineering **118** (2016) 012013 doi:10.1088/1757-899X/118/1/012013

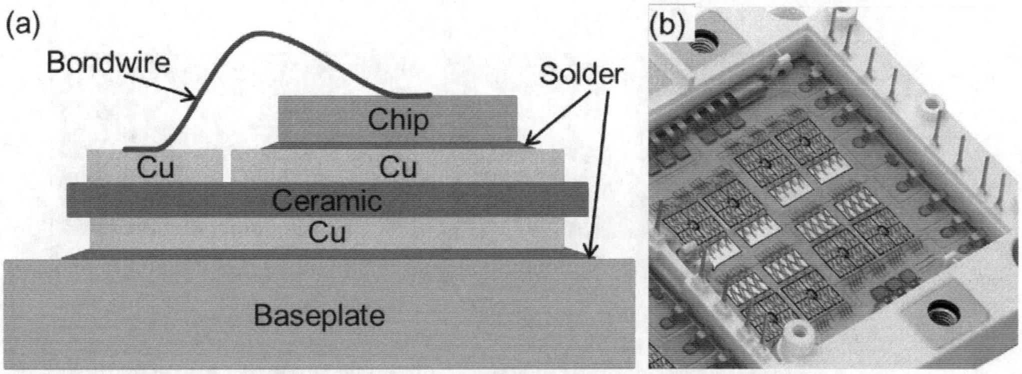


**Figure 1.** (a) Schematic illustration of a typical power module, (b) section of an Infineon HybridPack™ power module used for automotive applications.

The performance and lifetime of a power module strongly depend on its ability to dissipate the heat generated by the chips. The heat dissipation can be quantified by the thermal resistance $R_{th}$ [K/W] of the package, which is inversely proportional to the thermal conductivity $\lambda$ [W/m/K] of the materials used. One of the major obstacles for future generations of power modules are the increasing demands on system integration and power density. Both are accompanied by increased chip temperatures and hence decreased product lifetimes. The heat dissipation limit of conventional materials like Cu or Al is nearly reached. Hence for future devices novel materials with increased thermal conductivity need to be investigated. 'Annealed pyrolytic graphite' (APG) [5], a highly ordered graphite phase created by annealing turbostratic graphite is a possible candidate. APG has a thermal conductivity of above 1500 W/m/K in two spatial directions and 10 W/m/K in the third.
Aim of this work is to evaluate the usability of annealed pyrolytic graphite as inserts in metal matrices as a replacement for state-of-the-art plain metal baseplates for power modules.

## 3. Comparison of different simulation modules & insert geometries

Several physical modules (constant heat transfer 'ht-const. h', external forced heat sink 'ht-efts' and non-isothermal flow module 'ntif'), suitable for different thermal boundary conditions, are investigated. By comparing three different physics modules and options of heat transfer we found non-isothermal simulations to be closest to reality for temperature distribution at the surface of the heat sink, but also the most time consuming. Reason is the modeling of an active flushing liquid heat sink. The benefit of this method is the calculation of the heat transfer coefficient at every single point of the heat sink surface instead of using a fixed heat transfer coefficient.
We optimized the arrangements of the graphite inserts in the metal matrix. Some examples of the results for insert optimization are given in figure 2. The result for simulation without APG shows the very low heat spreading. By arranging APG in-between the heat spots (APG 1) heat congestion occurs, the APG is not able to absorb and spread the heat efficiently. A better arrangement of APG-stripes is the location directly underneath the heat spots. The thermal energy is transferred into the APG and is spread in the baseplate. The examples APG 2 and 3 show these kinds of arrangements with continuous and discontinuous stripes.

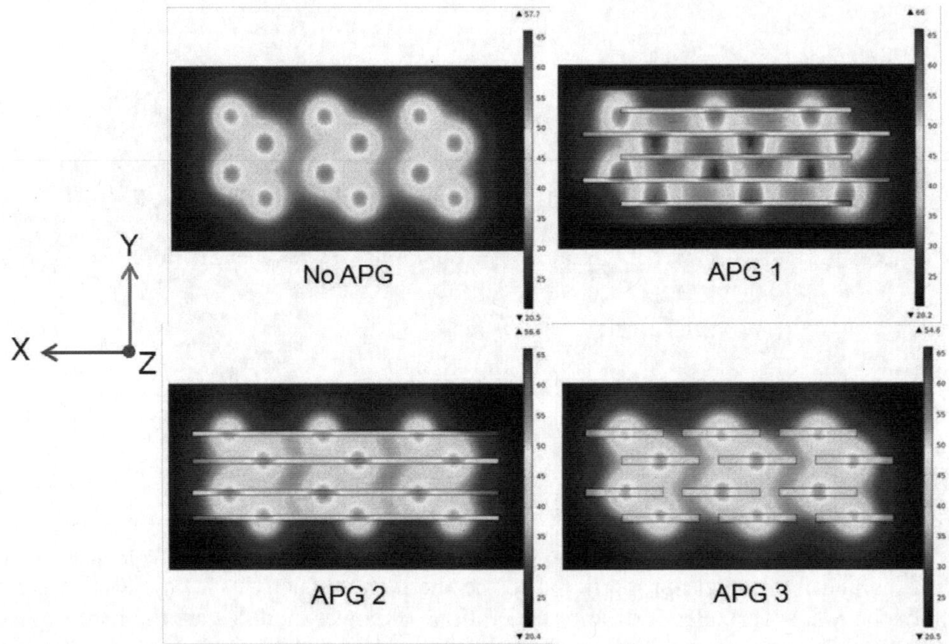


**Figure 2.** Selected results of temperature distribution at the heat sink surface (scale: 20 °C to 65 °C).

## 4. Influence of thermal contact conductance on heat dissipation

We evaluate the influence of the TCC between the Al matrix and the APG inserts on the heat dissipation of a power module using steady-state FEM simulations. We use a simplified structure of Infineon's HypridPack™1 power module as an example, allowing for general results applicable for most modules while still comparable to real-life products. In order to shift the simulation focus towards the baseplate and reduce calculation times we combine the thermal properties of the direct copper bonded substrate and both solder layers into one artificial substrate layer. The used model and the two different evaluated baseplate geometries are shown in figure 3. IGBTs are used as heat sources with an internal heat generation per volume resulting in 220 W per switch. A convection surface is attached to the bottom side; all other surfaces are perfectly isolated. The APG inserts are oriented with their high thermal conductivity in X and Z direction. All but the Al-APG interface have a perfect TCC.

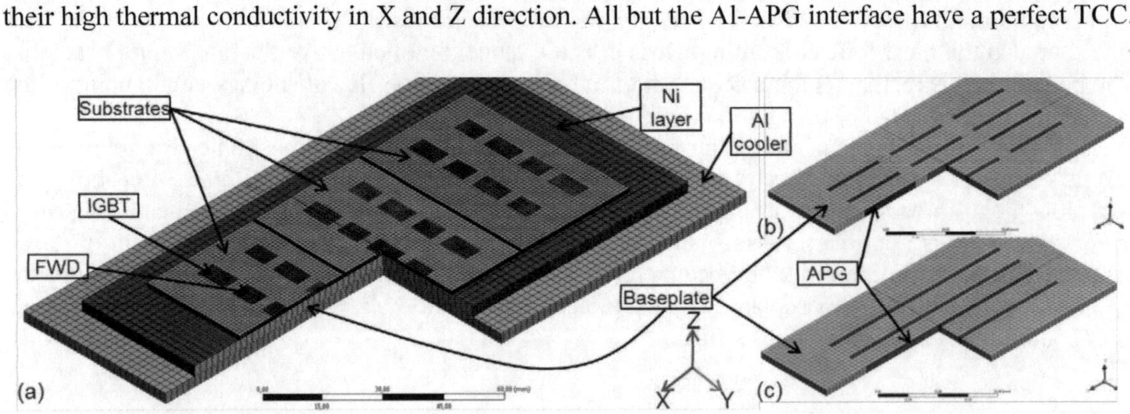


**Figure 3.** (a) Simulation model incl. mesh (scale: 60 mm); composite baseplate with 'short' (b) and 'long' (c) APG inserts (scale: 50 mm).

The maximum chip temperatures are calculated; the resulting temperature distributions and heat fluxes at the chip with the highest temperature are shown in figure 4. The highest calculated chip temperatures for both baseplate geometries depending on the TCC are given in table 1. Using a plain Al baseplate we calculate the maximum chip temperature to 133.8 °C. Compared to a plain Al baseplate the composite baseplate shows improved heat dissipation for a TCC as low as 200 kW/m²/K. For all TCCs the 'long' baseplate type performs better than the 'short' type. The inferior heat dissipation of the 'short' type can be explained with its reduced amount of APG and its interruptions within the APG inserts. This shows that heat dissipation in Z direction is not the only influencing factor. The continuous inserts allow for improved heat spreading in X direction and hence result in lower chip temperatures. The maximum chip temperature using a perfect thermal contact (infinite TCC) is more than 14 °C lower than that using a TCC of 10 kW/m²/K. This shows that the TCC has a major influence on the heat dissipation of module: if the TCC is too high the inserts act like pores and are detrimental to the heat dissipation.

**Table 1.** Maximum chip temperatures for both baseplate geometries depending on the thermal contact conductance between matrix and inserts. An infinite TCC represents perfect thermal contact.

| TCC | [kW/(m² K)] | ∞ | 800 | 400 | 200 | 100 | 10 |
|---|---|---|---|---|---|---|---|
| Max. chip temperature [°C] | 'short' | 128.1 | 129.5 | 130.5 | 132.5 | 135.3 | 142.4 |
| | 'long' | 127.6 | 128.9 | 129.8 | 131.6 | 134.3 | 141.8 |

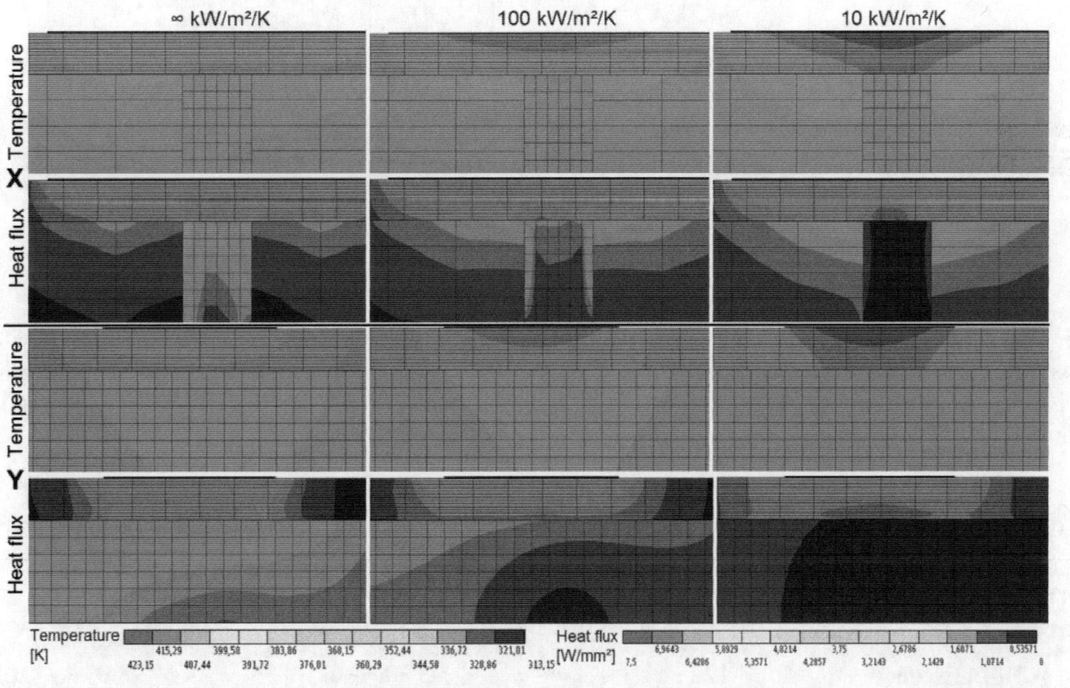


**Figure 4.** Simulation results using baseplate with 'long' inserts, cross-sections at chip with highest temperature. Temperature in row 1&3, heat flux in row 2&4. Viewing direction: rows 1&2 in X, rows 3&4 in Y direction. 1st column perfect TCC, 2nd column 100 kW/m²/K, 3rd column 10 kW/m²/K.

## 5. Measurement of cast composite baseplates

The lateral thermal conductivity of casted Al-alloy composite baseplates was investigated using a transient plane source technique (Hot Disk TPS 3500 by Hot Disk AB, Sweden, [6,7], sensor diameter 12.8 mm). We evaluated different sensor positions to investigate the influence of the TCC and the APG inserts on the heat dissipation as illustrated in figure 5. On the backside of the composite baseplates (the side the DCB would be soldered onto, not visible in figure 5) the inserts are not showing, instead a thin Al layer is between sensor and APG. For these backside measurements the sensor is positioned in X&Y direction just like the 'in-between' position on the front side. The average conductivities and standard deviations of the measurements are given in table 2. Measurements at the centered position show a higher conductivity than the position in-between the inserts, demonstrating that heat spreading in X direction occurs. The heat is dissipated anisotropically within the lateral dimensions beyond the marked areas in figure 5. Comparing the plain Al measurement with the Al-backside measurement reveals the detrimental effect of the TCC between Al and APG: Apparently the TCC is too low to allow heat absorption and dissipation by the inserts. Instead the inserts act like air pores and reduce the effective thermal conductivity. However if a heat source is positioned on top of an insert the effective thermal conductivity is improved significantly (more than 30%) compared to a plain Al baseplate.

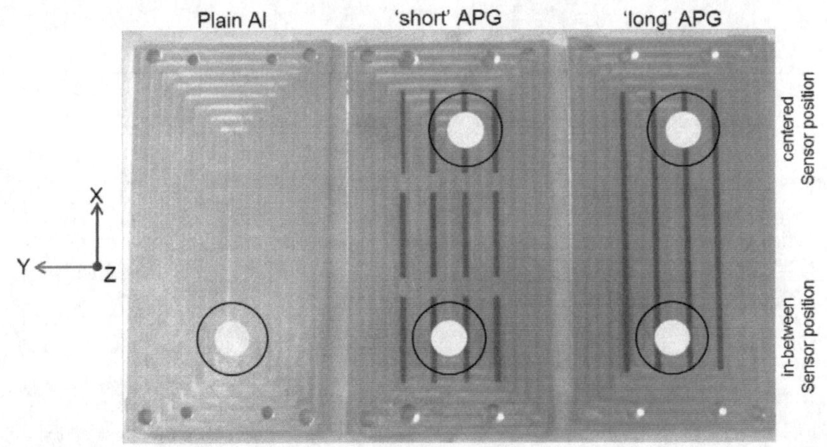


**Figure 5.** Image of the three different measured baseplate types. The yellow dots indicate the sensor position for the thermal conductivity measurements; the circle marks the maximum penetration depth for isotropic materials for the chosen measurement parameters based on the properties of the used Al-alloy.

**Table 2.** Lateral thermal conductivities for different sensor positions and baseplate types as illustrated in figure 5.

| λ [W/(m K)] | Plain Al | 'short' APG | 'long' APG |
|---|---|---|---|
| **Centered** | N/A | 162.2±0.7 | 153.8±2.0 |
| **In-between** | 122.3±1.1 | 113.5±1.6 | 98.0±0.7 |
| **Al-Backside** | N/A | 98.5±2.2 | 78.6±0.3 |

## 6. Summary and Outlook

Overall the performed simulations show that APG inserts in metal matrices can be used to improve the heat dissipation of high performance power semiconductor modules compared to plain Al alloy baseplates, if the insert arrangement in respect to the heat sources and their thermal contact conductance to the metal is optimized. These simulations have been confirmed by transient plane source measurements. During operation the baseplate of a power module bends due to thermal expansion. It is unclear whether the APG inserts break during the cyclic loads due to their brittle behavior. Hence in the next step we will evaluate the thermal conductivity of the cast composite baseplates after thermal cycling tests and create and evaluate a demonstrator power module.

## Acknowledgments

This work is supported by the ongoing project THERMACO (Smart Thermal conductive Al MMCs by casting), which receives funding from the European Union's seventh Framework Programme for Research and Technological Development under grant agreement no. 608978.

## References

[1] Hutsch T, Schubert T, Schmidt J, Weißgärber T and Kieback B Innovative Metal-Graphite Composites as Thermally Conducting Materials *PM2010 Proceedings - PM Functional Materials - Heat sinks* PM2010 World Congress

[2] Chai Y, Zhang K, Zhang M, Chan P C and Yuen M M 2007 Carbon nanotube/copper composites for via filling and thermal management *Electronic Components and Technology Conference, 2007. ECTC'07. Proceedings. 57th* (Reno, NV, USA: IEEE) pp 1224–9

[3] Fan W, Liu X and Mariner J 2012 Thermal pyrolytic graphite composite with coefficient of thermal expansion matching for advanced thermal management *Microwave Symposium Digest (MTT), 2012 IEEE MTT-S International* (Montreal, QC, Canada: IEEE) pp 1–3

[4] Zeidler H, Schubert A, Flemmig S, Penzel M and Essel S 2014 Graphene Aluminium Matrix Nano-Composites for Heat Transfer Applications *Proceedings of the 4th International Conference on Nanomanufacturing (nanoMan2014)* nanoMan2014 (Germany)

[5] Walker P L and Thrower P A 1973 *Chemistry and physics of carbon: a series of advances. Vol. 9* (New York: Marcel Dekker)

[6] Gustavsson M, Wang H, Trejo R M, Lara-Curzio E, Dinwiddie R B and Gustafsson S E 2006 On the Use of the Transient Hot-Strip Method for Measuring the Thermal Conductivity of High-Conducting Thin Bars *Int. J. Thermophys.* **27** 1816–25

[7] Ma Y, Gustavsson J S, Haglund Å, Gustavsson M and Gustafsson S E 2014 Pulse transient hot strip technique adapted for slab sample geometry to study anisotropic thermal transport properties of μm-thin crystalline films *Rev. Sci. Instrum.* **85** 044903

# Recent Developments for Ultrasonic-Assisted Friction Stir Welding: Joining, Testing, Corrosion – an Overview

**M Thomä[1], G Wagner[1], B Straß[2], C Conrad[2], B Wolter[2], S Benfer[3] and W Fürbeth[3]**

[1] Chair of Composite Materials, Institute of Materials Science and Engineering, Technische Universitaet Chemnitz, 09125 Chemnitz, Germany
[2] Fraunhofer Institute for Nondestructive Testing, 66123 Saarbrücken, Germany
[3] DECHEMA Society for Chemical Engineering and Biotechnology, 60486 Frankfurt am Main, Germany

marco.thomae@mb.tu-chemnitz.de

**Abstract.** Due to the steadily increasing demand on innovative manufacturing processes, modern lightweight construction concepts become more and more important. Especially joints of dissimilar metals offer a variety of advantages due to their high potential for lightweight construction. The focus of the investigations was Al/Mg-joints. Friction Stir Welding (FSW) is an efficient process to realize high strength joints between these materials in ductile condition. Furthermore, for a simultaneous transmission of power ultrasound during the FSW-process (US-FSW) a positive effect on the achievable tensile strength of the Al/Mg-joints was proven.

In the present work the industrial used die cast alloys EN AC-48000 (AlSi12CuNiMg) and AZ80 (MgAl8Zn) were joined by a machining center modified especially for Ultrasound Supported Friction Stir Welding. The appearing welding zone and the formation of intermetallic phases under the influence of power ultrasound were examined in particular. In order to identify optimal process parameters extensive preliminary process analyzes have been carried out. Following this, an ultrasound-induced more intensive stirring of the joining zone and as a result of this a considerably modified intermetallic zone was detected. At the same time an increase of the tensile strength of about 25 % for US-FSW-joints and for fatigue an up to three times higher number of cycles to failure in comparison to a conventional welding process was observed. Moreover, detailed corrosion analyzes have shown that especially the welding zone was influenced by the corrosive attack. To expand and deepen the knowledge of the US-FSW-process further material combinations such as Ti/Steel and Al/Steel will be considered in future.

## 1. Motivation

Aluminum and magnesium die cast alloys possess a high potential for industrial applications due to their low density and high specific tensile strength in combination with low manufacturing costs [1]. The fabrication of tailored components made of these two light metals is of central interest in many industrial sectors like the automotive sector and the aerospace industry. The main aim actually is to find alternative welding methods for these materials, because it is not possible to join aluminum and magnesium by fusion welding due to the strong formation of brittle intermetallics in the whole bonding zone. However Friction Stir Welding (FSW) as a low temperature pressure welding method is a suitable alternative to manufacture hybrid joints while largely avoiding intermetallic phases (IM-phases) [2], because these precipitates have a limiting influence on the mechanical properties of the realized Al/Mg-joints. But also for FSW the formation of IM-phases cannot be avoided completely. In the contact area between the dissimilar base materials two continuous IM-layers occur. To reduce their influence on the mechanical properties Ultrasonic Supported Friction Stir Welding (US-FSW) was developed with regard to disperse the IM-phases in the welding zone [3]. The present work will give an overview about the possibilities of US-FSW, the resulting microstructure as well as the monotonic and cyclic properties of Al/Mg-joints. Furthermore the corrosion behavior and suitable non-destructive testing methods will be discussed.

## 2. Experimental Procedure

The commercially available cast-alloys EN AC-48000 (AlSi12CuNiMg) and AZ80 (MgAl8Zn) were used for the investigations. For US-FSW the materials were butt-welded in die-casted condition using sheet geometry of 280 mm x 100 mm x 3.3 mm. For the welds a machining center DMU80T from Deckel Maho was used modified for FSW (Figure 1 a) [4].

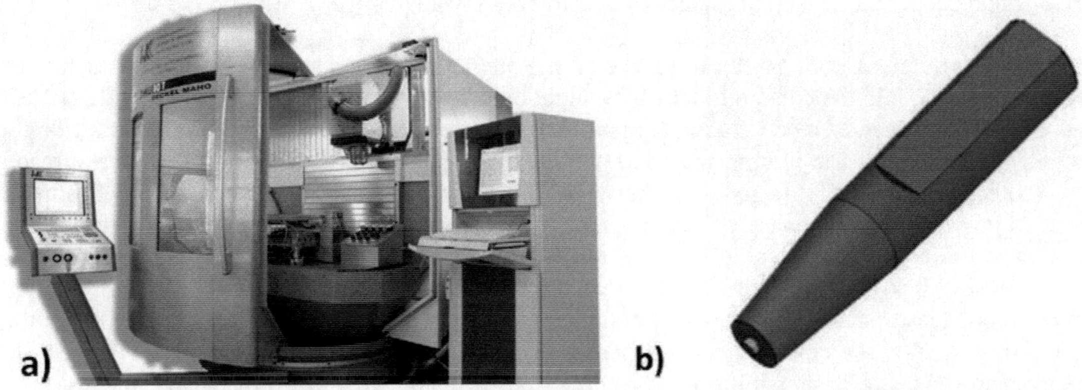


**Figure 1.** a) Machining center DMU80T modified for FSW, b) 3D-model of a welding tool [4]

The Al/Mg-hybrid-joints were produced with a welding tool made of hot-work steel 1.2344 (X40CrMoV51) (Fig. 1 b). It has a tool-shoulder of 16 mm diameter, a pin with a length of 3.1 mm and a metric thread (M4.5). For the input of the ultrasound a specially designed ultrasonic seam module type RM 20 from Schunk Sonosystems was used (Figure 2).

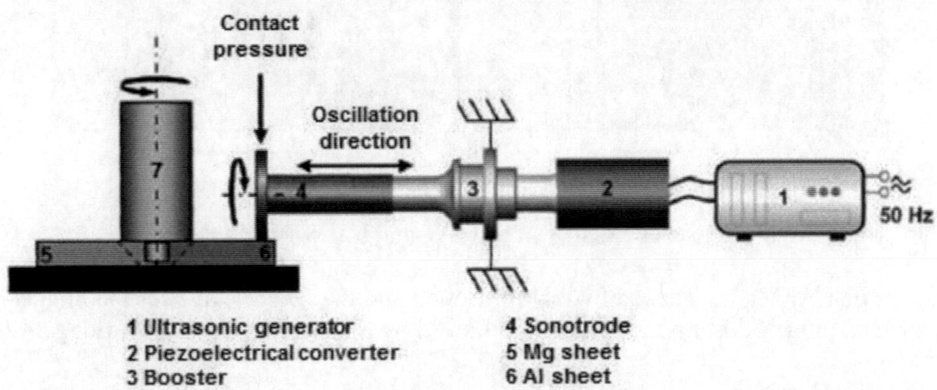


**Figure 2.** Ultrasonic assisted Friction Stir Welding, schematically [5]

The seam module works with a resonant frequency of 20 kHz, a maximal generator power of 3000 W and an amplitude of up to 35 µm. The optimized work distance between sonotrode and welding zone was 95 mm. The input of the ultrasound is controlled via LabVIEW and takes place in the aluminum sheet, which is also located as the Advancing Side.

## 3. Results and Discussion

Based on light-microscopic cross-section images a comparison between EN AC-48000/AZ80-FSW-joints and EN AC-48000/AZ80-**US**-FSW-joints was carried out regarding the influence of the ultrasound (Figure 3).

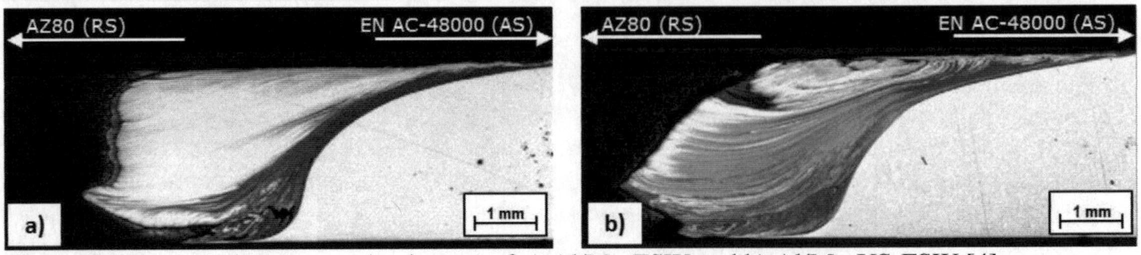


**Figure 3.** Microscopic cross-section images of a) Al/Mg-FSW and b) Al/Mg-US-FSW [4]

It is obvious that the ultrasound has an influence on the nugget shape and also on the microstructure of the welded joints. While the conventional FSW-nugget shows a vertical edge on the Mg-side (*Fig. 3 a*), the US-FSW-nugget is much flatter (Figure 3 b). In addition, the use of ultrasound strongly intensified the mixing of the joining-materials within the nugget – an observation that was already made by other researchers [6]. As a further part of the microstructural investigations Scanning Electron Microscopy (SEM) was carried out to gain a deeper understanding of the nugget-zone. Thereby two continuous intermetallic layers could be found at the interface of Mg and nugget. This effect for Al/Mg-FSW-joints had also been detected by Klag et al. [2]. The layers could be identified by EDX-Spot-Measurement as the intermetallic phases $Mg_{17}Al_{12}$ and $Al_3Mg_2$. These brittle precipitations impair the mechanical properties of the EN AC-48000/AZ80-FSW-joints. In comparison to this, the use of ultrasound-supported Friction Stir Welding leads to just one intermetallic $Mg_{17}Al_{12}$ layer (Figure 4).

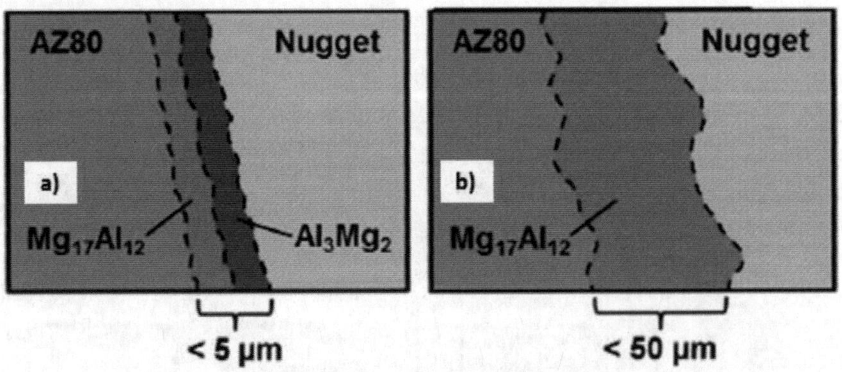


**Figure 4.** Interface Mg/nugget schematically of a) FSW and b) US-FSW [4]

The aluminum rich $Al_3Mg_2$ is cracked by the ultrasound and then dispersed over the nugget-zone [4]. A better understanding of this process could be achieved by EBSD-mappings of the nugget (Figure 5).

IOP Conf. Series: Materials Science and Engineering **118** (2016) 012014 doi:10.1088/1757-899X/118/1/012014

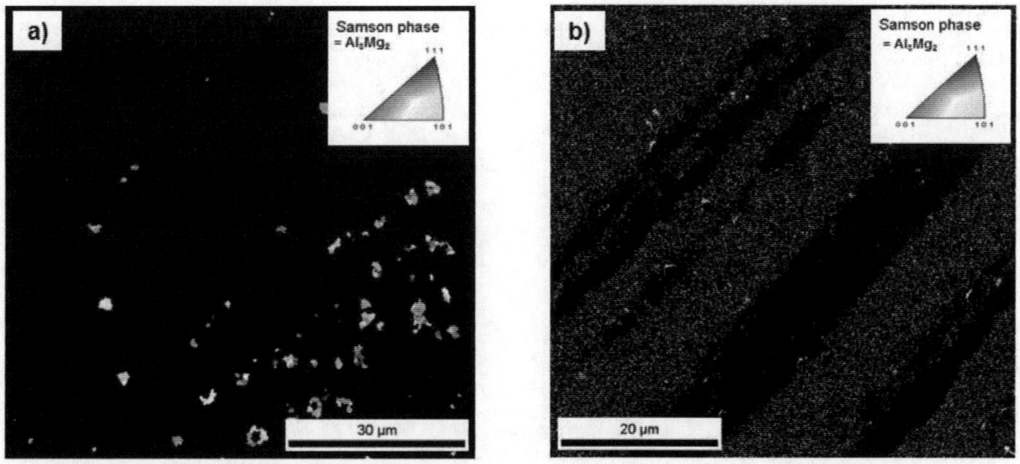


**Figure 5.** EBSD-analysis of $Al_3Mg_2$ in the nugget-zone: a) FSW and b) US-FSW [4]

The pictures show the described morphology of $Al_3Mg_2$ in the welding zone of EN AC-48000/AZ80-FSW-joints (*Fig. 5 a*) and in EN AC-48000/AZ80-US-FSW-joints (Figure 5 b). This advantageous impact of US-FSW was also discovered by Park [7].
Further investigations of the mechanical properties by static and dynamic testing were carried out in this work. Figure 6 represents a comparison of the tensile strengths of the two base materials, Al/Mg-FSW-joints and Al/Mg-US-FSW-joints.

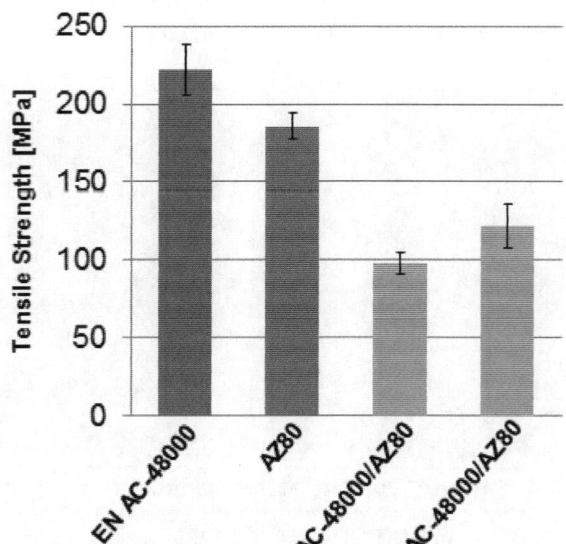


**Figure 6.** Comparison of the achieved tensile strengths of the base materials, the FSW-joints and the US-FSW-joints [4]

The conventionally welded joints achieve an average tensile strength of 98 MPa, which corresponds to 50 % of the strength of the weaker joining partner AZ80. The ultrasound-supported EN AC-48000/AZ80-joints show higher values of about 122 MPa, a tensile strength increase of 25 % in average.
The cyclic fatigue behavior of the welded materials was characterized by an S/N-curve (Figure 7). It can be seen that Al/Mg-US-FSW-joints attain significantly higher lifetimes for the same stress.

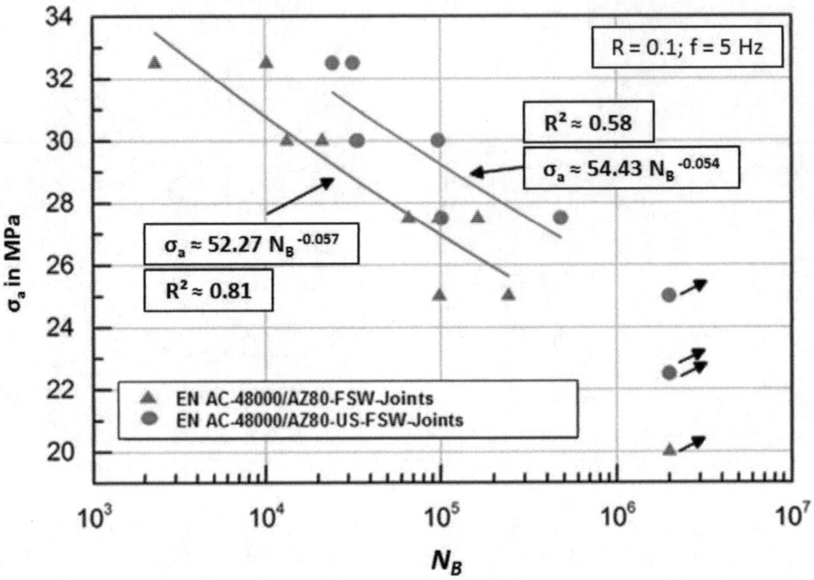


**Figure 7.** Comparison of S/N-curves for the EN AC-48000/AZ80-FSW-joints and the EN AC-48000/AZ80-US-FSW-joints [4]

Beside destructive testing non-destructive test methods, electro-magnetic-induced ultrasound (EMUS) and X-ray radiography were carried out on the Al/Mg-joints to verify their capability of detecting welding defects. Figure 8 shows the results of the EMUS-testing.

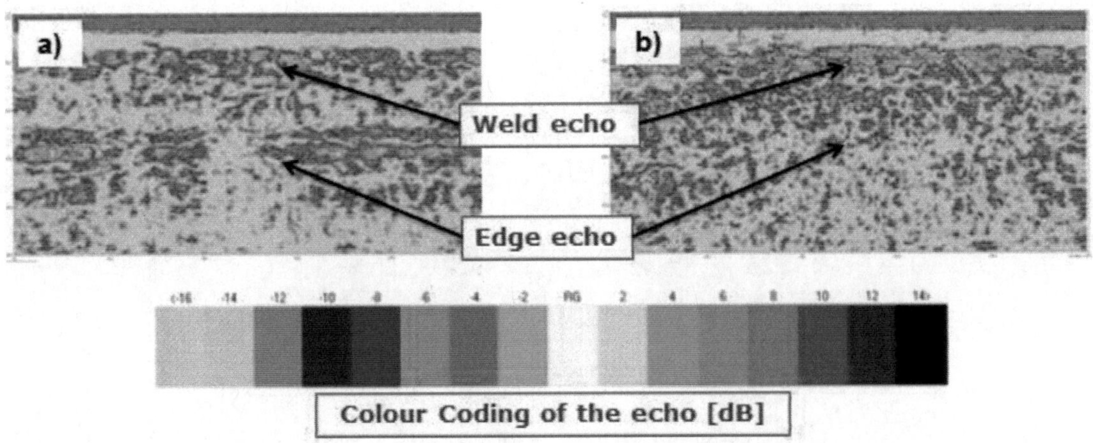


**Figure 8.** Non-destructive testing via EMUS at a sound FSW-weld (a) and at a flawed weld (b) [4]

A sound welding joint shows a weld echo and an edge echo by elevated intensities. The flawed weld exhibits only higher intensities only for the weld echo due to the present defects.
Also by using X-ray radiography differing images for conventional welded joints and ultrasound-supported welded joints could be observed (Figure 9).

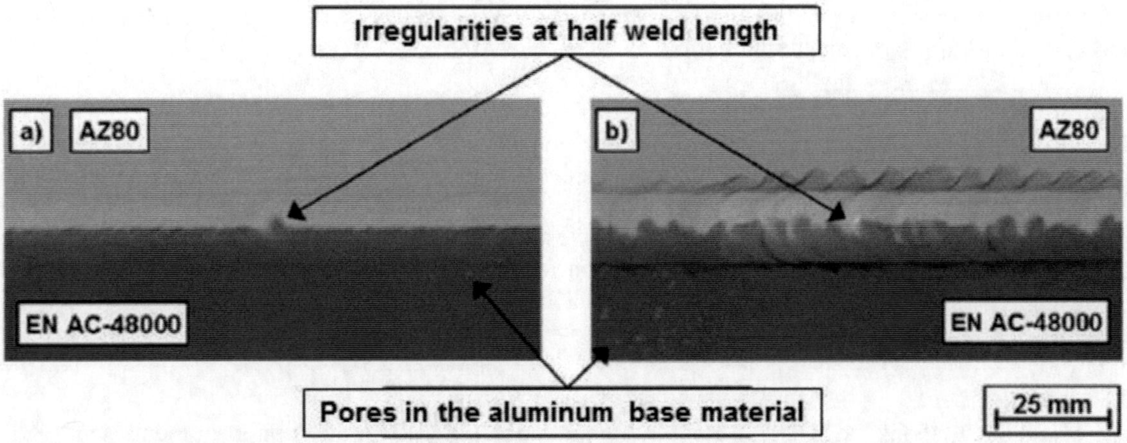


**Figure 9.** Non-destructive testing via X-ray radiography of an EN AC-48000/AZ80-FSW-joint (a) and at an EN AC-48000/AZ80-US-FSW-joint (b) [4]

Both images show a high weld quality and only one irregularity over the weld length, which could be attributed to the mold of the sheets. Additionally several pores in the aluminum base material could be detected. For US-FSW the more intensive stirring of the material in the joining area is clearly recognizable.

To attain more detailed results about corrosive processes, especially in the nugget area, spatially dissolved corrosion tests were carried out by means of Scanning Kelvin Probe microscopy. Figure 10 shows a comparison for the electrochemical potential difference for FSW- and US-FSW-joints.

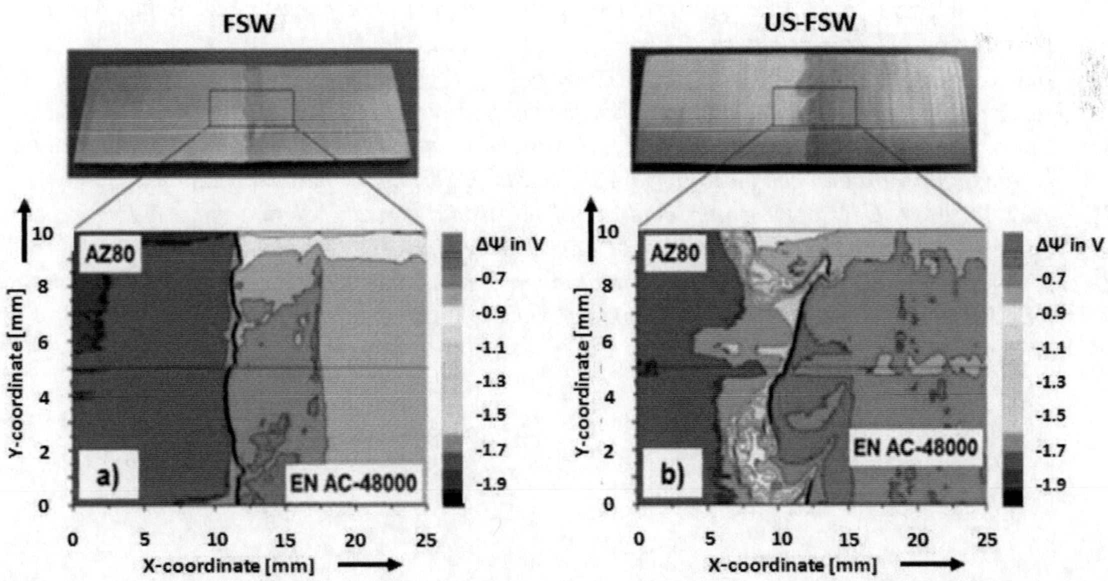


**Figure 10.** Comparative Scanning Kelvin Probe measurements of a) EN AC-48000/AZ80-FSW-joint and b) EN AC-48000/AZ80-US-FSW-joint [8]

Generally clear differences in corrosion behavior for the welding zone and the two base materials could be detected. In comparison to conventional FSW-joints an increase of the potential difference in ultrasound-supported joints can be observed, especially for the EN AC-48000-alloy and the nugget.

Furthermore a considerably higher gradation for US-FSW-joints can be observed in the welding zone and at its transition to the magnesium alloy.

## 4. Summary

The investigations have shown that ultrasound support during Friction Stir Welding positively influences the resulting microstructure and the achievable mechanical properties. The tensile strength increases up to 25 % and for fatigue the number of cycles to failure is three times higher for the same stress amplitude. Furthermore non-destructive testing methods have proven to be suitable to characterize joints. The important role of corrosion for joints made of dissimilar metals was presented by Scanning Kelvin Probe measurements, which have shown that the corrosive behavior of welded EN AC-48000/AZ80 sheets is determined by the microstructure of the developing nugget.

**Acknowledgement**

We would like to thank the "German Research Foundation (DFG)" for the support of the present work as a part of the Priority Program 1640 "Joining by Plastic Deformation".

## Literature

[1] Kainer K U and von Buch F, 1999. *Moderne Entwicklungen von Legierungen für den Leichtbau. Materialwissenschaft und Werkstofftechnik 30*, 159–167.

[2] Klag O, Gröbner J, Wagner G, Schmid-Fetzer R and Eifler D, 2013 *Microstructural and thermodynamic investigations on friction stir welded Mg/Al-joints. International Journal of Materials Research 105*. 145–155.

[3] Wodara J, 2004. *Ultraschallfügen und –trennen. DVS-Verlag GmbH. Düsseldorf.*

[4] Strass B, 2015. *Ultraschallunterstütztes Rührreibschweißen von Al/Mg-Verbunden: Prozessanalyse, mikrostruktureller, mechanischer und korrosiver Eigenschaftsvergleich. Dissertation TU Kaiserslautern.*

[5] Balle F, Wagner G and Eifler D, 2007. *Ultrasonic spot welding of aluminum sheet/carbon fiber reinforced polymer – joints. Materialwissenschaft und Werkstofftechnik 38*. 934–938.

[6] Liu X, Wu C and Padhy G K, 2015. *Characterization of plastic deformation and material flow in ultrasonic vibration enhanced friction stir welding. Scripta Materialia 102*. 95–98.

[7] Park K, 2009. *Development and analysis of ultrasonic assisted friction stir welding process. Dissertation Department for Mechanical Engineering University of Michigan.*

[8] N N, 2012–2015. *Persönliche Mitteilungen und experimentelle Untersuchungen des DECHEMA Forschungsinstitutes, Frankfurt/Main.*

IOP Conf. Series: Materials Science and Engineering **118** (2016) 012015 doi:10.1088/1757-899X/118/1/012015

# New approach to design of ceramic/polymer material compounds

**A Todt[1], D Nestler[1], Trautmann M[1] and G Wagner[1]**

[1] Chair of Composite Materials, Institute of Materials Science and Engineering, Technische Universitaet Chemnitz, 09125 Chemnitz, Germany

andreas.todt@mb.tu-chemnitz.de

**Abstract.** The damage tolerance of carbon fibre-reinforced ceramic-matrix composite materials depends on their porosity and can be rather significant. Complex structures are difficult to produce. The integration of simple geometric structures of ceramic-matrix composite materials in complex polymer-based hybrid structures is a possible approach of realising those structures. These hybrid material compounds, produced in a cost-efficient way, combine the different advantages of the individual components in one hybrid material compound. In addition the individual parts can be designed to fit a specific application and the resulting forces. All these different advantages result in a significant reduction of not only the production costs and the production time, but also opens up new areas of application, such as the large-scale production of wear-resistant and chemically inert, energy dampening components for reactors or in areas of medicine. The low wettability of the ceramic component however is a disadvantage of this approach. During the course of this contribution, different C/C composite materials with a specific porosity were produced, while adjusting the resin/hardening agent-ratio, as well as the processing parameters. After the production, different penetration tests were conducted with a polymer component. The final part of the article is comprised of the microstructural analysis and the explanation of the mechanical relationships.

## 1. Introduction

Carbon fibre-reinforced Carbon Composites (C/C) show great mechanical dampening behaviour, as is the case for ethylene vinyl acetate. This opens up new areas of application for these composites. Examples are energy absorbers for mechanically oscillating or impact related tensions and loads, for example areas of personal protection as a shielding component for projectiles or in an industrial environment as a dampening aid for oscillating masses [1, 2].

The ceramic based compounds currently utilised are characterised by a low energy absorption capacity in relation to their weight and an elaborate production. As a result of this, a new material concept of these compounds is required. There are already a high number of concepts based on metal components e.g. hybrid laminates. These compounds can be composed of one or preferably of multiple energy dissipating components, depending on the place of application and the type of stress placed upon the parts [1–3].

The Improvement of ceramic and polymeric composite materials to create hybrid material compounds enables the utilisation of the beneficial energy dissipating properties of the individual components in an integrated material system. These new materials open up new areas of application, for example as wear and tear resistant and chemically inert energy absorbing parts for reactors or in areas of medical engineering [1, 4].

In order to improve the penetration properties, the surface structure has to be enhanced. During this study, various C/C composites were produced with different resin/hardening agent ratios and the microstructural and mechanical properties were analysed.
The investigations was preferably performed by means of 3-point bending tests and the dynamic mechanic thermal analyses (DMTA). Finally, additional penetration tests were conducted on selected polymers. These were carried out in order to give an outlook on further developments regarding hybrid ceramic/polymeric material compounds.

## 2. Experimental

### *2.1. Production of the carbon fibre reinforced composites*

The C/C composites were produced by means of a multi-levelled process. During the first stage a multi-layered fibre/plastic composite (VF ≈ 60 Vol.-%), consisting of twelve layers of fibre fabrics and a phenolic Novolake resin (pre-polymer) each, was created by hand laminating. The plain weave carbon fibre fabric type Aero (style 447), 160 g/m² consists of warp threads and wefts of fibres of the type Tenax® HTA40 (200 tex), 3 k.
The phenolic novolac pre-polymer Prefere 943990 of Dynea (short: PN 943990) underwent a preliminary examination. During the analysis the material showed indicators of the formation of a porous foam matrix carbon. In the carbon fibre-reinforced plastic (CFRP) state, the polymer shows a low amount of pore-development, while the cross-linking in an autoclave at a pressure of > 10 bar can be characterised as medium. The chemical cross-linked agent chosen for the phenolic novolac pre-polymer is called 1,3,5,7- Tetraazatrizyclo[3,3,1,13,7]decane (colloquial: Urotropine). This hardener was selected due to the excellent development of open pores and was added to the PN 943990 in various weight amounts.
After the production of the laminates, they underwent a hardening process in an autoclave (type 4575, Parr Instruments). The temperature for the curing process was determined empirically and is calculated following the formula: P {T(t) [°C]| dT(t) [K/min]| t [min]}i=1 to 3; [T∞]. The empiric set of parameters was determined to be the following: T0 = 20 °C; {80|1|60}; {120|1|60}; {160|1|60}; T∞ = T0. The pressure parameters and resin/hardening agent ratios are shown in Table 1.

**Table 1.** Test matrix: curing of the CFRP-composites

| No. | resin content [wt%] | Hardening agent content [wt%] | pressure [$10^5$ Pa] |
|---|---|---|---|
| 1 | 96 | 4 | 10 |
| 2 | 96 | 4 | 30 |
| 3 | 88 | 12 | 10 |
| 4 | 88 | 12 | 30 |

In order to enable a complete curing reaction, the samples underwent a hardening and tempering process in a furnace at 320 °C for 4 hours and subsequently cooled freely. Prior to the pyrolysis a separation tool of the type Discotom-6 (Struers) was instructed to cut the CFRP samples. The individual sizes of the samples are discussed further in the chapter about the characterisation.
The second stage of the C/C-composite production consisted of a pyrolysis of the CFRP samples in an argon-filled tube furnace, type CTF 12/65 (Ströhlein). The whole process and program of the pyrolysis was determined and optimised during prior tests. During these tests the thermogravimetric mass degradation of matrix polymers and the thickness of the samples were examined. The whole process was separated in three temperature-time-regimes over 40 h at 300 °C and 1000 °C. Further details of the process will not be disclosed.

### *2.2. Process supporting examinations*
The open porosity of the C/C and CFRP composites was determined according to DIN EN 1389 version B [5].

### *2.3. Examination of the mechanical characteristics of the C/C composites*
The characterisation of the mechanical properties of the C/C-samples was carried out by a three point bending test (Fa. Kammrath & Weiss) according to DIN EN 658-3. The distance between the supporting pins was 40 mm and 9 mm between the loading pins. The testing speed was set to 2 µm/s. Five samples of each sample series were tested [6]. In addition, the mechanical dampening characteristics of the C/C samples were determined by utilising a dynamic mechanical thermal analysis (DMTA) according to DIN EN ISO 6721-1 [7].

### *2.4. Wetting testing of the C/C composites with ethylene vinyl acetate*
The EVA (Ethylenvinylacetat) shows an excellent specific mechanical dissipation factor [8]. For further tests the C/C samples were adjusted for the characterisation methods regarding their size and wetted with EVA. Prior tests showed EVA Greenflex ML 60 (ENI polimeri europa) to be suitable, as the rheological properties are advantageous. The polymer was applied to the C/C composites by means of injection moulding.
The injection pressure was set to 750 bar and the holding pressure to 200 bar. The maximum feed stock temperature was set to 190 °C, while the lowest temperature of the tool was 20 °C. Subsequently the hybrid material compounds were demoulded.

### *2.5. Analysis of the microstructure*
All of the samples were analysed by characterising the cross sections of the C/C composites and the EVA/(C/C) compound materials by utilising an incident light microscope GX51 (Olympus), a camera (Dig 3300) and the image analysis system Stream Motion 1.7.

## 3. Results

### *3.1. Results of microstructure analysis*

#### *3.1.1. Microstructures of modified CFRP composites*
Figure 1 shows the light-microscopic bright field images of the cross sections of the CFRP samples, which were produced from PN 943990, with 4 and 12 % of a hardening agent (HTMA) and varying the hardening pressure from 10 to 30 bar. The samples with less hardener content of 4 % show far more cracks when treated with 10 bar instead of 30 bar ((a) and (b)).
When comparing the samples (a) and (c), which only differ in the content of the hardening agent (4–12 %), one can see that the increase of HTMA content leads to the development of cracks and even the formation of extensive delaminations. A higher pressure during the hardening process decreases the risk of similar structural defects (as can be seen in the comparison of (a) and (b) or (c) and (d)).

IOP Conf. Series: Materials Science and Engineering **118** (2016) 012015 doi:10.1088/1757-899X/118/1/012015

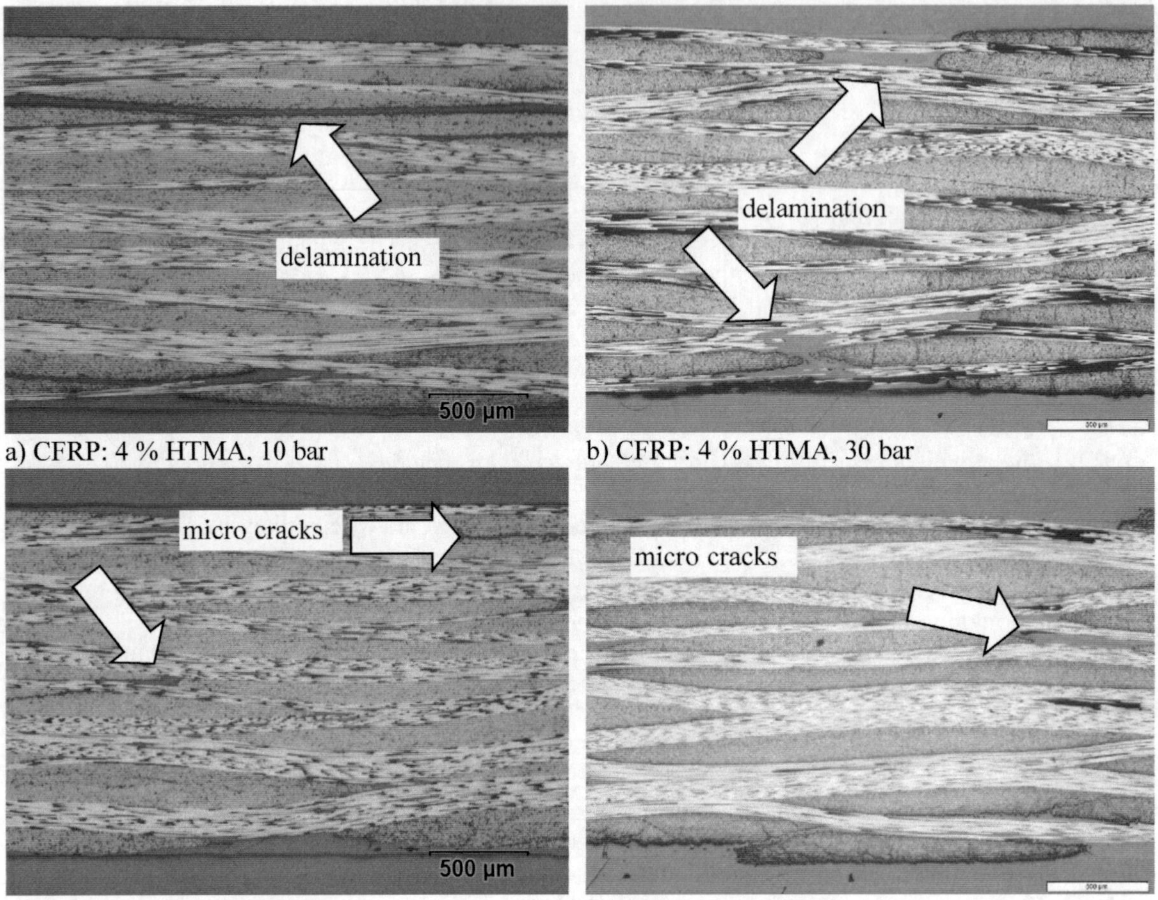


a) CFRP: 4 % HTMA, 10 bar
b) CFRP: 4 % HTMA, 30 bar
c) CFRP: 12 % HTMA, 10 bar
d) CFRP: 12 % HTMA, 30 bar

**Figure 1.** Light microscopic bright field images of the CFRP composites with a hardening agent content of 4 and 12 % HTMA, while varying the hardening pressure from 10 to 30 bar.

### *3.1.2. Microstructures of C/C composites*

Figure 2 shows the light microscopic bright field images of the cross sections of the C/C samples, which were produced from PN 943990, with 4–12 % HTMA and a varying pressure of 10 to 30 bar. The samples with a HTMA content of 4 % and a hardening pressure of 10 bar display more delaminations compared to the samples, which were pressurised under 30 bar ((c) and (d)). The comparison of the samples (a) and (c), which only differ in the increased HTMA content from 4 to 12 %, leads to the same result as it did with the CFRP composites, namely the samples with less hardening agent develop more structural defects than the ones with the lower HTMA content. The samples hardened under a pressure of 30 bar and 12 % HTMA show segmentation cracks (Figure 2c and d). Increasing the HTMA content to 12 % only leads to a decrease of the defects already created in the CFRP state ((c) and (d)).

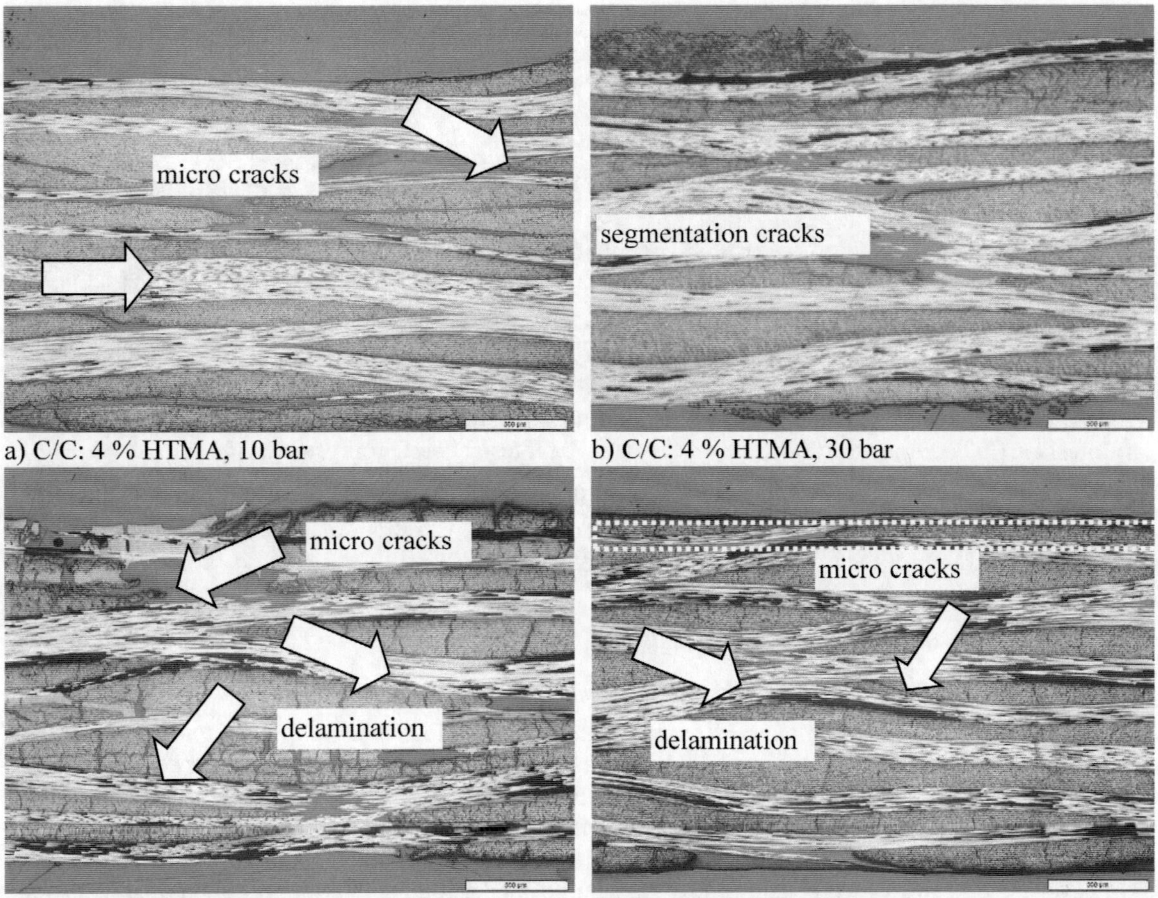


a) C/C: 4 % HTMA, 10 bar
b) C/C: 4 % HTMA, 30 bar
c) C/C: 12 % HTMA, 10 bar
d) C/C: 12 % HTMA, 30 bar

**Figure 2.** Light microscopic bright field images of the C/C composites with a hardening content of 4 and 12 % HTMA, while varying the hardening pressure from 10 to 30.

### *3.1.3. Microstructures of EVA/(C/C) Interface*

Due to the adequately high flexural strength values and above-average mechanical dampening characteristics C/C composites with a hardening content of 12 % and a previously applied pressure of 30 bar, were chosen for wetting tests of the EVA-polymer. A wide interface between the C/C composite and the Polymer can be seen in Figure 3. The image reveals the fact that the segmentation cracks are filled up completely.

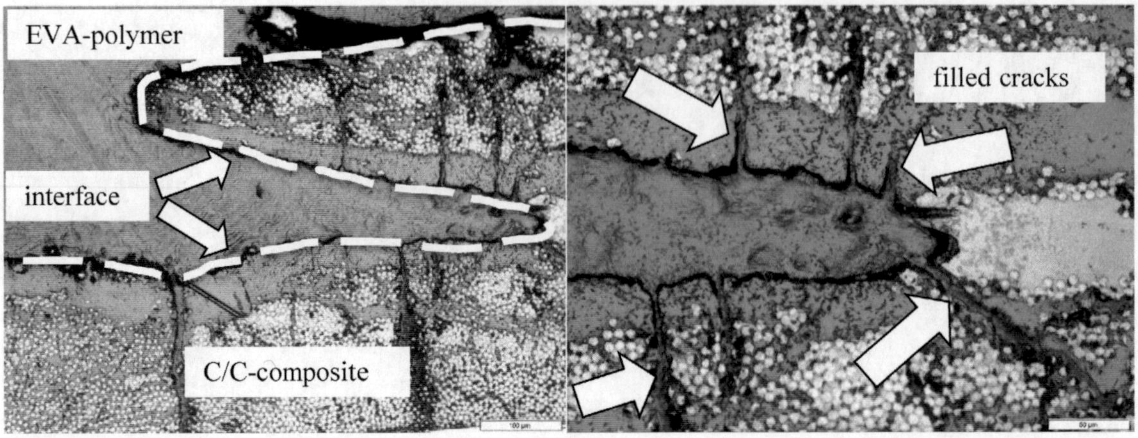


a) Interface C/C composite 100 X b) Interface C/C composite 50 X

**Figure 3.** Light microscopic bright field images of the C/C composites with a hardening content of 12 % HTMA and a hardening pressure of 30 bar. Complete infiltration of the EVA polymer in the segmentation cracks of the C/C composite at the interface.

*3.2. Results of Results of the mechanical tests*

*3.2.1. Flexural strength of modified C/C composites*

Figure 4 displays the average flexural strength values of the various C/C composites and the illustration of the fractures. The images show the development of unilateral shear defects while applying a pressure of 30 bar, whereas a hardening pressure of 10 bar only clarifies a shear fracture on both sides. As a result of these observations the hardening agent content does not seem to have a significant impact. The samples produced with 30 bar yield the higher flexural strength values compared to the other ones. However the best results were achieved by the samples undergoing a pressure of 30 bar and consisting of only 12 % hardening agent. These special samples indicate the generation of micro cracks as a result of the shear force; however these defects are not to be found within the entire sample volume.

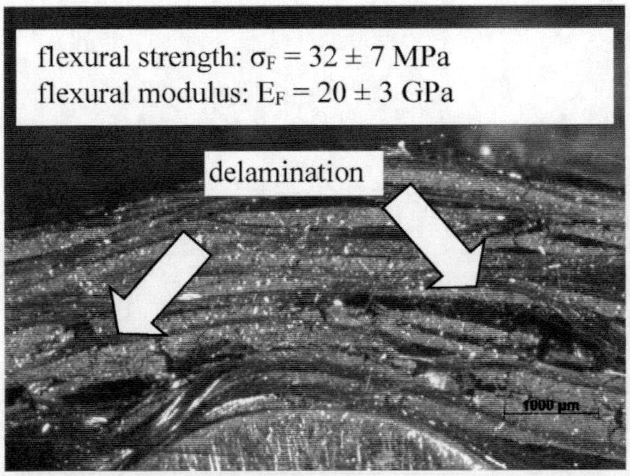


a) C/C: 4 % HTMA, 10 bar

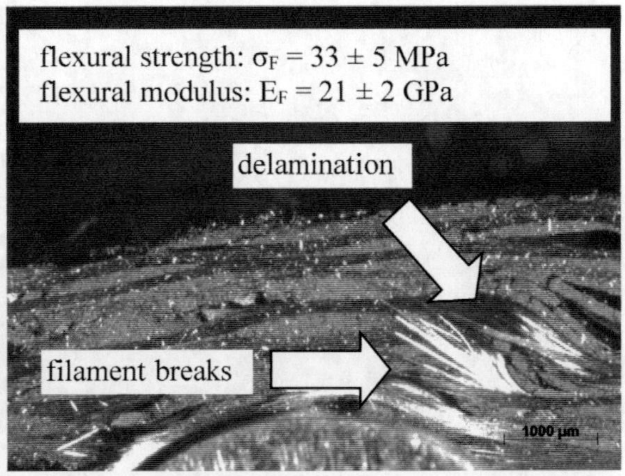


b) C/C: 4 % HTMA, 30 bar

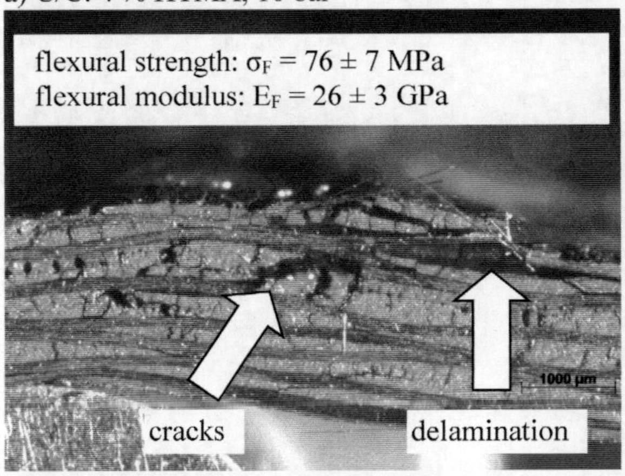


c) C/C: 12 % HTMA, 30 bar

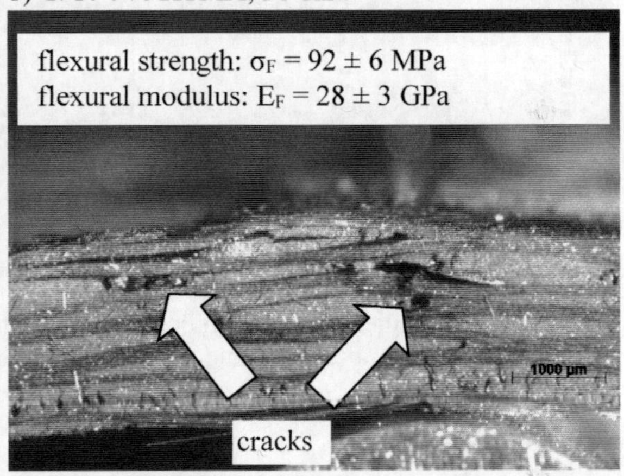


d) C/C: 12 % HTMA, 30 bar

**Figure 4.** Light microscopic in situ images of the C/C composites at the moment the maximum flexural strength is reached. The cracks and delaminations (marked by arrow) developing as a result of the shear force can be seen.

### *3.2.2. Mechanical damping behaviour of modified C/C-composites*

The results shown in Figure 5 display the development of a low open porosity and low mechanical dampening characteristics of the samples with a low hardening agent content undergoing a treatment with a high hardening pressure. The mechanical dampening behaviour is correlating with the behaviour of the open porosity at a hardening content of 4 %. Nevertheless, no significant influence can be shown. However, the samples with the lower content of the hardening agent, namely 4 %, display a significant impact/effect of a change in the pressure exerted upon them on the actual open porosity.

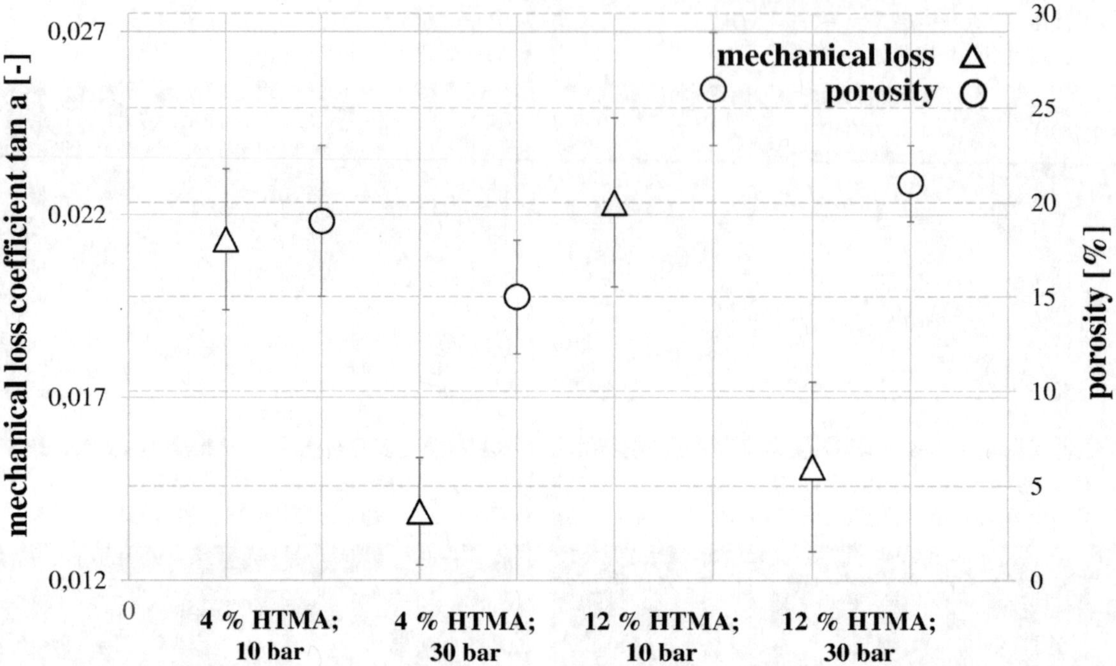


**Figure 5.** Mechanical dampening behaviour in depending of open porosity of the C/C composites; effect of open porosity on the mechanical loss coefficient.

## 4. Discussion

With the constant hardening pressure of 10 bar the samples consisting of 4 % hardening agent showed the highest number of structural defects, such as cracks and delaminations. These are further promoted by the intrinsic gas emission during the cross linking process. Once the pressure is raised to 30 bar, these defects, which are caused by the escaping reaction and decomposition gases, can only be found in the samples with 4 % HTMA. The characteristic C/C structural morphology forms within the sample containing 12 % of the hardening agent. This fact indicates the existence of a sufficient amount of adhesion between the fibre and the matrix to create reproducible mechanical properties. Due to the production of gases during cross linking reactions, the development of a specific pore-structure is presumable. However, the simultaneous creation of high cohesion forces within the amorphous carbon matrix prevents any delamination processes between the fibre bundles and the matrix. This explains the even development of segmentation cracks, which is a result of the fibres preventing the matrix from any shrinkage. The pre-existing structural defects of the CFRP composites significantly influence the system of the cracks and any delaminations during the pyrolysis. To investigate this further on, additional analyses and examinations of the pore structure are necessary.

The examination of the achieved flexural strength values indicates a connection to the macro structure. Samples containing prior defects displayed lower mechanical strength than their flawless counterparts. During the examination of this connection, an influence of the open porosity, which is accompanied by defects, on the processes was discovered. However, compared to pre-existing data [8] high mechanical dampening values were discovered, which showed an inverse development to the mechanical strength values. The elastic modulus reached the range of thermosetting SMC composite materials, whereas the mechanical dampening equalled the characteristic value of thermoplastic polymers [8]. The infiltration of indentations on the outer boundary surface indicates the possibility of a mechanical adhesion to the C/C composite for the production of hybrid ceramic/polymer material compounds.

## 5. Summary

In this investigation, porous C/C composites were created by utilising a special carbon precursor. The concentration of the hardening agent was varied between 4 and 12 %. During the cross linking reaction a pressure of 10 and 30 bar was applied. These processes were followed by a microstructural characterisation, which was performed by analysing the samples with a light microscope. The C/C composites were further investigated by means of a 3-point bending test and by performing DMTA measurements. Parallel to this process the open porosity was determined. A direct link between the open porosity and the mechanical dampening behaviour was found. The flexural strength correlates with the individually observed structural development during the pyrolysis. In summary the porosity and also the possibility of the development of structural defects increases depending on the concentration of the hardening agent, resulting in a reduction of the flexural strength. The development of a network of cracks suitable for a C/C composite with good mechanical characteristic values and an excellent mechanical dampening is realised with a hardening agent content of 12 % and a pressure of 30 bar. The wetting tests show the possibility of a PUR-infiltration of the outer boundary surfaces of C/C composites with a characteristic open porosity.

## 6. Outlook

The results show the potential of a targeted manipulation of the mechanical dampening behaviour by adjusting the parameters of the wetting process. Further studies and examinations regarding the pore development behaviour of CFRP composites and links to different parameters are necessary. The same thing can be said for the pyrolysis and the parameters attached to that process. Subsequently to these steps the properties of the fibre/matrix interface need to be characterised. In further experiments, the polymer/(C/C) interface is examined in more detail, in order to enable the development of hybrid material compounds from polymer and integrated C/C-based dampening components/elements. Due to the results, which indicate the possibility of a production of selected hybrid ceramic/polymer material compounds, processes and methods of a large-scale production of thermoplastic EVA-like or thermoelastic agents seem to be possible and feasible.

## 7. References

[1] Todt A, Nestler D, Roder, K, Wielage B, Kroll L, Spange St and Tröltzsch J, DE Patent 10,2014,200,510,A1. (2014).

[2] Kern GmbH, Datasheet, *Ethylen/Vinylacetat* (E/VA).

[3] Mucha H, *Untersuchungen zur Porositätsentwicklung von Phenolharzen als polymere-und Kohlenstoffspendermatrices in C-Faserverbundwerkstoffen*, Dissertation, TU Chemnitz, 2002.

[4] Wielage B, Weber D, Müller T, Steger H, *Thermomechanical Monitoring of Composite Materials during the Pyrolysis of C/C Composites*, in: Z. Key (Ed.), Engineering Materials **425** (2010), 95–105.

[5] DIN EN 1389:2004-03, *Hochleistungskeramik – Keramische Verbundwerkstoffe – Physikalische Eigenschaften - Bestimmung der Dichte und der scheinbaren Porosität.* (2004).

[6] DIN EN 658-3, *Advanced technical ceramics - Mechanical properties of ceramic composites at room temperature - Part 3: Determination of flexural strength.* (2002).

[7] DIN EN ISO 6721-1:2003-01, *Kunststoffe - Bestimmung dynamisch-mechanischer Eigenschaften - Teil 1: Allgemeine Grundlagen.* (2003).

[8] CES Edu Pack 2009, *Mechanical loss coefficient*, (2009).

**Acknowledgments**

This work was performed within the Federal Cluster of Excellence EXC 1075 "MERGE Technologies for Multifunctional Lightweight Structures" and supported by the German Research Foundation (DFG). Financial support is gratefully acknowledged. The authors thank Mr. Alexander Einenkel and Mr. Patrick Jahn, who produced the joints and Mrs. Heidrun Böhme, who made the bending tests and measurements of porosity. We thank Mr. Steilen for the translation into English.

IOP Conf. Series: Materials Science and Engineering **118** (2016) 012016 doi:10.1088/1757-899X/118/1/012016

# Microstructural characterisation of interfaces in magnetic pulse welded aluminum/aluminum joints

**S Sharafiev[1], C Pabst[2], M.F.-X Wagner[1] and P Groche[2]**

[1] Chair of Materials Science, Institute of Materials Science and Engineering, Technische Universität Chemnitz, D-09125 Chemnitz, Germany
[2] Institute for Production Engineering and Forming, Technische Universität Darmstadt, D-64287 Dramstadt, Germany

**Abstract.** Electromagnetic pulse welding provides novel and useful solutions in the field of multi-material structures. The joining occurs through the high-energy mechanical impact of metal sheets, which leads to a strong, cohesive, metallic bonding between the sheets. The high-velocity collision results in changes of the microstructure in the regions adjacent to the interface. The aim of this work is the microstructural characterization of the interface and the welding zone of Al/Al joints. One typical interface is analysed by optical and electron microscopy and using the electron backscatter diffraction technique. Our results show that the interface exhibits a well-known type of wavy weld geometry. Moreover, the formation of an intermediate layer that is different from the bulk aluminium structure is observed. The microstructure of this intermediate layer consists of ultrafine grains and in some regions exhibits a columnar grain structure. It is likely that this special microstructure (which plays an important role in determining the cohesive, continuous bonding between the metal sheets) is formed through a rapid melting and crystallisation process in the interface region.

## 1. Introduction

Electromagnetic pulse welding (EMPW) is a widely used technique to produce joints between metal sheets by plastic deformation. The joining is achieved by the collision of two partners after acceleration of the dynamic partner (flyer) towards the other, static partner (target). The bonding processes take place at very high strain rates up to $10^6$ $s^{-1}$. EMPW is increasingly used in industrial applications, even though there exists only limited knowledge of the bonding mechanisms. In most scientific studies, the characterisation of welding zones is limited to a macroscopic evaluation along the weld. For a more detailed understanding of EMPW it is necessary to analyze the microstructural changes in the interfacial region, where continuous metallic bonding leads to the adhesion between the metal joint partners. This may in turn allow to define useful processing regimes, such as impact velocities and angles that lead to proper joining along larger welding zones.

## 2. Experimental

In the present study the microstructure in an interface between two magnetic pulse welded pure aluminum sheets (EN AW-1050) was investigated. The sheet thickness was 2 mm. The EMPW process was carried out with an initial distance between plates of 2 mm. The electrical current ranged between 322 – 397 kA at 18.9 kHz. Further information on the industrial EMPW setup used for joining is presented in our previous work [1]. The structural changes in the welding zone were observed by optical (OM), scanning electron (SEM) and transmission electron microscopy (TEM) with selected area electron diffraction (SAED). In addition, electron backscatter diffraction (EBSD) measurements on the interface were carried out in the SEM. The samples were ground and polished for OM. Ion beam thinning and polishing were applied for the observations by TEM. A vibration polishing procedure was applied for the SEM and EBSD measurements.

## 3. Results and discussion

Focusing on the characterisation of the interfacial changes, which lead to the cohesive bonding between two aluminium sheets, only regions within the seam with wave formation and an intermediate layer are discussed here. For microscopic observation, the joints were cut along the cross-section in the weld centre, so that the cutting plane is perpendicular to the sheet plane and parallel to the direction of weld propagation. In this cross section the interface exhibited a typical wavy morphology, as can be seen in optical overview micrograph in figure 1. It should be noted that we do not address questions related to wave formation (such as wave length as a function of different impact conditions) in this short paper. This topic of wave formation during impact welding has been extensively discussed in other studies such as, for instance, in [2]. One important and general observation when studying EMPW-produced interfaces of aluminum materials is that even between similar aluminium/aluminium joints, some special microstructure can be always found in the weld zone. This interfacial microstructure is typically referred to as "intermediate layer", see e.g. [3]. The microstructure of this layer differs from that of the initial bulk material. Its thickness is about 5 – 15 µm. We found that this intermediate layer was unstable against the etching procedure typically used for aluminum alloys. Assuming that the so-called jet clears the sheet surface during the welding process [4] it is unlikely that the intermediate layer consists of surface aluminium and/or oxide particles. Instead, we propose that the changed microstructure is formed in the near-interface region during welding process and that it is characterized by a higher defect density, which makes it prone to removal during conventional etching.

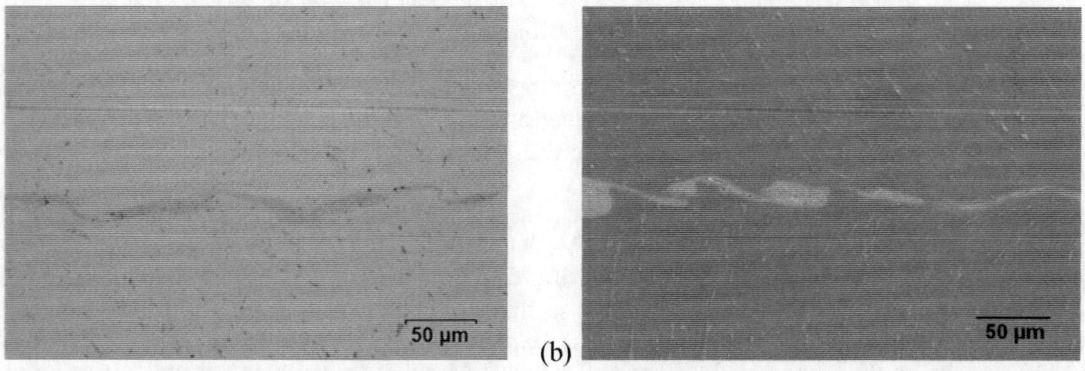


(a) (b)

**Figure 1.** Overview of the interface of an aluminum/aluminum joint produced by EMPW. The interface exhibits a characteristic wavy form and the intermediate layer differs from the initial bulk material: (a) OM image, (b) SEM micrograph.

For a detailed observation of the grain structure in the intermediate layer, the EBSD measurements were carried out. It should be noted that the preparation procedure for EBSD analysis needs to provide pristine surfaces, i.e., it requires a careful and time-consuming procedure to make sure that particularly the unstable microstructures near the weld zone are not affected. In figure 2, we present a representative SEM micrograph, and a corresponding EBSD image, of the welding zone. Compared to the initial grain structure, the much finer grain structure in the intermediate layer can be clearly seen in the EBSD image. Furthermore, mainly on the right part of the EBSD image, the grains exhibit a columnar shape. These columnar grains are elongated in the direction of the acceleration vector during EMPW. This specific feature of the weld zone region can be compared with the structures of grain growth in the presence of a strong temperature gradient. The microstructural features documented in figure 2 indicate that the material may well have been melted for a very brief period of time during impact welding: this local melting in a very thin surface region of the aluminum parts can result from the high impact energies during EMPW. This energy is expended on the deformation in the initial bulk material and (because of the very high strain rates) leads to near-adiabatic heating. Subsequent rapid crystallisation due to the high temperature gradient between the surface layer and the bulk material then leads to the elongated, columnar microstructures. We note that, considering the complex dynamics of impact welding, the tensile pulse directly after impact might also have an effect on the growth process; further work is required to analyse this effect in detail.

**Figure 2.** SEM micrograph and EBSD image of the welding zone; the green rectangle corresponds to the region analysed by EBSD. The intermediate layer exhibits a well-defined, fine-grained microstructure with columnar grains.

The microstructure of the welding zone and the intermediate layer can be studied in even greater detail with a much higher magnification in the TEM. TEM bright field images of the welding zone are shown in figure 3. The seam is oriented in vertical direction. The difference between the intermediate layer and the initial microstructure is evident, so that the intermediate layer boundaries can be very simply emphasized with thin white lines and arrows, as depicted in figure 3(a). The small part of the welding seam studied here is fully bonded. The TEM micrograph again clearly shows that the microstructure of the intermediate layer consists of very fine grains and with a correspondingly high density of grain boundaries. In figure 3(b), another position along the weld is shown with higher magnification. Only one side of the intermediate layer is shown, and a sharp boundary can be observed between the intermediate layer and the region of initial material (associated with a much coarser grain size). Within the layer, the columnar grains seem to be grown from the boundary in the interface direction (white arrows indicate the potential growth direction in figure 3(b)).

(a) (b)
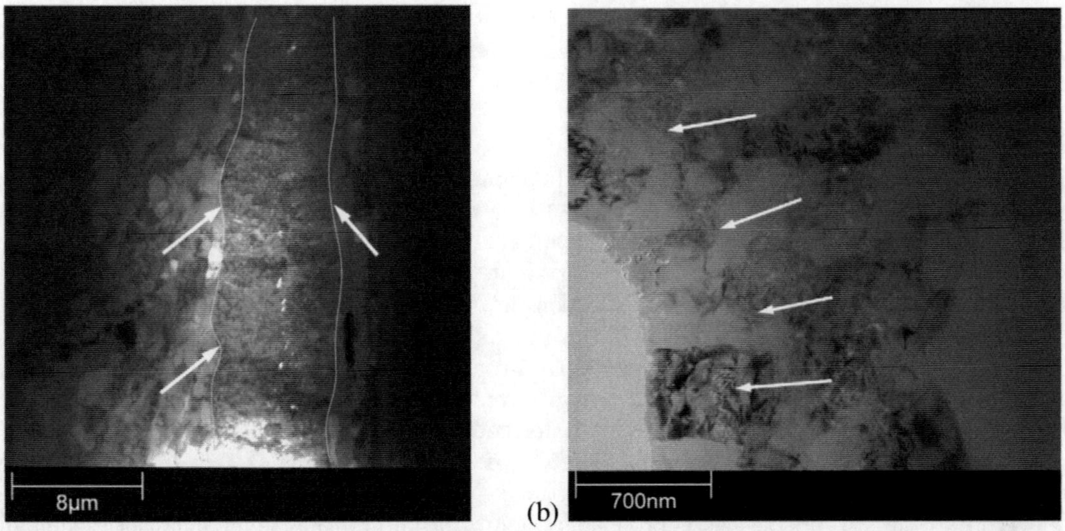


**Figure 3.** TEM bright field images of the intermediate layer. (a) The layer is highlighted by white lines and arrows. It exhibits a fine grain structure differing from the initial bulk grain structure. (b) One side of the intermediate layer (left) with grains in columnar form. The white arrows indicate the growth direction away from the coarse grained bulk material (right).

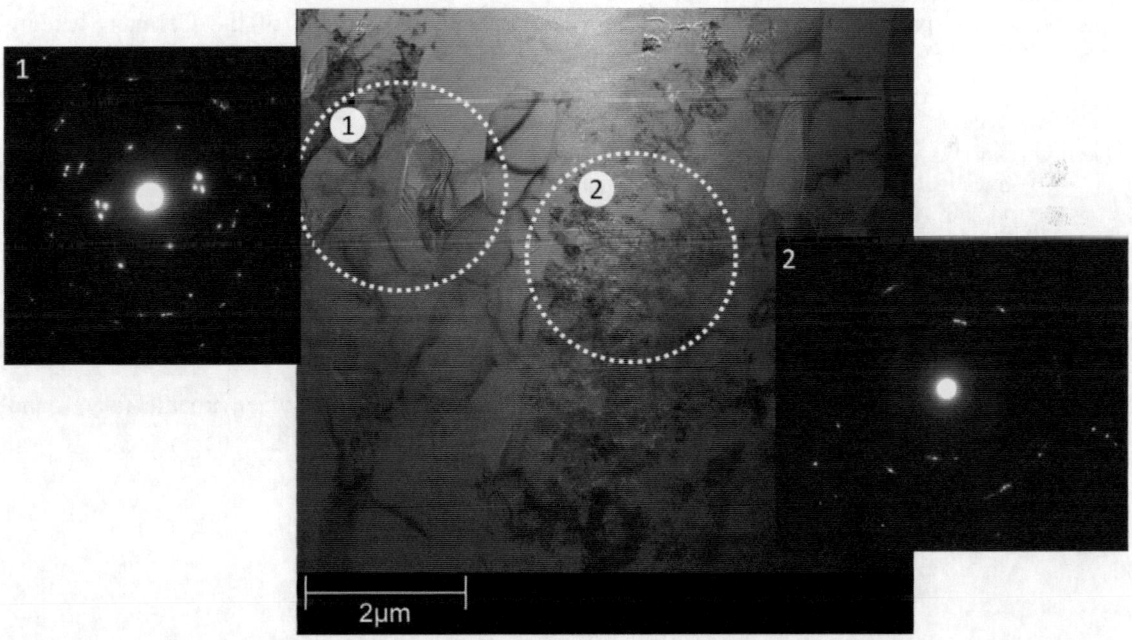


**Figure 4.** Bright field TEM image of the welding zone and SAED pattern of (1) thecoarse-grained near-interface region, and (2) the fine-grained intermediate layer. The low intensity of diffraction rings in (2) is due to the low thickness of the sample.

SAED patterns were recorded near the interface and inside the intermediate layer itself. Figure 4 shows the bright field TEM image; the regions where SAED patterns were recorded are highlighted by circles and the corresponding SAED patterns are shown on the left and right sides of figure 4, respectively. The first region (near-interface region) contains only a few coarse grains from the initial microstructure. The SAED pattern therefore only exhibits some individual diffraction spots. In

IOP Conf. Series: Materials Science and Engineering **118** (2016) 012016 doi:10.1088/1757-899X/118/1/012016

contrast, the SAED image of the intermediate layer is characterized by blurred diffraction spots and by concentric rings. These features clearly indicate the fine-grained microstructure in the intermediate layer.

## 4. Summary and conclusions

The interfacial microstructure of the aluminum/aluminum joints fabricated by magnetic pulse welding was characterised using optical and electron microscopy, and in particular by EBSD (in the SEM) and SAED (in the TEM). While this brief report only highlights typical results from our on-going microstructural investigations of interface regions in joints produced by EMPW, we highlight that the EBSD measurements and the TEM investigations are in good agreement; they clearly demonstrate that an intermediate layer is formed during EMPW, and that this layer is characterized by microstructural features (fine, and partly columnar grains) that strongly differ from the initial bulk material that can also be observed in the vicinity of the newly formed weld zones. Our microstructural observations further indicate that the intermediate layer is formed by local melting and rapid crystallisation of the near-interface regions. This process is closely related to the high impact energies during EMPW, which are concentrated on a small material volume and may hence lead to a considerable increase in temperature. This, in turn, results in pronounced temperature gradients in the aluminum sheets that are likely to lead to columnar grain growth in the thin intermediate layer during subsequent, rapid crystallization. Obviously, the local behaviour of the intermediate layer plays an important role in providing good adhesive bonding in the EMPW joints.

**Acknowledgements**

This work was performed in project A8, cooperating with project A5, of the German Research Foundation's (DFG) Priority Programme SPP 1640 "Joining by plastic deformation".

## 5. References

[1] Groche P, Wagner M.F.-X, Pabst C, Sharafiev S 2014 Development of a novel test rig to investigate the fundamentals of impact welding *J. Mater. Process. Technol.* **214** 10 pp 2009-17

[2] Ben-Artzya A, Stern A, Frage N, Shribman V, Sadot O 2010 Wave formation mechanism in magnetic pulse welding *Int. J. Impact Eng.* **37** 4 pp 397–404

[3] Lee K.-J, Kumai S, Arai T, Aizawa T 2007 Interfacial microstructure and strength of steel/aluminum alloy lap joint fabricated by magnetic pressure seam welding *Mater. Sci. Eng.* A **471** pp 95–101

[4] Kapil A, Sharma A 2015 Magnetic pulse welding: an efficient and environmentally friendly multi-material joining technique *J. Cleaner Prod.* **100** 1 pp 35–38

IOP Conf. Series: Materials Science and Engineering **118** (2016) 012017 doi:10.1088/1757-899X/118/1/012017

# On the development of an intrinsic hybrid composite

**R Kießling[1], J Ihlemann[1], M Riemer[2], W.-G Drossel[2], I Scharf[3], T Lampke[3], S Sharafiev[4], M Pouya[4] and M.F.-X Wagner[4]**

[1] Professorship of Solid Mechanics, Faculty of Mechanical Engineering, Technische Universität Chemnitz, D-09126 Chemnitz, Germany
[2] Fraunhofer Institute for Machine Tools and Forming Technology, D-09126 Chemnitz, Germany
[3] Professorship of Materials and Surface Engineering, Institute of Materials Science and Engineering, Technische Universität Chemnitz, D-09125 Chemnitz, Germany
[4] Chair of Materials Science, Institute of Materials Science and Engineering, Technische Universität Chemnitz, D-09125 Chemnitz, Germany

**Abstract.** Hybrid parts, which combine low weight with high strength, are moving into the focus of the automotive industry, due to their high potential for usage in the field of crash-relevant structures. In this contribution, the development of an intrinsic hybrid composite is presented, with a focus on the manufacturing process, complex simulations of the material behaviour and material testing. The hybrid composite is made up of a continuous fibre-reinforced plastic (FRP), in which a metallic insert is integrated. The mechanical behaviour of the individual components is characterised. For material modelling, an approach is pointed out that enables modelling at large strains by directly connected rheological elements. The connection between the FRP and the metallic insert is realised by a combination of form fit and adhesive bonds. On the one hand, adhesive bonds are generated within a sol gel process. On the other hand, local form elements of the metallic insert are pressed into the FRP. We show how these form elements are generated during the macroscopic forming process. In addition, the applied sol gel process is explained. Finally, we consider design concepts for a specimen type for high strain testing of the resulting interfaces.

## 1. Introduction

Hybrid composites are made of different classes of materials to combine the advantageous properties of the single components (see, e.g. [1, 2]). Due to the complex and expensive manufacturing process as well as difficulties in material testing and simulation, the application of these composites, e.g. in the field of crash relevant structures, is currently not considered in commercial production (see [3]). This contribution outlines the development of an intrinsic hybrid composite that can be manufactured robustly and reproducibly. The corresponding manufacturing process, based on the emergence of the compound and the simultaneous generation of the component geometry in a combined forming process, is devised. Furthermore, the material testing of the interfaces and the material modelling of the components of the hybrid composite at large strains are presented.

## 2. Materials of the components and the related material models

The investigated hybrid composite consists of a continuous fibre reinforced plastic (FRP) with an integrated metallic layer as shown in figure 1. The FRP is a unidirectional carbon fibre sheet composed of a PA 6 Ultramid B40 (BASF) matrix and Toray's T700S (50-K) carbon fibres. AA 1005 is employed for the metallic insert.

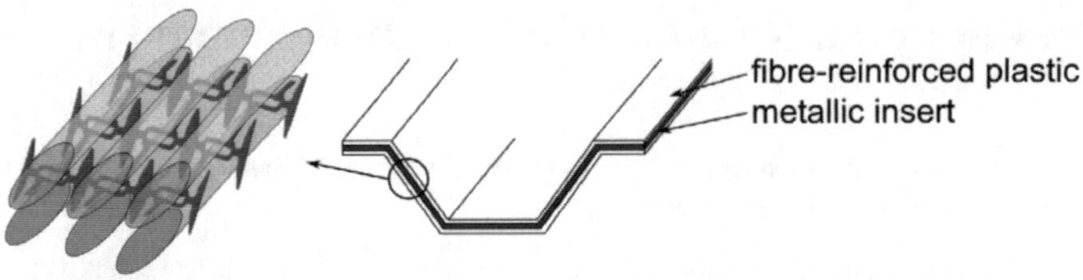


**Figure 1.** Macro and mesostructure of the investigated hybrid composite.

Finite element simulations are performed to estimate and optimise the crash behaviour of a structural part made of the hybrid composite. To simulate the corresponding mechanical behaviour of the deployed components at large strains, precise viscoplastic material models are required. The formulation and implementation of such a model requires a profound knowledge in the field of nonlinear continuum mechanics. Ihlemann [5] developed a general approach that enables material modelling at large strains based on the direct connection of rheological elements. Within this concept, rheological elements, representing elastic, viscous and plastic material behaviour, are combined in series and parallel connections. To deduce the required material model, only relations demanding the stress power equivalence of the rheological connections have to be evaluated. A numerical procedure by Kießling et al. [6] implements directly these connection relations. Consequently, a complex material model can be generated by the composition of a rheological model and the selection of provided rheological elements. Furthermore, it can be employed in commercial finite element codes by user subroutines, like UMAT for Abaqus. Therefore, this straight-forward and flexible approach of material modelling at large strains is deployed to develop the viscoplastic material models of the components of the intrinsic hybrid composite studied here.

The mechanical behaviour of the aluminium alloy is covered by the rheological model shown in figure 2(a). As demonstrated by Kießling et al. [6], the evaluation of the associated connection relations shows accordance with the material model presented by Shutov and Kreißig [4], which is able to simulate the viscoplastic behaviour of this material type. The rheological model, which is applied for the matrix material of the FRP, is given in figure 2(b). Furthermore, an extension of this rheological model, which covers anisotropic effects, is introduced. The resulting homogenised material model can be utilised for the FRP. Consequently, the mechanical behaviour of the applied components can be estimated as shown by the simulated stress-strain curves in figure 3 and utilised in finite element simulations of the intrinsic hybrid composite.

## 3. Connection between the fibre-reinforced plastics and the metallic insert

### *1.1. Form fit by local form elements*

One aim of this project is to identify basic mechanisms to establish a form fit between the laminate layers following the intrinsic concept. A general approach is to utilise in-plane deformations to trigger out-of-plane reactions of the geometric structure.

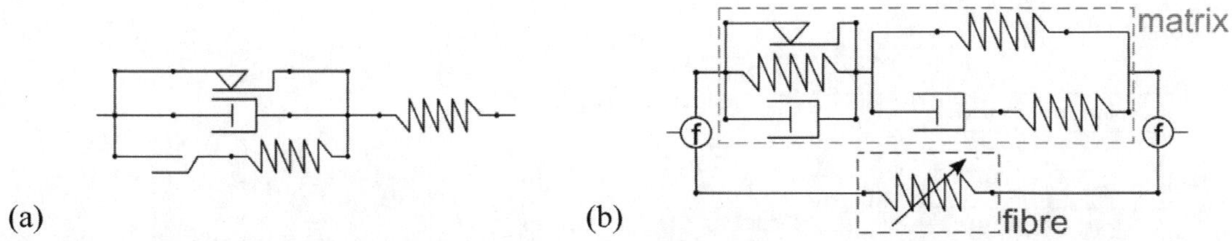


**Figure 2.** Rheological models: (a) aluminium alloy, (b) unidirectional FRP.

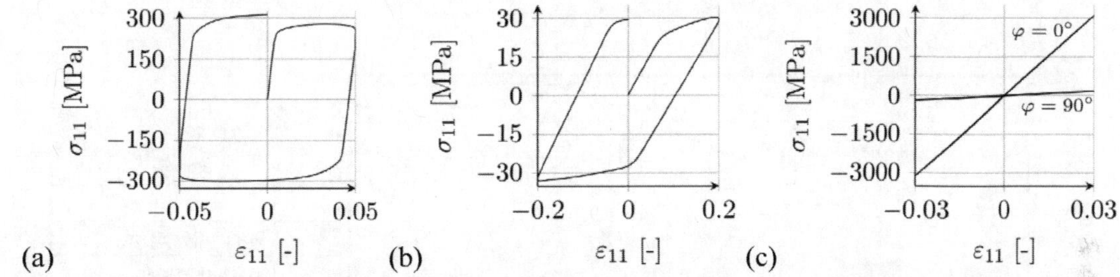


**Figure 3.** Results of isochoric tension tests with a sinusoidal time series (frequency 10 Hz): (a) AA, (b) PA 6, (c) unidirectional FRP with fibre orientation of 0° and 90°.

One promising effect, generating out-of-plane deformations, is local buckling [7]. To generate the required local compression stress, in-plane bending is utilised. By loading a curved beam, tensile forces can be converted into a bending moment. Figure 4(a) shows the basic geometric element with the width w, the length l and the eccentricity e of the applied load.

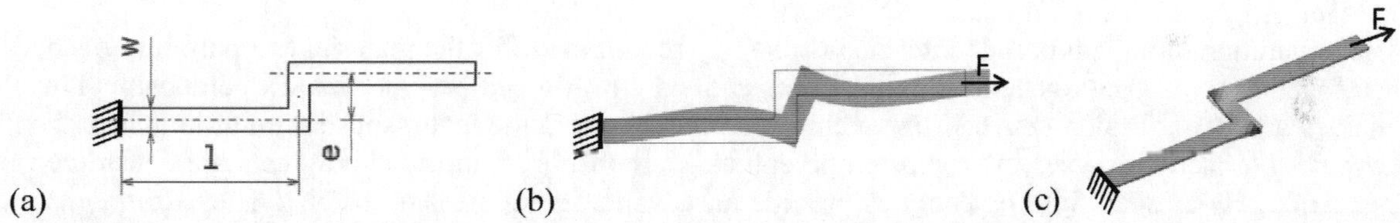


**Figure 4.** Principle for inducing out-of-plane deformations: (a) basic element, (b) in-plane bending, (c) out-of-plane deformation.

The applied force F generates the bending moment and thereby areas of the local compression state (cf. figure 4(b)). When the load exceeds a critical value, the structure suddenly starts to buckle and an out-of-plane deformation is realised (see figure 4(c)). The value of the critical load strongly depends on geometric ratios. Our investigations show that a large value of the parameter w favours the out-of-plane deformation. The length l has no significant influence on the critical load. In contrast, an increased eccentricity e leads to a reduced critical load. Taking the results of the numerical parameter studies into account, a specimen with an array of 4 form fit elements was designed (figure 5(a)). The linear buckling analysis of the specimen shows the out-of-plane deformation of the form fit elements under tensile load (figure 5(b)).

In a next step, physical specimens were manufactured by laser cutting and tensile tests were performed. Figure 6 illustrates a force-displacement curve from a representative specimen. From 0 to 3 mm traverse displacement, the elastic deformation of the specimen is observed. At the first peak of the curve, the critical buckling load is reached and the out-of-plane deformation occurs. This is combined with a significant reduction of the applied force. Beyond an extension of 10 mm, the applied force F continues to rise until fracture. The force which is achieved right before the failure of the form fit element is even slightly higher than the critical buckling load. This is an important relation aiming at a series combination of several elements.

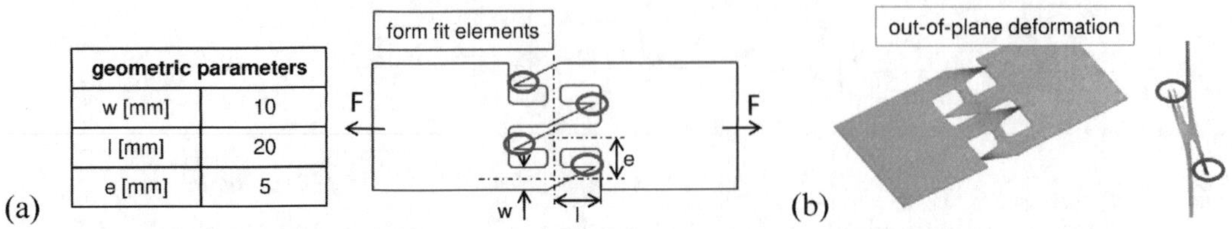


**Figure 5.** Geometry of the investigated specimen: (a) test specimen with applied load, (b) linear buckling analysis.

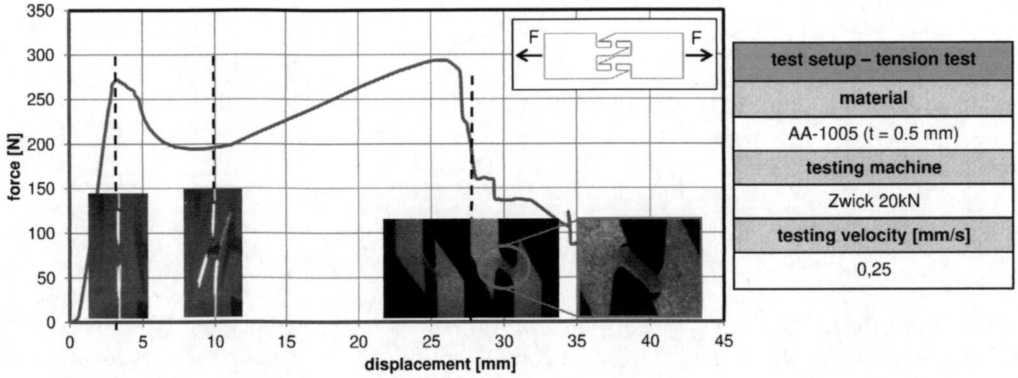


**Figure 6.** Force-displacement curve for tensile test.

*1.2. Sol-gel process generating adhesive bonds*

The generation of an appropriate interface is one key requirement for the material compatibility of the heterogeneous materials system of metal insert and fibre-reinforced polymer matrix composite. This enables a better adhesion between fibre-reinforced polymer and the metal substrate. Furthermore, the adapted interface decouples the corrosion potentials. Carbon fibre reinforced polymers are considered as particularly critical. Sol-Gel coatings provide an advantage as corrosion barrier. They form dense layers and provide good adhesion. The application of sol-gel layers prevents contact corrosion between the noble fibre-reinforced polymer (redox potential of + 0.74 V) and the less noble aluminium substrate (redox potential of - 1.67 V). It is challenging to develop a formulation for a sol-gel coating on aluminium before the forming process: The ductility of the coating has to be sufficient for this subsequent process.

Another function of the sol-gel coating is the optimisation of cohesive interactions between heterogeneous partners in the hybrid component to achieve the best possible connection. Therefore the coating is modified to their chemical surface composition to produce stronger interactions from the selected PA 6 matrix. For example, the properties of the sol-gel layer are modified by using hybrid sole [8], a mixture of chemically different silicates, for their production. This allows changing the wetting behaviour of the interface with the selected chemical composition. Furthermore, it is possible to determine the voltage profile in the layer with strains during the forming process by using structurally different organic silicates, so-called ORMOSILS. The wettability of the fully crosslinked surface of the sol-gel coating can be improved in a further step to achieve maximum intermolecular interactions to the polyamide. For example, alkylated silicates are used to modify the surface in the desired manner [9]. So the sol-gel layer gets a very oleophilic character. This is promoting the connectivity to thermoplastic polymers and adds a maximum material connection of both joining partners in the hybrid composite.

## 4. Mechanical characterisation of the interface

The interfacial properties of an intrinsic hybrid strongly affect the total composite strength. It is necessary to find a general approach for characterisation of the mechanical properties of hybrid structures with different interface concepts, and to understand the relevant strength and damage mechanisms in their interfaces, both under quasi-static and dynamic loading conditions. Typically, mechanical testing of composite structures is performed by conventional shear tests and/or tensile tests normal to the interface. However, information from such tests is often limited to critical forces (or nominal stresses) for complete failure. In order to characterise the microstructural developments and damage processes of metal-plastic interfaces in greater detail, experiments need to be interrupted (preferably at different, well-defined time increments). In our recent experiments, we use the digital image correlation (DIC) method for measurements of local strain distributions. This technique allows mapping of surface strains and provides considerably more information in terms of local strain/stress peaks and damage initiation compared to conventional mechanical testing. Such strain mapping methods are also useful for careful design of stopped experiments, which yield partially damaged samples that can then be further analysed, for instance with optical or electron microscopy.

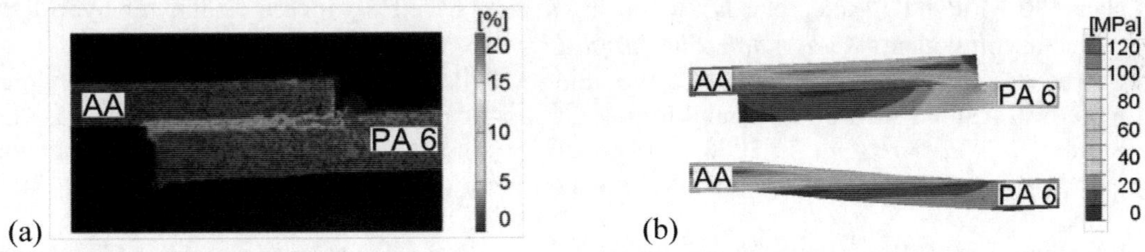


**Figure 7.** AA/PA 6 composite in a shear test: (a) DIC image representing the strain distribution, (b) FEM simulation showing the von Mises stress distribution for two different specimen geometries.

The deformation and damage mechanisms and hybrid response differ at quasi-static and dynamic conditions. The conventional specimen geometries for static shear tests cannot be used for dynamic tests. A hybrid composite exhibits local stiffness changes at the interface, and this effect leads to impedance jumps and thus to undesirable wave reflections during dynamic deformation; reflected elastic waves distort the measured pulse signals. Therefore, one key goal of this project is to develop new sample geometries that allow for a careful control of stresses and strains at hybrid interfaces during dynamic loading. In figure 7, we present first examples of our experimental and numerical research on such improved sample geometries. Figure 7(a) shows DIC data (the color code is related to the maximum principal strain) obtained during a quasi-static test of an AA/PA 6 composite sample. Damage and crack initiation in the interface is associated with large displacements, and the DIC algorithm interprets these data as locally strongly increased strains. Clearly, this method can be used to optimise experimental procedures for stopped tests. Moreover, the strain data can also be used to validate finite element simulations of similar experiments. Figure 7(b) shows two examples (also quasi-static) of sample geometries using the finite element method to search for samples that provide more homogeneous stress distributions in the interface: the conventional sample geometry is associated with an inhomogeneous distribution of von Mises stresses, whereas a sample with adapted cross sections of the AA and PA 6 parts clearly shows a much more homogeneous distribution. Current and future work is focused on using similar methods to experimentally and numerically study stresses and strains during dynamic deformation, and to analyse the resulting interface damage.

## 5. Concluding remarks

In summary, the experimental and numerical findings presented in this short paper represent the current state of the ongoing development of an intrinsic hybrid composite. Clearly, our project requires the combined expertise of the research centres of production engineering, materials engineering and applied mechanics. Current work is focused on combining the individual results in order to actually manufacture a demonstrator of the investigated hybrid composite.

**Acknowledgements**
This research was supported by the German National Science Foundation (DFG) within the Priority Program 1712.

## 6. References

[1] Zeilinger H 1985 *Verbund- und Hybridwerkstoffe – Definition, Zusammensetzung, Anwendung* VDI-Bericht 563 (Düsseldorf: VDI-Verlag)

[2] Haase R, Müller R, Landgrebe D, Scholz P, Riemer M 2015 Process design for hybrid sheet metal components *Acta Matall. Sin. (Engl. Lett.)* **28** pp 1518–24

[3] Carrado A, Faerber J, Niemeyer S, Ziegmann G, Palkowski H 2011 Metall/ polymer/ metall hybrid systems: towards potential formability applications *Compos. Struct.* **93** pp 715–21

[4] Shutov A V, Kreißig R 2008 Finite strain viscoplasticity with nonlinear kinematic hardening: phenomenological modelling and time integration *Comput. Method. Appl.* M. **197** pp 2015–29

[5] Ihlemann J 2014 *Beobachterkonzepte und Darstellungsformen der Nichtlinearen Kontinuumsmechanik* ISBN 978-3-18-334018-7 (Düsseldorf: VDI-Verlag)

[6] Kießling R, Landgraf R, Ihlemann J 2015 Direct connection of rheological elements at large strains: application to multiplicative viscoplasticity *Proc. Appl. Math. Mech.* **15** pp 313–14

[7] Iwase E, Hui P.-C, Woolf D, Rodriguez A W, Johnson S G, Capasso F, Loncar M 2012 Control of buckling in large micromembranes using engineered support structures *J. Micromech. Microeng.* **22** pp 1–7

[8] Darwich S, Scharf I, Engelhardt G, Lampke T (2011) Sol-gel coating for alluminium alloy: self-healing characteristics *In: J. E. P. T.* pp 3–13

[9] Mahadik S A, Parale V, Vhatkara S R, Mahadik D B, Kavale M S, Wagh P B, Gupta S, Gurav J 2013 Superhydrophobic silicia coating by dip coating method *Appl. Surf. Sci.* **277** pp 67–72

# Experimental evaluation of Bauschinger effects during tension-compression in-plane deformation of sheet materials

**M Härtel[1], C Illgen[1] and M F-X Wagner[1]**
[1] Chair of Materials Science, Institute of Materials Science and Engineering, Technische Universität Chemnitz, 09125 Chemnitz, Germany

markus.haertel@mb.tu-chemnitz.de

**Abstract**

An experimental analysis of the effect of in-plane strain path changes and Bauschinger effects during mechanical loading that involves compression is quite challenging for sheet materials because they are prone to buckling. In this paper, we present a newly developed method that allows to deform sheet metals in tension and in compression. Our experimental setup primarily consists of an anti-buckling clamping device that can be used in a universal mechanical testing machine. Surface strains are measured with a digital image correlation (DIC) system. To demonstrate the viability of our method, first tests are conducted on thin sheet specimens (1 mm thickness) of a DC06 steel. Several approaches to reduce friction between the clamping system and the specimen are investigated. The validity of the test results is established in monotonic tensile tests. Measurements with different amounts of maximum tensile strain and reversed strain values are also discussed, showing that the material exhibits a Bauschinger effect, which is further quantified using different Bauschinger parameters.

## 1. Introduction

The decrease of the yield strength after a load path change is known as Bauschinger effect [1]. To quantify the Bauschinger effect as a function of different amounts of maximum (previous) strain, various parameters (Bauschinger effect parameter, Bauschinger strain, Bauschinger strain parameter) have been proposed, see e.g. [2-4]. Those parameters can be determined in experiments with load path changes. Relatively simple experiments with load path changes are shear tests with a reversal of shear direction [5, 6] or tensile tests with subsequent compression tests [2, 7]. Compared to the established tension-compression testing of bulk materials [8-11], tension-compression testing of sheet materials is challenging: sheet materials are prone to buckling under compressive loading. To overcome this issue, comb-shaped devices are often used; these devices provide additional support from the sides of the sample to prevent buckling (see also figure 1d for an example of a comb geometry). Kuwabara et al. developed such supporting clamps. They used a conventional specimen and measured forces with a load cell and strains with a strain gage, but their experimental approach required the design of a dedicated load frame [12]. Tan et al. and Boger et al. suggested to use a custom-designed, non-standard dog bone specimen, hydraulic clamps and strain gages for strain measurements [13, 14]. To determine forces, they performed calculations to account for friction in the experimental setup, using experimental friction values from [15].

Piao et al. used a similar setup, but with an additional option to heat the clamps [16], and Lee et al. used a similar rig like Kuwabara with slightly different combs and also with the heated clamps. For strain measurements they used a laser extensometer [17]. Knoerr et al. employed special clamps fixed with springs on non-standardized specimens. They performed (out of plane) strain measurements using digital image correlation (DIC) [18]. These previous studies provided a useful basis for the development of our own experimental setup, which we built in the light of previous approaches (and the corresponding issues and limitations), and considering the following requirements: The experimental setup should

- allow for continuous, uninterrupted tension and compression cycling
- facilitate the use of standard specimens
- fit into a conventional universal tensile testing machine
- allow for direct, in-plane strain measurements.

This paper primarily reports on the development of our experimental setup dedicated to tension-compression testing. Moreover, with this new setup, tension-compression-tests were performed on steel sheet material. We compare the test results with those of conventional monotonic tensile tests. Measurements with different amounts of prestrain and reversed strain values are also discussed, showing that the material exhibits a Bauschinger effect. The magnitude of the Bauschinger effect is finally evaluated using different Bauschinger parameters, and compared to bulk material data.

## 2. Experimental

In all experiments presented in this paper, a steel sheet material with a thickness of 1 mm was used. The steel is a low-carbon, interstitial free DC06 with 0.02 wt.-% C, 0.25 wt.-% Mn and 0.3 wt.-% Ti and Nb. This alloy was chosen as simple reference materials because the low amount of alloying elements allows to exclude further effects associated with the formation of precipitates, additional heat treatments or impurities.

### *2.1 Development of a new experimental setup for tension-compression testing*

Figure 1 shows the new experimental setup for tension-compression testing that was developed to meet all requirements listed above. The setup consists of clamps (to hold the sample) and of four moveable, comb-shaped parts (which restrict out of plane deformation). This setup is installed between a die set that is adapted to fit into a conventional testing machine (ZWICK/ROELL Zmart Pro Universal 100 kN), figure 1a. A standardized specimen (following the regulations of [19]) with a gauge length of 50 mm and a width of 12.5 mm is placed between the two front parts and the two back parts, figure 1b. The comb-shaped clamps prevent buckling. For strain measurements, a DIC system is used, figure 1b. Because of the thickness of the clamps (16 mm, to ensure a high stiffness), proper illumination of the sample surface, which is visible between the comb-shaped fingers, a ring light is installed around the DIC camera, figure 1c. This results in excellent illumination without any shadows on the surface of the specimen that is painted with speckle patterns, figure 1d. During the experiment, the DIC system tracks the displacements of small regions on the sample surface, and thus allows for accurate, local measurements of surface strains.

**Figure 1:** Tension-compression setup with comb-shaped fingers (a) installed between a die set in a standard testing machine. The standardized tensile specimen is placed between the two front and the two back parts of the tool (b). For sufficient illumination between the combs' fingers, a ring light is placed around the DIC camera (c); the illuminated surface of the specimen contains speckle patterns for DIC measurements (d).

While the new setup in principle meets our design criteria, certain issues arise during application: on one hand, friction occurs between the clamps and the specimen, and this effect cannot simply be neglected; on the other hand, even when friction is minimized by appropriate experimental measures, the force data cannot be used directly to calculate stresses (and thus material strength values). One typical approach is to perform complementary FEM simulations to account for friction in the setup (see [14, 16, 18]), but because this approach is tedious, one goal of our study was to accurately determine stresses without additional simulations. Below, we describe strategies that allow to conveniently handle these challenges.

### 2.2 *Reducing friction in the experimental setup*

The issue of friction is quite common (and most challenging) in tension-compression testing. When no special measures are taken to reduce friction, one frequently observes strain localization: the specimen is then primarily deformed in the zone near the "fingers" of the comb-shaped clamps. For a given nominal strain, this effect ultimately leads to higher stresses compared to conventional tests. This is obvious from the stress-strain curves presented in figure 2, where we compare the benchmark curve measured without the comb-shaped clamps with an experiment with clamps (but without trying to reduce friction: curve labeled "with clamps only") and with experiments where different attempts were made to reduce friction. The common approach to reduce friction is to place Teflon (PTFE) sheets between the specimen and the anti-buckling device to reduce friction [12-14, 16-18]. We observed, however, that this approach is not sufficient when applied to our experimental setup: putting PTFE foil on one or both sides of the sample, and even combining PTFE foil and PTFE spray, does reduce friction, but the corresponding stress-strain curves still do not agree very well with the reference measurement.

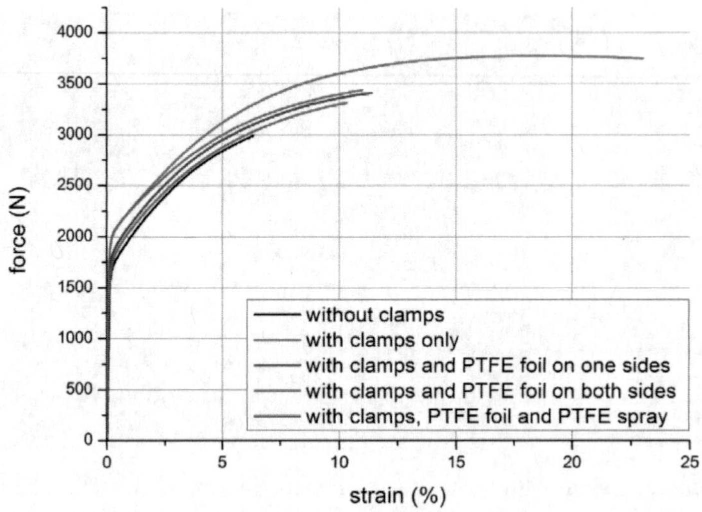


**Figure 2:** Force-strain-curves determined while using different methods to reduce friction in the experimental setup. The benchmark is the curve "without clamps". The measurement without any reduction of friction ("with clamps only") exhibits the strongest deviations. Compared to the results of the literature, none of the methods using Teflon is in sufficiently good agreement with the benchmark measurement.

To improve on these first results, we also tried applying other lubricants used commonly in metal forming, such as Beruforge (produced by Bechem Lubrication Technology) or $MoS_2$, but this also did not work. Knoerr et al. suggest the use of springs to press on the anti-buckling device [18], but in our experimental setup, this method lead to large variation in the resulting force-strain-curves, especially at the point of load path change (i.e., at maximum strain, prior to load reversal).

We finally came up with a new idea to prevent friction by machining a notch in sheet thickness (1 mm) into the clamps, figure 3. This allows the specimen to move freely and without additional friction during the initial tensile deformation. Despite this increased degree of freedom, after the load path change (and during subsequent compressive loading) the clamps immediately and effectively avoid buckling. The stress-strain curves in figure 4 demonstrate that this approach is quite effective: there is a very good agreement between tests performed with and without the modified clamps.

IOP Conf. Series: Materials Science and Engineering **118** (2016) 012018 doi:10.1088/1757-899X/118/1/012018

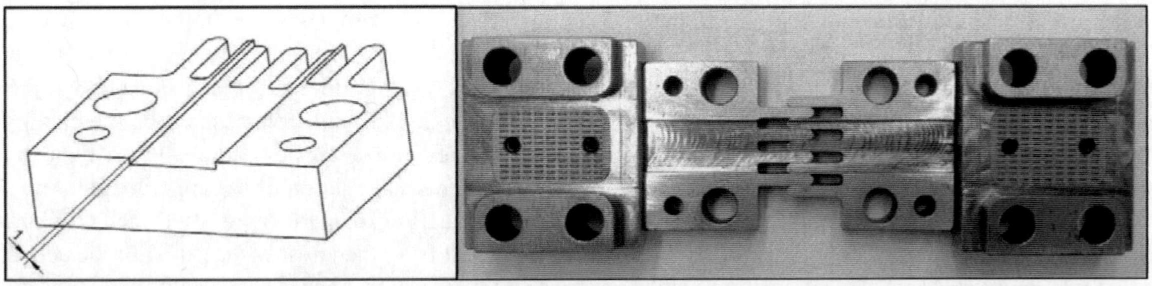

**Figure 3:** A successful strategy to considerably reduce friction in our tension-compression device is to machine a notch in sheet thickness (1 mm) into the clamps (shown schematically on the left). The photograph on the right shows the pair of clamps with the machined notch.

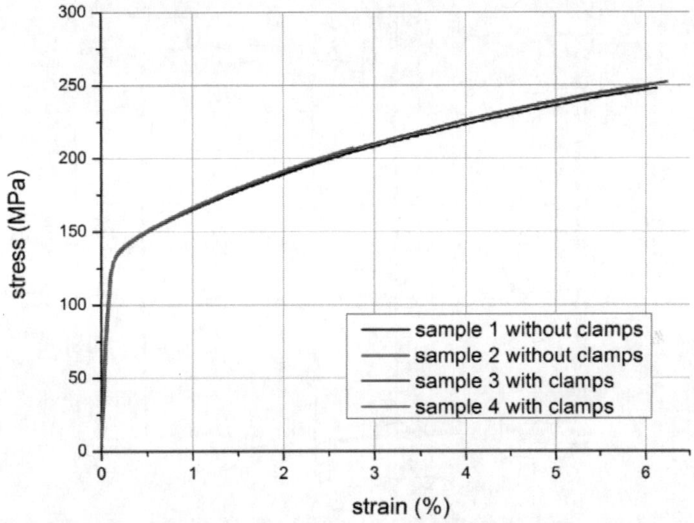


**Figure 4:** The comparison between four different samples demonstrates that machining a notch in the clamps is an effective method to reduce friction: there is very good agreement between samples tested without clamps (as a reference) and those with clamps with the machined notch. Both at the yield point and at strains up to 6 %, all curves agree very well.

### 2.3 *Measuring forces and determining accurate material strength data*

In all experiments reported here, forces are measured using the load cell of the conventional testing machine. The clamping device is installed in this testing machine between a die set, see also figure 1a. Unfortunately, this die set is also subjected to internal friction, which results in increased force values, and thus in artificial "residual stresses" when data from the load cell are used to calculate stresses. One option to measure those "residual stresses" is to perform an additional measurement after the sample has been removed, using only the empty setup after each test, and following the same machine path. Figure 5 shows forces as a function of nominal strain during measurements without samples for strain paths with four different maximum strains. These forces, which simply result from internal friction, are not constant, but in all measurements they are in a similar range. The measured strain values are much higher than the maximum strains during the actual tension-compression tests because strains are measured on the clamps and no material resists the deformation. We also note that forces under tensile load are lower than the corresponding forces under compression. This indicates that the internal friction of the experimental setup is higher under compressive loads. Such calibration measurements were performed after each tension-compression test, and the "residual stresses" were then subtracted from the stresses measured in the actual tension-compression test. This procedure results in stress-strain curves that agree very well with data from conventional tensile tests (i.e., using a standardized sample shape in a uniaxial testing machine, and no special experimental setup) performed on the same material.

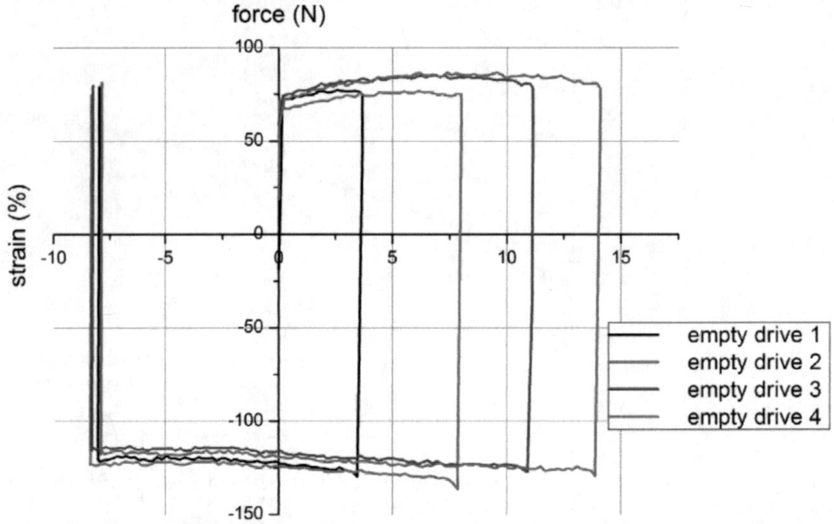


**Figure 5:** Force-strain curves measured with the empty drives for different maximum strains. Force values in tension are lower lower than forces under compression. These force data are used to determine accurate stress data from the actual experiments.

## 3. Results and Discussion

While the first part of this paper (and in particular the two preceding sections) have been focused on establishing and validating a detailed testing protocol for our new experimental setup for tension-compression testing of thin sheet materials, we now present first results (on the DC06 sheet material) that were obtained following our experimental procedure. In figure 6, our first results of tension-compression tests using maximum strains of 0.5, 1, 1.5, and 2 %, respectively, are shown. Note that the loading parts of all stress-strain curves fall on top of each other – this demonstrates that our experimental setup provides reproducible as well as reliable (as demonstrated by comparing to conventional tensile data) results. The yield strength of all samples in tension is 122 MPa, which is also in excellent agreement with the conventional tensile test. Note that, in all cases, the stress at the onset of plastic deformation after load reversal is lower than the maximum tensile stress prior to load reduction: the material clearly exhibits a Bauschinger effect. Furthermore, we observe that the flow curve after load reversal does not exhibit a distinct point separating the elastic from the plastic region. Therefore, the yield strength under compression load cannot be determined exactly. Tan et al. observed the same phenomenon [13]; we follow their suggestion to use $R_{p0.2}$ as a reasonable approximation for the yield strength under compression.

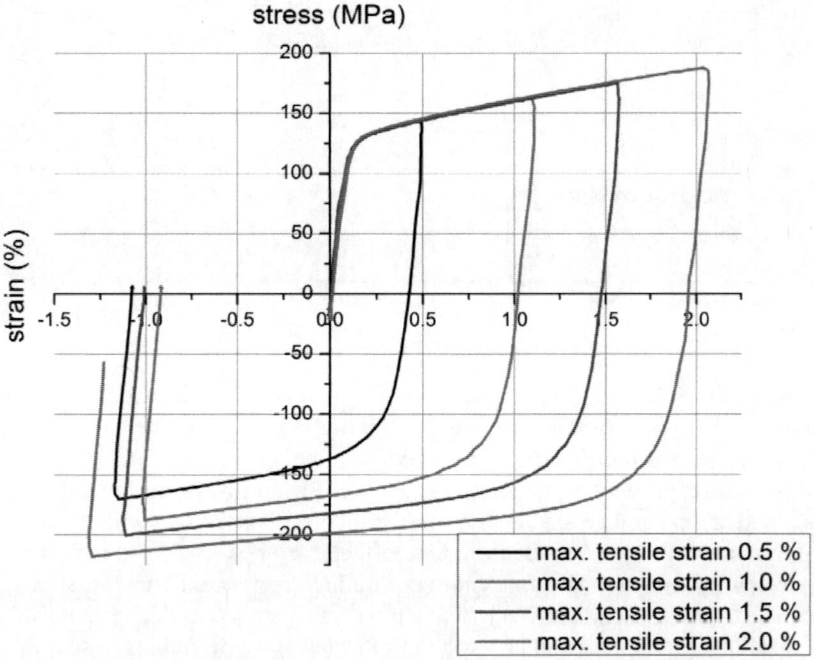


**Figure 6:** Stress-strain curves of tension-compression experiments with different maximum tensile strains (0.5, 1, 1.5, and 2 %). The yield strength under tensile load (122 MPa) agrees well with data from conventional tests. After load reversal, all specimens exhibit reduced yield strengths under compression.

Different parameters have been proposed to quantify the Bauschinger effect. In [3, 4]), a Bauschinger effect Factor was introduced. This factor, $\alpha$, is defined as the ratio between the yield strength after load reversal (i.e., in compression) $\sigma_R$ and the maximum tensile stress $\sigma_F$, figure 7. Abel and Muir proposed two parameters that are focused on strain values: The first one is the Bauschinger strain $\beta$ [2] which can be described as the strain that is necessary (after load reversal) to reach the same stress as the maximum tensile stress prior to load removal, figure 7. The second parameter is the Bauschinger strain parameter $\beta_\varepsilon$ [2]. This parameter is defined as the ratio between the Bauschinger strain $\beta$ and the maximum tensile strain $\varepsilon_P$, figure 7.

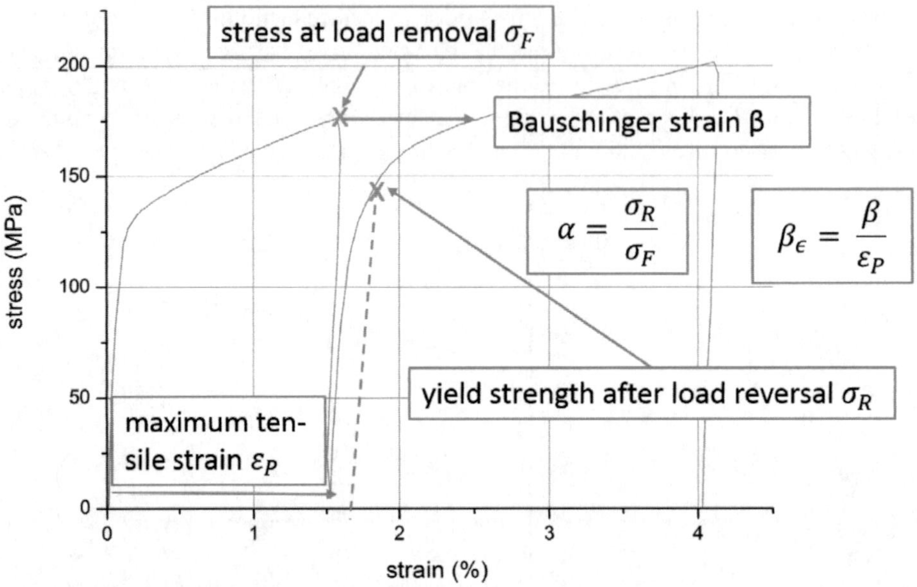


**Figure 7:** Schematic representation of a stress-strain curve of a tension-compression test (compressive stresses multiplied by -1 for better illustration). The specimen is deformed up to a maximum tensile strain $\varepsilon_P$, and a maximum stress $\sigma_F$ , respectively. Then the load is reduced and reversed so that the sample is loaded in compression. The material's subsequent yield strength in compression is $\sigma_R$. The Bauschinger effect factor $\alpha$ is defined as the ratio between $\sigma_R$ and $\sigma_F$. The Bauschinger strain $\beta$ corresponds to the strain that, after load reversal, is necessary to reach the same stress as at load removal. The Bauschinger strain parameter $\beta_\varepsilon$ is the ratio between Bauschinger strain $\beta$ and maximum tensile strain $\varepsilon_P$.

To quantify the Bauschinger effect in the DC06 tested here, for a direct comparison with data from the literature, and as a first step towards further research on Bauschinger effects during multi-axial deformation, we calculated the different Bauschinger parameters as a function of different maximum tensile strains, figure 8. The Bauschinger effect factor $\alpha$ (which, like all Bauschinger factors discussed here, would be equal to 1.0 in the absence of a Bauschinger effect) decreases slightly with increasing maximum tensile strain: for a maximum strain of 0.5 , $\alpha$ is 0.91; for a maximum strain of 2 %, $\alpha$ is 0.82. This trend of the Bauschinger effect factor has also been observed for different sheet [15] and bulk materials [3].

In contrast, the Bauschinger strain $\beta$ increases with increasing tensile strain: for 0.5 %, $\beta$ is 0.56, and for 2 %, $\beta$ is 1.18. Again, this is in excellent agreement with results on bulk material with a similar chemical composition reported in [2]. Finally, when we consider the ratio between Bauschinger strain $\beta$ and the maximum tensile $\varepsilon_P$ (which is the so called Bauschinger strain parameter $\beta_\varepsilon$) one interesting result is that, for 0.5 %, $\beta_\varepsilon$ is higher than 1. This demonstrates that a higher strain after load reversal is necessary to reach the same (absolute) stress value as at load reversal. Abel and Muir observed the same effect for bulk material [2], but only for small tensile strains. At higher strains, $\beta_\varepsilon$ decreases rapidly: for 1 %, $\beta_\varepsilon$ is 0.76, for 1.5 % it is reduced to 0.66, and for 2 % it is 0.58. This observation also agrees well with [2].

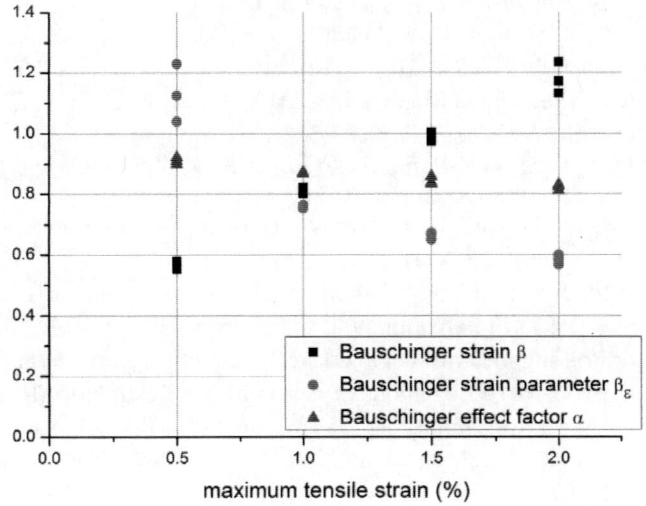


**Figure 8:** Evolution of the different Bauschinger parameters as a function of maximum tensile strain.

## 4. Summary and Conclusions

This investigation focused on the development of a new experimental setup for in-plane tension-compression deformation of sheet metals. Our design meets several requirements; it (i) allows using a conventional/ standardized specimen, (ii) can be directly used for testing in a conventional testing machine and (iii) it designed for in-plane strain measurements with digital image correlation. One key issue for tension-compression tests is friction between parts of the setup and the specimens. This issue was successfully tackled by machining a notch in sheet thickness into the clamps. Moreover, we discussed how the amount of friction in the die set can be determined by performing empty runs after the mechanical tests themselves. By subtracting the characteristic force values from the measured forces during the tension-compression experiments, stress-strain curves that are in very good agreement with data from conventional tensile tests can be obtained. Using our optimized experimental setup, we performed first investigations on samples of a DC06 steel, and we discussed the evolution of different parameters that allow quantification of the Bauschinger effect as a function of different maximum tensile strains. Our experimental results show that the steel sheet DC06 clearly exhibits a Bauschinger effect, and they fully agree with literature data on bulk material with a similar chemical composition.

## Acknowledgements

The authors gratefully acknowledge recognition of this research by the Doerrenberg award committee.

**References**


[1] Bauschinger J 1886 *Mitteilung XV aus dem Mechanisch-technischen Laboratorium der Königlichen Technischen Hochschule in München* **13** 115

[2] Abel A and Muir H 1972 *Philos. Mag.* **26** 489

[3] Gupta S P and Kodali S P 1976 *Scr. Metall.* **10** 111

[4] Richards M D, Van Tyne C J and Matlock D K 2011 *Mater. Sci. Eng. A* **528** 7926

[5] Lewandowska M 2003 *Mat. Chem. Phys.* **81** 555

[6] Vincze G, Rauch E F, Gracio J J, Barlat F and Lopes A B 2005 *Acta Mater.* **53** 1005

[7] Bate P S and Wilson D V 1986 *Acta Metall.* **34** 1097

[8] Norm DIN EN ISO 6892: Metallic materials - Tensile testing - Part 1: Method of test at room temperature. 2009. Berlin: Beuth Verlag

[9] Norm ASTM E8M: Standard Test Methods for Tension Testing of Metallic Materials. 2015. West Conshohocken: ASTM International

[10] Norm DIN 50106: Prüfung metallischer Werkstoffe; Druckversuch. 1978. Berlin: Beuth Verlag

[11] Norm ASTM E9: Standard Test Methods of Compression Testing of Metallic Materials at Room Temperature, 2009. West Conshohocken: ASTM International

[12] Kuwabara T, Nagata K and Nakako T 2001 Proc. of Int. Conf. Advanced Materials and Processing Technol. 407

[13] Tan Z and Magnusson C 1995 *Mater. Sci. Eng. A* **183** 7926

[14] Boger R K, Wagoner R H, Barlat F, Lee M G and Chung K 2005 *Int. J. Plast.* **21** 2319

[15] Balakrishnan V 1999 *Measurement of in-plane Bauschinger effect in metal sheet* (Columbus, Ohio: The Ohio State University)

[16] Piao K, Lee J K, Kim J H, Kim H Y, Chung K, Barlat F and Wagoner R H 2012 *Int. J. Plast.* **38** 27

[17] Lee M G, Kim J H, Kim D, Seo O S, Nguyen N T and Kim H Y 2013 *Expe. Mech.* **53** 1039

[18] Knoerr L, Sever N, McKune P and Faath T 2014 Proc. of AIP Conference **654** 654

[19] Norm DIN 50125: Prüfung metallischer Werkstoffe – Zugproben. 2009. Berlin: Beuth Verlag

IOP Conf. Series: Materials Science and Engineering **118** (2016) 012019 doi:10.1088/1757-899X/118/1/012019

# Experimental and numerical investigation on cold flat rolling processes of DC04 sheets with special focus on residual stresses

**A Bauer[1,2], T Mehner[3], C Binotsch[2], M Sieber[3], B Awiszus[2] and T Lampke[3]**

[1] Corresponding author, email: a.bauer@mb.tu-chemnitz.de
[2] Chair of Virtual Production Engineering, Institute for Machine Tools and Production Processes, Technische Universität Chemnitz, 09125 Chemnitz, Germany
[3] Materials and Surface Engineering Group, Institute of Materials Science and Engineering, Technische Universität Chemnitz, 09125 Chemnitz, Germany

**Abstract.** The process of cold flat rolling is a widespread industrial technique to manufacture semi-finished products, e.g., for the automotive or homewares industry. Basic knowledge of the process regarding dimensioning and adjustment of defined characteristics is already state of the art. However, a detailed consideration and analysis with respect to local inhomogeneous residual stresses in several process steps mostly remains disregarded. A broad understanding of the process due to the distribution of residual stresses in the workpiece and the direction of the stress tensors allows for a definition of the characteristics of the workpiece even before the actual manufacturing process. For that purpose, it is necessary to perform numerical investigations by means of the finite element analysis (FEA) of cold flat rolling processes.
Within this contribution, several approaches for the calibration of the FEA with the real flat rolling process will be addressed and discussed. To ensure that the numerical consideration provides realistic results, this calibration is indispensable. General parameters such as geometry, height reduction, rolling temperature, process time, and the rolling speed are considered as well as a photogrammetric survey, and calculated residual stresses with results of X-ray diffraction (XRD) will be compared.
In the course of the experiments, a good agreement between the stress results of the FEA and the XRD was found in the center of the specimen. In combination with the allocation of the stress orientations, the agreement close to the edges is also fine. Some issues that cause differences between the FEA and the experiment are dis-cussed.

## 1. Introduction

The process of cold flat rolling is a widespread industrial technique to manufacture semi-finished products, e.g., for the automotive or homewares industry. The normal procedure for the cold strip production is divided in the following six main components: pickling, cold rolling, annealing, skin pass rolling, refinement and adjustment. One of the major tasks of cold flat rolling is the defined adjustment of measurements, material homogeneity, surface quality like roughness or waviness, and the mechanical properties [1]. The quality of the hot rolling strip has a big influence on the achievable results in the cold rolling process because cold rolling is mostly situated at the end of the process chain. With a good understanding of the process parameters, it is possible to influence the mechanical properties in a defined way, even at this time of the process. Ideally, a defined rolling state without any following post treatment (e.g., heat) can be achieved. Therefore, it is necessary to acquire a good knowledge of the effects of different parameters on the cold rolling process. A good way of taking most of those parameters into account is a well calibrated finite elements analysis (FEA). With the help of the FEA, it is possible to analyze the process numerically under variation of different parameters. Some of the most important results related to the mechanical properties are residual stresses. These residual stresses are a major part of this consideration. A deeper understanding of the evolution, direction and expression of these stresses mostly remains unconsidered. Just a few authors analyzed residual stresses in forming processes in a more detailed way [e.g., 2, 3, 4, 5, 6], but even they do not analyze the orientation of the stress tensors and the basic cold flat rolling processes.

## 2. Experimental and simulative setup

The material used for the experiments is a general DC04 sheet metal (*Max Baum Stahl Service GmbH*). It is very important to choose an ordinary steel with a small proportion of alloying elements to gain understanding of the basic processes without phase transformations or further micromechanical problems. The material composition is given with 0.042 % C, 0.242 % Mn, 0.010 % P, 0.013 % S, 0.035 % Al, 0.011 % Si, and Fe balance [7]. The roughness average of the used DC04 is about $R_a = 0.99 \pm 0.09$ µm in rolling direction. Prior to rolling, the metal sheets were cut into samples with the dimensions of 100 mm × 14 mm × 2 mm by water jet cutting. Water jet cutting was used to avoid additional shear or tensile stresses induced by the cutting process.

After cutting, the samples were rolled at room temperature on a *Duo/Quarto EW 105x100* strip rolling mill by *Bühler & Co GmbH*. This rolling mill is especially designed for experimental purposes. The roll speed can be varied from 3.0 rpm to 24.3 rpm with a roll diameter of 105 mm. The defined consistent roll speed in the experiments was selected to be 12.12 rpm. The constant rolling force within the rolling mill is specified with a maximum of 100 kN while using the duo mode with 2 working rolls. The samples were loaded manually until they overcame the roll feed and proceeded through the rotary movement of the working rolls. The distance of the rolls can be adjusted with a step size of 10 µm.

While processing the samples, a thickness reduction of 0.5 mm occurs. Thus, the thickness of the analyzed specimen is 1.5 mm. The evenness of the samples was ensured because bends across the length of the specimen cause a distinct variation of the measured stresses and may even alter the sign of the residual stresses.

The numerical analysis of the flat rolling process was set up in the finite element simulation program *simufact.forming (SF) 13.0*, which is especially appropriate for forming processes. More precisely, the *general purpose (GP) tool* was used, which includes additional functions of importance for rolling processes. With the *simufact.forming general purpose tool* (SF GP), it is possible to generate a defined mesh over for the specimen, which can be manually positioned.

Remeshing of the sample is possible but was not applied. Related special characteristics concerning the process are discussed in detail in chapter 4. In general, the numerical model was generated with 8-node-hexahedral elements with the size of 0.500 mm × 0.250 mm × 0.125 mm. Furthermore, the simulative model was set up as 3D mechanical simulation. In the numerical analysis with SF GP, thermal effects were generally not taken into account because only minor temperature increases of about 15–20 °C occurred. But for calibrating the results, a coupled analysis of mechanical and thermal effects was performed once. The material model for the used DC04 was defined with a flow curve received from and adjusted by the *Autoform (Autoform Engineering GmbH)* material library. The tools were defined as rigid bodies and the kinematics were transferred from the real flat rolling process. Between the working rolls and the sample, a combined friction coefficient of $\mu = 0.35$ was added. Furthermore, an additional pusher was needed to simulate the mechanical force, which is required to overcome the roll feed. So called "fences" were used to prevent the specimen from bending (figure 1).

During the numerical analysis, the cold flat rolling process is very susceptible to bending due to the contact values and therefore, it is necessary to integrate those fences with $\mu = 0$. As solver, the iterative sparse calculation was used. Within a second process step, the working rolls release the sample to achieve results with no remaining loads on the surface.

Photogrammetric analyzes were performed with *ARGUS* (*Gesellschaft für optische Messtechnik mbH*), which is capable of detecting the shape of an object with only a couple of photos taken from different positions. Therefore, a pattern of circles with a defined diameter of 0.5 mm and a distance of 1 mm was printed electrolytically on the surface of the specimen. Subsequently, the samples were processed in the rolling mill. The pattern elongates during the rolling process. To evaluate the results, the rolled sample was photographed with the ARGUS camera and the deformation of the sample was detected through the deformation of the defined circles. The evaluation was performed with the *ARGUS-Forming-Ananlysis-GOM*-v6.2.0 software. Results are, e.g., the major strain, the minor strain, the *von-Mises* strain, and the reduction of the thickness. The results can be illustrated with a color scale and the associated distribution over the specimen. For the rolling processes, the interesting values are the major and minor strain as well as the *von-Mises* strain.

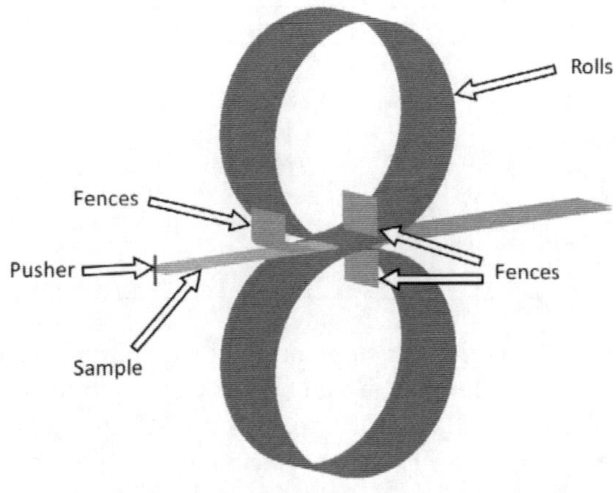


**Figure 1.** Rolling setup in the simulation model.

Residual stresses were measured using a diffractometer *D8 Discover* (*Bruker AXS*) with cobalt $K_\alpha$ radiation for the X-ray diffraction (XRD) analysis. A pinhole aperture with a diameter of 0.5 mm was used to confine the X-ray beam. Line scans along the y direction were performed starting from the center of the sample measuring 10 positions with a step size of 0.7 mm. The $\sin^2\psi$ method was applied: $\sin^2\psi$ was varied with a step size of 0.1 with both positive and negative $\psi$ values. Three rotation angles around the sample normal were used ($\varphi = 0°, 45°, 90°$). The lattice planes {211} of α iron (*Young*'s modulus 220 GPa, *Poisson*'s ratio 0.28 [8]) were measured and evaluated by means of the program *Leptos* (*Bruker AXS*).

## 3. Calibration of the finite elements model and results

According to the VDI standard 3633, the FEA is a tool, which defines the reproduction of a real process. Due to the fact that the numerical calculation is always just an approximation to the reality, there are also many parameters, which can be adjusted within a numerical model. Those parameters have to be set so that the simulation can generate results, which can be used as a prediction for real processes. To ensure that the numerical model fulfills these criteria, a calibration needs to be performed with the associated process. Without any comparison to the reality, it is very hard to interpret the calculated results in the right way. The significance of simulations without reference to test stands or related processes is not very meaningful. To guarantee that the numerical model shows results similar to the checked problems, there are different options, which are pointed out in the following sections.

### *3.1. Calibration with standard process parameters*

The rolling technology includes some parameters, which are quite easy to prove and calibrate. Among these classic parameters are, for example, geometric measurements. The examined rolling specimen was 100 mm in length (x-axis), 14 mm in width (y-axis) and 2 mm in thickness (z-axis). This geometry was exactly rebuilt in SF GP. During the rolling process, the geometry became elongated, and the y/z ratio of 7 indicated a small widening in the y direction. The sample also expands in the numerical analysis (table 1).

**Table 1.** Comparison of measurements, temperature and rolling period for the calibration of the simulation with experimental values for the first rolling pass (2.0 mm to 1.5 mm).

| Parameter | Experiment | Simulation | Deviation (%) |
|---|---|---|---|
| Sample geometry | | | |
| Length (mm) | 127.5 | 128.0 | 1 |
| Width (mm) | 14.5 | 14.4 | < 1 |
| Thickness (mm) | 1.50 | 1.50 | < 1 |
| | | | |
| Rolling period (s) | 2.3 | 2.1 | 9 |
| Temperature (K) | 311 | 310 | < 1 |

Other parameters are the time for one rolling step and the temperature, which was inducted into the specimen (table 1). The time was stopped with a digital chronometer and the temperature was measured with an industrial thermometer on the surface of the specimen directly after each rolling pass. For the calculation of the temperature within the FEA, it is necessary to activate the coupled process for mechanical and thermal effects. This function mostly remains deactivated in the context of cold rolling because it requires additional calculation time. Therefore, it is sufficient to realize this calculation once for every rolling pass if it should be required for the results.

In addition to these standard ways of calibrating numerical simulations, there are some special approaches to achieve well matched processes. These particular tools need a good agreement between reality and simulation. Some methods such as detecting the grain structure for mechanical stress analyses are not appropriate for every kind of numerical analysis. Thus, it must be carefully considered which type of analysis fits the simulation and leads to good parameters for the calibration of the process.

The von-Mises strain measured with the ARGUS GOM system is shown in figure 2. The local influences on the different measuring points within the photogrammetric analysis cause a more irregular distribution of values than in the FEA, but in general, they are at almost the same level. Therefore, it can be inferred that the calibration shows a good agreement. The values of the numerical analysis are at about 0.29–0.30 in the center of the specimen, while the photogrammetric survey lies at about 0.27–0.28.

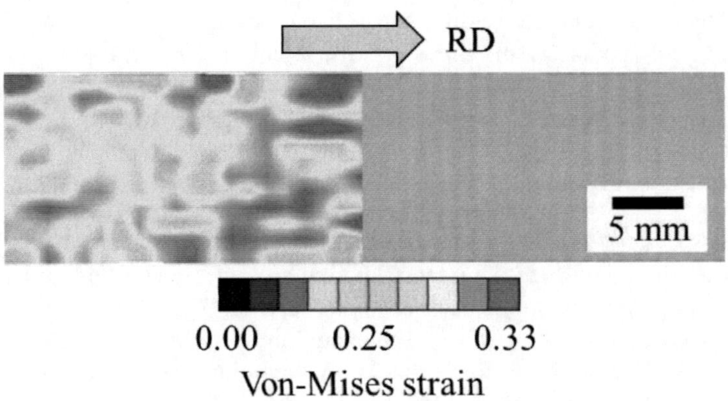


**Figure 2.** Comparison of von-Mises strain of the GOM-ARGUS (left) and FEA (right) results.

Other important calibration parameters are the residual stresses of the specimen, which are discussed in detail in the following section.

### *3.2. Calibration of residual stresses*

In contrast to hot rolling, during which recovery and recrystallization processes occur and residual stresses are reduced, hardening has to be considered when dealing with cold rolling processes. As a result of the micromechanical properties of cold processed steel, the displacements of the material during the forming process entail that stresses are induced into the specimen. Those stresses lead to a change of the mechanical characteristics of the workpieces. To influence the behavior of the workpiece in a suitable way, it is necessary to understand how and where stresses appear, and how they can be influenced by proper manufacturing. The numerical calculation can be a good tool to prove which factors influence the results in what way. To gain relevant significance, a good calibration between numerical analysis and the real process is needed. The more factors are compared, the better a calibration works. In addition, it is very important to adjust the calibration to the required results. For the consideration in this paper, the analysis of residual stresses constitutes a major part. For this purpose, the XRD method is among other reasons appropriate because of the penetration depth that is comparable with the size of the elements in the FEA. A well calibrated simulation can be used as a tool to predict influences of varying forming parameters.

The residual stresses within the considered specimen evolve in a special way. The flow resistance in y direction is larger compared to the flow resistance in x direction. This circumstance can be ascribed to the geometrical shape in flat rolling because of the distribution of contact surfaces within the rolling gap. Further, there is a bigger amount of contact in width than lengthwise. For that reason, the material principally flows in rolling direction. The bigger the ratio between width (y) and pressed length (in the rolling gap), the lower is the increase in y direction. Another important factor is the relation between width and height of the workpiece. From a y/z ratio > 10, it can be assumed that the rolling process is free of the increase in width [9]. In the case of y/z = 7, the analyzed specimens are situated in the upper areas and are just subject to a minor increase in width of approximately 0.4 mm (rolling step 2.0 mm to 1.5 mm). Through the reduction in thickness, the relation increases to a factor of approximately 9.5 in further rolling passes, which leads to a lower increase in width with every following rolling step.

As a result of the chosen sample geometry, there is a special development of stresses because the material flow in width (y) direction is hindered by the geometry in the rolling gap. Therefore, significantly differing stresses arise at the edges of the sample compared to the center areas. At this point, the knowledge of the general distribution of stresses and of the orientation of the stress tensors is necessary. To achieve an exact calibration of the actual process with the numerical analysis, both the orientation of the residual stresses and their amount has to be analyzed. Through this concept, some conclusions can be drawn in relation to the material flow and the stress development. Hence, an aimed impact on the stresses through the geometry of the rolled material and the rolling tools is possible. In the following, a detailed comparison of the numerical analysis and the XRD will show the accordance and the details of the numerical results.

The typical distribution within the following specimen evolves through the combination of different phenomena. As described above, the material is hindered to flow in y direction due to the geometrical dimensions of sample and rolling gap. Nevertheless, the sample is reduced in thickness as a result of the increasing infeed of the working rolls. Thus, the rolling direction is the only possible direction for the material flow. As a result of the forced material flow in x direction, tensile stresses develop. The tensile stresses are denoted as positive and compressive stresses as negative values. The allocation of stresses takes place in such a way within the SF GP so that the most positive value is called "maximum residual stresses" and the most negative one "minimal residual stresses". In this way, the maximum residual stresses could be the smallest, according to the amount, if the intermediate and the minimum residual stresses are negative [10]. The intermediate residual stress is typically not taken into consideration because it is often close to zero. Within the XRD, the stresses were given and arranged as follows: The residual stresses are arranged in the order of their amount. In other words, the principal stress with the highest amount is denoted as $\sigma_1$, and the one with the lowest amount is called $\sigma_3$. According to this principle, the order of stresses between the XRD and the numerical analysis has to be considered additionally. Therefore, the measured values have to be compared and analyzed in advance. To rate the observed differences between the results of the XRD and the FEA, the measurement uncertainties of both techniques have to be considered. The uncertainties within the XRD account for about ± 10 MPa. For the FEA, uncertainties are more difficult to define due to imperfections of the underlying material model, inaccurate initial material states, and numerical errors that may increase in subsequent simulations. What follows is that only absolute numbers can be estimated. The stress results of the simulation are shown in figure 3, and the comparison with the experimental values can be found in table 2.

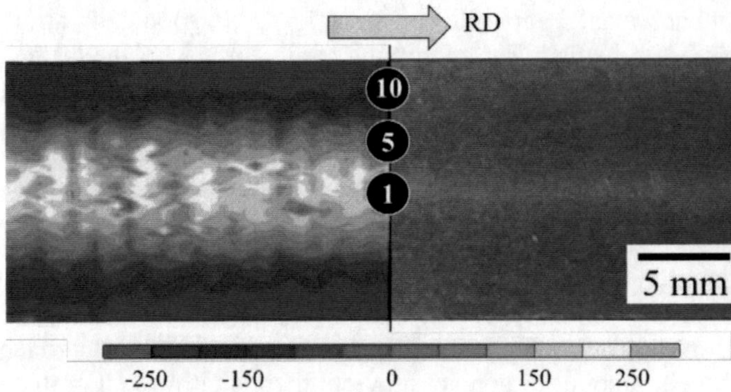


**Figure 3.** FEA results of $\sigma_{xx}$ with the marked positions (1–10) for the XRD line scan.

**Table 2.** Comparison of principal stresses for the calibration of the simulation with experimental values for the rolling pass 2.0 mm to 1.5 mm.

| Parameters | Experiment (MPa) | Simulation (MPa) | Difference (MPa) |
|---|---|---|---|
| $\sigma_1$ center | 240 | 219 | 21 |
| $\sigma_1$ edges | 132 | 94 | 38 |
| $\sigma_2$ center | 18 | 13 | 5 |
| $\sigma_2$ edges | -73 | -100 | 27 |

The orientation of the stress tensors can be evaluated in addition to the scalar stress values. Therefore, the described distribution of the stresses can be seen and easily understood. The stress tensors of the maximum principal stresses in the center of the specimen are generally orientated in rolling direction. With increasing distance to the center, the stresses of the specimen are progressively orientated to the edges of the sample and perpendicular to the stresses in the center (figure. 4). This phenomenon occurs because of the above-mentioned experimental procedure in which the material flow in rolling direction (x) is free, while the flow in width direction (y) is hindered. Thus, the described tensile stresses develop. According to the amount, the tensile stresses at the edge of the specimen are clearly lower than the stresses in the center of the sample. Within the minimal principal stress value, the stress behavior is opposite. The simulations reveal that the minimal principal stresses in the center of the sample are almost zero and orientated to the edges of the specimen, the stresses at the edges are large with an almost antiparallel orientation to the rolling direction (RD). Those stresses are orientated in the direction opposite to the rolling direction (figure. 4) because of the prevented material flow in width direction. The corresponding values on the basis of which the development of the tensor direction can be reconstructed are listed in table 3.

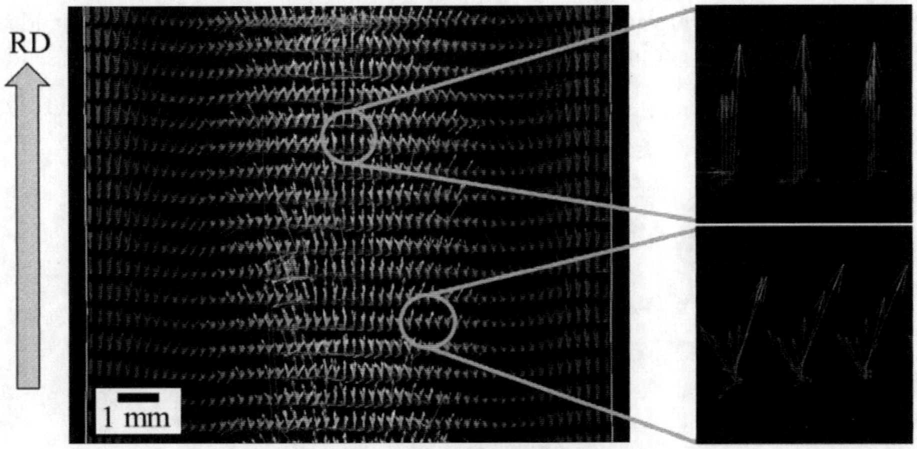


**Figure 4.** Combined illustration of the orientation of the principal stress tensors of $\sigma_1$ and $\sigma_2$ (upper inset: orientation in rolling direction; lower inset: increasing deviation of the stress tensors to the rolling direction).

**Table 3.** Comparison between experimental (exp) and simulative (sim) stresses in ($\sigma_{xx}$) and perpendicular ($\sigma_{yy}$) to the rolling direction (RD) as well as the orientation of $\sigma_1$ and $\sigma_2$ for the rolling pass 2.0 mm to 1.5 mm.

| Position (figure 3) | Experiment $\sigma_{xx}$ (MPa) | Simulation $\sigma_{xx}$ (MPa) | Orientation $\sigma_1$ (exp/sim) (° to RD) | Experiment $\sigma_{yy}$ (MPa) | Simulation $\sigma_{yy}$ (MPa) | Orientation $\sigma_2$ (exp/sim) (° to RD) |
|---|---|---|---|---|---|---|
| 1 | 179 | 183 | -7/ 0.0 | -23 | -39 | 82/ 90.0 |
| 2 | 170 | 172 | -2/ 0.0 | -70 | -35 | 88/ 90.0 |
| 3 | 179 | 167 | 5/ 0.0 | -56 | -45 | 96/ 90.0 |
| 4 | 155 | 116 | 4/ 0.0 | -71 | -30 | 94/ 90.0 |
| 5 | 137 | 18 | 4/22.5 | -73 | -50 | 94/112.5 |
| 6 | 116 | -33 | 8/22.5 | -82 | -52 | 98/112.5 |
| 7 | 98 | -122 | 25/45.0 | -70 | -51 | 105/135.0 |
| 8 | 67 | -170 | 29/45.0 | -69 | -49 | 109/135.0 |
| 9 | -40 | -184 | 29/67.5 | -126 | -43 | 110/157.5 |
| 10 | -120 | -192 | 45/67.5 | -119 | -29 | 135/157.5 |

Given the difficulties in quantifying the right angles, the orientation of the stress tensors within the FEA is set up in steps of 22.5°. Within the XRD, the corresponding angles are provided together with the results. Additionally, it should be mentioned at this point that the step size of the positions within the FEA is not the same as used within the XRD. The available distances solely can be changed in steps of 0.25 mm depending on the used distance between the nodes within the FEM (0.7 mm in XRD). Despite of the sometimes differing values at the measuring positions, it can be seen that the stresses within the FEA and the XRD develop in an equivalent way. For example, the stresses with the same orientation in table 3, like $\sigma_{xx}$ in position 10 (experiment) and position 7 (simulation), both with an angle of 45° to the rolling direction, also correspond with the values of stress: 120 MPa (experiment) and 122 MPa (FEA). This also applies to angels of around 22.5° and to the results in rolling direction (around 0 degree), see figure 5. The stresses and angles which do not occur for $\sigma_{xx}$ within the XRD are those with a value between 45° and 90°. This can be caused by the XRD beam diameter of approximately 0.7 mm and thus, the measured values are averaged over the corresponding area. Close to the edges of the specimen, the XRD beam is wider than the sample and thus may cause measurement uncertainties. Additional differences can have an origin in the evolution of the microstructure within the rolled specimen. Furthermore, the sharp and ideal edges of the specimen within the FEA are an idealization of the reality, which causes a bigger increase of stresses close to the edges.

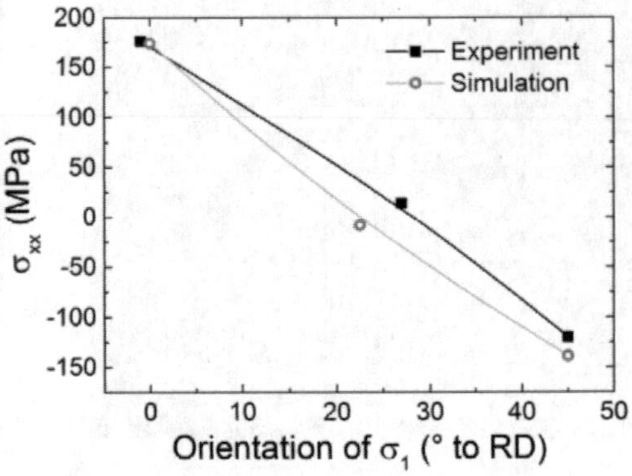


**Figure 5.** The residual stresses in rolling direction ($\sigma_{xx}$) in dependence of the orientation of the 1st principal stress component. The alignment of the points with similar orientations of $\sigma_1$ results in a very good agreement between experiment and simulation.

## 4. Summary and conclusions

Generally, it can be concluded that the principal stresses cannot be analyzed separately. To get an overview of the effective stresses in the specimen, the whole stress tensors given in the results section have to be considered. In this example, the stress state in FEA consists of tensile stresses in the center of the sample and compressive stresses at the edges. This stress characteristic was also measured experimentally. The decay of the tensile stresses as well as the change of the stress orientation from the center of the specimen to the edges numerically occurs at lower distances from the center compared to the experimental values. These differences can be attributed to the idealized sharp edges in the simulation that allow for less deformation compared to the experiment. Generally, the agreement of the residual-stress values is good despite of experimental and numerical uncertainties.

Overall, there are different factors, which have to be considered for cold flat rolling processes, to achieve realistic results within the FEA. Most importantly, no remeshing of the sample should take place within the FEA. During the remeshing process, existing residual stresses can be eliminated and spread over the sample on a new level. In addition to the new distribution of the mesh, the values of the stresses were quite lower than the real stresses or the given stresses of the numerical analysis without remeshing. Hence, the analysis without remeshing shows a much better agreement with the experiment. However, the lack of remeshing causes other problems in further rolling steps such as problematic ratios within the edge lengths of the elements, which possibly reduces the accuracy of the simulation.

**Acknowledgement**
The authors gratefully acknowledge the Deutsche Forschungsgemeinschaft (DFG) for supporting this work (projects LA 1274/27-1 and AW 6/27-1).

**References**

[1] Hoffmann H, Neugebauer R and Spur G (Eds.) 2012 *Handbuch Umformen* vol 1 (München: Carl Hanser Verlag)

[2] Kato B and Aoki H 1978 Residual stresses in cold-formed tubes *J. Strain Anal.* **13** 193

[3] Razny W, Fischer F D, Finstermann G, Schwenzfeier W and Zeman K 1996 The influence of some rolling parameters on the residual stresses after rolling *J. Mat. Proc. Technol.* **60** 81

[4] Dixit U S and Dixit P M 1997 A study on residual stresses in rolling *Int. J. Mach. Tools Manufact.* **37** 837

[5] Cruise R B and Gardner L 2008 Residual stress analysis of structural stainless steel sections. *J. Constr. Steel Res.* **64** 352

[6] Moen C D, Igusa T and Schafer B W 2008 Prediction of residual stresses and strains in cold-formed steel members *Thin-Walled Struct.* **46** 1274

[7] Max Baum Stahl Service GmbH, DC04 material data sheet with test results of the used batch.

[8] Eigenmann B and Macherauch E 1996 Röntgenographische Untersuchung von Spannungszuständen in Werkstoffen. Teil III *Matwiss. Werkstofftechn.* **27** 426

[9] Kopp R and Wiegels H (Eds.) 1999 *Einführung in die Umformtechnik* vol 2 (Aachen: Verlag Mainz)

[10] Simufact Engineering GmbH 2015 manual simufact.forming Anwendung

# Semi-solid processing and its as yet unexplored potential

**D Aišman, K Rubešová and B Mašek**

University of West Bohemia in Pilsen, Research Centre of Forming Technology – FORTECH, Univerzitni 22, 30614 Pilsen, Czech Republic

daisman@vctt.zcu.cz

**Abstract.** Thanks to the available advanced control technology, manufacturing processes which have been described in the past but their hidden potential remained untapped are now continuing to be developed. As a result, even conventional materials which have been around for years can be manipulated to obtain unusual microstructures with specific mechanical and physical properties. Semi-solid processing belongs to the above-described group: it had been studied in the past but, due to complicated process control, it gradually lost its appeal. However, advanced techniques of temperature field control enable engineers to control this complex process accurately. One of the innovative methods of semi-solid processing is mini-thixoforming. As it focuses on very small-size products, it offers very steep heating curves and extremely high solidification and cooling rates, when compared to conventional thixoforming. The capabilities of this process were tested on X210Cr12 ledeburitic tool steel. After the optimum processing conditions were found, additional materials were tried, ranging from low-carbon microalloyed steels through medium-alloy steels to high-alloy tool steels. The microstructure evolution upon the mini-thixoforming process is an issue of its own. The final microstructure of X210Cr12 consisted of more than 90 % of austenite and chromium carbides. Semi-solid processing of a steel with a high vanadium content led to a microstructure comprising MA matrix and globular vanadium carbides. In a low-alloy steel, martensitic microstructure was obtained.

**Keywords:** semi-solid state, thixoforming, rapid solidification, tool steel

## 1. Introduction

Increasing demands on materials properties drive the development of new processing routes for both metallic and non-metallic materials. Thanks to these routes and innovative procedures, attractive mechanical properties can be obtained, even in materials which used to be processed mostly by conventional techniques. One of the interesting paths to obtaining attractive properties is semi-solid processing [1]. This technique offers wide opportunities for producing non-traditional microstructures and enables some difficult-to-form materials to be processed. The forming process involves either one or a very few forming operations, and therefore delivers significant time savings and contributes to cost-effective production. In addition, it can produce intricate-shape products. Two routes are available when thixotropic behaviour is desired in a material [2,3]. These comprise thixoprocessing and rheoprocessing [4]. Thixoprocessing involves materials which are first melted, and then cast and let solidify. After that, they are reheated to the semi-solid range and formed. In rheoprocessing, the material is cooled down to the semi-solid state and then formed. Rheoprocessing is thus more energy-efficient, as it does not require reheating to the semi-solid state. On the other hand, it places much greater demands on the process technology and control, which is why it is less common than thixoprocessing. The shared characteristic of both processing routes is the thixotropic behaviour of the metal. Thixotropy is a deformation behaviour which is defined by a decrease in viscosity with increasing shear rate and time in semi-solid state. The shear stress and deformation in the semi-solid material during forming cause the material's viscosity to decrease [5]. The required forming force is relatively low when compared to the forces needed for conventional processes. Optimum thixotropic behaviour following heating to the semi-solid state should occur when the solid fraction is between 10 and 60 %, depending on the particular semi-solid processing route [6]. For this reason, alloys which are less sensitive to the solid fraction are better candidates for thixoforming, as they are less likely to solidify undesirably while they cool during die cavity filling. With them, the forming temperature is easier to control, which is a key aspect. Over the years, various semi-solid processing techniques have been developed. They became known as SSM (Semi-solid Metal Forming) processes. Among them, there are rheocasting, thixocasting, thixomoulding, thixoforging, thixoextrusion and thixorolling. In recent years, the research centre developed a semi-solid forming process which is termed mini-thixoforming [7]. Although the volume of the metal processed by this technique is small, the inherent problems with heating, namely the temperature field uniformity and temperature control accuracy, have been eliminated, thanks to an unconventional solution [8]. It led to successful processing of steels whose processing had been impossible to control due to narrow temperature intervals and high forming temperatures. In earlier experiments, X210Cr12 tool steel had been employed. The material is widely used for experimental semi-solid processing thanks to its wide freezing range which lies lower than those of other steels. In addition, a specific microstructure has been obtained thanks to particular thixoforming conditions in this otherwise ledeburitic steel. This microstructure consisted of polyhedral austenite grains embedded in carbide network. The austenite fraction exceeded 90 % [9]. Nevertheless, the question remained of how the microstructure would evolve upon semi-solid processing in other steels which are less commonly used for thixoforming. The purpose of the present experiment was to explore the capabilities of the thixoforming process and the microstructural evolution in a wider range of material from low-carbon high-alloy steels to high-alloy powder tool steels.

## 2. Experimental Programme

### *2.1. Selection of experimental materials*

Based on literature search and calculations, X5CrNiCuNb16-4 precipitation-hardenable martensitic stainless steel stabilized with Nb was chosen as the experimental material. This material is a representative of the group of low-carbon high-alloy materials (Table 1). The steel chosen from the medium-carbon group was the microalloyed 30MnVS6 (Table 1). This steel, alloyed predominantly with manganese and silicon, is typically used for making forged parts. Another experimental material was the high-alloy CPM 15V steel. It contains a large amount of hard vanadium and chromium carbides which can be expected to remain stable and present in the microstructure during processing. The high-alloy steel CPM REX 121 was the last experimental material chosen. Of all materials selected, this had the highest carbon level. In addition, it contained almost 10 % vanadium. It had high tungsten and cobalt levels and an increased molybdenum content. As it has high levels of alloying elements and is manufactured by powder metallurgy, it belongs to materials which are difficult-to-form by conventional techniques.

**Table 1.** Chemical composition of the experimental materials

| | C | Si | Cr | Ni | W | Co | Cu | Mn | V | Mo | Nb | P | S | Al | Ti |
|---|---|---|---|---|---|---|---|---|---|---|---|---|---|---|---|
| **X5CrNiCu Nb16-4** | 0.024 | 0.27 | 15.9 | 4.47 | - | - | 3.20 | 0.89 | - | 0.21 | 0.2 | 0.023 | 0.018 | - | 0.11 |
| **30MnVS6** | 0.31 | 0.62 | 0.20 | 0.02 | - | - | 0.03 | 1.50 | 0.098 | 0.007 | - | 0.018 | 0.024 | 0.02 | 0.026 |
| **CPM 15V** | 3.4 | 0.9 | 5.25 | - | - | - | - | 0.5 | 14.5 | 1.3 | - | - | - | - | - |
| **CPM REX 121** | 3.42 | - | 3.99 | - | 9.75 | 8.71 | - | 0.51 | 9.18 | 5.21 | - | - | - | - | - |

In order to develop a particular semi-solid processing route, one needs to map a number of materials parameters across the entire temperature range of the process. In this case, the required parameters were calculated using the JMatPro software because the exact values are very difficult to determine experimentally [11]. The dependence of liquid fraction on temperature plays a key role (Figure 1). The melting curves clearly show the differences among the materials in the semi-solid state. The most favourable properties for thixoforming were found in the CPM REX 121 steel, where the increase in the melt fraction is almost linear. In addition, its freezing range is at lower temperatures than in the other steels. Based on the preliminary calculation, the suitable forming range for the CPM REX 121 steel was identified as 1200 – 1230 °C. The X5CrNiCuNb16-4 steel offers a relatively favourable freezing range as well. According to the calculated data, the forming range is between 1390 and 1415 °C. The K190 steel exhibits more problematic characteristics, as its increase in liquid fraction is steep. The suitable temperature interval is 1240–1255 °C. The most difficult one is the CPM 15V steel. Here, an increase in temperature of a mere 18 °C results in 90 % of melt in the material. The temperature interval suitable for forming is narrower than 10 °C, which places extreme demands on the temperature field control.

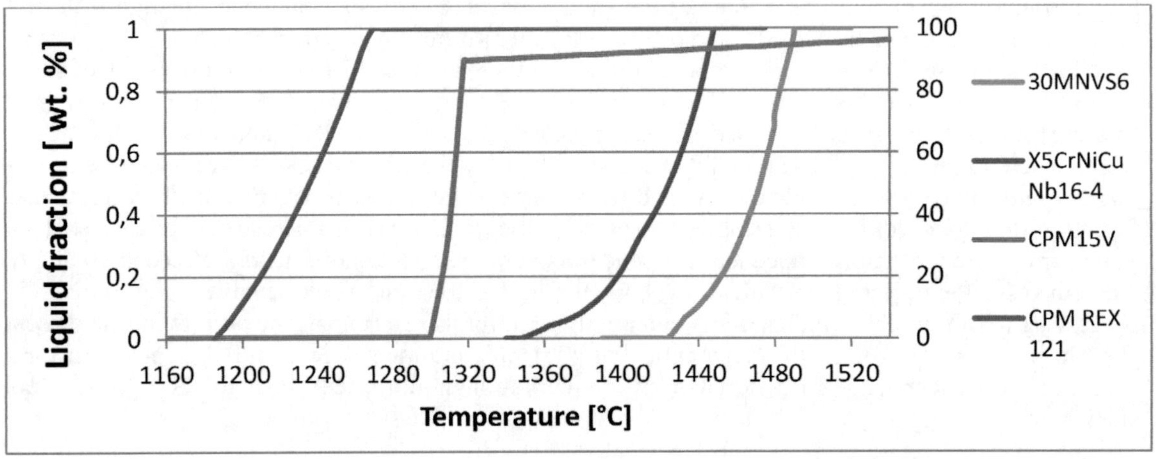


**Figure 1.** Calculated theoretical liquid fraction vs. temperature

Semi-solid processing was carried out using the mini-thixoforming process which involves forming in a closed die cavity and cross-extrusion [10]. The cross-section of the die cavity was 1.9×5 mm and its length was 20 mm. From all steels, stock of 48 mm length and 6 mm diameter was prepared. Their ends were cone-shaped in order to facilitate the passage of the heating current and to align the piece inside the shot sleeve. The stock was heated directly in the shot sleeve. The need for handling a melted workpiece was thus eliminated. High deformation speed is important to the semi-solid forming process, as the material must not solidify before it fills the die cavity. Therefore, the mean stroke velocity was 1 m/s. In the course of the forming process, the temperature was continuously measured with a thermocouple and regulated. As the initial forming temperatures were only determined on the basis of the materials' chemical composition, the actual appropriate forming temperatures for all materials had to be found experimentally by iteration.

## 3. Results and discussion

Based on calculation, a forming temperature of 1400 °C was chosen for the X5CrNiCuNb16-4 material. The feedstock was heated to this temperature in 55 seconds. It was then held for 5 seconds to homogenize the temperature. The first experiments with the calculated parameters did not lead to complete filling of the cavity. The proposed temperature of 1400 °C was too low. The shape and other features of the feedstock showed that the liquid fraction was less than 10 %, the percentage normally reported as sufficient for the thixotropic behaviour to occur. In subsequent trials, the temperature was increased in steps of 5 °C. Finally, forming in the range of 1435 – 1445 °C proved optimal. In this temperature interval, the die cavity filled completely.

For the 30MnVS6 steel, the chosen heating temperature was 1460 °C. In this case, the calculation proved very accurate, as the optimum forming range found by experiments was 1450–1460 °C.

CPM 15V had the narrowest available temperature interval of all the experimental materials. Its initial chosen temperature was 1305 °C. At this level, the liquid fraction was expected to be approx. 30 %. The feedstock, however, melted in the process, which showed that the temperature was too high. Each next trial temperature was therefore 5°C lower than the last. In this manner, the heating temperature was decreased in steps to 1265 °C. After the heating trials with zero axial force, schedules with pre-defined amounts of compressive deformation were tested as well. Based on deformation force plots, the optimum forming range for the CPM steel was found to be 1265 – 1270 °C.

The first processing temperature chosen for trials on the CPM REX121 steel was 1215 °C. The outcome showed that at this temperature the liquid fraction required for adequate fluidity was not reached. By raising the heating temperature in steps of 5 °C, the liquid fraction increased and the entire die cavity was filled eventually. The resulting product had the desired surface quality. The die cavity only filled completely at the temperature of 1235 °C.

The effects of semi-solid processing were explored using metallographic analysis. Optical and electron microscopic observation showed that the single-phase microstructure of ferrite, a solid-solution phase, was retained after semi-solid processing of the X5CrNiCuNb16-4 material (Figure 2). In order to confirm this, diffraction phase analysis was carried out. According to the measured data, the ferrite fraction was 90 %. Another phase identified was austenite. Its amount of 5.5 % was just above the reported 5 % detection limit of the method. The remaining two phases, CrSi and Cr3Si, were present in amounts corresponding to the measurement error. Chemical homogeneity of the products was checked using EDS microanalysis because diffusion in semi-solid state is very rapid due to high temperature and under such conditions, phase or chemical segregation might occur. Chemical microanalysis only proved a slight increase in Cu concentration along grain boundaries. It can be attributed to its migration from the solid phase to the melt upon partial melting. The distributions of other elements were uniform. This is in agreement with the hardness readings of 340±2.3 HV which were identical before and after the processing. The above findings show that the rapid removal of heat did not lead to precipitation which is otherwise common in steels of this type at relatively low temperatures.

In the 30MnVS6 steel, the initial microstructure consisted of ferrite and pearlite. The processing produced a hardening microstructure, a mixture of martensite and bainite which is a relatively uncommon microstructure in semi-solid processed materials (Figure 2). As a result, the hardness of the material increased from 235 HV to 760 HV.

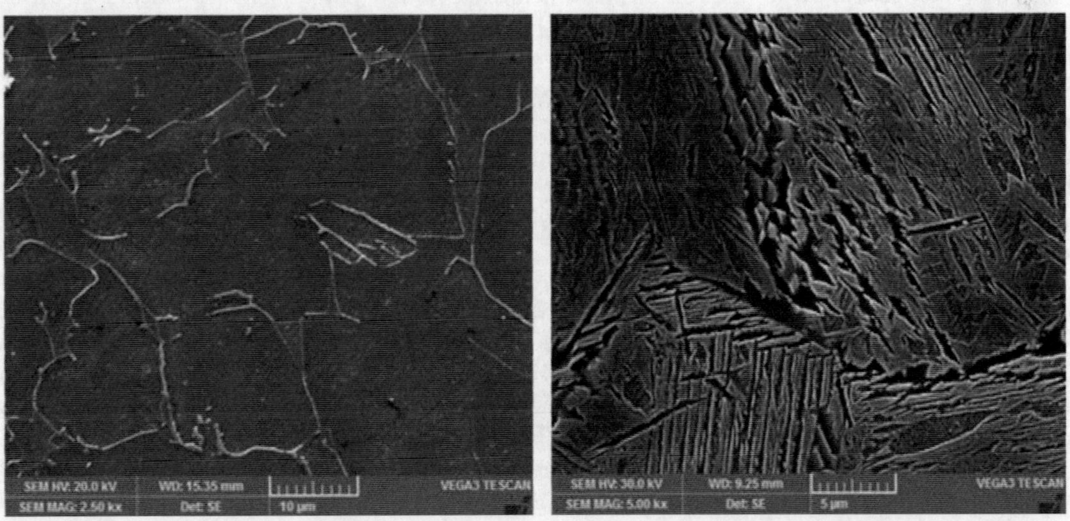


**Figure 2.** Left: resulting microstructure of X5CrNiCuNb16-4 steel; right: 30MnVS6 steel

The mini-thixoformed CPM 15V contained globular vanadium carbides embedded in an austenitic-martensitic matrix. This is in agreement with the results of diffraction phase analysis. The resulting microstructure contained 50 % austenite and 30 % dispersed martensite. High-stability $V_8C_7$ carbides resisted melting during semi-solid processing and made up 20 % of the structure. Carbides with greater chromium content melted completely and their elements formed a eutectic, predominantly on grain boundaries (Figure 3). In the initial condition, the average measured hardness was 298 HV10. Upon thixoforming, its hardness was 728 HV10, which is a 2.5-times higher value. This increase can be attributed mainly to the formation of martensite within the matrix and to the chromium precipitates in the form of network.

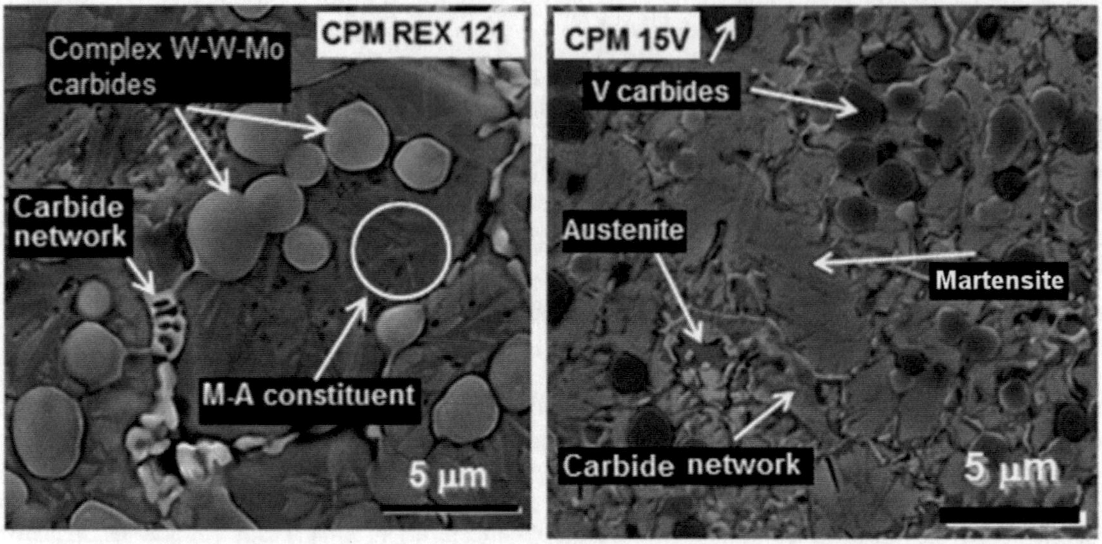


**Figure 3.** Left: resulting microstructure of CPM REX 121 steel; right: CPM 15V

The microstructure of the product of CPM REX 121 consisted of complex carbides and austenitic matrix and a small fraction of martensitic needles (Figure 3). Primary chromium carbides with sizes below 1 μm partially melted and transformed into a eutectic located at austenite grain boundaries. Its volume fraction increased slightly with the processing temperature. Complex V-W-Mo carbides, on the other hand, retained their initial character. In some cases, multiple carbides coalesced into larger particles. Changes at boundaries of large carbides took place. Those became more spherical than in the initial condition which is evidence of their partial melting. Using diffraction analysis, a 35 % fraction of austenitic matrix and 10 % fraction of martensite were found. The balance consisted of carbides. The largest share of those (35 %) had the $M_8C_7$-type carbides, in which the dominant element was vanadium. Minority fractions were represented by $Mo_2V_4C_5$ and $Cr_2FeV_2$. By contrast, CPM REX 121 showed the highest initial hardness, 407 HV10, of all materials. It had a high carbon content and, consequently, high volume fraction of martensite upon processing. This led to a notable increase in hardness to 830 HV10, which is the highest final value among all the materials compared. A contribution to this increase can be attributed to high levels of alloying elements, and thus a high proportion of undissolved carbides in the microstructure.

## 4. Conclusion

The present experiment involved mini-thixoforming of steels with various levels of carbon and alloying elements. Thanks to a close control of the entire process, the die cavity filled completely with all the materials. It was found that alloyed steels with higher carbon levels have microstructures which consist of the M-A constituent matrix and carbides whose shapes and distributions in the matrix depend predominantly on their chemistries. Owing to their high melting point, vanadium carbides retain their initial form. Chromium carbides, on the hand, melt partially and precipitate in the form of a eutectic. Steels with high chromium and low carbon levels exhibit single-phase structures with dissolved chromium. Thanks to the high level of chromium in solid solution, they can be expected to exhibit excellent corrosion resistance. By contrast, materials with lower level of alloying elements form hardening microstructures. Using this route, finished products with final microstructures and high strengths can be produced.

**Acknowledgement**
This paper includes results achieved under project nos. LO1502 Development of Regional Technological Institute and SGS-2015-028 Semi-solid Processing and New Structures without Carbide Net. The projects were subsidised by the Czech Ministry of Education from specific resources of the state budget for research and development.

**References**

[1] Džugan J and Nový Z 2015 *Advanced Materials Research* vol 1127 p 99
[2] Kanga C G, Seob P K and Jeonb YP 2015 *Journal of Materials Processing Technology* vol 160 p59
[3] Bauer W H, Collins E A and Eirich FR 1960 Rheology: *Academic Press* vol 3 p 158
[4] Rosso M 2012 Journal of Achievements in Materials and Manufacturing Engineering vol 54 p110
[5] Fleming M C 1991 *Metallurgical Transactions* vol 22 p 957
[6] Hirt G, Khizhnyakova L, Baadjou R, Knauf F and Kopp R 2009 Thixoforming: Semi-solid Metal Processing (WILEY-VCH Verlag GmbH)
[7] Aišman D, Jirková H and Mašek B 2009 Proceedings of the 20th International DAAAM Symposium Intelligent Manufacturing & Automation: Theory, Practice & Educationvol 20 p 1895
[8] Ronešová A amd Mašek B Digital Generator of Phase Shift Modulation Patent USA (2009), No. 7,480,155
[9] Sen I, Jirkova H, Masek B, Böhme M and Wagner M. F.-X. 2012 *Metallurgical and Materials Transactions* vol 43 p 3034
[10] Mašek B, Aišman D, Rubešová K and Jirková H 2014 *Materials Science Forum* vol 783pp. 801
[11] JMatPro, The Java-based Materials Property Simulation Package, Version 6.1, Sente Software Ltd., Surrey Technology Center, UK

IOP Conf. Series: Materials Science and Engineering **118** (2016) 012021 doi:10.1088/1757-899X/118/1/012021

# Graded High-Strength Spring-Steels by a Special Inductive Heat Treatment

**A Tump[1,2] and R Brandt[2]**

[1] Mubea Fahrwerksfedern GmbH, Dr.-Karl-Heinz-Muhr-Straße , 99631 Weißensee, Germany
[2] Lehrstuhl für Werkstoffsysteme für den Fahrzeugleichtbau, Universität Siegen – Fakultät IV, Am Eichenhang 50, 57076 Siegen, Germany

**Abstract.** A method for effective lightweight design is the use of materials with high specific strength. As materials e.g. titanium are very expensive, steel is still the most important material for manufacturing automotive components. Steel is cost efficient, easy to recycle and its tensile strength easily exceeds 2,000 MPa by means of modern QT-technology (Quenched and Tempered). Therefore, lightweight design is still feasible in spite of the high density of steel. However, a further increase of tensile strength is limited, especially due to an increasing notch sensitivity and exposure to a corrosive environment. One solution is a special QT-process for steel, which creates a hardness gradient from the surface to the core of the material. This type of tailored material possesses a softer layer, which improves material properties such as fracture toughness and notch sensitivity. This leads to a better resistance to stress corrosion cracking and corrosion fatigue. Due to this optimization, a weight reduction is feasible without the use of expensive alloying elements. To understand the damage mechanism a comprehensive testing procedure was performed on homogeneous and gradient steels. Some results regarding the fracture mechanic behavior of such steels will be discussed.

## 1. Introduction

The application of materials with a high specific strength is a common method to reduce the weight of components that demand a high fatigue strength. Thus, high-strength steels are still important for the manufacturing of automotive components. Steel is cost efficient, easy to procure and easily recyclable compared to other lightweight-materials, e.g. titanium. Steel can have a tensile strength of more than 2,000 MPa that allows components to be designed with considerable weight reduction. However, the use of a high-strength steel requires a certain corrosion protection for components when they are exposed to a corrosive environment.

Alloying elements like Ni, V, Nb, Ti, W or Cu [1], [2], [3] are applied in order to increase the fatigue limit of steels. These elements are able to increase the toughness (Ni), to refine the grain size (V, Nb, Ti), and enhance the corrosion resistance (Cu) [3]. Furthermore, the fatigue limit is enhanced by means of casehardening, nitrating or carburizing [4]. However, increasing the hardness leads to a decrease in ductility, an increase in notch sensitivity [5], [6] as well as the hydrogen embrittlement sensitivity [7]. A new approach is the SLM-technology (Surface Layer Modification) for coil springs that increases the fatigue strength of high-strength spring steels which are exposed to a corrosive environment [8]. The SLM-technology aims to reduce the hardness within a thin surface layer of the spring wire, thus leading to a graded material, which exhibits a locally increased notch sensitivity and corrosion resistance [6].

For a better understanding of the microstructural mechanisms, a fracture mechanical investigation was conducted for homogeneously hardened as well as graded specimens.

## 2. Characterization of the material

For this study, a standard spring steel (54SiCr6) was used. Its composition is given in Table 1.

**Table 1.** Chemical composition of 54SiCr6 (arithmetic mean out of three specimen)

| C | Si | Mn | P | S | Cu | Al | Cr | Ni | V | As |
|---|---|---|---|---|---|---|---|---|---|---|
| 0.535 | 1.400 | 0.700 | 0.008 | 0.005 | 0.050 | 0.023 | 0.250 | 0.125 | 0.005 | 0.005 |

The initial wire rod diameter of $d_{wire} = 20$ mm was firstly reduced to $d_{wire} = 17$ mm by means of a drawing process. Then the drawn rod was tested for local cracks and hole-like flaws in both longitudinal and transverse direction by means of eddy current inspection (Circograph® and Defectomat® from the Institute Dr. Foerster GmbH & Co. KG).

The drawn wire was then inductively heated at a temperature of 900 °C and subsequently water quenched to room temperature. Finally the hardened wire was tempered at different temperatures (minimum 400 °C) in order to have five types of specimens with different tensile strengths ranging from $R_m$ = 1,700 MPa to $R_m$ = 2,200 MPa. The samples to be graded, representing the SLM technology, were treated by an additional inductive heating procedure (Figure 1) within the tempering unit in such a way that an overall tensile strength of $R_m$ = 2,000 MPa was reached.

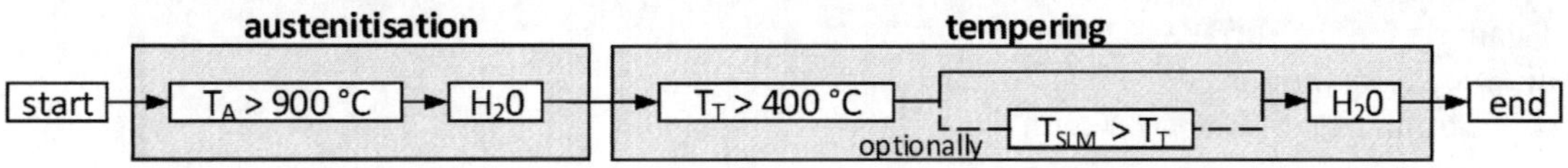


**Figure 1.** Schematic heat treatment process for homogeneous and gradient wire

The resulting martensitic structure contained retained austenite < 4 % (measured with X-ray diffractometer D8 Advance™ from Bruker Corporation). The prior austenite grain size was approximately 10 µm (Table 2).

All four-point-bending specimen (SENB – Single Edge Notched Beam) for fracture mechanical testing were milled and ground out of the QT-wire. Finally, the notches were eroded (Figure 2 – a, b) to a depth of 1.5 mm.

The crack propagation rates da/dN were measured according to ASTM E 647–08 [9].

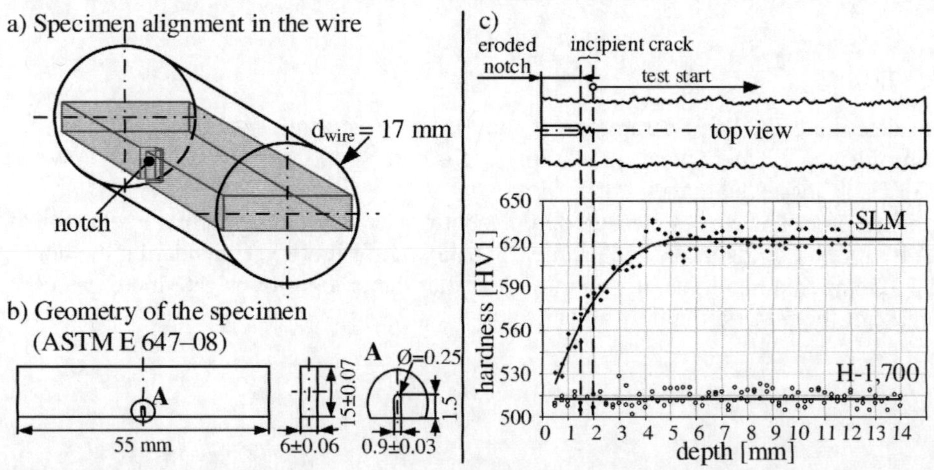


**Figure 2.** a) Specimen position in the wire, b) Geometry of the specimen, c) characteristic hardness profile of the graded material (SLM) and in comparison to the homogeneous material H-1,700

The graded material exhibits a characteristic hardness profile (Figure 2 - c). At the surface the hardness is 530 HV1, and it steadily increases up to 620 HV1 at a depth of approximately 5 mm. From there on the hardness remains constant.

The average material properties, like tensile strength $R_m$, yield strength $R_{p0.2}$, reduction of area Z, Vickers hardness HV1 and grain size class G of all tested specimens are listed in Table 2.

**Table 2.** Material properties of the tested specimen

| Material | tensile test wire [a] [MPa] | | | hardness SENB [b] [HV1] | | | ASTM grain size number [b] | |
|---|---|---|---|---|---|---|---|---|
| | $R_m$ | $R_{P0.2}$ | Z | surface [c] | core [d] | integral | surface [c] | core [e] |
| 1. H-1,700 | 1,690 | 1,462 | 46 | | 512 | | G11.0 | G11.0 |
| 2. H-1,900 | 1,876 | 1,633 | 43 | | 573 | | G10.5 | G10.0 |
| 3. H-2,000 | 1,980 | 1,717 | 41 | | 605 | | G11.0 | G11.0 |
| 4. H-2,100 | 2,091 | 1,810 | 31 | | 635 | | G11.0 | G10.5 |
| 5. H-2,200 | 2,207 | 1,894 | 24 | | 664 | | G10.5 | G10.5 |
| 6. SLM | 2,022 | 1,735 | 45 | 581 | 623 | 615 | G11.0 | G10.5 |

[a] arithmetic mean out of five specimens
[b] arithmetic mean of three specimens
[c] distance from the surface = 2 mm
[d] distance from the surface = 5–15 mm
[e] depth from the surface = 7.5 mm

## 3. Fracture mechanical testing method

The tests were performed at a constant stress ratio of R = 0.2 with a test frequency of f = 120 Hz using a resonant testing machine (Cracktronik with Fractomat from Russenberger Prüfmaschinen AG). Sharp starting cracks were established at a stress intensity level of $\Delta K = 18\ MPa\sqrt{m}$. The crack length was measured with crack measuring films by indirectly measuring the change in electrical resistance. The fracture-analysis were performed by means of scanning electron microscopy (Sigma 300 from Carl Zeiss Microscopy GmbH).

## 4. Results

The characterization of stable crack growth is commonly described by the Paris-Erdogan-Law (1).

$$\frac{da}{dN} = C \cdot (K)^m \tag{1}$$

In order to determine the Paris-parameters (C and m) at least three specimens of each material were tested. Using the measurement points of all samples the corresponding regression curves were calculated by means of Gauß'least square method.

All Paris-Erdogan-law curves measured, representing the stable crack growth behavior of the materials with different hardness levels and profiles, are shown in Figure 3. Considering the homogeneously hardened and tempered materials, it becomes clear that the crack growth rate increases corresponding to the increase of the material hardness level.

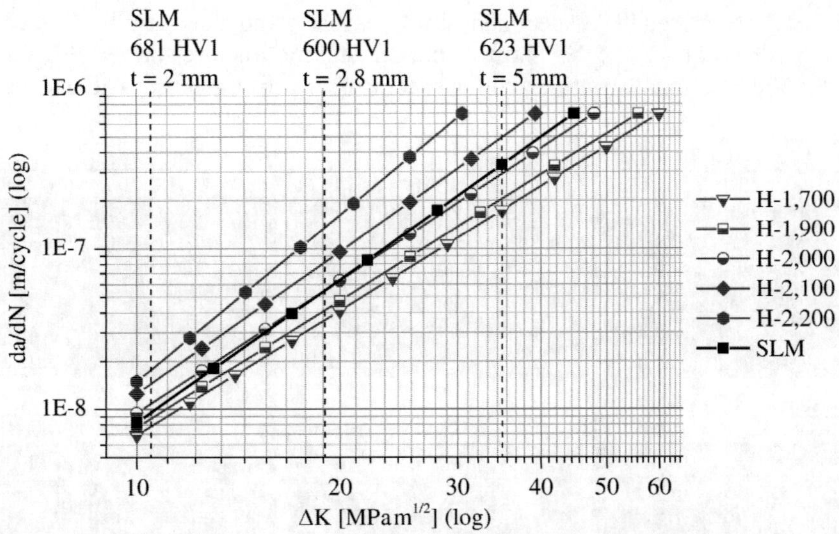


**Figure 3.** Regression curves for each material in the state of stable crack propagation

It is obvious that the curves are shifted upwards with increasing material hardness and the gradient becomes steeper, too (Figure 3). The numerical values of the Paris-Erdogan-law parameter C and m are given in the table of figure (Figure 4. – b).

a)

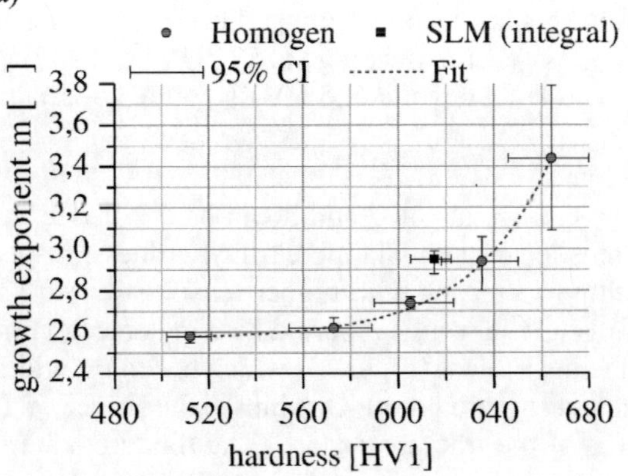


b)

| material | m | C | $R^2$ [%] | s |
|---|---|---|---|---|
| H-1,700 | 2.58 | 1.81E-11 | 96.1 | 0.14 |
| H-1,900 | 2.62 | 1.85E-11 | 96.0 | 0.14 |
| H-2,000 | 2.74 | 1.72E-11 | 94.8 | 0.19 |
| H-2,100 | 2.94 | 1.44E-11 | 75.4 | 0.23 |
| H-2,200 | 3.44 | 5.39E-12 | 68.9 | 0.16 |
| SLM | 2.95 | 9.15E-12 | 95.7 | 0.14 |

with: $R^2$: coefficient of determination
s: standard deviation
CI: confidence interval

**Figure 4.** a) Relation between the hardness and the Paris-Parameter m, b) Paris-Erdogan-law parameters of the different materials

For a better understanding of the change of the growth exponent m with the hardness of the material, the growth rate is plotted vs. the hardness in Figure 4a). At comparatively low hardness levels of 510 HV1 to almost 600 HV1 only a minor increase of m is discovered. However, as soon as the hardness exceeds a level of 600 HV1 the growth exponent m shows a considerable increase with materials hardness. Compared to the homogeneous material, the graded material has a growth exponent value of m = 2.95 which is close to the value of the homogenously hardened material H-2,100. A corresponding symbol for SLM, at its integral hardness, is also plotted in the diagram (Figure 4. – a). It is clearly located above the curve representing the homogeneous materials which means that the gradient is higher. However, the site of the Paris-Erdogan-law curve of SLM lies significantly below the curve representing H-2,100 (Figure 2.).

The fractography of the fractured SLM specimens discloses that homogeneously hardened and tempered materials up to a hardness of ≈ 575 HV1 shows a transcrystalline fracture surface (Figure 5. – a). At a magnification of 2,500 individual striations with a spacing of nearly 1 µm are visible (Figure 5. – a.2).

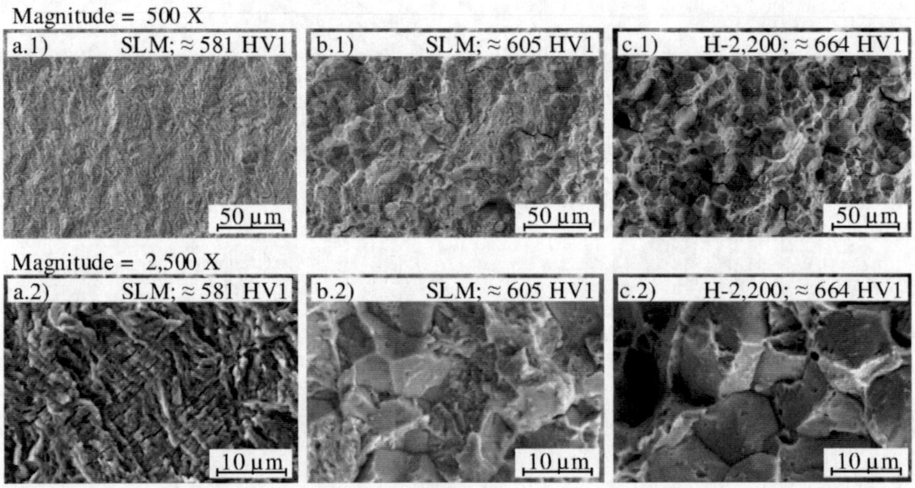


**Figure 5** a) Transcrystalline striations at 581 HV1 (SLM-specimen), b) Mixed fracture patterns of striations and intercrystalline fracture at 605 HV1 (SLM-specimen), c) Nearly complete intercrystalline fracture at 664 HV1 (H-2000-specimen)

The portion of intercrystalline fracture sites increases as the cracks becomes longer, i.e. the core material becomes harder. More and more intercrystalline fracture sites occur in the range of 575 HV1 to 628 HV1 and intercrystalline fracture clearly dominates at a hardness level beyond 628 HV1 (Figure 5. – c.).

## 5. Discussion and concluding remarks

Stable crack propagation of high strength spring steels were investigated and it can be concluded that the crack propagation rate accelerates with increasing tensile strength and increasing hardness, respectively. This result is not unexpected because the ductility decreases with higher tensile strength and similar results can be found in the literature [5]. In contrast to the results reported here indicate not only a mere shift of the Paris-Erdogan-curve but also a change in gradient (Figure 3) which becomes evident by plotting the growth exponent m vs. the material hardness (Figure 4 – a). One hint to the source of the changing gradient was found in the study of the patterns of the fracture surface. The fracture patterns show a change from transcrystalline striations, with a spacing of approximately 1 µm between them, to a nearly pure intercrystalline state with increasing hardness (Figure 5).

Furthermore a method for a fracture mechanical testing procedure for SLM material has been developed. By means of this method, it was possible to analyze different phases of crack growth within a graded material. The crack propagation rate of the graded material is "slow" in the beginning, similar to the rate of the homogeneous material H-1,900. This can be understood by comparing the local SLM-hardness (581 HV1) at a depth of 2 mm (Figure 2 – c) with the average hardness of H-1,900 (573 HV1). The continuously increasing hardness of SLM with increasing crack length leads to a faster increase of the crack propagation rate. At a crack-length of 2.8 mm the crack propagation rate is still below that of the material H-2,000 which has approximately the same tensile strength or the same integral hardness, respectively (Table 2). This implies that SLM-technology can delay the time to failure, considering cracks starting from the material surface or from inclusions within a surface layer of 2.8 mm thickness.

Overall, the gradient steel was subject to fracture mechanical testing under dry conditions and the crack propagation behavior could be characterized. The effect of a corrosive environment was not taken into account and will be a subject for future investigations

**Acknowledgements**
Special thanks to the "Lehrstuhl für Materialkunde und Werkstoffprüfung" of the University of Siegen for conducting the fracture mechanical tests and to the "Europäischer Fonds für regionale Entwicklung" (EFRE) for the financial support.

**Literature**

[1] Masumoto K.; Kaiso M.; Ibaraki N.: Developments of Ultra-high Strength Wire Rods for Valve Springs with Excellent Fatigue Life, Kobe Steel Engineering Reports 2009 – Band 59 Heft 1, pp.67-70, 2009

[2] Perrad F.; Charvieux F., Languillaume J.: A new spring steel with improved ductility dedicated for high strength parabolic leaf springs, 2nd International Conference Super-High Strength Steels, Associazione Italiana di Metallurgia (AIM), 2010

[3] Perrad F.; Mendibide C.; Yoshihara N.; Namimura Y.; Ibaraki N.: High strength spring steels with improved ductility and corrosion resistance, International Conference on Steels in Cars and Trucks (SCT), pp. 106-113, Stahleisen GmbH, 2008

[4] Nishimura T.; Hagiwara Y.; Yamada K.; Ozone T.; Itoh Y.; Hayashi H.; Homma S., The Valve Springs Carbo-Nitrided at High Temperature for High Speed Engines; SAE Technical – Paper Series, no. 890777, 1989, ISSN 0148-7191,

[5] Tange A.; Akuto T.; Takamura N.: Relation between shot-peening residual stress distribution and fatigue crack propagation life in spring steel, Transactions of Japan Society of Spring Engineers – Vol. 1991 No. 36, pp. 47–53, NHK Spring Co.,Ltd., 2010

[6] Tump A.; Brandt R.; Klapprott S.: Gradientenstahl für Achsfedern, Tagungsband des Ilmenauer Federntags; 2013

[7] Kobayashi K.; Yamahuchi A.; Komazaki S.-i.; Misawa T.; Kohno Y.; Fukuzumi T.: Effekt of tensile strength on pitting corrosion resistance and hydrogen desorption profile in automobile spring steels, 11th International Congress on Fracture, 2005

[8] Neubrand J.; Hartwig M.; Patent term – DE 10 2009 011 118 A1, Muhr und Bender KG, 2010

[9] ASTM E 647–08, Standard Test Method for Measurement of Fatigue Crack Growth Rates, ASTM International

# Development of microstructure in high-alloy steel K390 using semi-solid forming

**K Opatova[1], D Aisman[1], K Rubesova[1], K Ibrahim[1] and S Jenicek[1]**

[1]University of West Bohemia in Pilsen, Univerzitni 22, 30614 Pilsen, Czech Republic

opatovak@rti.zcu.cz

**Abstract**. Semi-solid processing of light alloys, namely aluminium and magnesium alloys, is a widely known and well-established process. By contrast, processing of powder steels which have high levels of alloying elements is a rather new subject of research. Thixoforming of high-alloy steels entails a number of technical difficulties. If these are overcome, the method can offer a variety of benefits. First of all, the final product shape and the desired mechanical properties can be obtained using a single forming operation. Semi-solid forming can produce unusual powder steel microstructures unattainable by any other route. Generally, the microstructures, which are normally found in thixoformed steels, consist of large fractions of globular or polygonal particles of metastable austenite embedded in a carbide network. An example is the X210Cr12 steel which is often used for semi-solid processing experiments. A disadvantage of the normal microstructure configuration is the brittleness of the carbide network, in which cracks initiate and propagate, causing low energy fractures. However, there is a newly-developed mini-thixoforming route which produces microstructures with an inverted configuration. Here, the material chosen for this purpose was K390 steel, in which the content of alloying elements is up to 24 %. Its microstructure which was obtained by mini-thixoforming did not contain polyhedral austenite grains but hard carbides embedded in a ductile austenitic matrix. This provided the material with improved toughness. The spaces between the austenite grains were filled with a eutectic in which chromium, molybdenum and cobalt were distributed uniformly. After the processing parameters were optimized, complex-shaped demonstration products were manufactured by this route. These products showed an extraordinary compressive strength and high wear resistance, thanks to the hardness of their microstructure constituents, predominantly the carbides.

**Keywords:** Semi-solid state, high-alloy steel, K390, thixoforming

## 1. Introduction

Semi-solid processing of powder steels is a new and technically interesting technique which enables complex components with advanced properties to be produced using a single forming operation. It allows engineers to combine the advantages of complex-shapes of castings and mechanical properties of forgings in materials which – due to their high alloy levels – are difficult to form by conventional methods, not to mention making small and intricate-shaped parts by forming [1]. The purpose of the present experiment was to explore the potential of the above-mentioned material for mini-thixoforming. The process parameters were gradually optimized in order to obtain complex-shaped products. The forming process took place in a closed cavity of a die which had been specially developed for the mini-thixoforming technique. Using this tool, thin-walled products with the wall thickness of approximately 1 mm can be made.

Thixoforming involves heating a material to the region between the solidus and liquidus temperatures and forming it in a partially melted state [2]. Relevant data for this experiment was predicted with the aid of calculation in JMatPro software. Based on this data, initial conditions were specified for this steel. The most important of those was the temperature, at which the material contains 15 % liquid fraction. The preliminary calculation clearly showed that achieving the desired liquid fraction would be difficult. The reason is that this parameter increases very steeply with temperature. The temperature range available for semi-solid forming thus becomes narrow: approximately 60 °C (Fig. 1). For this reason, the process parameters were subsequently modified, namely the heating profile and temperature.

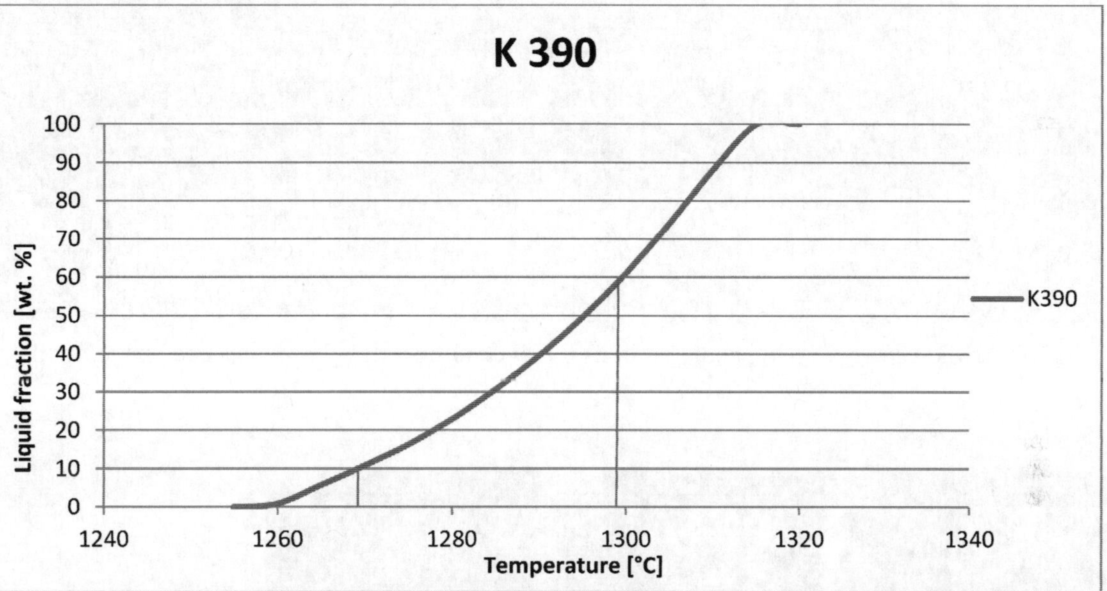


**Figure 1.** The actual temperature interval for effective thixoforming – liquid between 10 % and 60 %.

## 2. Experimental programme

In order to obtain the desired "inverted" microstructure, the chemical composition of the material had to be altered adequately, as it was substantially different from the chemistries of those steels which are normally used for thixoforming experiments [3]. The objective was to preserve carbide particles in the material in semi-solid state. After the process, these would remain embedded in metastable austenite. The formation of carbide network would be suppressed. Based on calculations, the high-alloy K390 powder steel was chosen as the experimental material. It has high vanadium (up to 8.94 %) and carbon (up to 2.48 %) levels (Tab. 1). Given its high hardness and wear resistance, the K390 steel is typically used for making tools.

Its initial microstructure contains a ferrite matrix with high levels of, molybdenum and chromium, and with uniformly distributed carbides. Its hardness is 255 HV30. The carbides are mainly vanadium carbides with sizes between tenths of micrometres and 2 μm. Image analysis shows that their proportion in the microstructure is approx. 35 % (Fig. 2). Distributions of alloying elements were mapped using microscopic observation and EDS analysis (Fig. 3).

**Table 1.** Chemical composition of experimental material.

| Amounts of alloying elements [wt. %] | C | Cr | Cu | Co | Mn | Mo | Ni | Si | V | W |
|---|---|---|---|---|---|---|---|---|---|---|
| **K390** | 2.48 | 4.20 | 0.10 | 1.95 | 0.40 | 3.64 | 0.19 | 0.57 | 8.94 | 0.85 |

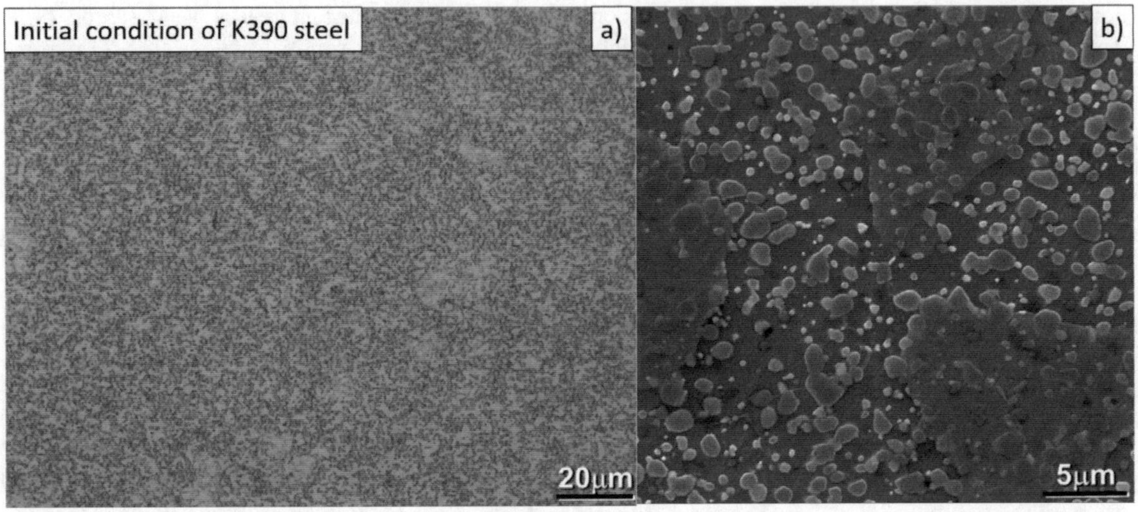


**Figure 2.** Microstructure of the material in its initial condition: a) light microscope image, b) SEM image.

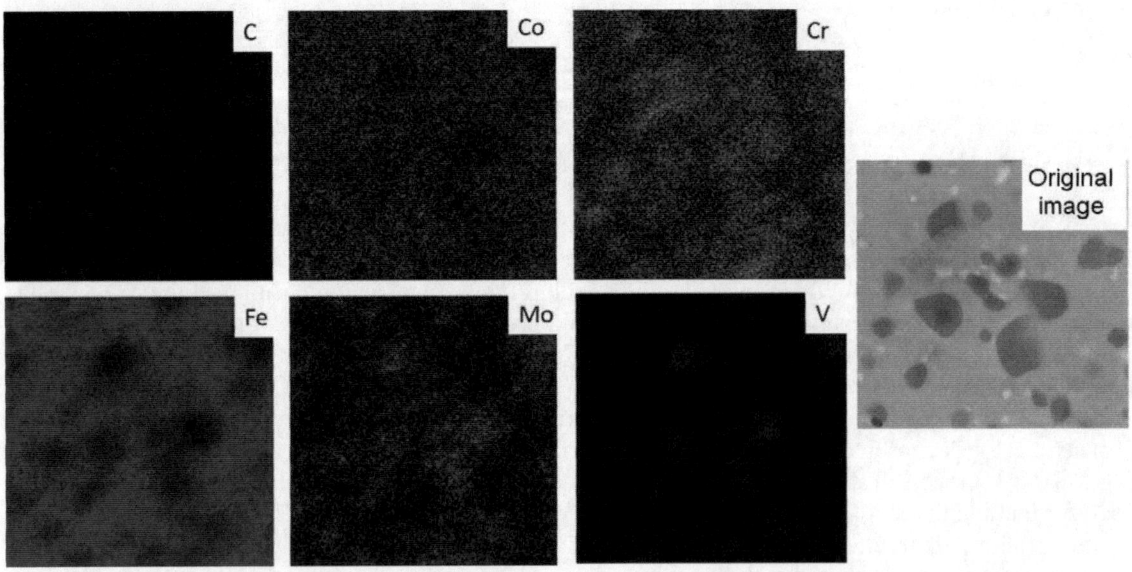


**Figure 3.** Maps of distributions of chemical elements in the initial condition of the material.

## 3. Choice of parameters and semi-solid processing

The experiment was carried out in a specially developed device which comprised a titanium die for mini-thixoforming. Cylindrical stock of 6 mm diameter and 50 mm length with truncated cone ends was used. The purpose of the conical ends was to align the stock to the axis of the die and provide contact surfaces for energy supply via copper grips. In order to provide a wider range of product shapes, the die had an exchangeable shape insert. The stock was heated by electrical resistance and induction while placed in the die. The design of the forming device provides high rates of heating, solidification and cooling. With a digital high-frequency control system, the entire process could be controlled accurately, including the forming route and the heating profile [4].

As this process is impossible to describe and calculate to adequate accuracy using numerical methods, it was optimized iteratively by experiment. The forming process parameters were optimized, until the die cavity filled completely with semi-solid metal and the product's final microstructure was uniform. An important aspect of mini-thixoforming is accurate temperature control for achieving the desired liquid fraction. It was found that the K390 steel begins to melt at 1255 °C and becomes fully melted at 1315 °C. The available range for semi-solid forming is therefore approximately 60 °C. The actual temperature interval for effective thixoforming is, however, narrower because the liquid fraction should be kept between 10 % and 60 % [5].

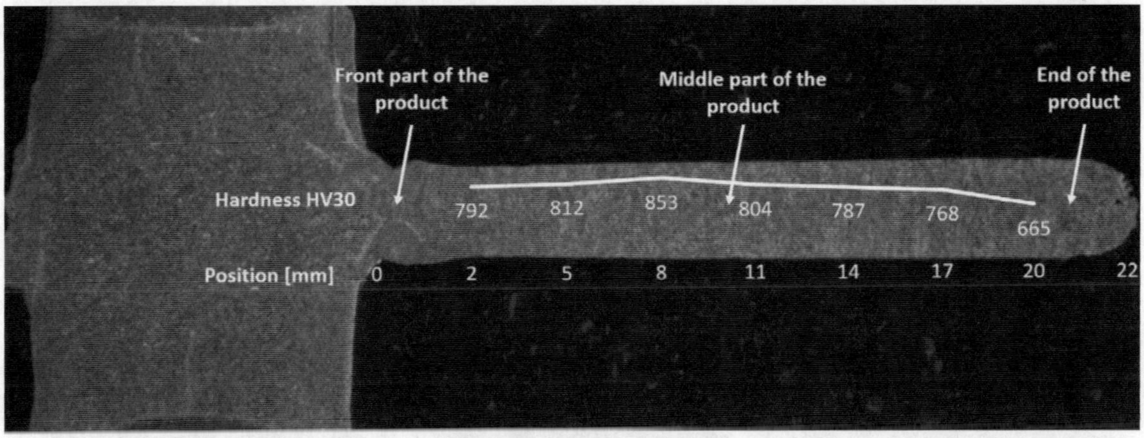


**Figure 4.** Hardness and microstructure analysis on axial cross-section through the product.

## 4. Results and discussion

The first forming run took place at a temperature at which, according to the preliminary calculation, the material should contain 15 % liquid fraction. The stock was heated to 1270 °C over 56 seconds and then formed using a maximum force of 6.5 kN. The deformation itself took less than 0.3 seconds. This temperature proved too high, as the stock lost shape stability owing to excessive melting. The material made uncontrolled local contact with the die wall, and the temperature field was thus disturbed. As a consequence, the die cavity was not filled. In the next runs, the temperature was reduced step-by-step, until the appropriate temperature for complete filling of the die cavity was found. In each case, the material flow through the die cavity was subsequently studied by metallographic observation of transverse sections through the product [6]. Besides optical microscopic and electron microscopic analysis, hardness measurement was undertaken in various locations within the product (Fig. 4). Amounts of individual microstructural constituents were determined by X-ray diffraction analysis (using a diffractometer with a Co-Kα source).

Heating to 1253 °C was found to be the most suitable option. At this temperature, the stock becomes melted to a sufficient extent, while retaining shape stability. It is then easily extruded into the product die cavity.

Microstructures of the products were studied. In accordance with the expected use of high-stability carbides of vanadium, an "inverted" microstructure was produced. The chemical composition of the material, the high heating rate and rapid cooling caused the initial ferrite matrix to transform into an austenite-martensite mixture. Chromium-vanadium eutectic formed as well, and the initial vanadium carbides were retained. The temperature at which the latter dissolve is 1280 °C. The processing temperature was thus lower enough than this dissolution temperature. These retained carbides were more spherical than their initial form, due to diffusion and mechanical effects, similar to the newly-formed ones [7]. The chromium-molybdenum-vanadium eutectic was present along austenite grain boundaries. Scanning electron microscopic observation and EDS studies, performed on Zeiss microscope with Oxford Instruments detectors, showed that a majority of alloying elements redistributed during thermomechanical processing and that the nature of the matrix changed profoundly (Fig. 5). The microstructure was homogeneous along the entire product axis. Only the region near the entry opening of the die cavity contained less globular and non-uniformly distributed carbides (Fig. 6). This was caused by the short path they travelled and by the less intensive mechanical interaction with the melt and other solid particles. These joint effects also slowed down diffusion which normally greatly contributes to spheroidization.

As the nature of the matrix changed and part of the super-saturated austenite transformed to martensite, the hardness increased to a level as high as 853 HV30.

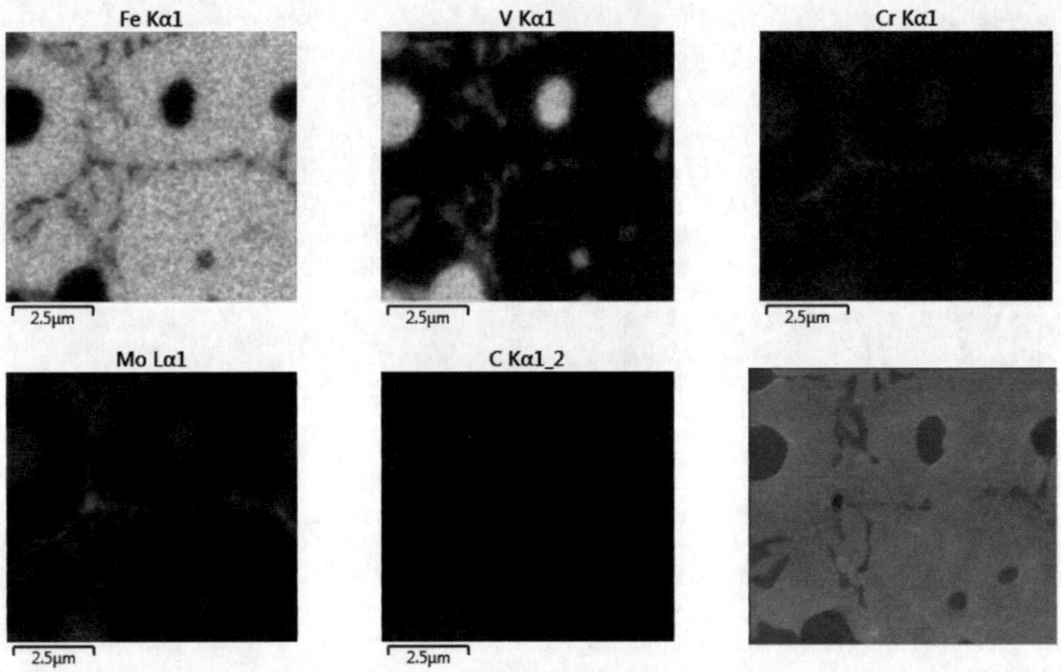


**Figure 5.** EDS Maps of redistributions of chemical elements after mini-thixoforming.

X-ray diffraction analysis was employed to determine the fractions of microstructural constituents in the product. The matrix consisted of austenite grains, carbide network along grain boundaries and a low fraction of martensite needles (Fig. 6 c)). The presence of martensite could also be derived from hardness values (Fig. 4). The amounts of individual phases in the microstructure of K390 steel are in agreement with its chemistry. Given the high amount of vanadium, a majority of carbides contained this element. The structure also includes chromium and molybdenum carbides which are less frequent (Tab. 2).

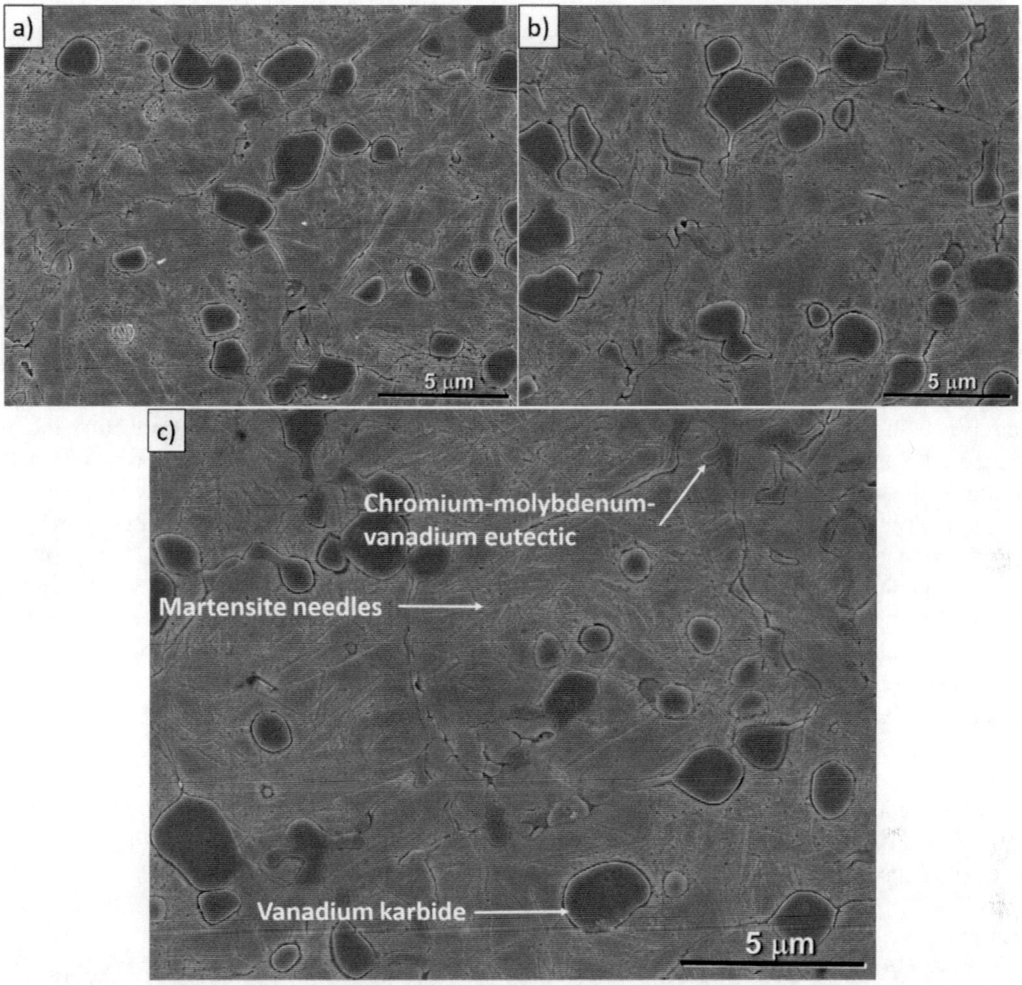


**Figure 6.** Microstructure of K390 steel upon mini-thixoforming; a) front part of the product; b) end of the product; c) middle part of the product.

**Table 2.** Fractions of microstructural constituents in K390 steel upon semi-solid processing, as determined by X-ray diffraction analysis. Measurements were performed on an automated powder diffractometer AXS Bruker D8 Discover with planar, positionally sensitive detector HI-STAR and a cobalt-ray lamp ($\lambda_{K\alpha}$ = 0.1790307 nm).

| | Phase | Content [wt. %] |
|---|---|---|
| | Austenite | 31.3 |
| | $Fe_2VSi$ | 16.9 |
| | VC | 15.5 |
| **K390** | Martensite | 10.8 |
| | $Fe_7V_3$ | 9.1 |
| | $V_4WC_5$ | 7.6 |
| | $Mo_2C$ | 4.5 |
| | $Fe_7Co_3$ | 4.3 |

## 5. Conclusions

The primary objective of this research was to produce an "inverted" microstructure which is profoundly different from all microstructure types known so far in semi-solid processed materials. When high-alloy tool steels are thixoformed, their final microstructure is typically strong but brittle, containing polygonal grains of super-saturated metastable austenite embedded in a carbide-austenite network. The ductile austenitic constituent deforms readily but, on the whole, the carbide-austenite network makes the material very brittle. For this reason, an alternative-type microstructure was sought which would eliminate or, at least, suppress these deficiencies. One of the solutions found involved the use of the high-alloy tool steel K390, which contains 8.94 % vanadium, for semi-solid processing. It is the high vanadium content which causes the highly-stable vanadium carbides to remain undissolved in the semi-solid state and enables them to greatly contribute to the alteration of the microstructure. To translate this concept into practice, a semi-solid processing technique was proposed which is now known as mini-thixoforming. Besides rapid heating, it also benefits from rapid solidification and cooling processes. Its parameters were experimentally optimized step by step. With the aid of calculations and experiments, a suitable heating temperature in the semi-solid range was found. At this temperature, the material can be thixoformed. Very small intricate-shaped products were obtained whose microstructures were substantially different from the microstructures obtained previously by semi-solid processing. These experimentally-produced microstructures consisted of a matrix of metastable and highly super-saturated austenite, in which globular vanadium carbides were embedded. The austenite also contained martensite and a small fraction of carbide-austenite eutectic. The best results were obtained using the route which involved heating to 1253 °C over 60 seconds and lateral extrusion into the shape die cavity for the product. Using these conditions, products with good surface quality and the desired "inverted" microstructures were obtained repeatedly.

**Acknowledgments**
This paper includes results created within the project LO1502 Development of Regional Technological Institute. The project is subsidised by the Ministry of Education, Youth and Sports from specific resources of the state budget for research and development.

## References

[1] Jirkova H, Aisman D, Sen I, Wagner M F – X, Behulova M, Kusy M and Masek B 2013 Mini-thixoforming of a steel produced by powder metallurgy *Solid State Phenomena* **192-193** 500-5
[2] Hirt G, Bleck W, Bührig-Polaczek A, Shimahara H, Püttgen W and Afrath. C 2006 Semi solid casting and forging of steel *Solid State Phenomena* **116-117** 34-43
[3] Mikmekova S, Masek B, Jirkova H, Aisman D, Mullerova I and Frank L 2013 Microstructure of X210Cr12 steel after the forming in semi-solid state visualized by very low energy SEM in ultra high vacuum *Applied Surface Science* **275** 403-8
[4] Masek B, Aisman D, Behulova M and Jirkova H 2010 Structure of Miniature components from steel produced by forming in semi-solid state *Transactions of Nonferrous Metals Society of China* **20** 1037-41
[5] Fleming M C 1991 Behavior of metal alloys in the semisolid state, *Metallurgical Transactions, A* **22** 957-81
[6] Püttgen W, Bleck W, Hirt G and Shimahara H 2007 Thixoforming of Steels – A Status Report *Advanced Engineering Materials* **9** 231-45
[7] Vancura F, Masek B, Aisman D, Jirkova H, Wagner M F – X and Bohme M 2014 Modification of metastase microstructure of CPM15V steel by heat exposure after treatment in semi-solid state *Journal of Alloys and Compounds* **586** 159-64

# Capabilities of Unconventional Processing of Multiphase AHSS Steels

**H Jirková, B Mašek*, C Štádler and Š Jeníček**

*masekb@kmm.zcu.cz

University of West Bohemia, CZ-30614 Pilsen, Czech Republic

**Abstract**. Today, new types of materials and procedures are sought continuously in order to achieve lower manufacturing costs, reduced energy consumption, shorter production times and other savings. In terms of the materials, TRIP steels are an attractive choice, as they provide an excellent combination of strength and ductility. They also offer good energy absorption in crash scenarios. Their main use is in the production of automotive body parts. One can expect that well-chosen processing parameters and unconventional forming routes would enable a wider range of thin-walled products to be made of these steels. Those could include thin-walled hollow products with excellent mechanical properties imparted by effective manufacturing routes at relatively low costs. If these materials are to be employed in real-world forming processes, an appropriate forming route must be chosen, integrated into an appropriate production chain and then optimized in terms of its parameters. This article describes a study of a rotary spin extrusion process. In the first stage, the impact of strain magnitude on microstructural evolution was studied in CMnSi steel using physical modelling of thermomechanical treatment. Subsequently, trials of a real-life technology chain, which efficiently combined incremental forming and heat treatment, were carried out on low-alloy CMnSi and CMnSiNb steels. The resulting products were stepped hollow parts of various diameters. Their strength was close to 1000 MPa and their elongation level exceeded 20 %.

**Keywords**. TRIP, spin extrusion, incremental forming, hollow parts, AHSS

## 1. Introduction

One of the available ways of improving the efficiency of engineering production involves using low-alloy steels, i.e. low-cost materials, and processing them by innovative techniques to obtain very complex shapes of products with excellent mechanical properties. The aim of this experiment was to conduct trials of rotary spin extrusion and reduction rolling on low-alloy multiphase TRIP steels, and thus produce stepped hollow products from solid bar stock without drilling axial holes by metal cutting methods.

TRIP steels possess multiphase microstructures which consist of ferrite, bainite and retained austenite. Their great asset is an excellent combination of high strength and good ductility [1]. Today, they are mostly used for making sheet components in automotive industry. However, recent research efforts have been focused on several novel combinations of multiple processes based on thermomechanical treatment or heat treatment and incremental warm and cold forming of bulk stock [2, 3].

## 2. Experimental programme

The above reasons inspired the authors of this study to use multiphase TRIP steel stock for trials of unconventional production of rotation-symmetric hollow products by forming. The technology chain comprised warm rotary spin extrusion, in-process annealing in the intercritical range, and reduction rolling. Stepped hollow thin-walled demonstration products were obtained, with outer diameters ranging from 37 to 40 mm and with a wall thickness of approximately 4 mm.

The experimental programme was divided into two stages. The first one comprised physical modelling of thermomechanical treatment in a thermomechanical simulator. Its purpose was to optimize the process variables. These findings were then used for the actual production of thin-walled products.

The fraction of retained austenite in the material was determined by either X-ray diffraction analysis or image analysis. The latter was also used for finding the fraction of ferrite and the ferrite grain size on metallographic sections which had been prepared by two-step etching with 3 % nital and 10 % aqueous solution of $Na_2S_2O_3$.

### *2.1. Experimental materials*

Two TRIP steels were employed: C-Mn-Si and C-Mn-Si-Nb (Tab. 1). The conventional chemistry of the C-Mn-Si steel is suitable for obtaining the desired final microstructure. The other steel only differs in having niobium as a microalloying addition, which contributes to its slightly higher strength.

**Table 1.** Chemical composition of the experimental materials.

| | C | Mn | Si | P | S | Cr | Ni | Cu | Al | Nb | Mo | W |
|---|---|---|---|---|---|---|---|---|---|---|---|---|
| T1 | 0.206 | 1.418 | 1.849 | 0.007 | 0.005 | 0.007 | 0.071 | 0.059 | 0.006 | 0.002 | 0.02 | 0.02 |
| T2 | 0.21 | 1.449 | 1.797 | 0.008 | 0.005 | 0.008 | 0.072 | 0.058 | 0.006 | 0.059 | 0.02 | 0.02 |

### *2.2. Physical modelling of thermomechanical sequence*

Physical modelling of the thermomechanical sequence was carried out in a thermomechanical simulator with high frequency resistance heating. The samples for the experiment had 8 mm diameter and active length of 16 mm. The crucial aspects are the temperature profile and the strain intensity. At this stage, the focus was on the latter. Besides its role in imparting the required shape to the workpiece, deformation has a substantial impact on grain refinement. In TRIP steels, deformation also promotes the formation of ferrite. Deformation in the intercritical range between 730 and 800 °C increases the volume fraction of ferrite by accelerating its nucleation process [4, 5]. It also accelerates the nucleation of bainite but retards the subsequent bainite growth, thus reducing the size of bainitic regions and the resulting volume fraction of bainite [4, 5]. At the same time, it leads to larger fractions of retained austenite which thus exhibits finer grain and contains higher carbon levels. This austenite is typically found in granular form which facilitates the TRIP effect better than the acicular retained austenite. Deformation applied in the intercritical region therefore stabilizes retained austenite both chemically and mechanically [5].

The thermomechanical treatment involved heating to 900 °C and soaking for 20 seconds. After soaking, several different schedules were applied which involved various numbers of deformation steps and various levels of total strain (Fig. 1). High number of deformation steps is characteristic of the rotary spin extrusion process. In this process, three rollers introduce strain into the workpiece, change its shape and dictate the microstructure evolution. The workpiece was held at the temperature of bainitic transformation of 425 °C for 600 s and then cooled in air. In the real-world process, this holding step takes place in a furnace.

The deformation schedules took place in 900-720 °C range. The first schedule involved twenty deformation steps. In the subsequent 40-step schedules, two different deformation step magnitudes were used (Tab. 2). Thus, two different true strain levels were applied. The same deformation step magnitude was used in the 60-step deformation schedule. (Tab. 2).

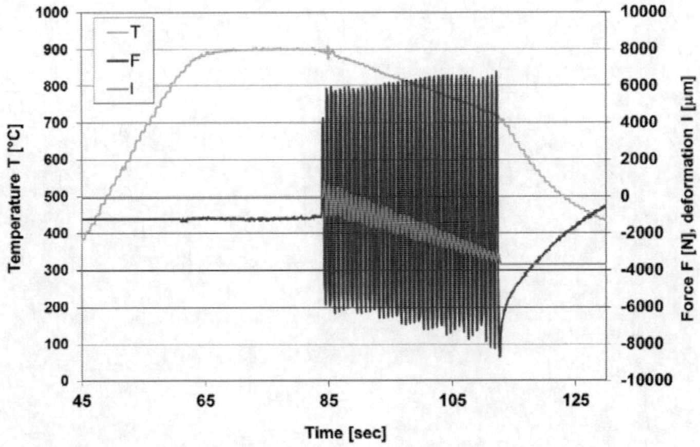


**Figure 1.** Example of thermomechanical treatment – 900 °C/20 s - 40-step incremental deformation – 425 °C/600 s

The true logarithmic strain φ imparted in the twenty-step schedule was 2.8. The resulting ferritic-bainitic microstructure contained fine ferrite grains and consisted of more than 50 % ferrite and about 10 % retained austenite. The size of bainite particles was under 10 µm.

The true strain imparted in the 40-step schedule with smaller deformation steps was 5.8. This is more than twice as much as in the 20-step deformation schedule. It led to very fine ferritic-bainitic microstructure with 55 % ferrite and 9 % retained austenite (Fig. 2). The ferrite grain size of 2.1±0.9 µm is close to the known physical limit of grain refinement by means of thermomechanical treatment.

The next 40-step schedule comprised deformation steps which were twice as large as in the previous case. The true strain was 10.4, almost double the value in the previous schedule. The microstructure configuration was very similar to the previous case. The deformation promoted the formation of ferrite whose fraction was up to 60 % (Fig. 3).

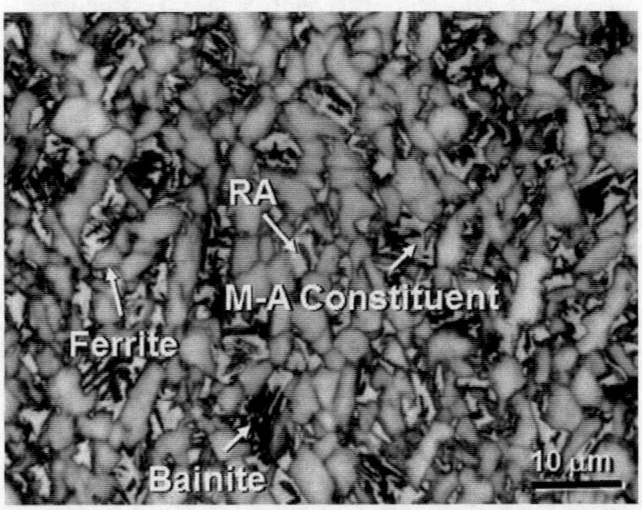


**Figure 2.** Micrograph of TRIP steel after 40-step incremental deformation with small deformation steps applied during continuous cooling from 900 to 720 °C. Two-step etch

**Figure 3.** Micrograph of a TRIP steel after 40-step incremental deformation with larger deformation steps applied during continuous cooling from 900 to 720 °C. The total strain level was 10.4. Nital etch

**Figure 4.** Detail of bainite sheaf, nital etch

The detail in the scanning electron micrograph shows a bainite sheaf with very fine laths free of any notable carbide precipitates, with polyhedral proeutectoid ferrite grains, and distinct retained austenite regions located between ferrite grains, as well as between bainitic ferrite laths (Fig. 4).

Carbide precipitation during bainitic transformation was suppressed and ferrite formation promoted by the additions of silicon and manganese. Silicon, an element insoluble in cementite, prevents or at least retards carbide precipitation during bainitic transformation, and promotes the diffusion of carbon into retained austenite. Manganese, an austenite stabilizer, increases the solubility of carbon in austenite and expands the region available for cooling because it delays the formation of pearlite. [1]

After the 60-step schedule with the total strain level of 13.4, the microstructure was similar to those obtained in the previous cases. The fraction of ferrite has not increased and its grain has not become finer than before. In the next schedule, the total strain level was increased further and the workpiece temperature was decreased but the material's ability to undergo plastic deformation proved insufficient and the specimen failed.

In one of the thermomechanical treatment variants which may prove viable in practice, first deformation steps are applied even before the end of the austenitizing hold, i.e. before the temperature starts to decrease. This was tried in a schedule which therefore comprised both isothermal and non-isothermal forming. The first 10 deformation steps were applied isothermally at 900 °C. They were immediately followed by another non-isothermal 10 steps, with the deformation sequence ending at 720 °C. The resulting fine ferritic-bainitic structure was almost identical to that obtained after the non-isothermal 20-step schedule in the 950-720 °C range.

**Table 2.** Impact of total strain on ferrite evolution

| Temperature interval of deformation [°C] | Number of deformation steps | φ [-] | Microstructure | Ferrite grain size [µm] | Ferrite fraction [%] | Retained austenite fraction [%] |
|---|---|---|---|---|---|---|
| 900 – 720 | 20 | 2.8 | ferrite + bainite +retained austenite | 2.3±1 | 51 | 11 |
| | 20* | 2.8 | | 2.3±1.2 | 56 | 11 |
| | 40 | 5.8 | | 2.1±0.9 | 55 | 9 |
| | 40 | 10.4 | | 2.1±0.9 | 60 | 10 |
| | 60 | 13.4 | | 2.1±1.1 | 59 | 10 |
| | 60 | 15.8 | | Specimen failure | | |

* 10 deformation steps at the austenitizing temperature of 900 °C and 10 deformation steps during cooling from 900 to 720 °C

### 2.3. *Manufacturing process*

In the preceding experiment, it was found that the material can sustain severe plastic deformation. A manufacturing route was therefore proposed for making thin-walled hollow products by incremental forming, and its trials were carried out on C-Mn-Si and C-Mn-Si-Nb steels. In the first step, bar stock of 58 mm diameter was converted by warm rotary spin extrusion to semi-finished products of 51 mm diameter with approx. 9 mm wall thickness (Fig. 5). The products were annealed in an intercritical temperature range. Their ferritic-pearlitic microstructure then transformed to multiphase ferrite-bainite-retained austenite microstructure with a good plasticity and a potential for the TRIP effect. The semi-finished products were then cold reduction-rolled. This operation produced two steps on the outer diameter of the product. The first stepped-down diameter produced was 42 or 44 mm and the second one was 37 mm.

**Figure 5.** Hollow product after rotary spin extrusion at 680-700 °C. The hollow portion is on the left, the middle part is the input stock and the right portion of the workpiece is the grip hold

*2.3.1. Rotary spin extrusion.* Both experimental materials were extruded after heating in a furnace to 750 °C. The forming operation was monitored with a thermal imaging camera. By analyzing the temperature field within the stock, it was found that the extrusion process started at 680-700 °C. Just after the end of the forming operation, the temperature within the semi-finished product was still relatively uniform: approximately 650 °C.

In the semi-finished products of both materials, the microstructures consisted of ferrite and pearlite (Fig. 6). The microstructure was studied in a region between the point of entry of the mandrel and a depth of approximately 40 mm. Down to the depth of 5-10 mm, ferrite grains and regular pearlite colonies were found. In the extrusion direction, this continuously changed into a deformed microstructure. The ferrite grains were heavily distorted and the pearlite colonies disintegrated.

*2.3.2. Intercritical annealing.* The temperature of the subsequent intercritical annealing (austenitizing) was 810 °C and the annealing times were either 15 or 30 minutes. The products were then cooled to 420 °C to prepare for bainitic transformation which was to take place in a salt bath. The salt bath was used instead of a furnace. The holding time was 8 minutes and was followed by water cooling.

In the C-Mn-Si steel, the microstructure upon intercritical annealing consisted of retained austenite islands in a ferritic matrix (Fig. 7). Only a very small volume fraction of bainite was found. Martensite-austenite (M-A) constituent was detected in several retained austenite islands in carbon extraction replicas. No differences between the experimental steels or between the outcomes of 15-minute and 30-minute austenitizing holds were found. This was confirmed by observation of carbon extraction replicas (Fig. 3). Several cementite lamellae were found which, given their amount, should not have any substantial impact on mechanical properties of the material. The difference between the holding times had no impact on the amount of retained austenite. In the niobium-free steel, the fraction of retained austenite was 15 % after the 15-minute hold and 17 % after the 30-minute hold. In the niobium-alloyed steel, the amounts of retained austenite were 16 % and 15 % (Table 3).

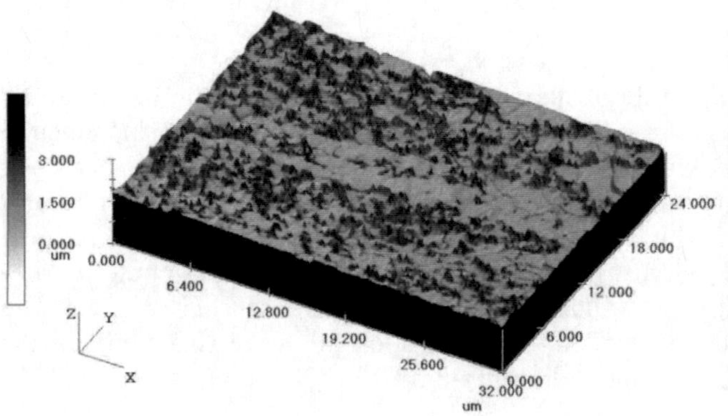


**Figure 6.** 3D-image of an etched metallographic section through the C-Mn-Si steel (T1) upon extrusion at 750 °C

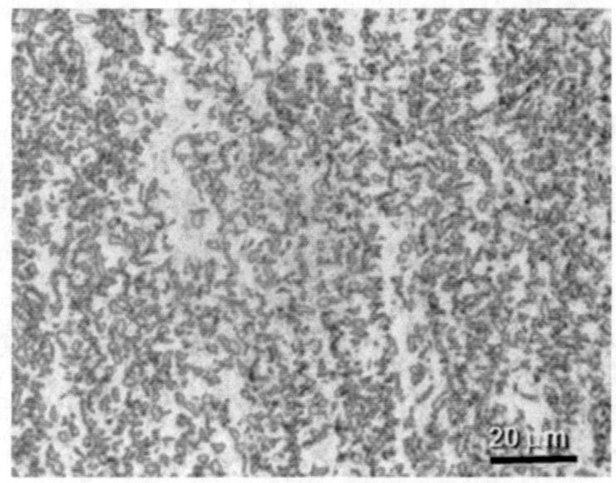


**Figure 7.** Micrograph of Mn-Si specimen (T1) upon intercritical annealing for 15 minutes

Mechanical properties were measured using miniature tension test (Table 3). The samples were located in the wall of the hollow product with the longitudinal axis in the same direction. The samples were taken from the part with uniform structure and its size was 2 x 1.5 x 5 mm. Test pieces of the niobium-free steel upon 15-minute holding showed an elongation of 30 % and an ultimate strength of 904 MPa. Test pieces which had been austenitized for 30 minutes showed an elongation of 32 % and an ultimate strength of 927 MPa. The holding times had only a minor impact on mechanical properties. The difference between the ultimate strength levels was a mere 23 MPa. Higher strength of the niobium-free steel may be due to the slightly larger fractions of hardening structure and retained austenite.

The niobium-containing steel which had been held at temperature for 30 minutes had an ultimate strength of 994 MPa. After the austenitizing time was extended from 15 to 30 minutes, the ultimate strength increased by 40 MPa but elongation levels remained unchanged: 28 % in both cases. The strength of the C-Mn-Si-Nb steel was higher than that of the niobium-free steel upon both schedules. High elongation level of the C-Mn-Si-Nb steel remained unchanged. The higher strength was caused by niobium carbides which greatly affect solid solution strengthening.

**Table 3.** Intercritical annealing schedules and mechanical properties

| Steel | Schedule | RA [%] | $R_m$ [MPa] | | $R_e$ [MPa] | | $A_{5mm}$ [%] | |
|---|---|---|---|---|---|---|---|---|
| Mn-Si (T1) | Furnace 810 °C/15 min. - salt bath 420 °C/8 min - water | 15 | 914 / 893 | 904 | 460 / 397 | 429 | 29 / 31 | 30 |
| | Furnace 810 °C/30 min. - salt bath | 17 | 924 | 927 | 454 | 451 | 33 | 32 |

| | | | | | | | | |
|---|---|---|---|---|---|---|---|---|
| | 420 °C/8 min - water | | 930 | | 447 | | 31 | |
| Mn-Si-Nb (T2) | Furnace 810 °C/15 min. - salt bath 420 °C/8 min - water | 16 | 1031 / 855 | 943 | 511 / 436 | 473 | 24 / 31 | 28 |
| | Furnace 810 °C/30 min. - salt bath 420 °C/8 min - water | 15 | 993 / 994 | 994 | 543 / 553 | 548 | 27 / 29 | 28 |

Based on the above-described results, semi-finished test products were heat treated using intercritical annealing with a 15-minute hold at 810 °C and subsequent bainitic transformation in a salt bath at 420 °C for 8 minutes.

*2.3.3. Reduction rolling*. The semi-finished products were reduction-rolled to obtain hollow products with a wall thickness of 4 mm. The reduction was 6 %. Two steps were formed on the product. The first was a step from 50 mm to 42 mm diameter, and the second had the smaller diameter of 37 mm. Samples for mechanical testing and metallographic observation were taken from both stepped-down areas.

The resulting microstructure consisted of deformed ferritic matrix with disintegrated particles of MA-constituent (Fig. 8). Final material properties and fatigue behaviour will be the topics of future research.

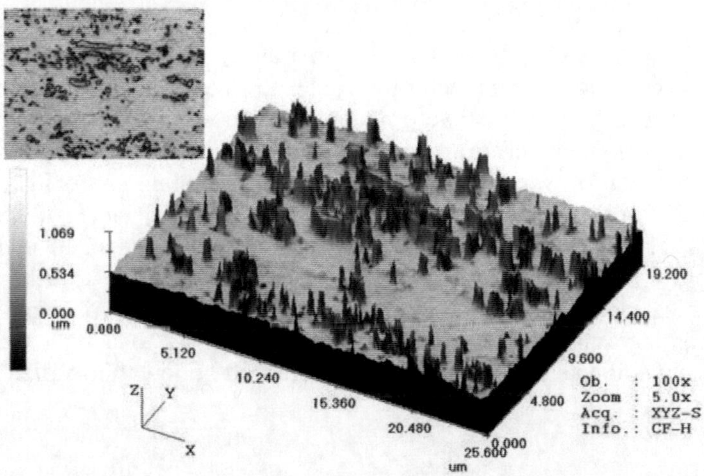


**Figure 8.** Three-dimensional representation of an etched metallographic section through C-Mn-Si steel (T1) upon cold reduction rolling, nital etch

With regard to the very good results obtained thus far, another schedule with twice as large reduction was undertaken. Instead of using the 6% reduction, stepping down to the diameter of 44 mm was carried out using 12% reduction. Another step to 37 mm was formed in the same fashion. (Fig. 9).

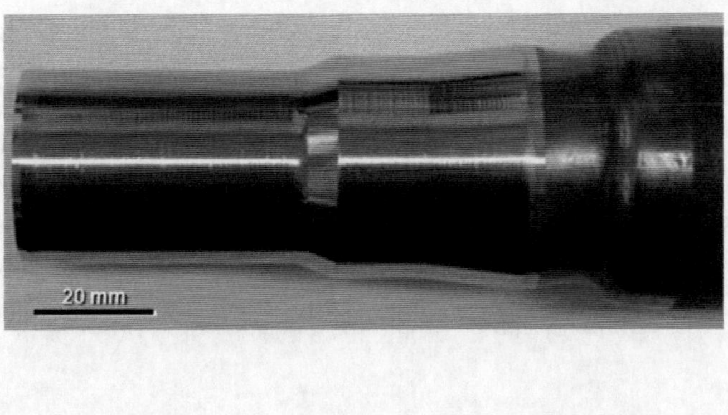

**Figure 9**. Example of twice stepped-down product from TRIP C-Mn-Si-Nb steel upon cold reduction rolling

Mechanical properties were measured by tension testing. The geometry of samples was 2 x 1.5 x 2 mm. Following the reduction to 42 mm, the C-Mn-Si steel (T1) had a strength of 891 MPa and $A_{5mm}$ elongation of 19 % (Tab. 4). Subsequent reductions to the diameter of 37.3 mm did not cause any substantial increase in ultimate strength. The elongation level remained about 20 %.

Upon the smaller reduction of diameter, the C-Mn-Si-Nb steel (T2) had a strength around 950 MPa and a similar $A_{5mm}$ elongation of approx. 20 %. The subsequent reduction to the final diameter of 37 mm led to an increase in ultimate strength to 965 MPa.

**Table 4.** Final mechanical properties after reduction rolling

| Steel | Diameter reduction [%] | $R_m$ [MPa] | $A_{5mm}$ [%] | Z [%] |
|---|---|---|---|---|
| C-Mn-Si (T1) | 14.4 | 891 | 19 | 57 |
| | 25.4 | 947 | 18 | 48 |
| C-Mn-Si-Nb (T2) | 16 | 953 | 21 | 45 |
| | 26 | 965 | 21 | 54 |

## 3. Conclusion

Based on previous experience and results of material-technological modelling of incremental thermomechanical processing, a manufacturing route for making hollow stepped products was proposed and tested successfully. A flexible production chain was used which comprised a sequence of warm rotary spin extrusion, intercritical annealing and cold reduction rolling to the final shape. The unconventional combination of warm forming, intercritical annealing and cold forming produced an appropriate combination of technological properties in low-alloy TRIP steels of CMnSi and CMnSiNb types. The process of microstructure evolution was designed to meet the requirements for technological properties of the materials throughout the manufacturing sequence and to enable the desired shapes and final properties to be obtained with as low energy consumption as possible and without the need for the final heat treatment operation. Thanks to a new material-technological strategy, a zero-waste manufacturing route for making stepped hollow thin-walled products was tested successfully.

**Acknowledgements**

This paper presents results achieved under the project LO1502 Development of Regional Technological Institute. The projects are subsidised by the Ministry of Education of the Czech Republic from specific resources of the state budget for research and development.

**References**

[1] Bleck W 2002 *International Conference on TRIP – Aided High Strength Ferrous Alloys* Using the TRIP effect – the down of a promising group of cold formable steels Prof. Bruno C. De Cooman Aachen

[2] Jirkova H, Rezek M, Meyer L W and Masek B 2009 Influence of austenitization temperature and number of incremental steps on structure development of TRIP-steel, *Proceedings of the 20th International DAAAM Symposium Intelligent Manufacturing & Automation: Theory, Practice & Education* **20** 1461-2

[3] Masek B, Stankova H, Novy Z, Meyer L W and Kracik A 2009 The Influence of thermomechanical treatment of TRIP steel on its final microstructure *Journal of Materials Engineering and Performance* **18**

[4] Basuki A and Aernoudt E 1999 Influence of rolling of TRIP steel in the intercritical region on the stability of retained austenite *Journal of Materials Processing Technology* **89 – 90** 37 – 43

[5] Godet S and Jacques P J 2004 *2nd International Conference on Thermomechanical Processing of Steels* Thermomechanical processing of TRIP-assisted multiphase steels Belgium

IOP Conf. Series: Materials Science and Engineering **118** (2016) 012024 doi:10.1088/1757-899X/118/1/012024

# Obtaining a TRIP microstructure by thermomechanical treatment without isothermal holding

**B Mašek[1], H Jirková[1], L Kučerová[1] and M F-X Wagner[2]**

[1]University of West Bohemia in Pilsen, Research Centre of Forming Technology, CZ-30614 Pilsen, Czech Republic
masekb@kmm.zcu.cz
[2] Chair of Materials Science, Institute of Materials Science and Engineering, Technische Universität Chemnitz, D-09125 Chemnitz, Germany
martin.wagner@mb.tu-chemnitz.de

**Abstract**. The contemporary development of technological processes for the production of modern multiphase steels can be characterized by the need for precise control of their technological parameters. The design of modern technological processes that allow sophisticated microstructures to be obtained usually cannot be carried out on real production equipment for technical as well as economical reasons. Therefore, new processes and test devices are continuously being developed to make it possible to simulate and model thermomechanical treatments on small specimens with precise control and monitoring of process parameters. A simulator for experimental modelling of thermomechanical processes has been developed at the University of West Bohemia. In this paper, to demonstrate the feasibility of simulating thermomechanical treatments with this setup on a lab scale, we discuss the thermomechanical treatment of TRIP steels without isothermal holding – a processing route that is difficult to handle and thus poses several technological as well as economic problems. The realistic processing of wire rolling with different cooling strategies is tested on the TRIP CMnSiNb steel. Our results show that the processing route without isothermal holding allows to obtain multiphase microstructures with a tensile strength of up to 835 MPa and a ductility $A_{5mm}$ = 25 %.

**Key words**: TRIP steel, thermomechanical treatment, isothermal holding, multiphase structure

## 1. Introduction

Rising prices of raw materials and energy, combined with the high workload of production devices, make it generally impossible to test or optimize new manufacturing processes and material treatments directly under production conditions. Material-technological modelling of microstructural evolution is a very efficient way of developing and optimizing new procedures. Material-technological modelling represents the establishment of such an environment that resembles the real time conditions for material processing as closely as possible. The process conditions in forming technologies, in thermomechanical treatments, and potentially in heat treatments can be characterized by a series of selected parameters. In material-technological modelling it is necessary to know the effects of these parameters and to adjust the model parameters accordingly. With properly selected parameters, a good agreement between the modelled and the real process can be expected [1-3].

For such a material-technological modelling approach, a thermomechanical simulator is available in the FORTECH Research Centre at the University of West Bohemia in Pilsen (Fig. 1). This custom-built simulator allows for the temperature and some selected deformation parameters to be controlled very carefully, so that they correspond closely to those of a real process – or those that can be expected in a real process in the case of the development of new technologies, materials or structures.

**Figure 1.** Thermomechanical simulator (overview: left; heated sample during a thermomechanical experiment: right).

This device also specifically allows rapid changes of both the temperature and the deformation parameters. Therefore even complex conditions associated with industrial/technological processes can be simulated accurately. For instance, the simulation of the deformation component can reach a speed of 3 m/s with the possibility of repeating several arbitrary, precisely driven deformation steps within a few seconds depending on the deformation regime. Concerning temperature behaviour, controlled changes of over 100 °C and 250 °C per second can be obtained both during heating and cooling of steel samples, respectively. Exact process monitoring is obviously guaranteed by the simulator's built-in high precision sensors and can further be improved and extended by attaching external monitoring devices to the control system. Examples of additional monitoring devices include an optical pyrometer with automatic emissivity correction or a high-speed camera capable of capturing fast deformation processes in detail, which may help to analyse, for instance, crack formation.

Material-technological modelling can be used not only for the development of new technologies but also for testing new materials such as multiphase steels. With these steels, it is possible to obtain very good mechanical properties, but the microstructural evolution must be carefully controlled during processing. Using the thermomechanical simulator presented here, it is possible to optimize the processing parameters [4, 5]. In this paper, we briefly discuss a challenging example for our experimental approach using the thermomechanical simulator: we study the thermo-mechanical treatment of a TRIP steel without isothermal holding.

## 2. Experimental program

For the experimental program, a low-alloyed CMnSiNb TRIP steel (see Tab. 1) was chosen. TRIP steels are multiphase steels and their microstructure consists of ferrite, bainite and small amounts of retained austenite [6]. Because of their high capacity for energy absorption and good fatigue behaviour they have recently been used in the automotive industry for safety components such as seat structures, cross-members, long post reinforcements, aprons and fender reinforcements [7].

The standard treatment of TRIP steels consists of isothermal holding in the bainitic temperature range. During this holding period, the formation of bainite is promoted and retained austenite is stabilized. The need to hold the material at constant temperature, however, is difficult to control and thus causes several technological as well as economic problems. The material-technological modelling approach was used to examine a new technology without isothermal holding during cooling, and to compare it to a processing route with isothermal holding. The material was treated using a regime which simulates normal rolling mill processing, to determine whether or not this innovative and economically beneficial material is suitable for treatment in a specific rolling mill and what material properties can be expected in the final product.

**Table 1.** Chemical composition of the TRIP steel.

| C | Mn | Si | P | S | Cr | Ni | Cu | Al | Nb | Mo | V | W |
|---|---|---|---|---|---|---|---|---|---|---|---|---|
| 0.21 | 1.45 | 1.80 | 0.008 | 0.005 | 0.008 | 0.072 | 0.058 | 0.006 | 0.059 | 0.02 | 0.005 | 0.02 |

## *2.1 Thermomechanical treatment*

For the experiment, a thermo-mechanical regime was applied that simulated a rolling strategy for a rolling track with heating to 1050 °C and passing through twenty rolling mills in 2.8 s (Fig. 2). The tensile and compressive deformations represent the individual passes. During the design of this deformation regime, time intervals between individual rolling mills were taken into account. The deformation took place during a temperature decrease within an interval from 1050 °C to 830 °C. During this deformation a true strain of 1.66 with various strain rates from 0.3 to 20 $s^{-1}$ was reached. In the final part of the simulated regime the influence of various cooling rates and isothermal holding times on the resulting microstructures was examined.

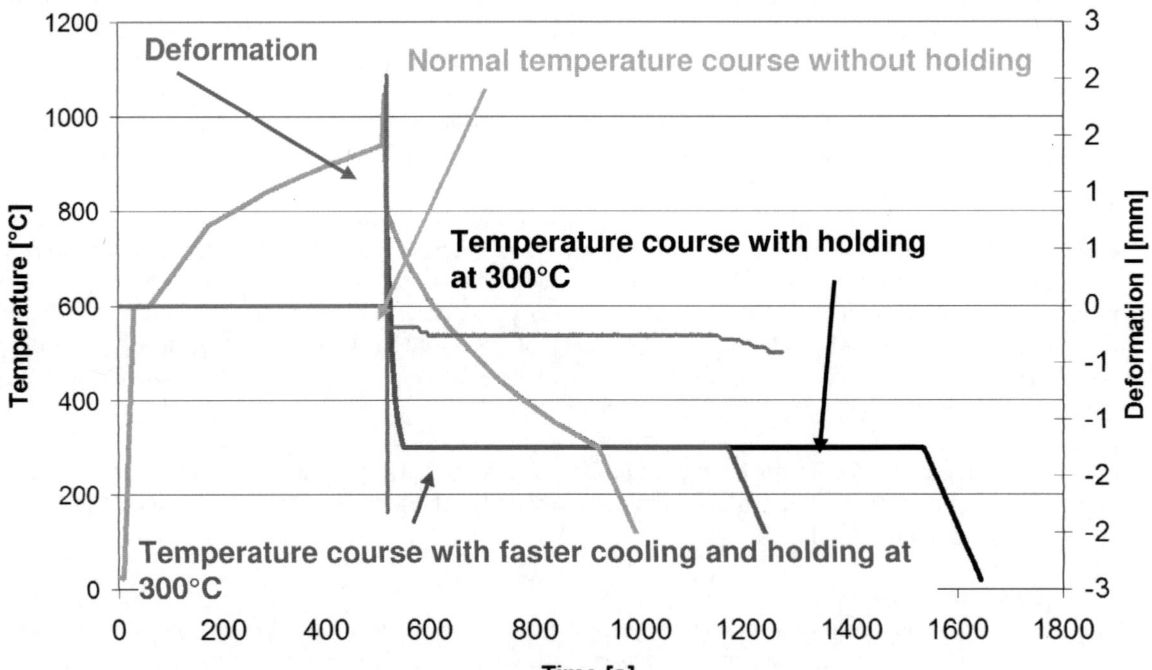


**Figure 2.** Temperature behaviour in regimes simulating the rolling mill process with various cooling strategies

Three different cooling strategies were tested (Tab. 2). In the first simulated regime, the deformation was followed by cooling at a rate of 1 K/s to room temperature. This slow cooling represents one of the realistic processing alternatives. In the next strategy, cooling was interrupted at 300 °C (with a holding time of 600 s). This isothermal holding time is characteristic for TRIP steels and it serves to stabilize the retained austenite. Stabilized austenite contributes to a good combination of strength and ductility during a subsequent cold deformation. To avoid pearlite formation, the cooling rate in the third strategy was increased up to 16 K/s.

The obtained microstructures were investigated by optical, laser confocal and scanning electron microscopy. The fraction of retained austenite was measured by X-ray diffraction analysis and the mechanical properties were documented by performing tensile tests on mini tensile test specimens with a gauge length of 5 mm.

**Table 2.** Parameters and results (volume fraction of retained austenite: RA; ultimate tensile strength: Rm; fracture strain: $A_{5mm}$) of material–technological modelling of a rolling mill process with different cooling strategies.

| Cooling time after deformation | $T_B$ [°C] | $t_B$ [s] | $R_A$ [%] | $R_m$ [MPa] | $A_{5mm}$ [%] |
|---|---|---|---|---|---|
| 830-200°C / 1 K/s | without bainitic holding | | 15 | 835 | 25 |
| 830-300°C / 1 K/s | 300 | 600 | 16 | 835 | 26 |
| 830-300°C / 16 K/s | 300 | 600 | 12 | 1043 | 29 |

## 3. Results and discussion

After the first simulated regime with a cooling rate of 1 K/s and without isothermal holding, the microstructure consisted of ferrite, bainite and M-A constituent with a small amount of pearlite (Fig. 3a). The presence of pearlite and the transformation of retained austenite to martensite in the M-A constituent could be confirmed by scanning electron microscopy (Fig. 3b). Pearlite is undesirable in TRIP microstructures because it decreases the carbon content in the retained austenite and therefore reduces its chemical stability. The ferrite volume fraction was 59 % and the volume fraction of retained austenite measured via X-ray diffraction analysis reached 15 %. The ferritic grain size was determined to be approximately 3.5 μm.

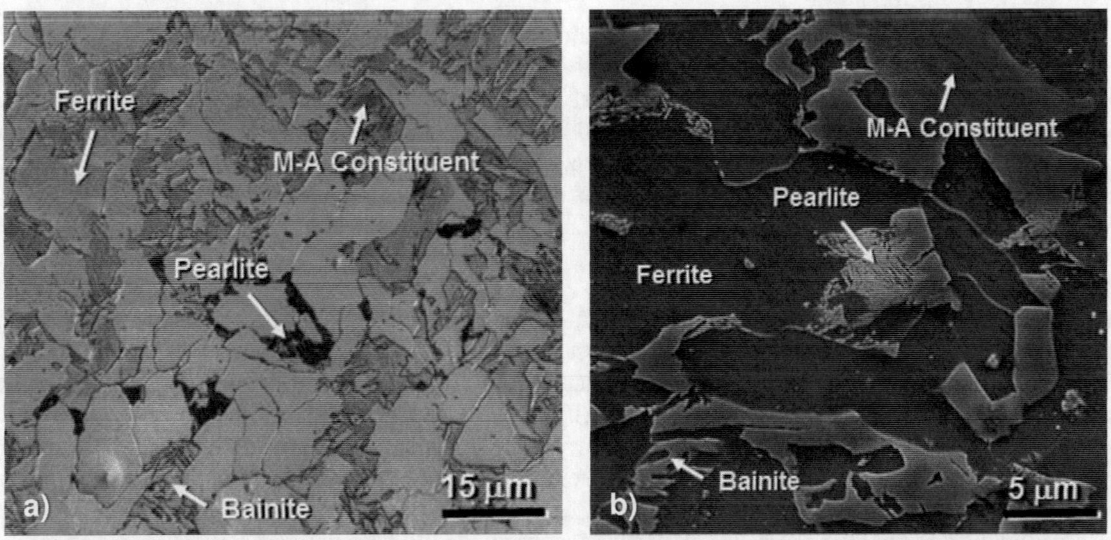


**Figure 3.** Microstructures obtained following a regime with usual behaviour for rolling mill. a) laser confocal microscope, b) scanning electron microscope – detail of pearlite islands and M-A constituent.

In the next strategy with a holding time at 300 °C a ferrite-bainite microstructure with 58 % of ferrite and 16 % of retained austenite was achieved. The grain size remained unchanged. In comparison to the strategy without holding, no pearlite was detected (Fig. 4a).

Further, the cooling rate between the last deformation step and holding at 300°C was increased up to 16 K/s. This treatment resulted in a ferrite-bainite microstructure with large bainitic blocks (. a) laser confocal microscope, 4b). Only laths of bainitic ferrite were observed and no polyhedral ferritic grains were detected. The structure contained 12% of retained austenite. The retained austenite typically occurred between the laths of bainitic ferrite.

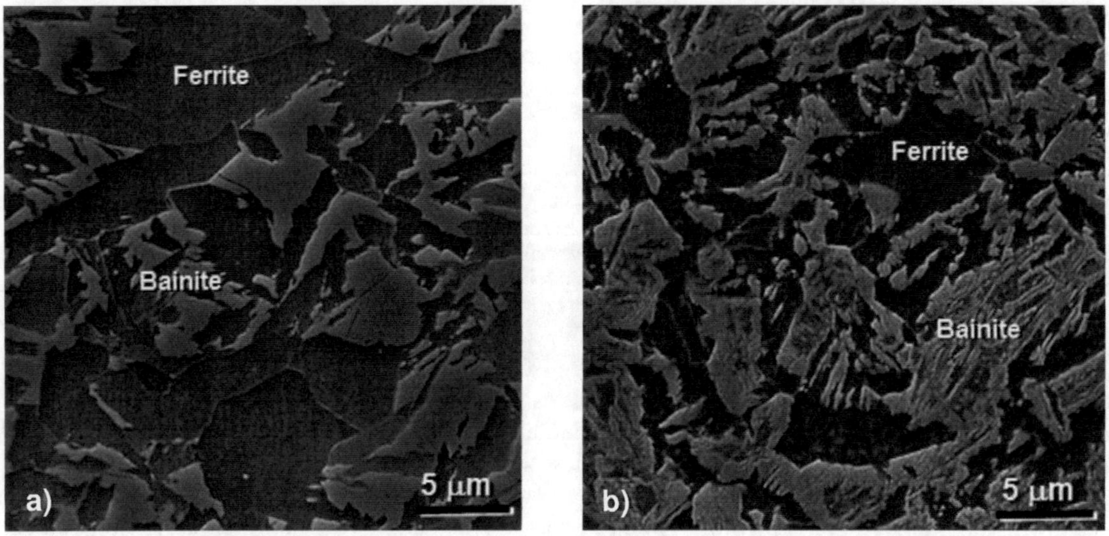


**Figure 4.** Microstructures produced from regimes with holding time at 300 °C.
a) Cooling rate of 1 K/s, b) cooling rate of 16 K/s.

The results of the mechanical tests performed on the different samples are summarized in Table 2. Because of the relatively small amount of thermo-mechanically treated material, mini-tensile samples were used for testing. The mechanical data in Tab. 2 shows that the best results were achieved for the regime with fast cooling and with holding at 300 °C. With this regime, the tensile strength was determined to be 1043 MPa and ductility is characterized by a fracture strain of $A_{5mm}$ = 29 %. Both strategies with longer cooling times gave rise to lower tensile strength (by approximately 170 MPa lower), but at a similar ductility.

## 4. Summary and conclusions

The behaviour of low-alloyed CMnSiNb steel under various treatment strategies was studied with the help of material-technological modelling on a custom-built TMB simulator. The thermomechanical process consisted of thermal and deformation phases. The deformation regime ran at a high strain rate, which corresponds to the real conditions in a rolling mill. The first results discussed here very briefly demonstrate the high sensitivity of microstructural development to changes of processing simulation parameters. The required microstructures were achieved through material-technological modelling on real material specimens, which allowed material and mechanical properties to be measured. The verified parameters of the thermomechanical process can be used for accelerated application of the innovative treatment of CMnSiNb steels with a tensile strength of over 1000 MPa and ductility A5mm = 29 %. Finally, it was found that even when isothermal holding is removed, the required microstructures with very small fractions of pearlite, and with reasonably good mechanical properties, can be achieved. This may well provide the basis for a wider application of this type of steels.

**Acknowledgements**
This paper includes results obtained within the project LO1502 Development of Regional Technological Institute. The project is subsidized by the Ministry of Education, Youth and Sports from specific resources of the state budget for research and development. We also acknowledge funding by the German Research Foundation (Deutsche Forschungsgemeinschaft, DFG) in project WA2602/10-1.

**References**

[1] Mašek B, Jirková H, Malina J and Ronešová A 2009 *Proceedings of the 5th European Rolling Conference* New Possibilities in Material-Technological Modelling for Real Rolling Technology Production Chains England
[2] Mašek B, Jirková H, Malina J and Ronešová A 2009 *Proceedings of the XXVIII. Verformungskundliches Kolloquium* Material-technology Modelling the Development of Structures in Wires during Controlled Cooling on Industrial Cooling Conveyors Östereich
[3] Mašek B, Jirková H, Malina J and Jeníček Š 2010 Advanced Material-Technological Modelling of Complex Dynamic Thermomechanical Processes *Materials Science Forum* **654-656** 1594-1597
[4] Mašek B, Staňková H, Malina J, Skálová L and Meyer L W 2007 Physical modelling of microstructure development during technological processes with intensive incremental deformation *Key Engineering Materials* **345-346** 943-946
[5] Mašek B, Staňková H, Klauberová D and Skálová L 2008 *3rd International Conference on Thermomechanical Processing of Steel* The Most Recent from Physical-Material Modelling of UHSS Structure Development on a Thermomechanical Simulator Italy
[6] Bleck W 2002 *Proceedings of International Conference on TRIP–Aided High Strength Ferrous Alloys* Using the TRIP effect – the dawn of a promising group of cold formable steels Belgium
[7] www.arcelorauto.com/v_ang/produits/fiches/trip3.html
[8] Staňková H, Skálová L, Meyer L W and Mašek B 2007 Influence of aluminium on retained austenite stabilization during thermomechanical treatment of low-alloyed TRIP steel *Acta Metallurgica Slovaca 13* **1** 208-213

IOP Conf. Series: Materials Science and Engineering **118** (2016) 012025 doi:10.1088/1757-899X/118/1/012025

# Plasma electrolytic oxidation of Titanium Aluminides

**R Morgenstern[1], M Sieber[1], T Grund[2], T Lampke[1] and B Wielage[2]**

[1] Materials and Surface Engineering Group, Institute of Materials Science and Engineering, Technische Universität Chemnitz, 09125 Chemnitz, Germany
[2] Professorship of Composite Materials, Institute of Materials Science and Engineering, Technische Universität Chemnitz, 09125 Chemnitz, Germany

**Abstract.** Due to their outstanding specific mechanical and high-temperature properties, titanium aluminides exhibit a high potential for lightweight components exposed to high temperatures. However, their application is limited through their low wear resistance and the increasing high-temperature oxidation starting from about 750 °C. By the use of oxide ceramic coatings, these constraints can be set aside and the possible applications of titanium aluminides can be extended. The plasma electrolytic oxidation (PEO) represents a process for the generation of oxide ceramic conversion coatings with high thickness. The current work aims at the clarification of different electrolyte components' influences on the oxide layer evolution on alloy TNM-B1 (Ti43.5Al4Nb1Mo0.1B) and the creation of compact and wear resistant coatings. Model experiments were applied using a ramp-wise increase of the anodic potential in order to show the influence of electrolyte components on the discharge initiation and the early stage of the oxide layer growth. The production of PEO layers with technically relevant thicknesses close to 100 µm was conducted in alkaline electrolytes with varying amounts of $Na_2SiO_3{\cdot}5H_2O$ and $K_4P_2O_7$ under symmetrically pulsed current conditions. Coating properties were evaluated with regard to morphology, chemical composition, hardness and wear resistance. The addition of phosphates and silicates leads to an increasing substrate passivation and the growth of compact oxide layers with higher thicknesses. Optimal electrolyte compositions for maximum coating hardness and thickness were identified by statistical analysis. Under these conditions, a homogeneous inner layer with low porosity can be achieved. The frictional wear behavior of the compact coating layer is superior to a hard anodized layer on aluminum.

## 1. Introduction

The reduction of moved masses, which can be achieved by the use of modern lightweight materials, leads to high savings of energy and costs in aerospace technology. The application of conventional titanium alloys is limited to about 600 °C. At higher temperatures, nickel-base alloys can be substituted by titanium aluminide alloys, which are based on intermetallic phases in the titanium-aluminum binary phase system. Because of their high creep resistance, low density and comparatively high oxidation resistance, alloys with a chemical composition close to the γ-TiAl phase are viewed as particularly promising. Large research efforts were conducted in order to improve the fracture toughness of these alloys during the last 25 years. This aim was reached by the creation of a homogeneous and fine-grained microstructure through thermomechanical treatments [1, 2].
A further enhancement of the operating temperature of titanium aluminides above 750 °C requires an improvement of the high-temperature oxidation and wear resistance (e.g. against erosion). This cannot be achieved by the variation of the alloy composition without losing the desirable mechanical

properties. Thus, the application of surface technologies is necessary. As already shown by Li et al. [3], protective oxide ceramic coatings can be generated by the plasma electrolytic oxidation (PEO). This technique enables the production of high coating thicknesses over 100 µm, which is a benefit compared to thin PVD or CVD layers that tend to cracking and spallation under high wear stress as the supporting effect of the substrate is not sufficient [4]. Furthermore, the PEO process is typically conducted at room temperature and requires short times, which leads to a higher energy and cost efficiency in comparison with thermal oxidation [5] or thermochemical processes like nitriding or carburizing [6]. Also, there is no thermal modification of substrate microstructure and mechanical properties.

The influence of electrolyte components on the electrochemical passivation or dissolution at anodic potentials during PEO was primarily investigated at aluminum, titanium and magnesium based substrates [7, 8]. Thus, the addition of silicates or phosphates to the electrolyte results in a stronger passivation of aluminum and titanium. The latter is important to avoid strong substrate dissolution at rising anodic potentials. Starting from a certain electric field strength, water vapor is ionized at the anode and the passive layer is broken through by impact and tunnel ionization. Within the discharge channel, substrate material gets molten and anodically oxidized and forms oxide ceramic deposits. During the PEO process, many small discharges or microarcs generate a continuous oxide layer at the substrate surface. Multiple remelting and oxidizing leads to the densification of coating areas close to the substrate and a growing coating thickness. Thus far, the impact of electrolyte components on the initiation of the PEO process as well as on coating microstructure and properties has not been reported in the literature for TiAl substrates.

## 2. Material and methods

The general effect of single electrolyte components was investigated in a small electrolyte volume of 200 ml at a circular surface of 0.79 cm² using TNM-B1 (Ti43.5Al4Nb1Mo0.1B) gamma titanium aluminide alloy. The samples were polarized as anode against a stainless steel cathode. The current-potential characteristic was recorded with an electrochemical workstation (IM6+CVB120, Zahner, Germany) with the potential increasing from 0 V to 80 V at a rate of 1 V/s. Afterwards, the potential was kept constant at 80 V for 120 s in order to observe the discharge phenomena and passivation effects. The experiments were conducted in an alkaline solution containing 70 (50) g/l of potassium hydroxide (KOH, Merck, purity: for analysis) and varying amounts of sodium metasilicate ($Na_2SiO_3{\cdot}5H_2O$, Aldrich, purity: for analysis), sodium metaphosphate ($Na(PO_3)n$, Carl Roth, purity: for analysis), potassium pyrophosphate ($K_4P_2O_7$, chemPUR, purity: for analysis), citric acid ($C_6H_8O_7$, VWR BDH Prolabo, purity: for analysis) and deionized water.

In order to produce PEO coatings with high thicknesses up to 100 µm, cylindrical samples of TNM-B1 alloy were embedded in epoxy resin and electrically connected by an aluminum wire. One circular surface with a diameter of 25 mm was exposed to the electrolyte. The plasma anodized coatings were produced in an alkaline electrolyte consisting of 10 g/l KOH. Contents of $Na_2SiO_3{\cdot}5H_2O$ and $K_4P_2O_7$ were varied according to a design of experiments between 10–20 g/l and 0–10 g/l respectively in aqueous solution. The coating process was carried out under a symmetrical rectangular bipolar pulse current regime with a pulse amplitude of 2 A and a pulse length of 10 ms in both anodic and cathodic half-cycle. The temperature of the electrolyte was kept constant at 20 °C during the process. The principal arrangement of the laboratory plant for the plasma anodizing process is shown elsewhere [9]. Reference samples for wear testing with the same dimensions consisting of Al 99.5 material were hard anodized in 10 vol% $H_2SO_4$ aqueous solution at 5 °C, 2 A/dm² for 1 h.

For scanning electron microscopy (SEM), cross sections were prepared by cutting, hot mounting in electrically conductive resin, diamond grinding, polishing and finally polishing with a silica oxide polishing suspension. Carbon coating was used to avoid sample charging. Secondary electron (SE) and backscattered electron (BSE) images were obtained at SEM LEO 1455VP using 25 kV excitation voltage. The elemental composition was simultaneously recorded by EDS data collection (EDAX Genesis). The X-Ray diffraction (XRD) phase analyses were performed using an X-ray diffractometer

D 5000 (Siemens) with Co anode. The phase identification was done with the help of the PDF-2 2010 database (International Centre for Diffraction Data).
Hardness values HM0.25/30/30 were obtained by registrating hardness measurement with Fischerscope HM2000 XYm at the cross-section of the coating. A Vickers diamond was used as the indenter. Per sample, a series of 5 measurements over the coating thickness was conducted. Oscillating dry sliding wear behavior was evaluated by a ball-on-disc arrangement (SRV, Optimol Instruments) at room temperature using a frequency of 20 Hz, 10 N normal force, WC-Co-ball with a diameter of 10 mm and an oscillation amplitude of 1 mm.

## 3. Results and discussion

### *3.1 Modell experiments on the influence of electrolyte components*

The influence of electrolyte components on the discharge initiation was determined in the potential range of 0–80 V in highly concentrated KOH solutions. In many cases, a current peak can be observed at the passivation potential. Within the subsequent passivation region, the current density stays at a low level. The discharge is initiated at the breakthrough potential, which is characterized by a sudden increase in current density. Above the initiation potential, a stable discharge process is marked by a constantly high current density. In that state, thin oxide ceramic layers consisting of an adherent base layer and an easily removable outer layer develop on the substrate. If the sample surface is completely covered by a dense oxide layer, the discharges might temporarily or permanently break down, which is characterized by a constantly low current density. An increase of the KOH concentration leads to lower discharge initiation potentials and higher discharge current densities. By using an aqueous solution of 70 g/l KOH, the discharge initiation potential amounts to 50 V, which enables the observation of potential shifts due to the addition of passivating electrolyte components within the regarded potential range. While retaining the KOH concentration, the other electrolyte components were added in the concentrations of 5, 50 and 100 g/l.
Generally, the effect of the investigated additions increases with higher concentrations. For a clearer presentation of the effects, only the curves for 100 g/l additive concentration are displayed in figure 1. It can be seen that the discharge initiation potential is shifted towards about 60 V and the discharge current density is slightly decreased by the addition of 100 g/l $K_4P_2O_7$. No remarkable change of the current-density time behavior was observed at $K_4P_2O_7$ concentrations of 50 g/l and below. The components $Na(PO_3)n$ and $Na_2SiO_3 \cdot 5H_2O$ contribute to the development of a thin and compact passive layer with the result that the discharge is temporarily interrupted or permanently prevented. Also the discharge initiation potential was shifted towards higher values by silicate addition (70 V at 100 g/l $Na_2SiO_3 \cdot 5H_2O$). A significantly stronger passivation is achieved by the addition of 50 or 100 g/l citric acid. This electrolyte component seems to be unsuitable as discharges are completely prevented within the examined potential range.

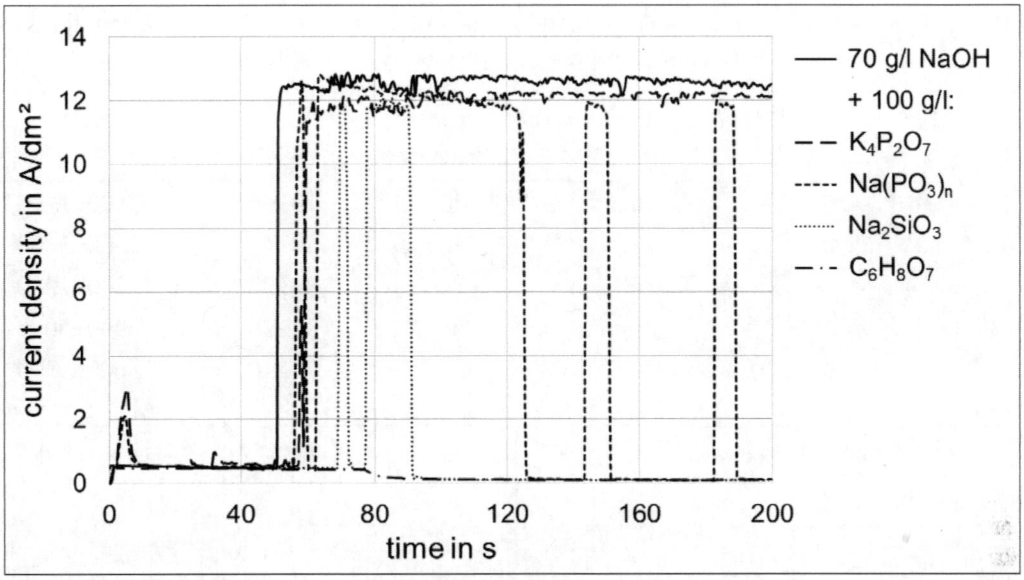


**Figure 1.** Influence of electrolyte component addition (100 g/l) on the current density time behavior at discharge initiation between 0 and 80 V (sweep rate 1 V/s) and early oxide coating growth at 80 V of a 70 g/l aqueous NaOH solution

### *3.2 Properties of PEO coatings with industrially relevant thickness*

The production of PEO coatings with industrially relevant thickness of up to 100 μm was conducted at constant KOH concentration of 10 g/l and temperature under pulsed current conditions. As shown by the model experiments, phosphates and silicates are suitable additives for influencing the discharge initiation and the PEO process itself. As $Na_2SiO_3$ and $K_4P_2O_7$ showed the strongest and weakest effect, respectively, both components were varied according to a statistical design of experiments. Coating thickness and hardness served as dependent variables. Also the process time has to be adjusted according to the process parameters in order to generate homogeneous coatings. Due to a local concentration of the electric field lines, the discharges were first initiated at the edge of the circular samples and continued towards the center. In case of compact PEO layers, the process ended as soon as the sample surface was completely covered. Otherwise, in case of a higher coating porosity, the process was aborted, when larger discharges started to destroy the PEO coating locally. The average hardness is related to the average coating porosity, whereas the standard deviations of coating thickness and hardness correlate with the structural homogeneity of PEO layers.

An overview of electrolyte compositions, process times and coating properties is provided by table 1. The analysis of variance reveals that both the pyrophosphate and the metasilicate concentration exert a significant influence on the coating thickness as shown by figures 2a and b. The error probability for considering the effect of the former and the latter on the coating thickness as significant is smaller than 2 % and 1 %, respectively. However, it cannot be excluded that the maximum thickness at 5 g/l $K_4P_2O_7$ and 15 g/l $Na_2SiO_3$ is also affiliated to the high process time of 240 minutes. Meanwhile, as displayed by figures 3a and b, the hardness is not significantly affected by the electrolyte composition. The error probability for considering the effect of pyrophosphate and metasilicate concentration as significant is 44 % and 48 % respectively. This is caused by the high deviation of the hardness because of the considerable amount of flaws in the coatings.

IOP Conf. Series: Materials Science and Engineering **118** (2016) 012025 doi:10.1088/1757-899X/118/1/012025

**Table 1.** Process time, coating thickness and hardness obtained for different concentrations of potassium pyrophosphate and sodium silicate

| No. | c(KOH)/ g/l | $c(K_4P_2O_7)$/ g/l | $c(Na_2SiO_3)$/ g/l | Process time /min | Coating thickness /µm | Coating hardness /$Nmm^{-2}$ |
|---|---|---|---|---|---|---|
| 1 | 10 | 5 | 15 | 240 | 103±25 | 3100±1100 |
| 2 | 10 | 0 | 10 | 230 | 30±4 | 2500±1200 |
| 3 | 10 | 10 | 10 | 90 | 82±22 | 3700±1000 |
| 4 | 10 | 10 | 20 | 90 | 92±29 | 2900±500 |
| 5 | 10 | 0 | 20 | 90 | 95±10 | 3600±1300 |

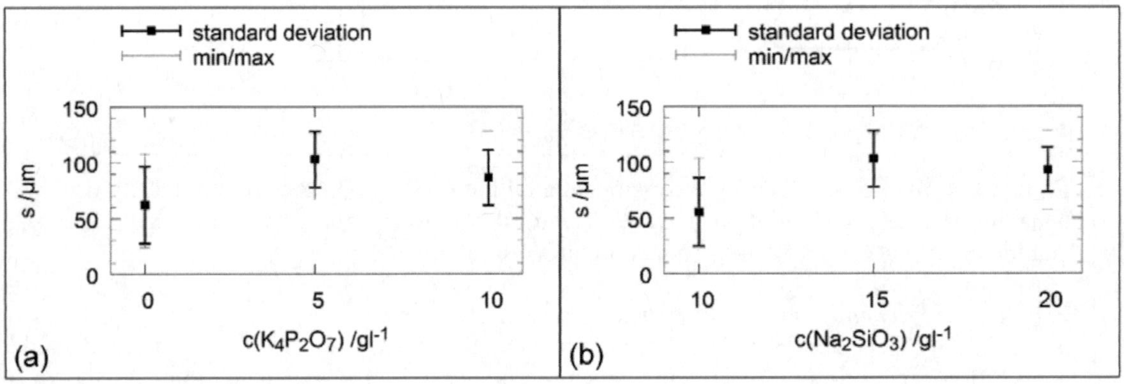


**Figure 2.** Influence of a) $K_4P_2O_7$ and b) $Na_2SiO_3$ concentrations on the coating thickness

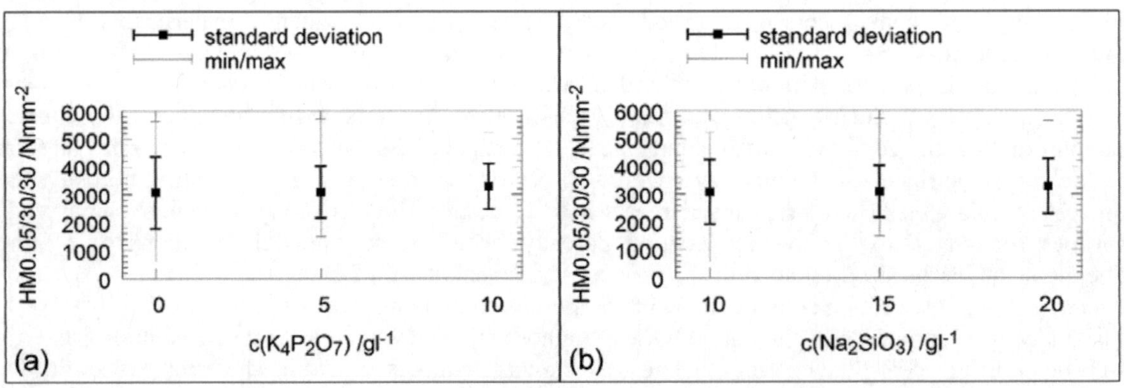


**Figure 3.** Influence of a) $K_4P_2O_7$ and b) $Na_2SiO_3$ concentrations on the coating thickness

The subsequent BSD images illustrate the influence of the electrolyte components on the coating microstructure. Low coating thickness of about 30 µm and high porosity are obtained in electrolyte 2 containing 10 g/l KOH and 10 g/l $Na_2SiO_3$ (figure 4a). A significant enhancement of coating thickness (figure 4b) is achieved by raising the $Na_2SiO_3$ concentration to 15 g/l and adding 5 g/l $K_4P_2O_7$ (electrolyte 1). Under these conditions, a compact inner coating area with low porosity can be generated (figure 4c). As shown by the results of EDS measurements in table 2, almost no constituents of the electrolyte (e.g. Si) are incorporated within the inner compact layer (area 3). However, higher concentrations of Si and K are detected within the outer porous layer (area 4) or PEO layers with a generally higher porosity (areas 1 and 2). This correlates to the less intense remelting and compaction of these areas. XRD measurements generally proved the existence of crystalline phases of $TiO_2$ (rutile, anatase, monoclinic) in PEO coatings on titanium aluminides. There is no clear evidence for the existence of $Al_2TiO_5$ as many characteristic reflection peaks are superimposed by those of $TiO_2$ phases. In case of higher Si contents in the outer porous layer (area 4) or porous PEO coatings (areas 1 and 2), significant fractions of unknown amorphous phases have to be expected. Generally, the thin porous PEO coating in figure 4a exhibits a lower Ti and higher Si content compared with the thicker and more compact coating in figure 4b. This might be explained by the presence of amorphous aluminum silicates. In both cases, an undisturbed interface between oxide and substrate can be observed. Cracks that proceed near the substrate-coating interface can be attributed to the materialographic preparation.

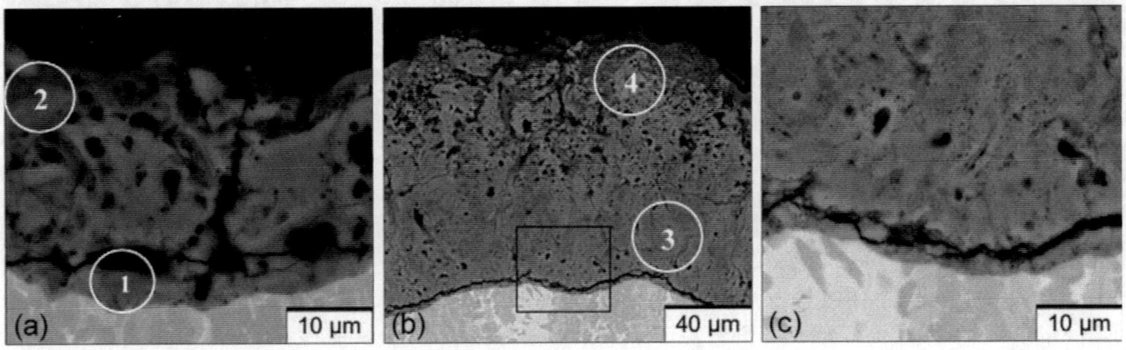


**Figure 4.** BSD images of PEO coatings generated from electrolytes a) 2 and b) 1; c) shows the framed detail of b) at higher magnification

**Table 2.** Results of EDS measurements at different areas of figure 4; chemical composition is given in at%

| No. | Ti | Al | O | Si | K |
|---|---|---|---|---|---|
| 1 | 13,7 | 14,2 | 64,4 | 6,4 | 0,4 |
| 2 | 6,7 | 6,1 | 67,3 | 19 | 0,5 |
| 3 | 18,6 | 14,9 | 65 | 0 | 0,3 |
| 4 | 13,9 | 10,7 | 61,1 | 4,7 | 6,5 |

As shown in figure 5a, PEO coatings with high thickness and low porosity can also be created without $K_4P_2O_7$ addition at a $Na_2SiO_3$ concentration of 20 g/l (electrolyte 5) at a significantly lower process time. However, in comparison with figure 4b, the coating appears to be more heterogeneous with respect to phase distribution and microstructure. A remarkable Si content of 4.5 at% is already detected within the inner compact layer. Thus, the presence of amorphous phases is to be expected. Areas with a significantly higher Si and K contents directly border to the inner compact layer. Those appear darker in figure 5a and continue towards the outer areas of the coating. The further addition of 10 g/l $K_4P_2O_7$ (electrolyte 4) leads to higher porosity and heterogeneity of coating microstructure and coating thickness, as can be seen at Fig. 5b. This correlates to a lower average hardness and a higher standard deviation of coating thickness in table 1. Also the coating appears to be less adherent to the substrate and more brittle as a higher number of larger cracks can be seen. The EDS measurements 4 to 6 show a more gradual increase in Si and K contents with increasing distance from the substrate compared to measurements 1 to 3 at figure 5a. Furthermore, higher amounts of Na and P were detected within the coating. Unexpectedly, the Na content is generally about twice the P content

although the $K_4P_2O_7$ concentration in the electrolyte was raised. As indicated by XRD measurements, small amounts of the crystalline phase $NaAlSi_3O_8$ might be present within the PEO coating.
Due to the substrate alloy composition, a small quantity (about 1 at%) of Nb was incorporated into all coatings. Generally, the Nb content decreases with increasing distance from the coating-substrate interface. Thus, no influence of the process parameters on the Nb content could be observed and no statement can be made, whether the Nb content influences the coating properties.

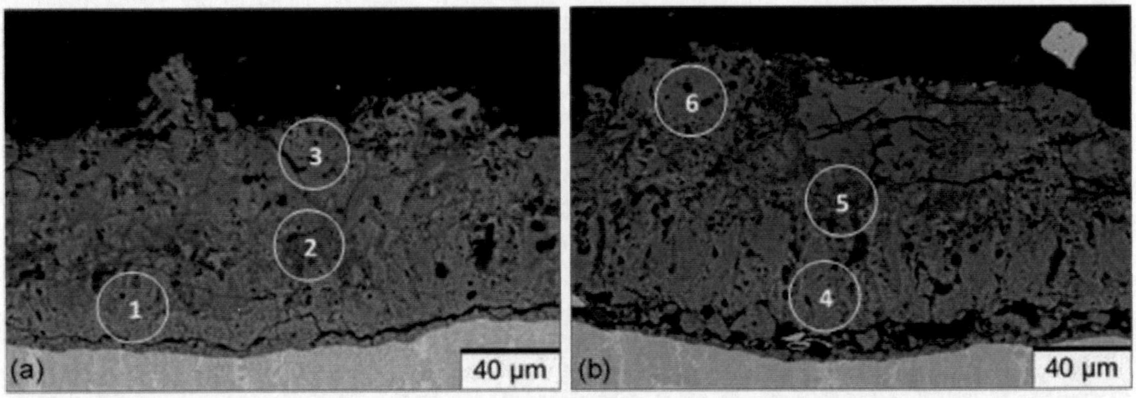


**Figure 5.** BSD images of PEO coatings generated from electrolytes a) 5 and b) 4

**Table 3.** Results of EDS measurements at different areas of figure 5; chemical composition is given in at%

| o. | Ti | Al | O | Si | K | Na+P |
|---|---|---|---|---|---|---|
| 1 | 14 | 13 | 66,6 | 4,5 | 0,9 | 0 |
| 2 | 9,1 | 10,7 | 59,7 | 9,3 | 8,5 | 0 |
| 3 | 8,4 | 8 | 60,9 | 9,4 | 9,4 | 0 |
| 4 | 16 | 11,5 | 67,3 | 1,8 | 1,5 | 0,8 |
| 5 | 10,2 | 10,6 | 61,8 | 7,9 | 6,5 | 2,3 |
| 6 | 6,4 | 7,7 | 60,9 | 10,2 | 11,2 | 3,3 |

With regard to high coating thickness and reasonable process time, PEO coatings of electrolytes 4 and 5 are promising candidates for practical application. Thus, they were chosen for oscillating dry sliding wear investigations. As shown in figure 6, both PEO coatings reveal high wear rates at the beginning of the test. This can be explained by the fast wear of the brittle outer porous coating areas. Due to high frequency and relative motion, a crack network is formed between pores and coating material is removed. As PEO coatings generated from electrolyte 4 are characterized by higher heterogeneity and porosity, higher initial wear and a high variation of the wear depths can be observed. When the outer porous layer is removed and the inner compact layer is reached, the wear rate is significantly reduced to a comparable low constant value for both PEO coatings. This can be attributed to the higher hardness of the crystalline phases and the fine cracks and pores that dissipate fatigue cracks. At this stage, three-body abrasion and fatigue wear are the dominating wear mechanisms. In comparison, a hard-anodized Al 99.5 sample with a comparable coating hardness of approximately 3000 MPa shows lower initial wear because of lower surface roughness and a more homogeneous porosity across the coating thickness. As the diameter of the ordered pore structure slightly decreases from the outer surface towards the substrate coating interface, the wear rate is also slightly decreasing during the wear process. However, the wear rate of the hard anodized coating remains higher than those of the compact inner layers of the PEO coatings on titanium aluminides. The wear tests on the hard anodized coating were aborted as soon as the coating thickness of approximately 50 µm was reached. Because of the higher wear stock and the lower wear rate of the inner compact layer, PEO coatings are more suitable as a long-term wear protection.

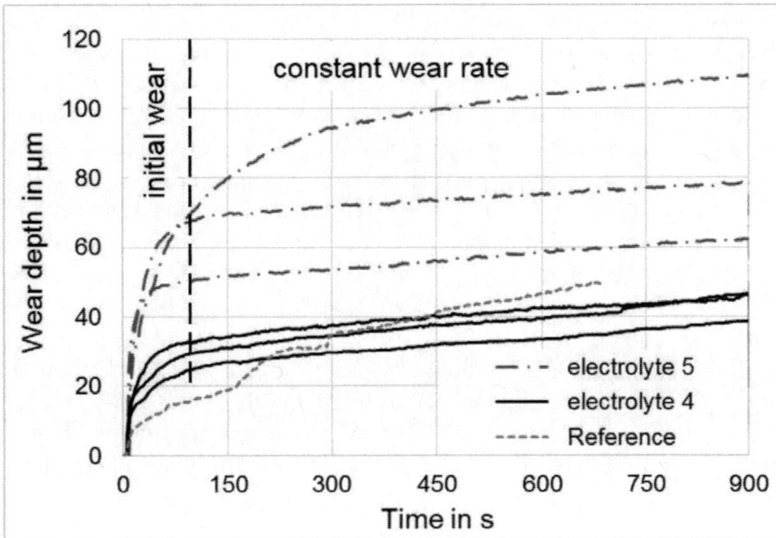


**Figure 6.** Progress of wear depth for PEO coatings generated from electrolytes 4 and 5 and hard anodized reference coating during oscillating dry sliding wear tests

## 4. Conclusions

It was the aim of the work to identify the influences of different electrolyte components on the plasma electrolytic oxidation of titanium aluminide alloy TNM-B1. Therefore, model experiments were conducted primarily in order to simulate the discharge initiation. It became apparent that an increasing KOH concentration leads to the reduction of the initiation potential and an increase of the discharge current density. The substrate passivation was slightly increased by the addition of $K_4P_2O_7$, whereas high concentrations of $Na(PO_3)n$ and especially $Na_2SiO_3$ led to a strong passivation. This resulted in a significant enhancement of the discharge initiation potential and interruptions or the termination of the discharges after short times. Furthermore, $C_6H_8O_7$ was proved to be an unsuitable additive due to strong passivation and the suppression of discharges. Hence, the applicability of the newly developed model process was demonstrated.

Based on the findings of the model experiments, PEO coatings with an industrially relevant thickness of about 100 µm were produced successfully. The statistical analysis of the experimental results showed that the coating thickness is significantly enhanced by the addition of both $K_4P_2O_7$ and $Na_2SiO_3$. The most homogeneous coating with the lowest porosity was obtained for 10 g/l KOH (supporting electrolyte) and the addition of 5 g/l $K_4P_2O_7$ and 15 g/l $Na_2SiO_3$ at a high process time of 240 min. Similar coating hardness and thickness can be achieved by the addition of 20 g/l $Na_2SiO_3$ without $K_4P_2O_7$ at a significantly lower process time of 90 min accompanied by a higher heterogeneity. Both parameter sets can be seen as optima regarding coating hardness, thickness and homogeneity due to a suitable combination of coating growth promotion and low incorporation of elements from the electrolyte. A reduction of the $Na_2SiO_3$ content to 10 g/l and the omission of $K_4P_2O_7$ leads to porous coatings with a significantly reduced thickness without a pronounced inner compact layer. On the other hand, the incorporation of elements from the electrolyte increases at 20 g/l $Na_2SiO_3$ and 10 g/l $K_4P_2O_7$. This leads to a higher portion of amorphous phases, higher porosity and brittleness of the coatings and a disturbed substrate-coating interface. Furthermore, oscillating dry sliding wear tests showed that the initial wear increases with increasing thickness of the outer porous layer. As soon as the outer porous layer is removed and the inner compact layer is reached, the wear rate is significantly reduced. The wear resistance of the inner compact layer proved to be superior in comparison with a hard anodized reference coating.

It is to be expected that the total wear volume of PEO coatings on titanium aluminides can be reduced by removing the outer porous layer. However, it has to be taken into account that the inner coating layer is not damaged by the machining of the porous layer. A further reduction of the coating porosity and an enhancement of the coating hardness can be achieved by increasing the alumina content of the PEO coating. This might be achieved by applying aluminum layers on the surface or the incorporation of alumina particles from the electrolyte. Furthermore, high-temperature properties of PEO coatings have to be investigated as titanium aluminides are predominantly used at high temperatures. These include coating adhesion and thermal shock resistance as well as high temperature oxidation and wear resistance.

**Acknowledgement**
The authors gratefully acknowledge funding by the German Research Foundation (Deutsche Forschungsgemeinschaft, DFG) in the framework of the research project LA 1274/31-1. The technical assistance of Pierre Kneipel, Christel Gläser, Karla Muhr, Dagobert Spieler, Sigrid Kaufmann and Gabriele Tauchmann is also gratefully acknowledged is also gratefully acknowledged.

## 5. References

[1] Wu X 2006 *Intermetallics* **14** 1114
[2] Schwaighofer E, Schloffer M, Schmoelzer T, Mayer S, Lindemann J, Guether V, Klose, J and Clemens H 2012 *Prakt. Metallogr.* **49** 124
[3] Li X, Wu X, Xue W, Cheng G, Zheng R and Cheng Y 2007 *Surf. Coat. Technol.* **201** 5556
[4] Gebauer K 2006 *Intermetallics* **14** 355
[5] Xia J, Li X Y, Li C X an Dong H 2004 *Thin Solid Films* **458** 212
[6] Boonruang C, Thongtem T, McNallan M and Thongtem S 2004 *Mater. Lett.* **58** 3175
[7] Yerokhin A L, Nie X, Leyland A, Matthews A and Dowey S J 1999 *Surf. Coat. Technol.* **122** 73
[8] Yerokhin A L, Nie X, Leyland A and Matthews A 2000 *Surf. Coat. Technol.* **130** 195
[9] Sieber M, Mehner T, Dietrich D, Alisch G, Nickel D, Meyer D, Scharf I and Lampke T 2014 *Surf. Coat. Technol.* **240** 96

IOP Conf. Series: Materials Science and Engineering **118** (2016) 012026 doi:10.1088/1757-899X/118/1/012026

# On the effect of incremental forming on alpha phase precipitation and mechanical behavior of beta-Ti-10V-2Fe-3Al

**S Winter[1] and M F-X Wagner[1]**

[1] Chair of Materials Science, Institute of Materials Science and Engineering, Technische Universität Chemnitz, 09125 Chemnitz, Germany

sven.winter@mb.tu-chemnitz.de

**Abstract.** A combination of good ductility and fatigue resistance makes β-titanium alloys interesting for many current and potential future applications. The mechanical behavior is primarily determined by microstructural parameters like (beta phase) grain size, morphology and volume fraction of primary / secondary α-phase precipitates, and this allows changing and optimizing their mechanical properties across a wide range. In this study, we investigate the possibility to modify the microstructure of the high-strength beta titanium alloy Ti-10V-2Fe-3Al, with a special focus on shape and volume fraction of primary α-phase. In addition to the conventional strategy for precipitation of primary α, a special thermo-mechanical processing is performed; this processing route combines the conventional heat treatment with incremental forming during the primary α-phase annealing. After incremental forming, considerable variations in terms of microstructure and mechanical properties can be obtained for different thermo-mechanical processing routes. The microstructures of the deformed samples are characterized by globular as well as lamellar (bimodal) α precipitates, whereas conventional annealing only results in the formation of lamellar precipitates. Because of the smaller size, and the lower amount, of α-phase after incremental forming, tensile strength is not as high as after the conventional strategy. However, high amounts of grain boundary α and lamellar $\alpha_p$-phase in the undeformed samples lead to a significantly lower ductility in comparison to the matrix with bimodal structures obtained by thermo-mechanical processing. These results illustrate the potential of incremental forming during the annealing to modify the microstructure of the beta titanium Ti-10V-2Fe-3Al in a wide range of volume fractions and morphologies of the primary α phase, which in turn leads to considerably changes, and improved, mechanical properties.

## 1. Introduction

Excellent mechanical and corrosion properties are key reasons for the use of titanium alloys, for instance in structural applications in the aerospace industry [1, 2]. The properties of different alloys are strongly affected by the volume fractions of hexagonal close-packed (hcp) α-phase and body-centered cubic (bcc) β-phase. Compared to the α-phase, the β-phase is characterized by a lower strength, but higher formability. A combination of good ductility and fatigue resistance makes β-titanium alloys interesting for many current and potential future applications. The mechanical behavior of titanium alloys is primarily determined by microstructural parameters like (β-phase) grain size, morphology and volume fraction of primary / secondary α-phase precipitates, and this allows changing and optimizing their mechanical properties across a wide range [3-5].

A typical example of light-weight titanium structural elements, hollow shafts, are manufactured along various processing routes. This typically involves cold extrusion, or deep hole drilling - but since the latter is characterized by a material utilization of only 40 %, it is quite inefficient for high-grade materials [6]. Incremental forming at elevated temperatures has recently been developed as an alternative for the fabrication of hollow shafts. One such process, which makes use of a partially rotating forming technology to produce axially symmetric hollow shafts directly from solid semi-finished bars, has been termed spin extrusion [7]. The metastable beta-titanium alloy Ti-10V-2Fe-3Al (Ti-10-2-3) was used in first experimental studies with a focus on spin-extrusion at high temperatures and on its effect on microstructural evolution [6]. Ti-10-2-3 is a candidate material for aerospace applications, and it is well known for its excellent formability at higher temperatures [1, 2]. Conventional heat treatments of this alloy consist of a solution treatment and a two-stage annealing for the precipitation of primary and secondary α phases. Annealing at various temperatures in the primary α phase field (from about 600 to 790 °C) leads to different volume fraction and sizes of the precipitates, but does not affect their morphology. Thermo-mechanical processing, in contrast, is known as an effective method to control the size and morphology of different phases in various titanium alloys. Several studies [8-10] have, for example, shown that a change in the morphology of α layers from lamellar to globular α can occur during hot working of Ti-6Al-4V alloy, which considerably improves the mechanical properties of the material.

While the microstructural characteristics of Ti-6Al-4V alloy after thermo-mechanical processing have been well documented in the literature, there is to our knowledge no systematic report on the effectiveness of concurrent hot deformation and annealing in the primary α region on the formation of different microstructures in Ti-10-2-3. In this study, we incrementally deform Ti-10-2-3 samples at different temperatures in the primary α phase field to simulate spin extrusion on a laboratory scale. We analyze how different microstructural processes such as phase transformation and work hardening are activated simultaneously, and how they influence the kinetics of microstructural evolution and the resulting changes in mechanical properties.

## 2. Experimental

The β-titanium alloy Ti-10V-2Fe-3Al (Ti-10-2-3) was obtained in solution-annealed condition from *Otto Fuchs KG* (Meinerzhagen, Germany). To modify the microstructure (i.e., morphology and volume fraction of the primary α-phase), two different heat treatment strategies were used in this study: In the first one (referred to hereafter as "*conventional heat treatment*" strategy), the material was solution annealed above the β-transus temperature (830 °C) for 30 min to achieve a fully recrystallized microstructure. The material was then water quenched to produce material fully in the β-phase. To investigate the influence of the primary α-phase on mechanical properties, the material was then heated again up to 600 °C, 650 °C, 700 °C or 750 °C for 60 min followed by air-cooling. From this procedure, four different material conditions with different sizes and volume fractions of the primary α-phase were obtained for further microstructural and mechanical analysis.

In addition to this conventional strategy, a special *"thermo-mechanical processing"* (TMP) approach was studied. This processing route combines the conventional heat treatment with an incremental forming during the primary α-phase annealing (the procedure is shown schematically in Figure 1). The forming was also realized at 600 °C, 650 °C, 700 °C, and 750 °C with 250 forming increments in 60 min. Each increment consisted of 1 mm compression and 1 mm tension. The whole process (forming, annealing and (air) cooling) was performed in a hydraulic universal testing machine of type *SchenckPTT250* with electrical resistance heating of the samples (see Figure 2 for the geometry of the TMP sample, which most notably has a parallel length of 20 mm – further details on this special sample type have been explained in detail in [11]).

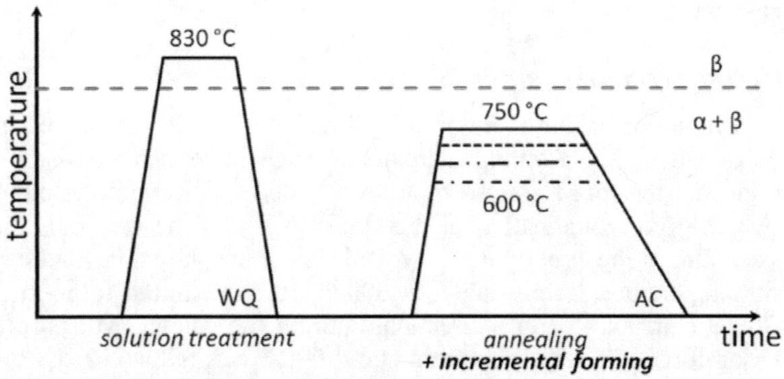


**Figure 1.** Schematic illustration of the applied heat treatment for the Ti-10V-2Fe-3Al alloy. First, a solution treatment is performed at 830 °C for 60 min with subsequent water quenching, followed by primary α phase annealing (at 4 different temperatures to obtain different primary α volume fractions: 600 °C, 650 °C, 700 ° and 750 °C) and air cooling. The thermo-mechanical processing includes an incremental forming during the α phase annealing.

Small ("mini") tensile specimens were cut out of the TMP samples by spark erosion. Depending on the forming temperature, 3 to 4 specimens could be taken out of one TMP-sample, as shown in greater detail in Figure 2. A systematic characterization of the stress-strain behavior as a function of the annealing temperature could be performed using a *UPM 1475 Zwick/Roell* tensile testing machine. Because of the small size of the mini tensile specimens, axial elongation was measured with an optical digital image correlation System (Aramis by GOM). This technique tracks the displacement fields of a speckle pattern on the specimen surface. The deformation field is recorded during testing and can subsequently be converted into strain data. The uniaxial tensile tests were carried out at an initial strain rate of $10^{-3}$ 1/s at RT.

Finally, the microstructures of all specimens were analyzed by optical microscopy and by scanning electron microscopy (SEM), using a *Zeiss Neon 40ESB* operated at 10 kV with a 4 quadrant backscatter detector.

**Figure 2.** The TMP sample of Ti-10-2-3 during the thermo-mechanical processing in the primary α phase field (600 °C - 750 °C) is deformed with 250 forming increments of ± 1 mm during a period of 60 min (left). After TMP, a "mini" tensile specimen is cut out by spark erosion (right). This specimen has a total length of 20 mm and a measuring length of 5 mm.

## 3. Results and discussion

### *3.1 Forces during thermo-mechanical processing*

We first focus on an evaluation of the thermomechanical behavior during processing. During TMP, each sample was deformed in 250 forming increments (each increment consisted of a total deformation of 1 mm in uniaxial tension, followed by a reverse deformation up to 1 mm in compression). Processing forces, which give an indication of thermal stability of the materials during incremental forming, were recorded during the experiments. Typical results are shown in Figure 3. Forming forces obviously depend on temperature. Interestingly, at 600 °C, the maximum force in each increment is nearly constant at 30 kN (for both loading directions) during the whole TMP procedure. In contrast, the incremental forming forces steadily drop from about 9 kN to 5 kN at 750 °C. It is to be expected that forming forces are lower at higher temperatures, but the transition towards softening during incremental forming at higher temperatures warrants a more detailed analysis. This softening effect can be directly related to the microstructures of the material at the different temperatures considered here (see also Figure 4 and the more detailed discussion in section 3.2): The samples deformed at 600 °C exhibit a high volume fraction of primary α precipitates (about 50 %, Table 1). Furthermore these precipitates are homogeneously distributed in the β grain structure. During the incremental forming process, some of the precipitates become oriented parallel to the direction of the highest shear stress, which is closely related to the formation of shear bands. Cyclic deformation then seems to occur primarily in those shear bands, and at constant stresses. In contrast, the samples deformed incrementally at 750 °C only have a primary α volume fraction of about 8 %; the precipitates are also distributed inhomogeneously. We did not observe the formation of shear bands and therefore the effect of decreasing forming forces is likely to be directly related to cyclic softening of the beta phase, which has been documented at high temperatures [12].

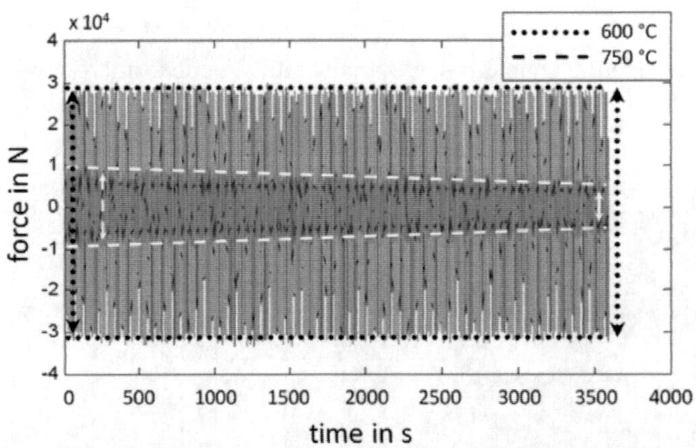


**Figure 3.** Force-time curves measured during incremental forming at 600 °C and 750 °C, respectively. At 600 °C the force remains nearly constant at about 30 kN (both in compression and in tension) for 60 minutes, whereas the processing force at 750 °C drops from 9 kN to 5 kN during the course of the incremental forming procedure.

### *3.2 Microstructural investigations*

To fully document the microstructural evolution in our Ti alloy during incremental forming, we first consider the characteristics of the initial microstructure in the undeformed Ti-10-2-3 samples after air-cooling. Figures 4a-d show the microstructures of conventionally heat-treated samples (i.e., with annealing between 600 °C and 750 °C, and without incremental deformation). In all heat-treated conditions the material is characterized by a large β-grain size of approx. 300 μm. It is obvious that most primary α phase has an acicular morphology, which is formed in the β grains. The volume fraction of the precipitates decreases with rising of the annealing temperature from 51 % at 600 °C to 19 % at 750 °C (Table 1). The size of the α needles and the interlamellar spacing between the α layers increase with increasing temperatures as well.

**Table 1.** Volume fraction (in %) of primary α after the conventional heat treatment vs. thermo-mechanical processing at different temperatures. The data was estimated from the SEM micrographs using simple image analysis algorithms.

| in % | 600 °C | 650 °C | 700 °C | 750 °C |
|---|---|---|---|---|
| **conv. HT** | 51 | 48 | 39 | 19 |
| **TMP** | 50 | 38 | 15 | 8 |

Grain boundary α (GB-α) could be observed in all samples after the conventional heat treatment (Figures 4a-4d, highlighted with arrows). In the SEM micrographs (most notably in Figures 4c and 4d) it is visible as a continuous thin layer with a thickness of approx. 1 µm. At 700 °C and 750 °C, the GB-α is surrounded by a zone without any precipitates. In contrast, the GB-α layers at 600 and 650 °C are considerably thinner, and are not associated with a precipitate-free zone. We highlight that this is an important observation because GB-α represents a preferred path for crack growth and can therefore lead to a lower amount of ductility. Finally, at lower annealing temperatures (Figures. 4a and 4b), the primary α precipitates form the well-known basket-weave microstructure.

In Figures 4e-h, we present the corresponding SEM micrographs obtained from samples of the TMP material. A comparison between the two heat treatment strategies shows that the morphology of the primary α phase is altered by incremental deformation. Instead of needles, the precipitates form a bimodal structure that consists of both globular and lamellar α-phase. While it seems reasonable to expect that the size of the primary α precipitates is reduced by incremental forming, our microstructural observations point out a different effect: compared to the conventional heat treatment without deformation, precipitate size seems to be unchanged at 700 °C and 750 °C to the conventional treatment; at 600 °C and 650 °C the precipitates in fact are significantly larger in comparison to those of the undeformed material, and no basket-weave structures could be observed. Moreover, the TMP samples show no tendency to precipitate GB-α at lower temperatures. Only at 750 °C, small amounts of GB-α could be observed. Most notably, mechanical loading during incremental forming does have an effect on precipitate orientation: all precipitates are orientated parallel to the direction of maximum shear stresses. Even at lower temperatures (600 and 650 °C) the primary α is related to the formation of clearly recognizable shear bands (see also Figures 5c and d). Comparing the conventional heat treatment and the TMP condition, we also observe that the conventionally annealed material systematically exhibits a higher volume fraction of primary α in comparison to the TMP samples. This difference increases with increasing temperature (see Table 1).

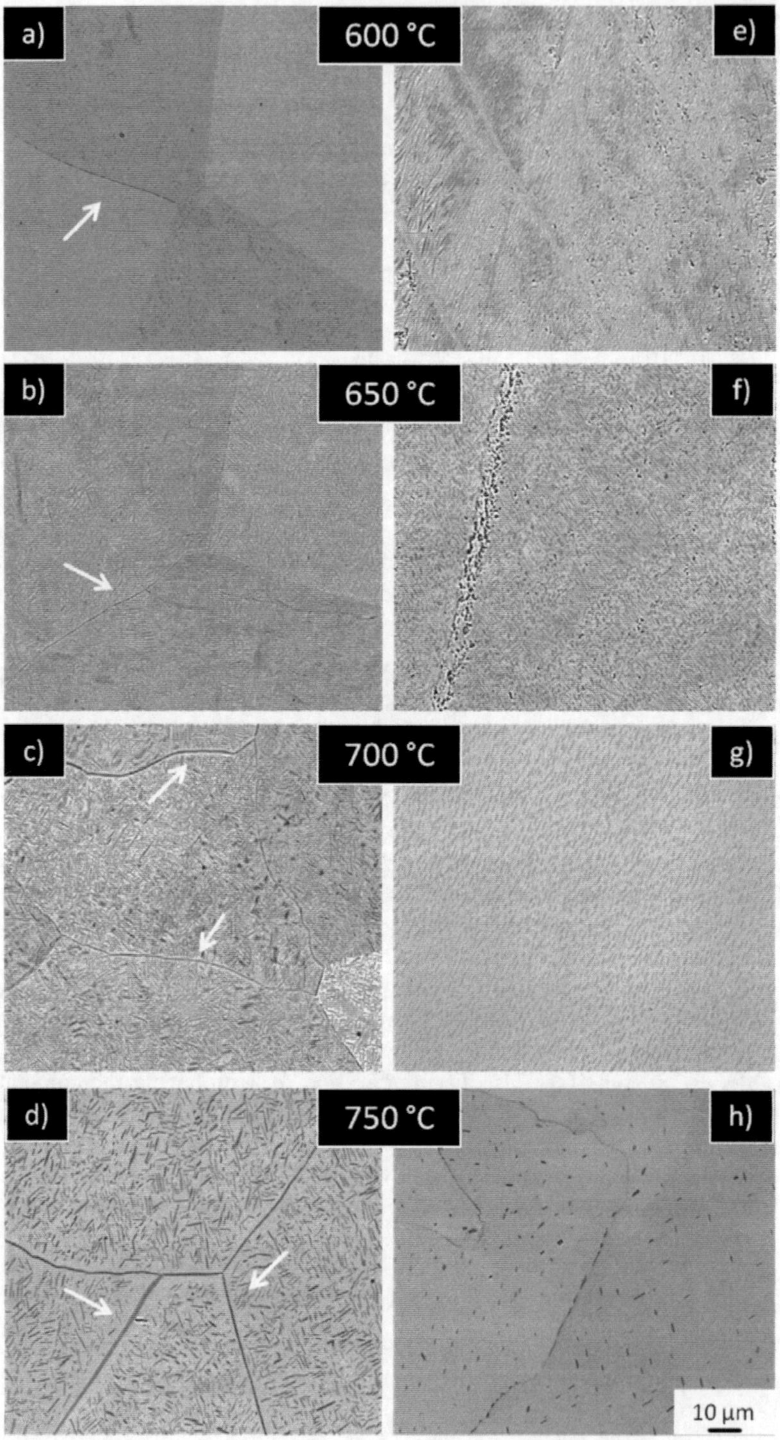


**Figure 4.** SEM micrographs of the conventional heat treated (a - d) and the TMP samples (e - h) after annealing at 4 different temperatures: The conventional heat treatment leads to finely dispersed lamellar $\alpha_p$ in the β-matrix. Lower temperatures result in higher $\alpha_p$ volume fraction. GB–α (highlighted by arrows) can be recognized at all temperatures. The TMP samples show a bimodal microstructure with globular and lamellar $\alpha_p$ - precipitates at 600 and 650 °C.

In Figure 5, we present some interesting microstructural features of the samples heat-treated at 600 °C and 650 °C, respectively, at higher magnification. On the left, (Figures 5a, b) SEM micrographs from undeformed conditions are shown, highlighting the basket-weave microstructures in both cases. At 650 °C, the α-phase volume fraction is slightly lower, so the precipitates have more space to grow which results in larger lamellae. In addition to the basket-weave structures, we also observed regions where the primary α precipitates are predominantly oriented in parallel and hence form no triangles (Figure 5a, dark areas, highlighted by arrows). Further research is required to fully analyze how these regions affect the mechanical behavior. The TMP samples (Figures 5c, d), in contrast, exhibit no basket-weave microstructures. Again, it is obvious from these micrographs that the $\alpha_p$-phase is oriented parallel to the direction of maximum shear stress (i.e., inclined by about 45° to the uniaxial loading direction). Incremental forming at 600 °C leads to the formation of distinct shear bands. Inside these bands, the α-phase is also strongly oriented parallel to the growth direction of the bands. The samples deformed at 650 °C exhibit a bimodal structure with both globular and lamellar α-phase; no significant occurrence of shear banding could be observed.

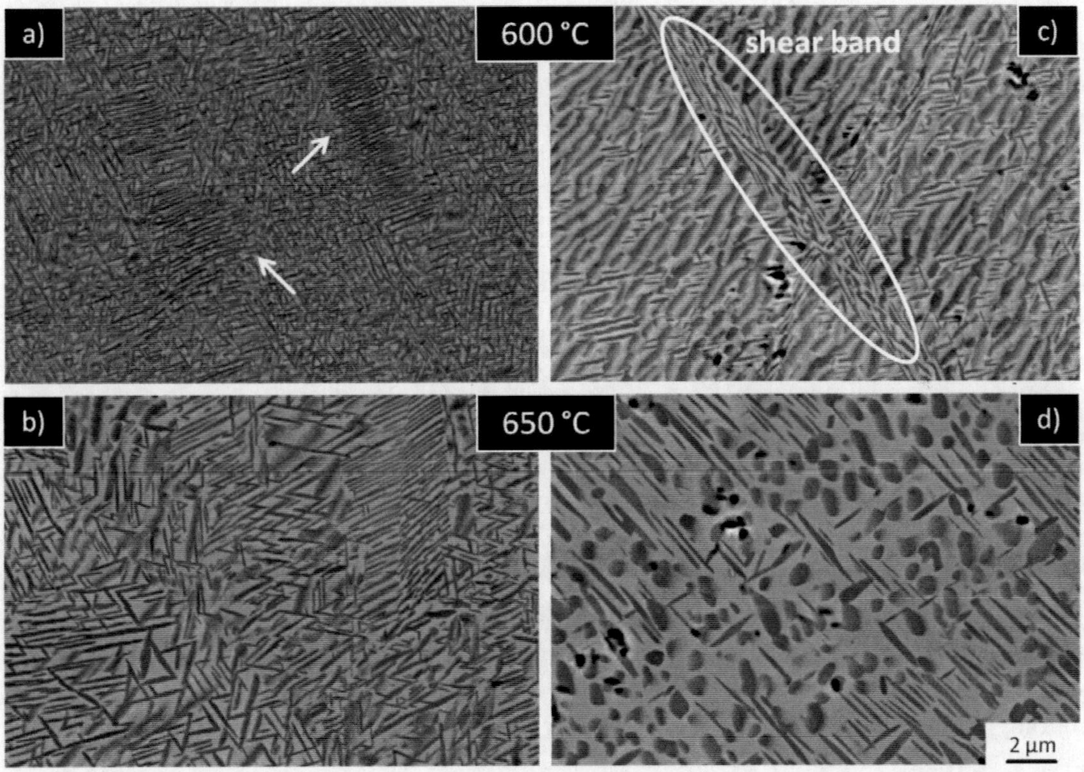


**Figure 5.** SEM micrographs of the conventionally heat-treated (a, b) and the TMP materials (c, d) after annealing at 600 °C (a, c) and 650 °C (b, d): The conventional heat treatment leads to a typical basket-weave structure and smaller precipitates. There are also regions where the primary α is primarily oriented in one direction (5a, dark areas, highlighted with arrows). The TMP samples exhibit a bimodal microstructure with globular and lamellar $\alpha_p$. Also the localization of deformation in shear bands can be clearly observed (as highlighted in c). The $\alpha_p$ - phase is oriented parallel to the orientation of maximum shear plane stress during incremental forming in all TMP samples.

### *3.3 Mechanical properties*

Stress-strain curves of the undeformed and the conventionally heat-treated materials are shown in Figure 6. Clearly, precipitation of primary α significantly increases the yield strength (YS) and the ultimate tensile strength (UTS) in comparison to the solution treated material (which only consists of the β-phase). After the conventional heat treatment the YS is increased from approx. 480 MPa (ST) to approx. 800 MPa after heating to 700°C or 750 °C, to 900 MPa after heating to 650 °C, and to a maximum value of 1050 MPa after heating to 600 °C. Ductility, however, decreases with lower annealing temperatures. The uniform elongation (UE) decreases from 39 % (ST) to 7 % (750 °C) and drops further to 2 % when the material is heat-treated at 600 °C. As expected, these results confirm that a large volume fraction of primary α precipitates increases the strength but decreases the ductility of Ti-10-2-3. We also observe that there is no significant strain hardening for the samples heat-treated at 700 °C and 750 °C samples, whereas those treated at lower temperatures (and the ST material) show some hardening.

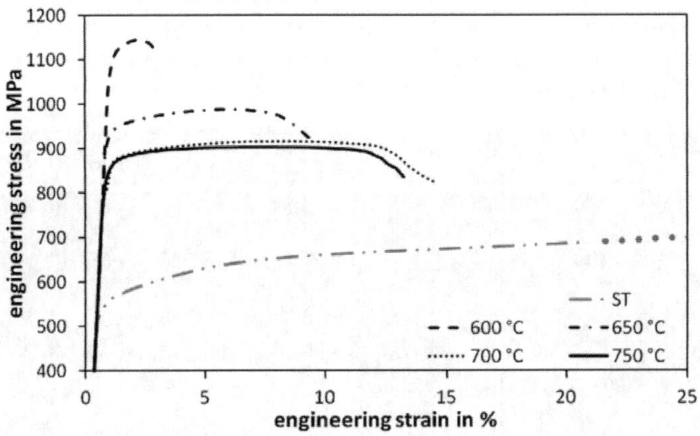


**Figure 6.** Engineering stress-strain curves of the samples heat-treated without incremental forming in comparison to the solution treated (ST) material. Yield strength and ultimate tensile strength are significantly increased in comparison to the ST material. Ductility decreases with lower annealing temperatures.

In comparison to materials subjected to the conventional heat treatment, the mechanical behavior of the samples subjected to incremental forming is different (Figure 7). After TMP the YS is increased from approx. 520 MPa (cyclic deformation at 750 °C) to approx. 650 MPa (700 °C), 700 MPa (650 °C) and to a maximum of 750 MPa (600 °C). The strength of all TMP conditions is lower than that of the corresponding materials that were heat-treated without incremental forming. This can be related to the larger amount of α-phase precipitates in the conventional heated samples (Table 1), which results in a stronger hardening of the β-matrix and thus leads to higher strength. In contrast, the ductility of the TMP samples is higher than for the conventional conditions. The samples that were cyclically deformed at 750 °C reach an UE of 22 %. The UE decreases with lower forming temperatures, but even the samples deformed during TMP at 600 °C still exhibit a UE of 6 %. This difference is mainly due to the fact that GB-α and the lamellar α morphology in the conventionally heated samples leads to lower ductility in comparison to the TMP samples that contain only globular or bimodal α precipitates. It is also known that the basket-weave structure, which occurs in the materials heat-treated at 600 °C and 650 °C, considerably reduces ductility. The strain hardening rate of the TMP samples is relatively low. Only those samples cyclically deformed at 750 °C exhibit a significant hardening behavior. Due to the small amount of primary α (about 8 %, Table 1), the strain-hardening behavior of this material condition is actually quite similar to the strain-hardening observed for the solution annealed material that is fully in the β-phase.

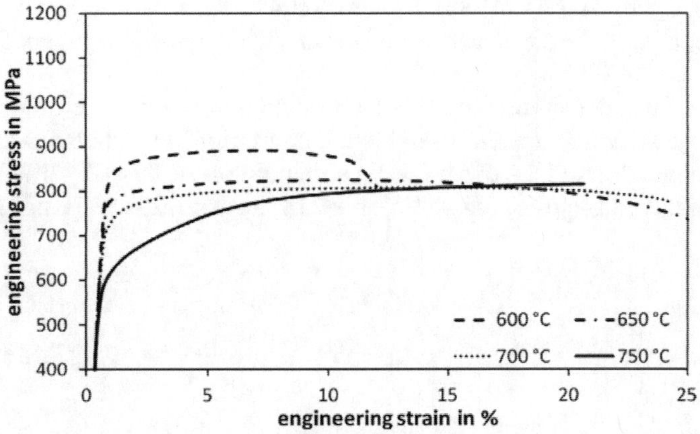


**Figure 7.** Engineering stress-strain curves of the TMP materials. The strength of all conditions is reduced compared to the corresponding materials produced without incremental forming, but ductility is is significantly increased.

## 4. Summary and conclusions

This paper presents first results of an ongoing investigation of the possibility to modify the microstructure of primary α, in particular with respect to precipitate shapes and volume fractions. In addition to the conventional strategy for the precipitation of primary α, a special route of thermo-mechanical processing was performed. This processing approach combines the conventional heat treatment with incremental forming during the primary α-phase annealing (performed in this study at 600 °C, 650 °C, 700 °C and 750 °C, each for 60 min). After the incremental forming, considerable variations in terms of microstructure and mechanical properties were observed for the different conditions. The microstructure of the deformed samples is characterized by globular and lamellar (bimodal) α precipitates, whereas conventional annealing leads to the formation of lamellar precipitates. Furthermore the primary α is oriented in the direction of maximum shear stresses during incremental forming, and forms no specific basket-weave structures. Due to the smaller size and the lower amount of α-phase after incremental forming, the strength lower than in materials treated with the conventional strategy. A high amount of grain boundary-α and lamellar $\alpha_p$-phase in the undeformed materials leads to a significantly lower ductility in comparison to the matrix with bimodal structures. Our study demonstrates that incremental forming of Ti-10-2-3 during the annealing step of the thermo-mechanical treatment can be used to modify the microstructures in a wide range of volume fractions and morphologies of the primary α-phase.

## References

[1] Donachie M J: *Titanium - a technical guide. ASM International, 2000.*
[2] Allen P: Titanium alloy development. *Advanced Materials & Processes.* **150** (4) (1996).
[3] Peters M and Leyens C: *Titan und Titanlegierungen.* WILEY-VCH, 2000, pp. 49-68.
[4] Kiese J and Wagner L: Fatigue behavior in Ti-10V-2Fe-3Al: Microstructural effects on crack nucleation and microcrack growth. *Fatigue '96*, Pergamon Press, 1996.
[5] Wang B et al: Mircrostructural evolution during aging of Ti-10V-2Fe-3Al titanium alloy. *Journal of University of Science and Technology* Beijing, **14** (4) (2007), pp. 335.
[6] Neugebauer R et al: Manufacture of a b-Titanium Hollow Shaft by Incremental Forming. *Production Engineering*, **5** (2011), pp. 227-232.
[7] Neugebauer R et al: Incremental Forming of Hollow Shapes. *Steel research int.*, (2005), **76** (2/3).
[8] Margolin H, Cohen P, Kimura H and Lzumi O: *Titanium 80*, Science and Technology, TMS-AIME, Warrendale (1980), pp. 1555-1556.

[9] Weiss I: *Metall. Trans.* A **17A** (1986), pp. 1935-1947.
[10] Welsch G, Lacombe P, Tricot R and Beranger G: *Sixth World Conference on Titanium*, France (1988), pp. 1289-1293
[11] Stănková H: Einfluss der inkrementellen Deformationen bei der thermomechanischen Behandlung auf die Eigenschaften von TRIP-Stählen. Dissertation, TU Chemnitz, 2008.
[12] Luquiau D, Feaugas X and Clavel M: Cyclic softening of the Ti-10V-2Fe-3Al titanium alloy. *Materials Science and Engineering A*, Volume **224**, Issues 1-2 (1997), pp. 146-156

IOP Conf. Series: Materials Science and Engineering **118** (2016) 012027 doi:10.1088/1757-899X/118/1/012027

# Multimetallic Electrodeposition on Carbon Fibers

**F BÖTTGER-HILLER[1], J KLEIBER[1], T BÖTTGER[1], T LAMPKE**

1 Materials and Surface Engineering Group, Institute of Materials Science and Engineering, Technische Universität Chemnitz, 09125 Chemnitz, Germany

**Abstract.** Efficient lightweight design requires intelligent materials that meet versatile functions. One approach is to extend the range of properties of carbon fiber reinforced plastics (CFRP) by plating the fiber component. Electroplating leads to metalized layers on carbon fibers. Herein only cyanide-free electrolytes where used. Until now dendrite-free layers were only obtained using current densities below 1.0 A $dm^{-2}$. In this work, dendrite-free tin and copper coatings were achieved by pre-metalizing the carbon fiber substrates. Furthermore, applying a combination of two metals with different sized thermal expansion coefficient lead to a bimetallic coating on carbon fiber rovings, which show an actuatory effect.

## 1. Introduction

The substitution of classic materials by carbon fiber reinforced polymers (CFRPs) provides advantages in materials properties. Impressive arguments for their use are high strength and flexibility, very low weight, positive behavior in important application criteria such as energy consumption, vibration resistance and damping, corrosion and aging. CFRPs show disadvantages like insufficient electrical and thermal conductivity. Furthermore, damages in CRFP parts are difficult to detect [1].

To overcome those weaknesses, several strategies which allow the functionalization of CFRP are under investigation. Therefore, three different strategies are preferred. On one hand it is possible to metalize the CFRP at its surface [2]. This way efforts consuming pre-treatment steps and the use of harmful chemicals. Furthermore, the layer is exposed to the environment what can cause contamination or oxidation. On the other hand it is possible to integrate the functionalization into the CFRP structure using metal wires [3]. In this case the millimeter thick metal wires forward weakness. In addition, metal wires are getting more and more expansive the smaller their diameters are.

The third preferred strategy is the production of metalized carbon fibers which are economic, environmentally friendly producible but currently not commercially available, except for nickelized carbon fibers [4–6]. A viable commercial transfer of metalized carbon fibers efforts the use of cyanide-free electrolytes and high current densities. In contrast to this requierements dendrites mainly grow while plating in example tin or copper at current densities above 1.0 A $dm^{-2}$ [4]. According to the plating metal, the coatings show more or less dendrites [7]. In this work, a conductive basic layer was used to prevent the formation of dendrites. Furthermore, two metals with different thermal expansion coefficients were combined to produce a bimetallic coating with actuatory effect for the first time.

IOP Conf. Series: Materials Science and Engineering **118** (2016) 012027 doi:10.1088/1757-899X/118/1/012027

## 2. Methods

### *2.1. Equipment and Materials*

The investigation of dendrite-free tin and copper coatings at carbon fibers were examined on carbon fiber woven fabrics (type Aero 160 from R&G Verbundwerkstoffe GmbH containing Toho Teijin Tenax HTA 40 E13 3K fibers). The woven fabrics were cut into segments of 40 mm x 60 mm to metalize the top layer.

Initial tests for the development of carbon fibers with an actuatory effect were examined on carbon fiber rovings of the type HTA 40 E 13 3 K (Toho Teijin Tenax). The fibers were cut into segments of approximately 70 mm length and fixed in a polypropylene holder.

The prepared substrates were immersed in a cyanide-free electrolyte for nickel, tin or copper at a temperature 25±1 °C. Electroplating experiments were examined by a laboratory power supply Beha Uniwat NG 304. Tempering and circulation of the electrolyte were ensured by using a laboratory stirrer IKA C MAG HS. The anodes were purchased from Galva-Metall GmbH.

Metallographic cross-sections were prepared to analyze the metal layers. Those samples were examined via optical microscope Olympus GX51 with Olympus XC30 camera.

### *2.2. Equipment and Materials*

Desizing of the substrates was carried out by immersing the carbon fibers for 60 minutes in an ultrasonic bath filled with acetone. The temperature was held at 60 °C during desizing.

The electrodeposition on carbon fiber rovings was carried out under variation of the process parameters like current density between 0.25 and 1.5 A $dm^{-2}$ and a plating time in a range from 5 to 60 minutes. The coated samples were rinsed with distilled water and ethanol. A subsequent drying process at 60 °C for 4 hours in a vacuum compartment dryer finalized the coating process of each layer.

The bi-metallic coating with an aspired actuatory effect requires an one-sided metal deposition at the carbon fiber surface. Thus, only one anode was used. In addition, it is necessary to protect the opposite side of the carbon fibers to prevent undesired metal deposition. Therefore, the bottom half of the rovings was embedded in an epoxy resin. After the successful metal deposition at the upper half, the samples were placed in acetone for 24 hours, to remove the protection layer.

## 3. Results

### *3.1. Dendrite-free copper plating*

Initially, the carbon fibers were nickelized in order to increase their conductivity and prevent the formation of dendrites. The coating process was observed by measuring the voltage in dependence of time. The resulting diagrams are similar to each other. Here, we present the deposition of the basic layer and a copper layer as representative example.

The initial voltage of the nickelization amounts 3.2 V and decreases to a constant value of 2.1 V after one minute deposition time (Figure 1). Copper plating takes place at a lower nearly constant voltage, which amounts to approximately 0.5 V.

IOP Conf. Series: Materials Science and Engineering **118** (2016) 012027 doi:10.1088/1757-899X/118/1/012027

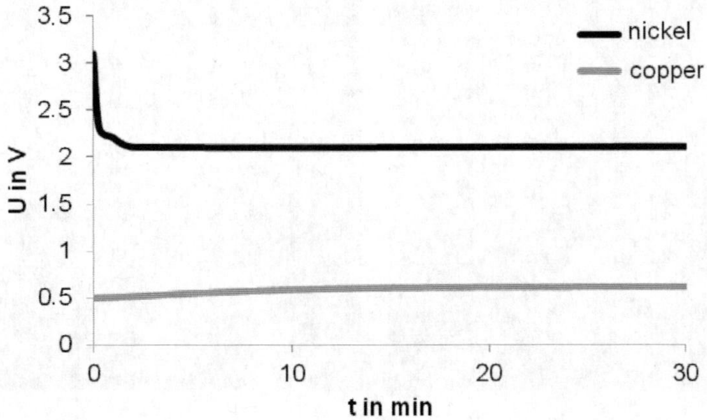


**Figure 1.** Time-voltage diagram recorded during the electroplating of a nickel basic layer deposited at a pure carbon fiber woven fabric and a copper layer deposited at a nickelized carbon fiber woven fabric.

The results of the coating processes can be observed by microscopy of cross-sectional views. The conductive basic layer is 1 μm thick and homogeneously distributed. Furthermore, the nickel layer shows a good penetration depth (Figure 2).

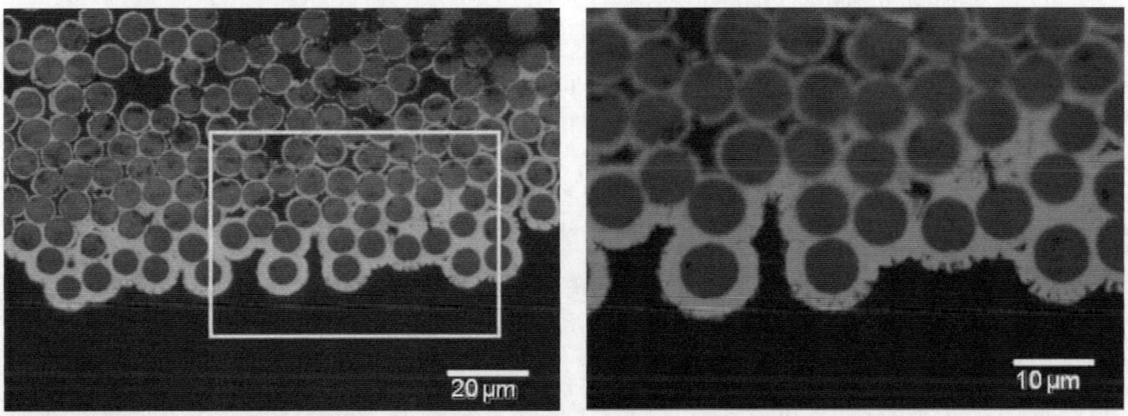


**Figure 2:** Cross sectional views of nickelized carbon fibers with a dendrite-free copper coating in different magnifications.

The electrodeposition of copper was carried out at a current density of 1.5 $dm^{-2}$ (Figure 1). The achieved copper layer covers the outer surface of the carbon fiber roving completely. In contrast to former coating experiments, it is possible to achieve very thin dendrite-free copper layers using the conductive basic layer (compare Figure 2 and [5]). The thickness of the copper layer amounts to approximately 2 μm.

### *3.2. Dendrite-free copper plating*

For the electrodeposition of dendrite-free tin layers, the carbon fibers were nickel plated too. The nickel-coating is 1 μm thick and homogeneously distributed and comparable to the conductive basis layer for dendrite-free copper. The penetration depth is as good as observed above (Figure 3).

IOP Conf. Series: Materials Science and Engineering **118** (2016) 012027 doi:10.1088/1757-899X/118/1/012027

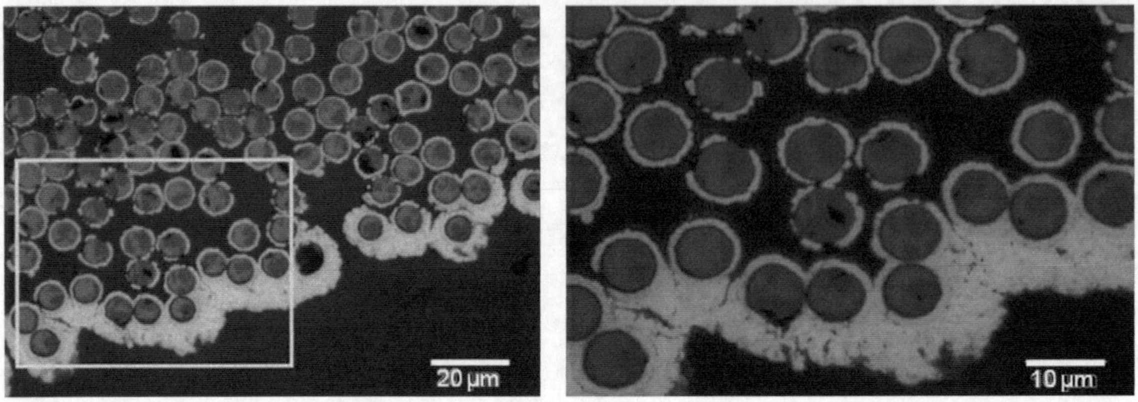


**Figure 3.** Cross sectional views of nickelized carbon fibers with a dendrite-free tin coating at different magnifications.

The tin layer was deposited at a current density of 1.5 A $dm^{-2}$ (Figure 3). The layer is located at the outer surface of the carbon fiber rovings and it does nearly cover their whole surface. The deposition of tin at a conductive basic layer leads to layers with an average thickness of approximately 7 µm. The coating is also dendrite-free.

### 3.3. *Bi-metallic coating with actuatory effect*

Two metals with preferably different expansion coefficients were chosen to create a bi-metallic coating with actuatory effect at a carbon fiber roving (Table 1). In the following example copper was plated as first and zinc as second layer.

**Table 1.** Coefficients of thermal expansion of copper, nickel, zinc, tin and carbon fibers [8].

| Material | $\alpha$ in $10^{-6}$ $K^{-1}$ |
|---|---|
| copper | 16,5 |
| nickel | 13,4 |
| zinc | 36 |
| tin | 27 |
| carbon fibers | –0,1 |

Due to the before applied protection layer using an epoxy resin, the coating is located at only one side of the carbon fiber roving. The protection layer could easily be removed after the electroplating processes.

The semi coated rovings show an actuatory effect (Figure 4).

**Figure 4.** Macro image of semi-coated carbon fiber rovings with a copper-zinc bi-metallic layer, showing an actuatory effect.

The samples are slightly deformed in the ground state (25 °C) due to residual stress. This deformation was completely removed by thermal post-treatment of the coated fibers up to 400 °C. Thus, the system is able to overcome a length of 6 mm.

## 4. Discussion

The carbon fiber woven could be nickelized easily, as known from literature [2]. This reproducible procedure can be used to create a conductive basic layer for further metallization. Due to the low current density a thin nickel layer was deposited. This shows a good penetration depth and coverage according to the greater convection and electrolyte used.

The voltage curve of the application of further layers is much lower than in case of direct carbon fiber metallization [see 2, 4–6,9], because an increasing conductivity of the layer leads to a decrease of voltage. The homogeneously conductive nickel layer provides a very good basis for the next layers. In case of copper, no additional seed is necessary and very thin coatings are possible while high current densities are applied. Caused by the nickel layer, the deposited copper layer is homogeneously distributed as well. Furthermore, the copper layer is located at the outer surface. The nickel layer shields the carbon fibers and prevents a further coating.

The conductive basic layer enables the dendrite-free deposition of tin, too. Remarkably, in comparison to literature (i.e. [5,9]), thick dendrite-free tin layers at carbon fibers could be described for the first time. In this work, a dendrite-free layer was obtained at a current density of 1.9 A $dm^{-2}$ while the tin deposition in [9] was no longer possible at current density of 1.0 A $dm^{-2}$.

Furthermore, a bi-metallic coating with actuatory effect was applied by the means of copper and zinc deposition. Both metals were chosen according to their coefficient of expansion. The one-sided coating provided a good opportunity to give the fibers actuatory elements. The deformation by heat reaction occurred relatively quickly. Just as quickly, the rebound occurred when removing the heat source. This is due to the small coating thickness, which also has only a small heat capacity. Figures
Each figure should have a brief caption describing it and, if necessary, a key to interpret the various lines and symbols on the figure.

## 5. Conclusion

In this work, dendrite-free copper and tin coatings at high current densities were achieved by pre-metalizing the carbon fiber substrates. Furthermore, applying a combination of two metals with different sized thermal expansion coefficients lead to a bimetallic coating on carbon fiber bundles, which shows an actuatory effect. To the best of our knowledge such results haven't been published yet. Structure-property relationships and cycling tests of the bi-metallic coating with actuatory effect are currently under investigation.

## 6. References

[1] Barlas T K and van Kuik G A M 2010 *Progress in Aerospace Sciences* **46** 1
[2] Böttger-Hiller F, Neumann S, Fehrmann F, Nier M, Böttger T, Scharf I, Nestler D, Böhme M, Hahn S, Nickel D, Wielage B, Lampke T 2014 *Mat.-wiss. u. Werkstofftech.* **45** 546
[3] Mäder T 2015 *Neuartige Sensoren zur Erfassung von Dehnungen in Faserkunststoffverbundwerkstoffen (Structural Health Monitoring)* Chemnitz (TU Chemnitz)
[4] Böttger-Hiller F, Nier M, Lampke T 2014 *International Surface Technology* **1** 44
[5] Nier M, Böttger T, Böttger-Hiller F, Nickel D, Scharf I, Nestler D, Wielage B, Lampke T 2013 *ICCM-19* 5820
[6] Böttger-Hiller F, Jahn P, Trautmann M, Lindner T, Nickel D, Lampke T 2015 *Mat.-wiss. u. Werkstofftech.* **46** 844
[7] Böttger-Hiller F, Nestler K, Zeidler H, Glowa G, Lampke T 2016 *AIMS Materials Science* submitted.
[8] Erbrecht R, König H, Martin K, Pfeil W, Wörstenfeld W 1999 *Das große Tafelwerk* (Berlin: Volk und Wissen)
[9] Dunzel H 2013 *Herstellung und Charakterisierung von metallischen Schichten auf Kohlenstofffasersubstraten als Funktionsschicht* Chemnitz (TU Chemnitz)

**Acknowledgments**
The financial support of the European Social Fund (ESF), and the BMWi (EXIST-Forschungstransfer – inca-fiber: 03EFE SN080) is greatly acknowledged. Moreover, the authors would like to thank Thomas Mäder for the fruitful discussions and experimental hints.

# Phase field simulation of a directional solidification of a ternary eutectic Mo-Si-B Alloy

**O Kazemi[1], G Hasemann[1], M Krüger[1] and T Halle[1]**
Institut für Werkstoff- und Fügetechnik, Otto-von-Guericke-Universität Magdeburg, P.O. Box 4120, D-39016 Magdeburg, Germany, E-Mail: omid.kazemi@ovgu.de

**Abstract**
We present a eutectic Phase-Field Model for a Mo-Si-B alloy at ternary eutectic composition (Mo-17.5Si-8B), under a constant thermal gradient. The process parameters like cooling rate and thermal gradient were obtained directly from the experimental procedure of zone melting. The equilibrium interface geometries and interface mobility were calculated using an isotropic model.
The phase equilibria and the other thermodynamic parameters are obtained by linearizing the Mo-Si-B ternary phase diagram.
We have investigated the effect of process parameters on the lamellar growth pattern and lamella pattern stability with respect to the Jackson-Hunt minimum undercooling spacing theory. In order to examine the generated results by the model, they were validated with experimental observed microstructures and measurements and showed to be in a good agreement with the experimental observations.

Keywords: Mo-Si-B alloy – Phase field simulation – Ternary Eutectic

## 1. Introduction

### *1.1. Phase-Field Method*

Investigating the advancement of interface pattern and formation of complex microstructures during solidification, have been attractive subjects for the scientific researches. Predicting solidification behavior and microstructure pattern in alloys have been always interestingly focused, since having a good sight of the microstructure helps to predict the final physical and chemical properties of the alloy. One of the most powerful tools for predicting the interface evolution pattern is the Phase-Field Method [1 - 6].

The main logic behind the phase transformations indicates that the free energy reduction is the principal driving force of the phase transformations. Taking this attitude into account, the phase-field method has been implemented to simulate microstructure evolution during solidification [2]. In the phase-field models (PFM's), in order to define the distribution of the contacting phases and interface in the whole system, a phase field variable shown by $\Phi$ is announced. The value of the $\Phi$ in liquid region is adjusted to $\Phi = 0$, and in solid region $\Phi = 1$.

In the whole PFMs, the interface is considered as a thin layer but finite in width. In the classical assumptions which are abolished, the interfaces were assumed to be a sharp layer. However, recent investigations apply diffusive interface in which the interface is considered to be a combination of the liquid and the present solid phases. The phase field and diffusion field equations are derived from a definition of the free energy functional of the system, which is decreasing with time [1, 3].

During the past decades there are several attempts to perform a phase field simulation of the solidification process based on the Spinodal decomposition. In this approach, it is assumed that the crea-

IOP Conf. Series: Materials Science and Engineering **118** (2016) 012028 doi:10.1088/1757-899X/118/1/012028

tion of new nuclei will happen when the distance between lamellae get large enough. However, the experimental investigations proved that in some cases the prediction may not be accurate because despite the simulation's prediction, due to the abrupt tilting and oscillating of the lamellae the distance between them would slightly reduce and the potential position for the nucleation would be vanished.

Recently, Pezzolla and Steinbach have rearranged the multiphase field method by introducing a new variable called interface field. Thank to interface field variable, the changes of phase-field at each distinct point would be decomposed into sum of changes of independent interface fields at the point [1, 11].

### *1.2. Mo-Si-B Alloys*

Newly Ni-based super-alloys have met their ultimate application limits in getting used in turbines and industry's further demands has induced lack of more efficient materials. An ongoing demand for the materials which are capable to operate in even increasing temperature has led to development of new ultrahigh temperature structural materials. A potential class of materials is Mo-Si-B alloys [7, 9].

A principal point for the current microstructural researches is to find appropriate alloy compositions and processing procedure to reach the optimal balance among the mechanical properties. In order to enhance the microstructures and recognize the correlation between microstructures and material properties, it is necessary to have a comprehensive knowledge of the phase equilibria and physical properties of the Mo–Si–B system [8].

In this paper we firstly derive isotropic phase-field and diffusion equations for eutectic solidification by using the interface-field method and the equal chemical potential condition and we perform a phase field simulation of Mo-Si-B ternary eutectic alloy's solidification. Consequently we present the results for conducting the simulation with the different operational parameters and then we investigate the instability of the lamellae and the interface in variation of lamella spacing and try to investigate the compatible range of lamellae spacing which adequately fits the Jackson-Hunt's minimum undercooling lamella spacing according to the given undercooling and growth velocity [13].

## 2. Eutectic Phase Field Model

In the eutectic solidification of Mo-Si-B alloy, there are three solid phases; $Mo_3Si$ having an A15 crystal structure, $Mo_5SiB_2$ which has T2 structure and a Mo solid solution forming from the liquid phase; L [8]. Phases are designated by $\phi_i$ (i=0,1,2,3). According to the principles of phase field, sum of the $\phi_i$ at each region must be equal to 1 [1]. The main principle in determining the phase evolution pattern is to investigate the evolution of Phase field variable in the pass of the time. The time derivative of the $\phi$ shows the development of each phase clearly:

$$\dot{\Phi}_s(X,t) = \mu[\sigma(\nabla^2\Phi_s - \frac{(1-\Phi_s)(1-2\Phi_s)\Phi_s}{\eta}^2) + \frac{1}{\eta}\Delta G\Phi_s(1-\Phi_s)] \qquad (1)$$

Where $\mu$ is interfacial mobility, $\eta$ is interfacial thickness and $\sigma$ denotes interfacial energy [1, 11].

The total free energy function of the system volume V is given by

$$F = \int_V [f^p + f^t + \lambda_L(\sum_i \Phi_i - 1)]dV \qquad (2)$$

In which the $\lambda_L$ is the Lagrange multiplier and $f^p$ is defined as follow:

$$f^p = \sum_{j>i}\sum_i[-\frac{\varepsilon_{ij}^2}{2}\nabla\Phi_i\cdot\nabla\Phi_j + \omega_{ij}\Phi_i\Phi_j] \quad (3)$$

Where $\varepsilon_{ij}$ is the gradient energy coefficient and $\omega_{ij}$ is the height of double well potential.

And $f^t$, which is thermodynamic potential could be defined by:

$$f^t = \sum_i \Phi_i f^i(c_i) \quad (4)$$

The $f^i(c_i)$ is the free energy density of i phase with composition of $c_i$ [1].

*2.1. Phase-field equation*

Kim et al. have defined a step function $S_i(x, t) = 1$ when $\phi_i > 0$ and $S_i(x, t) = 0$ otherwise. So the number of coexisting phases could be given by:

$$n(x,t) = \sum_{i=0}^{3} s_i(x,t) \quad (5)$$

Taking account to this concept the time derivative of the phase field variable could be shown as following:

$$\frac{\partial\Phi_i}{\partial t} = \frac{1}{n}\sum_{j\neq i} s_{ij}(\frac{\partial\Phi_i}{\partial t} - \frac{\partial\Phi_j}{\partial t}) \quad (6)$$

Thank to this consideration, the change of phase field takes place at the interface. It emphasizes that motions of the interface of each phase with the other participating phases will affect the changes of the phase field variable contributing to it. It should be noted that these equations are derived with keeping the assumption that all the coexisting phases have equal chemical potential [1].

*2.2. Diffusion equation*

The main concept in deriving the diffusion equation is mass conservation. Accompanying the free energy of the system it could be mentioned as following:

$$\frac{\partial c}{\partial t} = \nabla D \sum_i \Phi_i \nabla c_i \quad (6)$$

In which the D is diffusivity dependent on the phase field [1].

It is worth noting here that all of the calculations in this study were assumed to be in isotropic condition.

## 3. Methods and Parameters

The most important part of conducting a transition simulation is to identify the phase equilibria and recognize the thermodynamic description and material characteristics of the present phases. The Mo rich part of Mo-Si-B system has been the subject of many researches. It was firstly review by Nowotny et al. and they have performed the isothermal section at 1600 °C and have extracted the involved phases in assistance with XRD [10].

Nunes et al. have studied the liquidus projection of the Mo rich portion of Mo-Si-B system and reported that there are six principal reactions along the solidification. They found that a ternary eutectic reaction occurs in the system and have reported that it is as following:

L → $Mo_3Si + Mo_5SiB_2 + Mo_{(ss)}$ [8].

Performing the liquid projection process with the Factsage™ software has proved that the eutectic transition occurs at about 2260 °K. Figure 1 shows the prediction of the phase equilibria and temperature of the phase transitions.

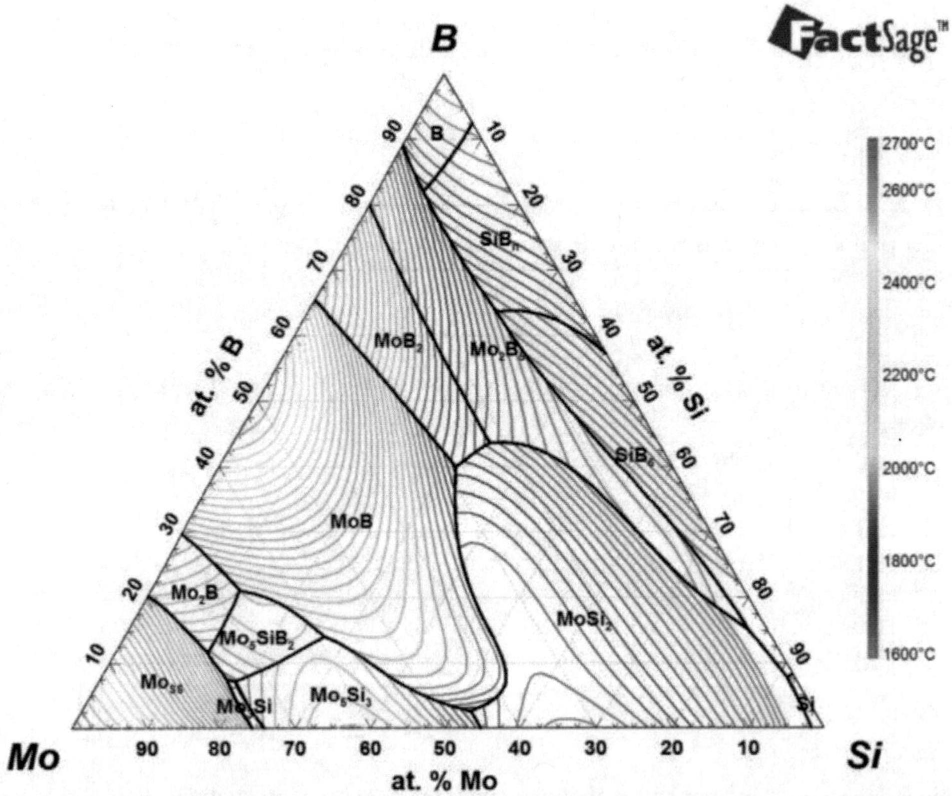


Figure 1: Liquidus projection of Mo-Si-B system, predicted by Factsage

In order to simulate the eutectic solidification process using the phase field method, a well cited scientific software called MICRESS was applied. To perform the simulation three major types of data were required: Material parameters such as diffusion data, phase interaction data like the mobility between phases, interface energy and thermodynamics data, process parameters like cooling rate and thermal gradient and the other numerical parameters for instance grid resolution and time step. The latter have to be adjusted to optimal values since although small values lead in more accurate results, but would increase the simulation time significantly. Otherwise the big enough amounts would result in unreliable consequences.

To carry out the simulation, a model containing the predicted phases was developed. The initial composition of the liquid was adjusted precisely to the ternary eutectic composition; Mo - Si 17 at % - B 7.7 at%. All the composition values are indicated in atomic percent in following.

Concerning the verification of the created results, they were compared with the experimentally observed outcomes. In this regard the cooling rate and thermal gradient were obtained from the experiments. Table 1 shows the parameter adjustments.

**Table 1**. Applied values for the required parameters

| Cooling rate [K/s] | Thermal Gradient [K/cm] | Surface energy between liquid and phases [J/cm$^2$] | Entropy of fusion in solid phases [J/cm$^3$K] | Entropy of fusion in liquid phases [J/cm$^3$K] | Diffusion coefficient in liquid [cm$^2$/s] |
|---|---|---|---|---|---|
| 200 | 10 | $10^{-4}$ | 0.1 | 0.9 | $2\times10^{-4}$ |

The simulations were performed with different interface thickness and an interface with three cells width has proved to result in the most compatible results.

## 4. Results and Discussion

Preliminary results showed to be in adequate accordance with the experimentally obtained ones. Figure 2 presents a schematic comparison between the initially achieved results of simulation and the observations of experiments.

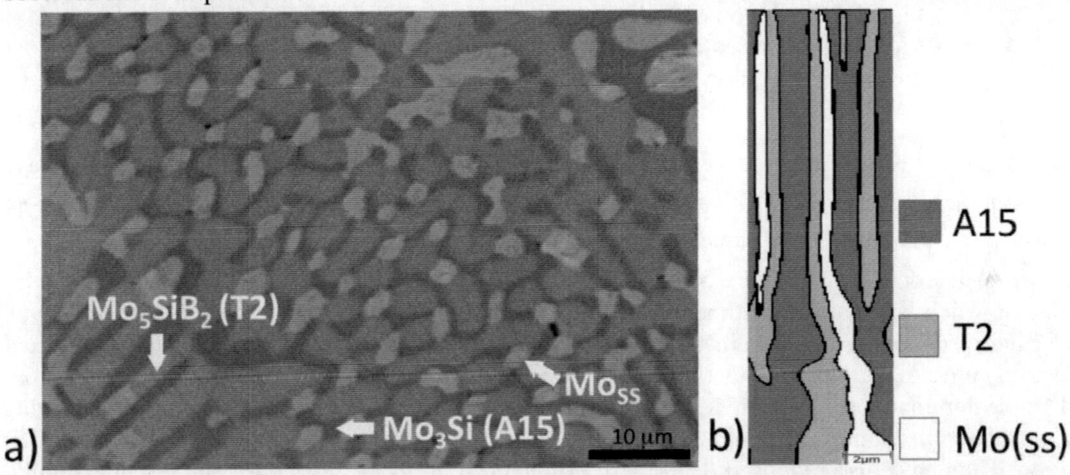


Figure 2. Eutectic lamellae observed in: a) experiments b) MICRESS simulation; the nuclei position was adjusted at the bottom so that from left the phases are presenting A15, T2 and $Mo_{(ss)}$ consequently. It is proved by simulations and experiments that almost 50 % of domain would be A15 and the other phases have almost equal volume fractions.

For the further studies we have investigated the instability of the ternary eutectic growth. For this purpose we have explored the Jackson-Hunt (JH) theory to address the minimum undercooling lamellae distances for the eutectic composition alloy. In this investigation the velocity of solid-liquid front was assumed to be constant and the simulation domain was containing three lamellae standing for three solid phases. Taking periodic boundary conditions into account, the generated results are capable to get mirrored for better consideration.

Choudhury et al. have proposed a generalized form of classical JH model [13] to examine the undercooling vs the lamella spacing in ternary systems [12]. We have followed their suggested procedure to study the changes of undercooling vs lamellae distance. Figure 3 shows the undercooling as a function of lamellae distance based on the theoretical assumptions in [12].

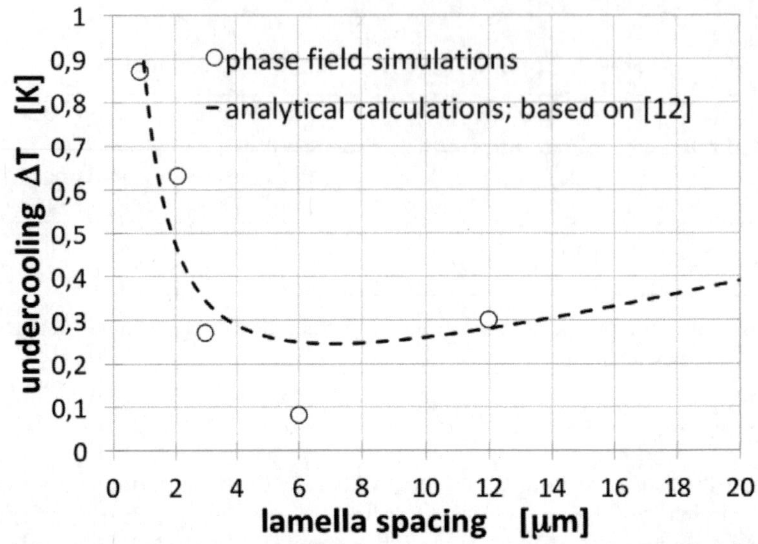


Figure 3. Variation of undercooling as a function of lamellae spacing for the Mo-Si-B ternary eutectic alloy, the lamellae spacing in this graph is showing the width of a unit containing three lamellae of solid phases

According to the Kim et al. we have defined a dimensionless variable $\Lambda$ which is $\lambda / \lambda_{JH}$ and have carried out the simulation with the different $\Lambda$ values to probe the possible instabilities in the system [1]. We have calculated the JH lamellae spacing for our system $\lambda_{JH} = 7$ µm connected to the minimum in undercooling (see Figure 3). This will be considered in following as a width of three lamellae in which lamellae are assumed to have equal width. For a ternary system, Choudhury et al. have suggested there are various instability patterns. Two mainly probed are patterns with different repeating units. According to Choudhury's work, if we call the three existing solid phases as A, B, and C, the dominant repeating units are ABC and ABAC [12]. It implies that along with the tilting and oscillating instability, the lamellae could shape different arrangement pattern during the solidification. They have also proposed that other permutations are possible. Confirming by the simulation and experimental investigations, the most predominant repeating pattern in Mo-Si-B eutectic alloy looks to be ABAC, however depending to the adjusted lamellae spacing, the phase field simulations proved the transition of the ABAC pattern to the other arrangements. This conversion happens following elimination of a lamellae or pinning or bifurcation of one lamella during evolution of the other lamellas. The growth pattern of the lamellae in Mo-Si-B ternary eutectic composition with constant proceeding velocity of solid phases into the liquid phase v = 0.05 µm/s, for $\Lambda = 0.12$, 0.17, 0.3, 0.85 and $\Lambda = 1.7$ is shown in Figure 4. According to the simulations, a transition to an intermittent oscillating growth pattern happens at $\Lambda = 1.7$.

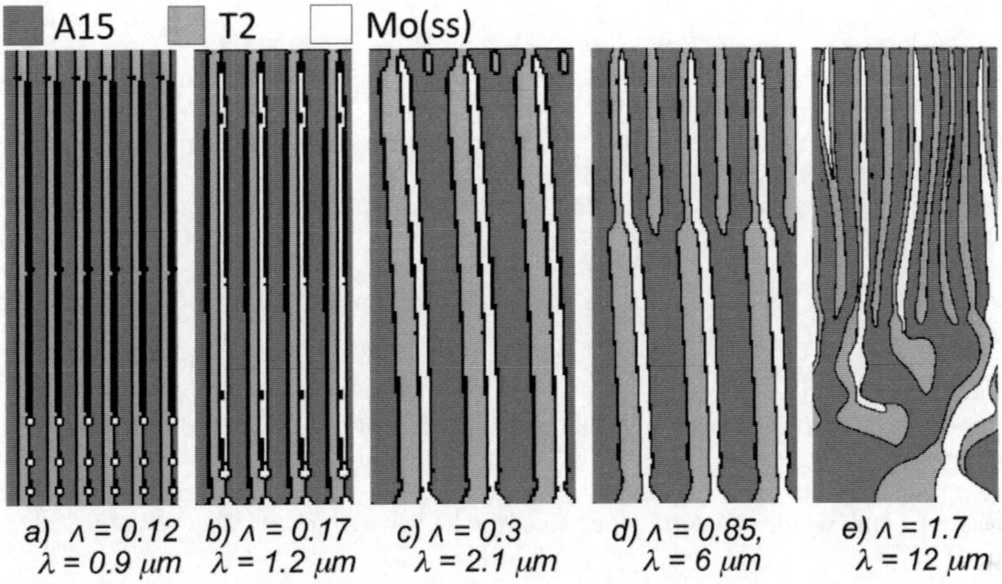


Figure 4. Growth pattern from the PFM computation at different $\Lambda = \lambda / \lambda_{JH}$

In the model proposed by Choudhury et al. it is assumed that all the solid phases have equal volume fraction regardless to the number of lamellae of each phase and width of each lamella. They have also considered that the undercooling at the interface of all solid phases with the liquid are equal to each other. Moreover, in their projected model it was assumed that all the phase diagram of binary composition of existing phases have similar characteristics such as liquidus slope. Despite to these assumptions the phase field simulation and model predictions are in good agreement with each other. Especially the stable lamellar spacing of 7 μm will be predicted in both models as the same.

## 5. Conclusions

Although several phase-field studies have been conducted for predicting a eutectic solidification's behavior, there is not adequate number of simulations for a ternary eutectic composition alloy. Several computations have been performed to predict the instability pattern in binary eutectic compositions but for the ternary eutectics there has not been a quantitative number of models. Despite of lack of sufficient verified models, we have derived the isotropic ternary eutectic model using MICRESS software with some major simplifications. Furthermore, we have investigated the minimum undercooling lamella distance for a fixed interface evolution velocity to probe the lamella spacing which results in a stable growth.

Performing the phase field simulation has shown a variety of growth patterns and instabilities in growing of lamellae which have been observed in the experimental investigations. Thanks to a generalized form of Jackson-Hunt theory for ternary alloys, the interval of lamellae spacing which results in a stable lamella growth was derived for the interface proceeding speed of 0.05 micrometer per second. The computation has shown that performing the simulation with the lamella spacing satisfactorily larger than the optimal JH lamella spacing would result in oscillating instability.

## References

[1] Kim S G, Kim W T, Suzuki T and Ode M 2004 Phase-field modeling of eutectic solidification. *Journal of crystal growth* **261** 135-58

[2] Apel M, Boettget B, Diepers H J and Steinbach I2002 2D and 3 D phase field simulation of lamella and fibrous eutectic growth *Journal of crystal growth* **237-9** 154-8
[3] Boettinger W J, Warren J A, Beckermann C and Karma A 2002 Phase-field simulation of solidification. *Annu. Rev. Mater. Res.* **32** 163-94
[4] Meng G, Lin X and Huang, W 2007 Lamellar spacing selection in regular eutectic solidification at low velocity. *Mater. Sci. Technol.* **23** 851-4
[5] Plapp M 2007 Three-dimensional phase-field simulations of directional solidification. *Journal of crystal growth* **303** 49-57
[6] Zhang R, Jing T, Jie W and Liu B 2006 Phase-field simulation of solidification in multicomponent alloys coupled with thermodynamic and diffusion mobility database. *Acta Materialia* **54** 2235-2239
[7] Lemberg J A and Ritchie R O 2012 Mo-Si-B Alloys for ultrahigh-temperature structural applications. *Adv. Mater.* **24** 3445-80
[8] Yang Y 2004 Thermodynamic modeling and experimental investigation of the Mo-Si-B-T quaternary system. Dissertation university of Wisconsin-Madison
[9] Hasemann G, Bogomol L, Schliephake D, Loboda P I and Krüger M 2014 Microstructure and creep properties of a near-eutectic directionally solidified multiphase Mo–Si–B alloy. *Intermetallics* **48** 25-33
[10] Yang Y and Chang Y A 2005 Thermodynamic modeling of the Mo-Si-B system. *Intermetallics* **13** 121-8
[11] Steinbach I and Pezzolla F 1999 A generalized field method for multiphase transformations using interface field. *Physica D* **134** 385-93
[12] Choudhury A, Plapp M and Nestler, B 2011 Theoretical and numerical study of lamellar eutectic three-phase growth in ternary alloys. *Cond-Mat. Mtr. Sci.*
[13] Jackson K A and Hunt J D 1966 *Trans. Metall. Soc. AIME* **236** 1129

IOP Conf. Series: Materials Science and Engineering **118** (2016) 012029 doi:10.1088/1757-899X/118/1/012029

# Prediction of Austenite Formation Temperatures Using Artificial Neural Networks

**P Schulze[1], E Schmidl[1], T Grund[1] and T Lampke[1]**

[1] Materials and Surface Engineering Group, Institute of Materials Science and Engineering, Technische Universität Chemnitz, 09125 Chemnitz, Germany

E-Mail: pierre.schulze@mb.tu-chemnitz.de

**Abstract.** For the modeling and design of heat treatments, in consideration of the development/ transformation of the microstructure, different material data depending on the chemical composition, the respective microstructure/phases and the temperature are necessary. Material data are, e.g. the thermal conductivity, heat capacity, thermal expansion and transformation data etc. The quality of thermal simulations strongly depends on the accuracy of the material data. For many materials, the required data - in particular for different microstructures and temperatures - are rare in the literature. In addition, a different chemical composition within the permitted limits of the considered steel alloy cannot be predicted. A solution for this problem is provided by the calculation of material data using Artificial Neural Networks (ANN). In the present study, the start and finish temperatures of the transformation from the bcc lattice to the fcc lattice structure of hypoeutectoid steels are calculated using an Artificial Neural Network. An appropriate database containing different transformation temperatures (austenite formation temperatures) to train the ANN is selected from the literature. In order to find a suitable feedforward network, the network topologies as well as the activation functions of the hidden layers are varied and subsequently evaluated in terms of the prediction accuracy. The transformation temperatures calculated by the ANN exhibit a very good compliance compared to the experimental data. The results show that the prediction performance is even higher compared to classical empirical equations such as Andrews or Brandis. Therefore, it can be assumed that the presented ANN is a convenient tool to distinguish between bcc and fcc phases in hypoeutectoid steels.

## 1. Introduction

For the modeling and design of heat treatment processes of steels, a large number of different material data is required. Such data comprises, e.g. thermal conductivity, heat capacity, thermal expansion and transformation data (TTT/CCT). For the transformation behavior, start- and finish-temperatures of the austenite formation are of major significance. These temperatures are relevant for the determination of the temperature required in different heat treatment processes and to distinguish between bcc and fcc lattice structure.

In figure 1, the different phases of a three-component steel (system of iron, carbon and metallic alloying element, with low content of the alloying metal M in accordance with [1]) are illustrated. The different phase areas and limits as well as the associated temperatures are distinctly marked. The start and finish temperatures of the austenite transformation during heating are described e.g. in time-temperature-austenitization diagrams (TTA), time-temperature-transformation diagrams (TTT) or continuous-cooling-transformation diagrams (CCT), which, for example, can be found in [2–8]. According to [4], the start and finish temperatures can be classified by dilatation curves as shown in figure 2.

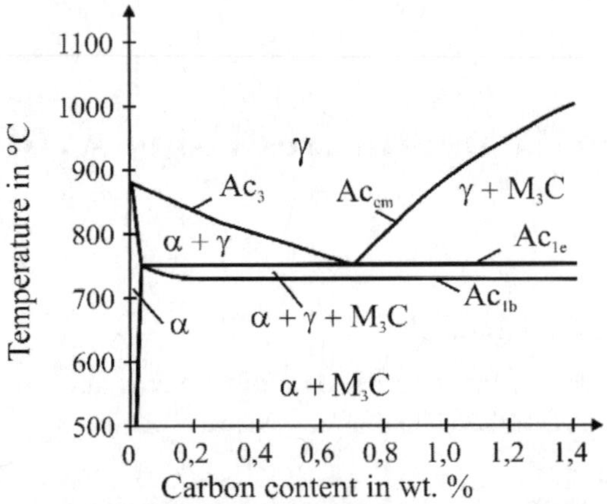


**Figure 1.** Section through a 3-component system iron-carbon-alloying element M at low and constant concentrations of M in accordance with [1].

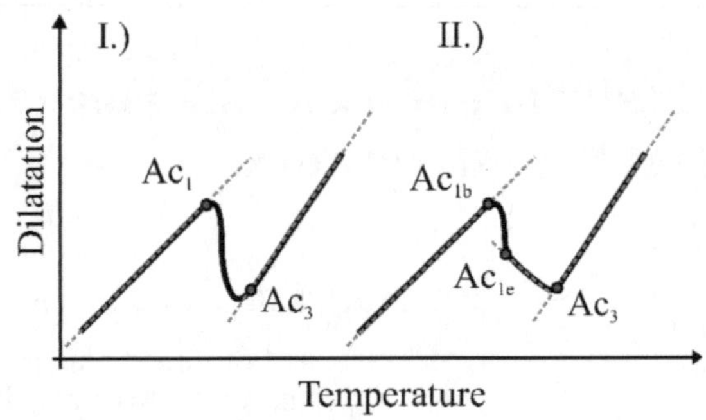


**Figure 2.** Schematic representation of the definition of the transformation points according to [4], I.) Hypoeutectoid steel II.) Hypoeutectoid steel with ferritic and perlitic initial microstructure.

The start of the austenite formation is usually denoted as $Ac_1$ or $Ac_{1b}$. A deviation from a linear dependency between temperature and dilatation occurs, when at least 1% of the volume has transformed from ferrite to austenite [4]. The temperature $Ac_3$ denotes the end of the ferrite/austenite transformation. If hypoeutectoid steels have a volume fraction of pearlite less than 60% the finish of the perlite transformation $Ac_{1e}$ is visible in the dilatation curve [4], figure 2. However, the transformation temperatures $Ac_{1e}$ and for high-alloyed and hypereutectoid steels $Ac_{cm}$ or $Ac_c$ are not considered in this paper due to the very limited accessible data situation.

As already mentioned above, material data for a large number of different steels are available in the literature. It is, however often not possible to find data for a very specific chemical composition or the found data show an insufficient quality. To calculate the temperatures $Ac_1$ and $Ac_3$ according to a given chemical composition, several empirical equations exist [9, 10]. In particular, the formulas of Andrews [9]

$$Ac_1 = 723 - 10.7\,Mn - 16.9\,Ni + 29.1\,Si + 16.9\,Cr + 290\,As + 6.38\,W \tag{1}$$

$$Ac_3 = 910 - 203\,\sqrt{C} - 15.2\,Ni + 44.7\,Si + 104\,V + 31.5\,Mo + 13.1\,W \tag{2}$$

and Brandis [10]

$$Ac_1 = 739 - 22\,C + 2\,Si - 7\,Mn + 14\,Cr + 13\,Mo - 13\,Ni + 20\,V \tag{3}$$

$$Ac_3 = 902 - 255\,C + 19\,Si - 11\,Mn - 5\,Cr + 13\,Mo - 20\,Ni + 55\,V \tag{4}$$

are often used to identify the start and finish temperatures of the ferrite/austenite transformation in steels. The symbols of the chemical elements represent the percentage by mass of these elements in the alloys. In [9] and [10], the limitations of the range of the chemical composition of these equations are described. According to [11], it is not certain whether the intrinsic relations between the chemical composition and $Ac_1$ or $Ac_3$ are considered by the empirical equations, which is why inaccuracies of the prediction can occur.

As it is shown in several publications [11–15], another more convenient mathematical approach to predict the material data is given by Artificial Neural Networks (ANN). ANN are advantageous in many cases, especially if the intricacy of the problem is immense and simplifications are not possible. Thus, ANN are used in materials science to predict effects and properties of welding processes, superalloys, fatigue properties, transformations, steel processing, machining and processing and many more [12]. The successful prediction of the austenite formation temperatures $Ac_1$ and $Ac_3$ using ANNs is already described in [11–15]. In the literature, two ANNs - each calculating one transformation temperature and one hidden layer with logistic activation function [13] or hyperbolic tangent transfer function [15] - were used. The prediction of both temperatures using a single ANN (but again only one hidden layer) is described in [11, 14]. The literature indicates that the architecture of the ANN and the activation functions influence the prediction accuracy of the neural network.

In this study a systematic procedure to train an ANN on the prediction of the austenite formation temperatures for a steel with given chemical composition is shown. The focus is on the prediction for hypoeutectoid steels using available literature data. The number of hidden layers as well as of different activation functions is varied, and the network with the highest prediction accuracy is compared to the accuracy of the classical empirical equations of Andrews and Brandis.

## 2. Modeling of Artificial Neural Network (ANN)

To model an ANN predicting the austenite formation temperatures of a given steel with high accuracy, data from a large number of steels with different chemical compositions were collected from the literature [2–8]. Often, elements such as N, Co, B, As, Sn, Zr, O, W, Ti were not measured but may influence the transformation temperatures. An element is used as input if it is considered in at least half of the surveyed data and if it is set to zero in the case of steels with no infor mation about the element content. If the experimental alloys have varying contents of alloying elements, which are not considered as input parameter, the alloying effects cannot be considered by the model either, and significant errors can occur [16]. Therefore, if a surveyed steel type had measured elements, e. g. boron, which were not considered as input data this steel would not be used for the ANN presented in this work. Because of the low variety of chemical compositions and due to the consideration of only unalloyed and low alloyed hypoeutectoid steels, the input data are reduced from 406 to 206 steel types with different chemical compositions. The elements considered, the content range, the mean content as well as the standard deviation (of the percentage of mass) of the 206 steels, which were used to model the neural network, are shown in table 1.

**Table 1.** Summary of the input and output variables of the ANN.

| | Variable | Minimum | Maximum | Mean | Standard deviation |
|---|---|---|---|---|---|
| Input in wt.% | C | 0.043 | 0.76 | 0.297 | ± 0.160 |
| | Si | 0.010 | 1.80 | 0.340 | ± 0.272 |
| | Mn | 0.040 | 2.11 | 0.744 | ± 0.420 |
| | P | 0.003 | 0.071 | 0.019 | ± 0.009 |
| | S | 0.004 | 0.090 | 0.018 | ± 0.009 |
| | Cr | 0.000 | 3.52 | 0.709 | ± 0.675 |
| | Cu | 0.000 | 0.98 | 0.112 | ± 0.123 |
| | Mo | 0.000 | 1.10 | 0.166 | ± 0.244 |
| | Ni | 0.000 | 4.28 | 0.432 | ± 0.887 |
| | V | 0.000 | 0.87 | 0.039 | ± 0.090 |
| | Al | 0.000 | 0.14 | 0.010 | ± 0.018 |
| Output in °C | $Ac_1$ | 657 | 800 | 733.3 | ± 23.07 |
| | $Ac_3$ | 740 | 954 | 823.1 | ± 39.60 |

To train the net the selected data from the literature were randomly divided ten times and arranged in groups of three: a training set (60% of the data), a validation set (20%) and a test/unseen set (20%). The test set was used to identify an over-trained network at the beginning as well as to compare the prediction accuracy between the ANN and the classical empirical equations of Andrews and Brandis afterwards (cf. section 3).

To find the best ANN with a minimum error between the experimental and ANN calculated temperatures, the hidden layer size and the number of neurons within one layer were systematically varied for each randomly grouped dataset (figure 3). Also the activation functions (logsig and tansig ) of the hidden layer neurons were changed. The activation function of the output layer was always set to be the identity function . The input and output data were normalized between zero and one.

A total of 8400 different ANNs were trained using the Levenberg-Marquardt algorithm which is implemented in the used MATLAB software. In so doing, the algorithm minimized the mean square error (MSE) between the predicted and experimentally measured temperatures adjusting the biases $b$ and the weights $w$ of all neurons. It was stopped, in the moment when the MSE of the training set was less than a predefined threshold value. At the end, the biases, weights and activation functions of the ANN with the lowest MSE, according to the validation set, were selected.

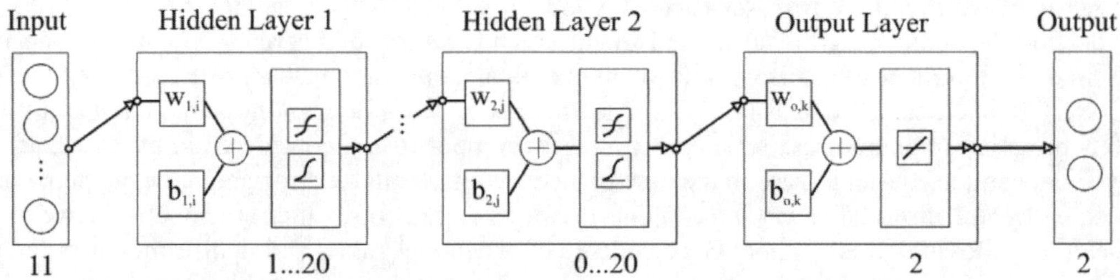


**Figure 3.** Schematic illustration of the trained feedforward network with the variation of the number of neurons and the activation functions in the hidden layers.

## 3. Results and Discussion

The mean square errors are illustrated in figure 4. They are represented as the radius of a sphere of all trained neural network configurations. A sphere with a small radius implies a low MSE. In figure 4 it can be seen that the MSE varies on a low level over the whole range of the number of neurons in the two hidden layers, except for an ANN with one neuron in the first hidden layer. As it is also shown in the literature [11–15], an ANN with only one hidden layer has an MSE which is in an acceptable magnitude. The ANN with the lowest MSE (further denoted as "best ANN") of all trained configurations has seven neurons in the first and sixteen neurons in the second hidden layer. The activation function of the hidden layers is the logsig function (see also black filled sphere in figure 4).

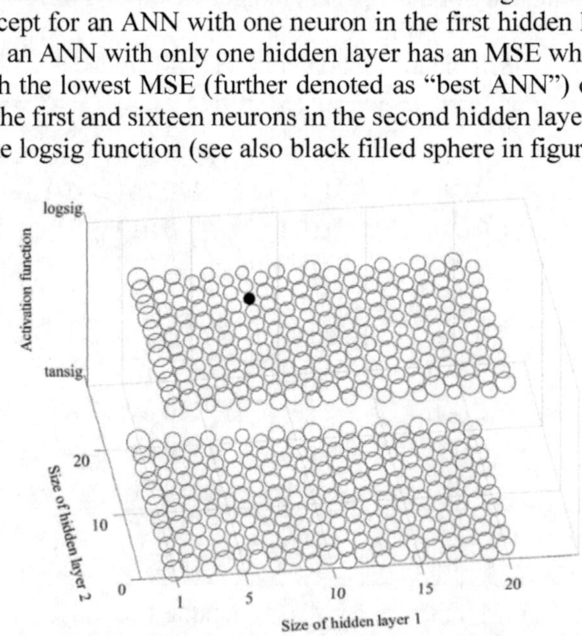


**Figure 4.** Mean square errors represented as the radii of a sphere of all trained ANN configurations. Black filled sphere: Configuration with the minimum MSE. Note: For reasons of clarity only the half of the configurations of the randomly selected dataset with the minimum MSE are illustrated.

To quantitatively estimate the prediction accuracy of the best ANN the mean relative estimation errors (MREs)

$$MRE_{1,2} = \frac{100}{n}\sum_{i=0}^{n}\left|\frac{Ac_{x_i}^{exp} - Ac_{x_i}^{ANN}}{Ac_{x_i}^{\text{exp}}}\right|, x \in \{1,3\} \tag{5}$$

(in which n denotes the number of different steel compositions and $Ac_x$ the $Ac_1$ or $Ac_3$ temperature) are compared to the corresponding values of the Andrews equations (1–2) and Brandis equations (3–4).

The predicted temperatures of the steel compositions used to train and validate the ANN show a higher accuracy than the values of the equations, figure 5. The average MRE of 1.24% of the best ANN (compared to 1.76% of the Andrews and 1.80% of the Brandis equations) confirms this statement, table 2. However, it is not unusual that an ANN has a very high prediction accuracy for trained data.

**Table 2.** Summary of the mean relative estimation errors of the best ANN and the empirical equations.

| Model | ANN | | | Andrews | | | Brandis | | |
|---|---|---|---|---|---|---|---|---|---|
| MRE | $Ac_1$ | $Ac_3$ | Ø | $Ac_1$ | $Ac_3$ | Ø | $Ac_1$ | $Ac_3$ | Ø |
| Train/Val | 1.21 | 1.27 | 1.24 | 1.57 | 1.95 | 1.76 | 1.55 | 2.05 | 1.80 |
| Unseen | 1.54 | 1.56 | 1.55 | 1.88 | 2.04 | 1.96 | 1.68 | 1.89 | 1.79 |

Therefore, the MRE of the test data, which were totally unknown during the training, is calculated. A qualitative comparison is illustrated in figure 6. Again, the ANN shows a higher prediction accuracy (average MRE of 1.55%) of the austenite formation temperatures compared to the equations of Andrews (1.96%) and Brandis (1.78%). In contrast to the equation models in which the prediction accuracy of the $Ac_3$ temperature is always noticeably lower than the accuracy of the $Ac_1$ temperature, the difference of the ANN is close to zero, table 2. Additionally, the difference between the MRE of the trained/validated and the unseen data of the best ANN is very low (about 0.3%). Hence, the modeled ANN should predict the temperatures of other composited steels with an equally high accuracy.

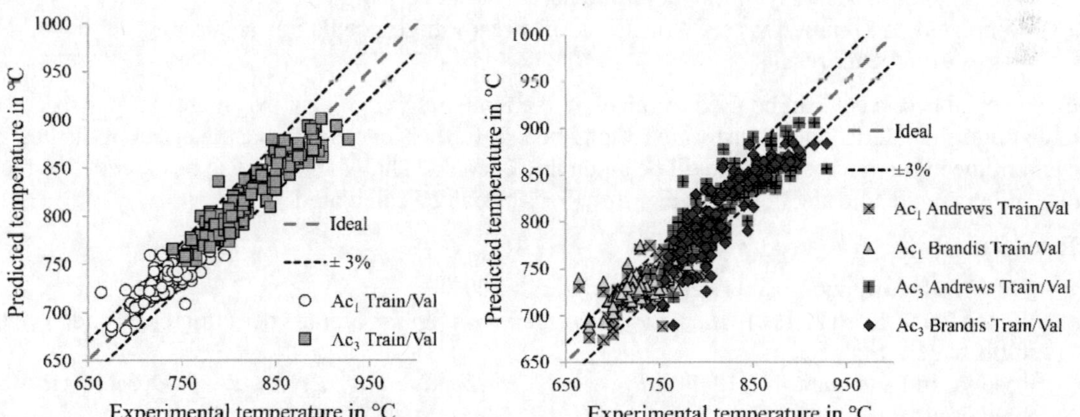


a) b)

**Figure 5.** Comparison between the predicted and experimentally measured temperatures of the a) best ANN for the training and validation data and b) classical equations of Andrews (1–2) as well as Brandis (3–4) for the training and validation data.

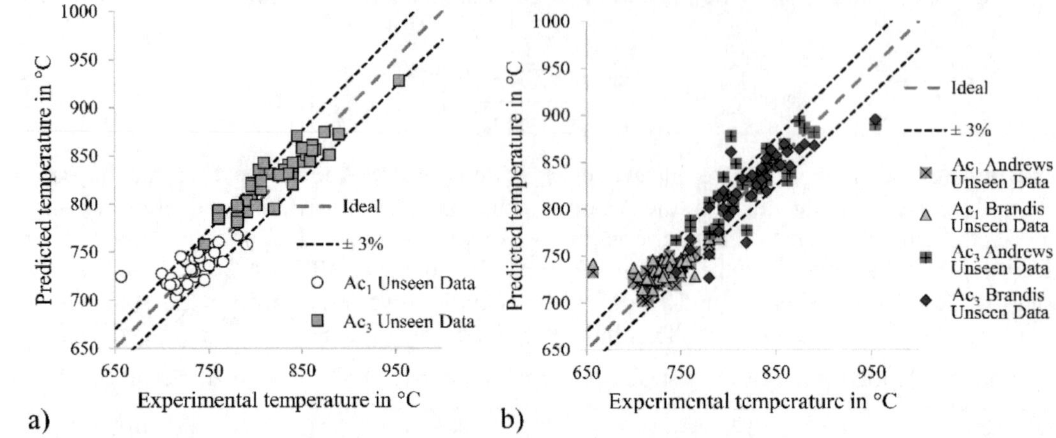


**Figure 6.** Comparison between the predicted and experimentally measured temperatures of the a) best ANN and b) classical equations of Andrews (1–2) as well as Brandis (3–4) for the test/unseen data

## 4. Summary and Conclusions

An artificial neural network predicting the austenite formation temperatures $Ac_1$ and $Ac_3$ in dependence of the steel composition was systematically trained to achieve a minimum MSE. The ANN with the minimum MSE of all trained configurations has seven neurons in the first and sixteen neurons in the second hidden layer. The activation function of the hidden layers is the logsig function. The following conclusions can be drawn:

(1) The MSE varies on a low level over the whole range of the number of neurons in the two hidden layers, except for an ANN with one neuron in the first hidden layer.
(2) The prediction accuracy of the identified best ANN is very high.
(3) Compared to commonly used equations of already high prediction accuracy, this best ANN shows even better results.

The developed best ANN can be used to calculate the temperatures of heat treatments (e. g. hardening) or to distinguish material data concerning fcc and bcc. In further investigations, the effect of the heating rate and of the initial microstructure will be taken into account. The ANN will also be extended to high-alloy and hypereutectoid steels, so that e.g. $Ac_{cm}$ or $Ac_c$ can be calculated.

## References

[1] 1984 Werkstoffkunde Stahl (Berlin: Springer) p 200
[2] Wever F and Rose A 1961 *Atlas zur Wärmebehandlung der Stähle* (berichtigter Nachdr.) (Düsseldorf: Verl. Stahleisen)
[3] Rose A and Hougardy H 1972 *Atlas zur Wärmebehandlung der Stähle 2* (Düsseldorf: Verl. Stahleisen)
[4] Orlich J, Rose A and Wiest P 1973 *Atlas zur Wärmebehandlung der Stähle 3* (Düsseldorf: Verl. Stahleisen)
[5] Orlich J and Pietrzeniuk H J 1976 *Atlas zur Wärmebehandlung der Stähle 4* (Düsseldorf: Verl. Stahleisen)
[6] 1987 *Zeit-Temperatur-Umwandlungsdiagramme* (Freital: Edelstahlwerk Freital)
[7] Seyffarth P 1982 *Schweiß-ZTU-Schaubilder* (Berlin: Verl. Technik)
[8] Seyffarth P, Meyer B and Scharff A 1992 *Großer Atlas Schweiß-ZTU-Schaubilder* (Düsseldorf: Dt. Verl. für Schweisstechnik, DVS-Verl.)
[9] Andrews K W 1965 *J. Iron and Steel Inst.* **203** 721–7
[10] Brandis H 1975 *TEW-Techn. Ber.* **1** 8–10
[11] You W, Xu W, Bai B and Fang H 2006 *Mater. Sci. Eng. A* **419** 276–82

IOP Conf. Series: Materials Science and Engineering **118** (2016) 012029 doi:10.1088/1757-899X/118/1/012029

[12] Bhadeshia H K D H 1999 *ISIJ Int.* **39** 966–79.
[13] Dobrzański L A and Trzaska J 2004 *J. Mater. Process. Technol.* **155–156** 1950–55.
[14] Rakhshkhorshid M and Teimouri Sendesi S A 2014 *J. Iron. Steel Res. Int.* **21** 246–51.
[15] Gavard L, Bhadeshia H K D H, MacKay D J C and Suzuki S 1996 *Mater. Sci. Technol.* **12** 453–63.
[16] Sha W 2011 *Mater. Sci. Eng. A* **528** 3275–76.

IOP Conf. Series: Materials Science and Engineering **118** (2016) 012030 doi:10.1088/1757-899X/118/1/012030

# Influence of thermally activated processes on the deformation behavior during low temperature ECAP

**S Fritsch[1], M Scholze[1] and M F-X Wagner[1]**

[1] Chair of Materials Science, Institute of Materials Science and Engineering, Technische Universität Chemnitz,
09215 Chemnitz, Germany

E-Mail: sebastian.fritsch@mb.tu-chemnitz.de

**Abstract.** High strength aluminum alloys are generally hard to deform. Therefore, the application of conventional severe plastic deformation methods to generate ultrafine-grained microstructures and to further increase strength is considerably limited. In this study, we consider low temperature deformation in a custom-built, cooled equal channel angular pressing (ECAP) tool (internal angle 90°) as an alternative approach to severely plastically deform a 7075 aluminum alloy. To document the maximum improvement of mechanical properties, these alloys are initially deformed from a solid solution heat-treated condition. We characterize the mechanical behavior and the microstructure of the coarse grained initial material at different low temperatures, and we analyze how a tendency for the PLC effect and the strain-hardening rate affect the formability during subsequent severe plastic deformation at low temperatures. We then discuss how the deformation temperature and velocity influence the occurrence of PLC effects and the homogeneity of the deformed ECAP billets. Besides the mechanical properties and these microstructural changes, we discuss technologically relevant processing parameters (such as pressing forces) and practical limitations, as well as changes in fracture behavior of the low temperature deformed materials as a function of deformation temperature.

## 1. Introduction

Severe plastic deformation (SPD) methods allow generating ultra–fine grained microstructures. One of the most interesting approaches to obtain high-strength materials in combination with good ductility is equal–channel angular pressing (ECAP). Enhanced properties such as very high strength can be provided by grain refinement into the sub micrometer range by deformation in the solution heat-treated condition, in combination with a suitable (post-ECAP) heat treatment. The reduction of grain size in SPD processes is based on the formation of low–angle grain boundaries in regions of high dislocation densities, which are then further transformed into high angle grain boundaries. While ECAP of many metals and alloys is well understood today, there are some difficulties with ECAP processing of high–strength, age hardening aluminum alloys such as AA7075: Because of the poor formability, the material tends to shear localization and cracking when ECAP is performed at RT. Higher deformation temperatures (120 °C) do in principle allow a successful production of homogenous billets. However, because of dynamic recovery processes and the formation of precipitates during the deformation, the mechanical properties of the resulting materials are hardly improved compared to the conventional peak aged condition.

One very promising approach to improve the mechanical behavior of hard-to-deform fcc metals is deformation at low temperatures. It is well known that forming at low temperatures (or even in the cryogenic regime) results in high dislocation densities because dynamic recovery processes are suppressed. This leads to an increase of strain hardening capability, which in turn impedes strain localization and crack formation. One important microstructural aspect is that, because of the limited mobility of defects at lower temperatures, cross slip processes are more strongly hindered in the cryogenic regime. This also results in reduced annihilation processes of dislocations and in more strongly pronounced strain hardening stages III and V: As long as additional dislocations can be stored, the material exhibits strain hardening, which is related to stable plastic deformation. Another limiting factor for ductility is dynamic strain aging, which can be simply related to the interaction of dislocations with solute atoms. Materials that are prone to the nucleation of deformation bands (which are observed as serrated flow in the stress–strain curves and can be classified in different types; most well–known: types A, B, C) are generally more difficult to deform homogeneously. This so-called Portevin–Le Chatelier effect (PLC effect) by its very nature is a thermally activated process and therefore strongly depends on temperature and strain rate. Both – an increased strain hardening rate and a reduced (or suppressed) PLC effect – can be related to a better formability of fcc metals at low temperatures.
The most widely established process that directly exploits the beneficial effects of forming at lower temperatures is cryogenic rolling. The main advantage of cryogenic rolling is that it is a relatively simple procedure that can be performed in conventional rolling mills. In contrast to the accumulative roll bonding process, the deformation takes place in small incremental steps to insure a limited increase of temperature. Various research groups have uses rolling mills for forming at cryogenic temperatures in constant rolling direction [1–6]. The billets were typically cooled in liquid nitrogen, then transferred to the rolling mill, and subsequently deformed. This procedure of cooling and subsequent deformation has to be repeated for each deformation step. The relatively high velocity during the rolling process leads to an increase of the billet temperature, which negatively influences the homogeneity of the billet and the resulting mechanical properties. Nevertheless, by accumulating deformation in small incremental steps, true plastic strains about three have been achieved in aluminum alloys [3, 7]. A suitable heat treatment after the deformation results in a strong increase of the initial flow stress in combination with a somewhat increased ductility. Moreover, the increased dislocation density results in a pronounced acceleration of precipitation kinetics, which shifts aging time to much lower values. The conventional process (constant rolling direction), however, adversely affects the useable residual volume of the billets: "uniaxial" rolling always results in very thin plates, which is a strongly limiting parameter for the use in technological applications. Other approaches for SPD of hard-to-deform age hardening aluminum alloys include multidirectional forging [8] and ECAP [9–13] at low temperatures. The main advantage of these processes is that the deformation conditions (dimensions of the billets, strain paths) can be carefully controlled. Especially ECAP is characterized by these positive features. Due to the constant cross section of the ECAP tool, and because the velocity and backpressure during the ECAP process can be adjusted, the deformation conditions can be exactly adapted to suit a particular alloy and large strains can be accumulated in a large useable (fully deformed) fraction of the total billet volume.
In the present study, the concept of ECAP at low temperatures is discussed in detail. We characterize the microstructure and the mechanical behavior of a coarse grained AA7075 alloy at different low temperatures. Furthermore we discuss how a tendency for the PLC effect and the strain hardening rate affect the formability during severe plastic deformation at low temperatures. We then discuss how the deformation temperature, pressing speed and backpressure influence the occurrence of PLC effects and the homogeneity of the deformed ECAP billets. Furthermore we characterize technologically relevant processing parameters and practical limitations, as well as changes in fracture behavior of the low temperature deformed materials as a function of the deformation temperature.

## 2. Experimental methods

For our investigations, the aluminum alloy 7075 in commercial purity (nominal chemical composition 5.9 Zn–2.3 Mg–1.3 Cu–0.2 Fe–0.1 Si wt.–%) as an extruded bar with a cross-section of 15 x 15 mm² was used. The material was solution heat treated at 475 °C for 2.25 h and then quenched to RT in water. To suppress natural aging processes, all specimens were immediately stored in a freezer after solutionizing. For mechanical characterization, tensile tests were performed in the solution heat-treated condition under quasi–static loading with an initial strain rate of $10^{-3}$ $s^{-1}$ (free gauge length and diameter of the cylindrical samples: 10.5 mm and 3.56 mm, respectively) at different temperatures (RT, -60 °C, -120 °C and -196 °C). Furthermore, we performed jump-tests (uniaxial tensile testing with a sudden change of strain rate of about one order of magnitude) in the range from $10^{-4}$ $s^{-1}$ to $10^{-1}$ $s^{-1}$ to characterize the strain rate sensitivity as a function of testing temperature. For statistically relevant data, three samples for each condition were tested. The investigations were performed in a mechanical tensile testing machine (Zwick/Roell) with a maximum force of 20 kN. To determine the strain values, digital image correlation systems (DIC) were used for RT measurements. Low temperature conditions were generated by different experimental methods: Testing at -60 °C was performed in a climate chamber installed at the testing machine. For mechanical testing at -120 °C and -196 °C, liquid nitrogen was filled in a special double–ring cooling device placed around the tensile specimen. Strain values at low temperatures were determined from the crosshead displacement data.

The material was deformed at -60 °C in a custom-designed ECAP tool that can be cooled to low temperatures and that includes moveable walls to minimize friction. More detailed information on the ECAP die can be taken from our earlier publication [9]. The equivalent plastic strain per ECAP pass was 1.1 [14]. ECAP was performed with varying pressing speeds (between 10 and 50 $mm/min^{-1}$); a backpressure of 80 bar was applied to prevent shear localization or cracking. To further reduce friction, the billets were covered with a lubricant (molybdenum disulphide, Aero Shell Grease 33MS) prior to forming.

The ECAP deformed billets and the tensile specimens were analyzed by optical microscopy (on the longitudinal plane) to characterize strain localization events due to the PLC effect. The samples were embedded in synthetic resin under vacuum at RT to avoid recrystallization and aging processes during sample preparation. After grinding and polishing, the samples were etched for 40 s at RT in an acidic solution (Graff-Sargent: 84 ml $H_2O$; 15.5 ml $HNO_3$; 0.5 ml HF and 3 g $CrO_3$). Optical micrographs were taken in phase contrast mode. The investigations of the fracture surfaces were performed with a field emission scanning electron microscope (Zeiss NEON40). The acceleration voltage was 5 kV with an operating distance of 4 mm. Besides the detectors for secondary and back–scattered electrons, an in–lens detector for enhanced contrast was used.

## 3. Results and discussion

As reported in our earlier publications [9, 15], deforming the solution heat-treated condition by SPD processes provides some advantages for potential improvements of the mechanical properties compared to ECAP of the conventional peak-aged condition (i.e., solution heat treated and aged to maximum hardness, T6). The solutionized alloys are free of precipitates and exhibit an improved ductility. This results in a much higher formability which is mandatory for SPD processes. Large amounts of plastic deformation lead to the formation of dislocation networks and (sub–)grain boundaries, which then also offers a possibility for the nucleation of precipitates. The concentration of dislocations is increased by increasing plastic strains. This provides a higher number of nucleation sites, which allows a more homogeneous formation of finely distributed precipitates during subsequent artificial aging. Furthermore the aging times are significantly shortened by the accelerated precipitation kinetics in ultra–fine grained microstructures [2, 3, 5].

Due to the high amount of foreign alloying elements in the investigated aluminum alloy, there is a strong susceptibility to deformation localization by the PLC effect [16].

Figure 1 shows the mechanical properties of AA7075 in the annealed condition as a function of testing temperature. The engineering stress–strain curve measured at RT clearly shows serrated flow and thus indicates occurrence of the PLC effect. A reduction of the testing temperature to -60 °C results in an increase of ductility (elongation to failure) of about 35 %. Lowering the temperature to -120 °C further increases ductility by ten percent and at -196 °C, elongation to failure is twice as high as at RT. Moreover, the strain range where the discontinuous flow is observed becomes much smaller at lower temperatures. While some effects of the PLC effect can still be observed at -60 °C, this behavior no occurs at -120 °C or at -196 °C. The uniform elongation is almost constant from RT down to -120 °C, but it exhibits a significant increase when tensile testing is performed at -196 °C. Tensile strength (as expressed by 0.2 % yield stress) also markedly increases from -120 °C to -196 °C, which can be related to less thermally activated cross slip of screw dislocations, which is the primary deformation mode in aluminum alloys [2, 3, 9, 17].

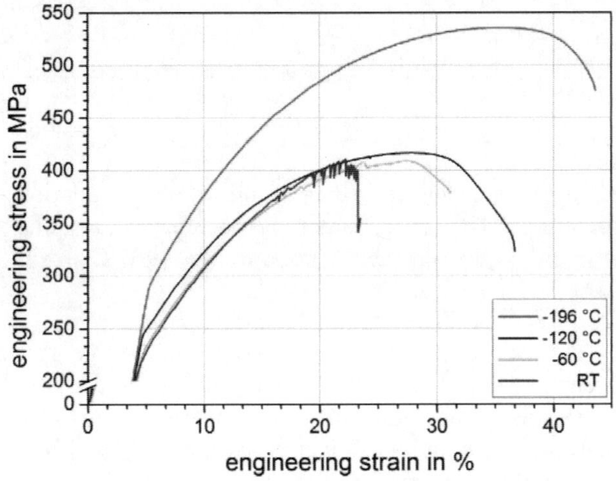


**Figure 1.** Engineering stress–strain curves of AA7075 under quasi–static loading at different testing temperatures. A reduction of testing temperature also leads to less serrated flow.

In figure 2, we present the strain hardening rate θ as a function of true stress as determined from the stress-strain curves measured at different temperatures. The raw data in figure 2a exhibit some scatter which results from the numerical calculation of the first derivative of the stress-strain data and which is particularly pronounced when serrated flow occurred. For a more precise presentation, these curves were therefore corrected using a smoothing function (figure 2b). While a decrease of testing temperature clearly leads to an improved ductility (see figure 1), the (almost) constant strain hardening rates between RT and -120 °C do not directly indicate a better formability at lower temperatures. Only when the testing temperature is reduced to -196 °C, the reduction of thermal activation results in a significantly higher strain hardening rate. From the data presented in figure 2, it is difficult to distinguish between strain hardening stages III and IV. This indicates that the characteristic microstructural processes – dynamic recovery (stage III) and interaction of an increasing amount of dislocations with (sub-)grain boundaries (stage IV) – take place simultaneously. At RT and at -60 °C, the tensile samples failed without reaching the conventional Considère criterion for the onset of necking ($\theta \geq \sigma_W$) [18], whereas at -120 °C and -196 °C, strain hardening stage V was reached and θ values fell below the Considère limit. This indicates that the sudden failure without significant necking (at RT and -60 °C) and the occurrence of a high strain hardening rate are related to dynamic strain aging. When local deformation in a PLC band exceeds the Considère limit, apparently premature failure of the material is observed from a macroscopic point of view.

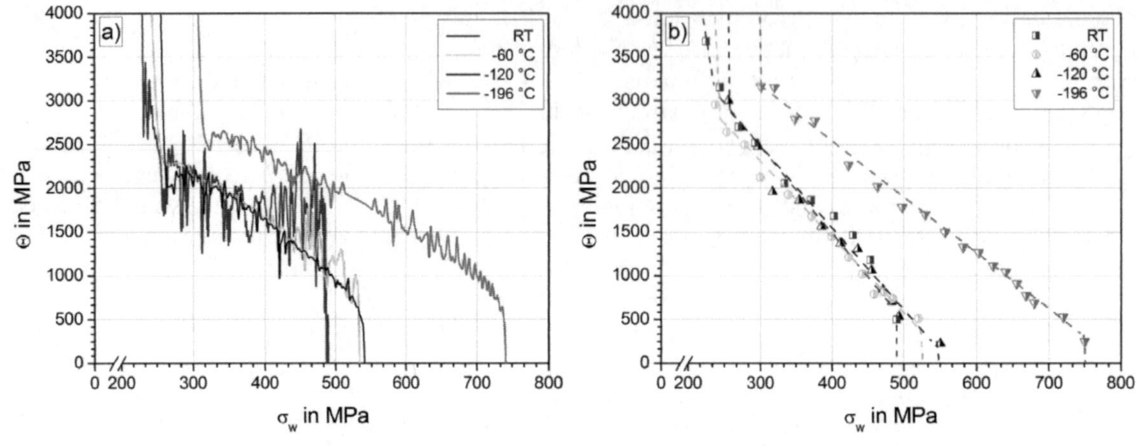


**Figure 2:** a) Full data and b) smoothed data of strain hardening rate as a function of true stress of AA7075 at different testing temperatures. The strain hardening rate curves are quite similar from RT down to -120 °C. Lower testing temperatures (cryogenic conditions: -196 °C) result in a marked increase of strain-hardening rate.

These results on the temperature-dependent mechanical properties of the age hardening aluminum alloy illustrate the important influence of the PLC effect. The formability of this alloy can be strongly increased by the suppression of dynamic strain aging at low temperatures. To further illustrate the increase of ductility with a decrease of deformation temperature from RT to -196 °C, optical micrographs (from the center and at the fracture surface of the tensile specimens) are shown in figure 3. Traces of deformation bands are apparent (from the deformation and the local displacement of large impurities that are present throughout this alloy and that are quite visible in differential interference contrast) at an angle of about 45° with respect to the loading direction (from left to right in all micrographs). These bands were also visible on the sample surfaces after testing at RT. In figure 3, these regions, where local shear deformation obviously exceeds the nominal one, have been artificially brightened for better visibility. Interestingly, in all cases where such deformation bands were present, fracture surfaces were oriented in parallel: because locally the critical shear stress was exceeded, shear failure occurred without necking at RT (figure 3b). Furthermore, parts of severed PLC bands were detected at the fracture surfaces, which confirmed the failure behavior within one band. This behavior is in a good agreement with a previous report by Chung et al., who described the local shear band formation as the primary failure mechanism of AA7075 at RT, [19].

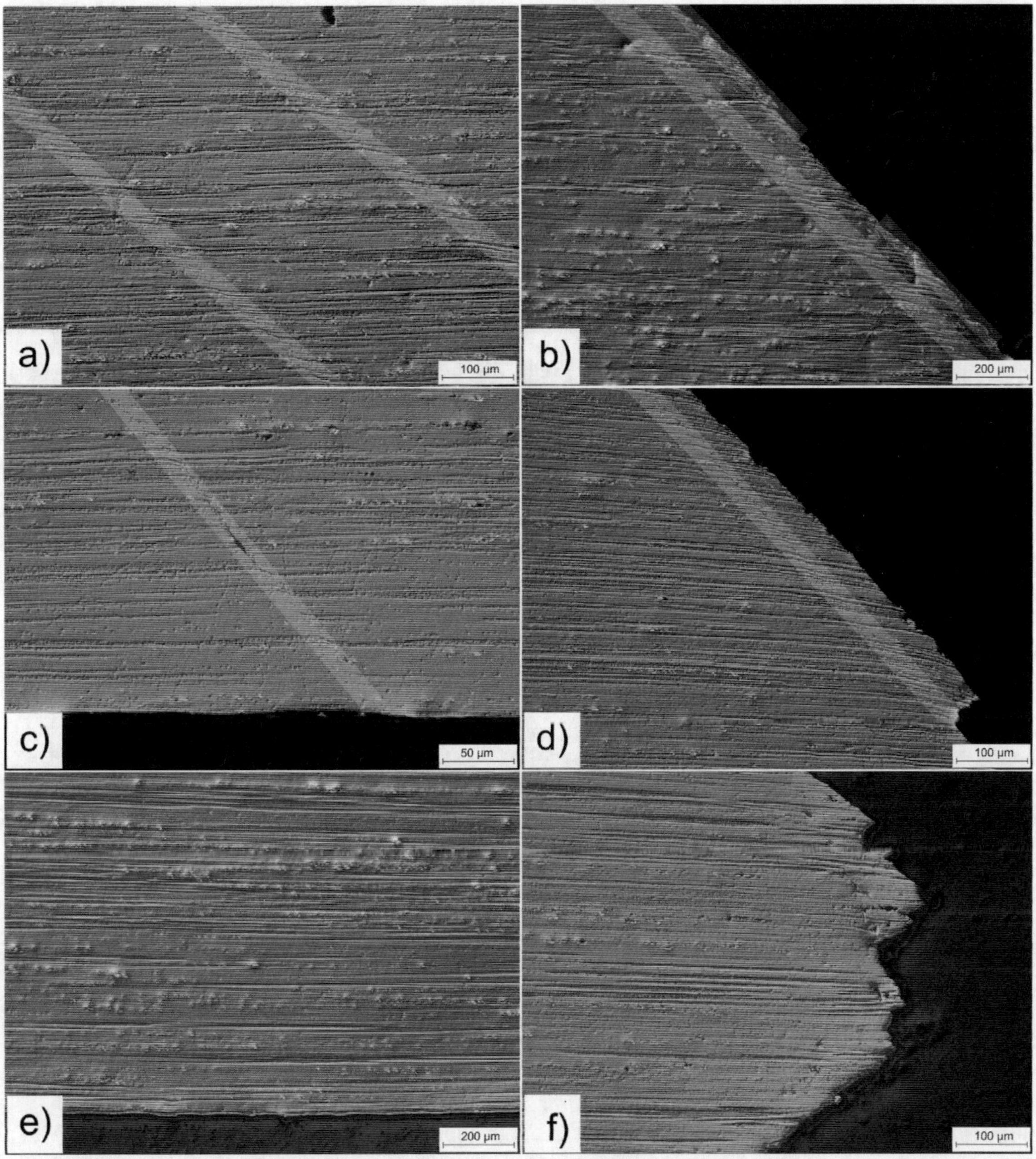


**Figure 3.** Optical micrographs in differential interference contrast of the solution heat-treated condition of AA7075. The micrographs were taken in the central section of the tensile specimens (orientation of the polished sections shown here: parallel to the deformation axis, which is from left to right in all micrographs) and in the region of fracture. PLC bands can be identified by analyzing the local displacements of the coarse impurities; the corresponding regions have been artificially brightened. a) and b) deformation at RT; c) and d) deformation at -60 °C; e) and f) deformation at -196 °C. Lower testing temperatures result in a transition from brittle (RT) to ductile (-196 °C) fracture.

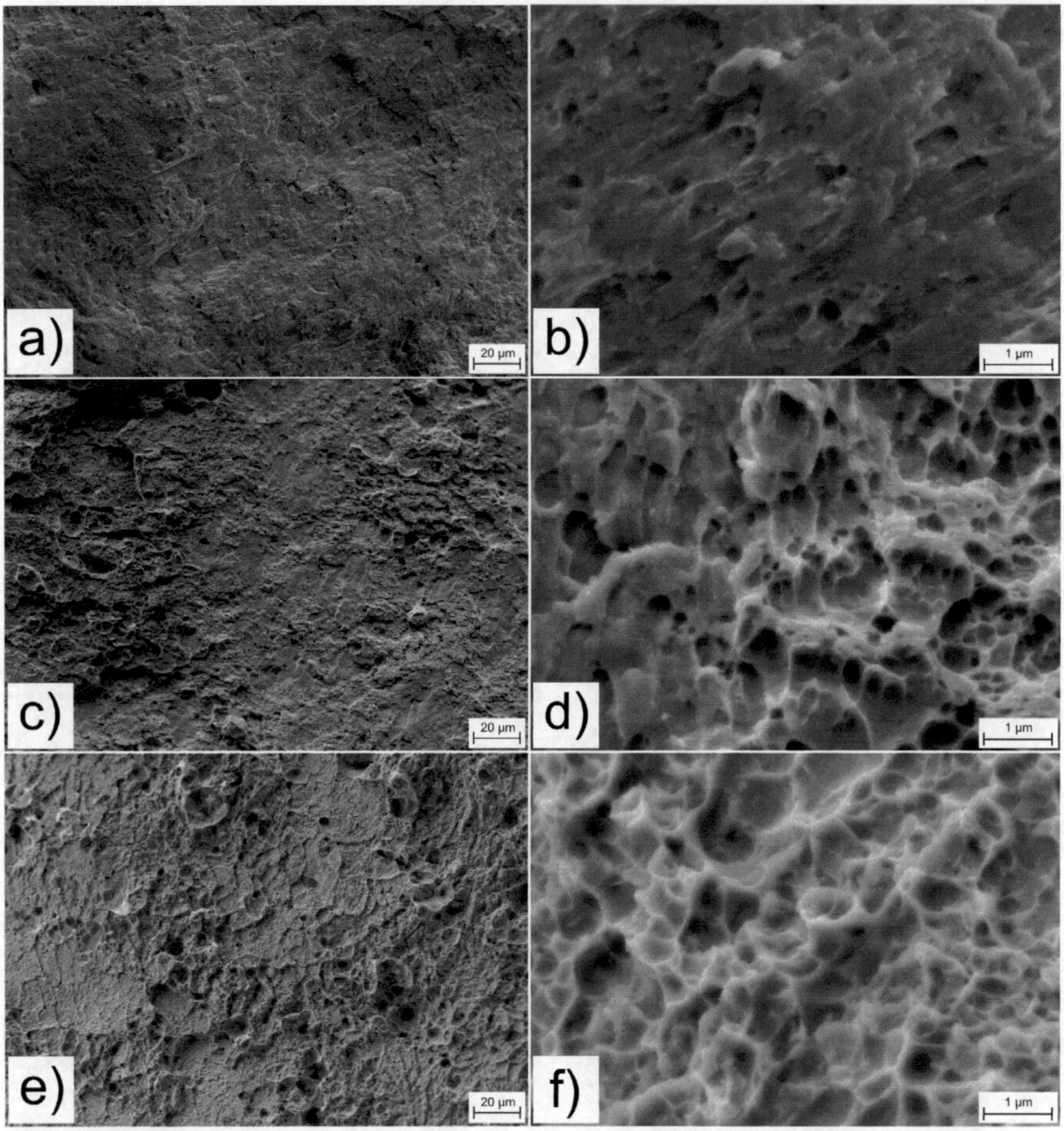


**Figure 4.** SEM micrograph of the fracture surface in solution heat treated condition of AA7075 with a transition from smooth shear fracture a) and b) at RT; to a mixture between brittle and ductile fracture c) and d) at -60 °C; to ductile fracture with dimples on the fracture surface e) and f) at -196 °C.

After deformation at -60 °C, PLC bands were no longer visible on the surface of the samples even though the microstructure showed a (somewhat lower) number of bands (figure 3c). Deformation at -60 °C results in smooth necking and the formation of shear fracture surfaces with a slightly elevated roughness (figure 3d). Crack nucleation of the samples occurred simultaneously along several PLC bands. This results from the overall reduction of the PLC effect and an increased ductility. Despite a significant reduction of serrated flow, the stress-strain behavior was considerably influenced by localization events (figure 1). This leads to an irregular microstructure and a premature failure along the deformation bands. Finally, cryogenic conditions lead to a distinct transition from brittle to ductile fracture; shear band fracture is suppressed (figure 3f), which agrees with literature [19]. At -196 °C the

IOP Conf. Series: Materials Science and Engineering **118** (2016) 012030 doi:10.1088/1757-899X/118/1/012030

PLC effect does no longer occur and deformation bands were not observed (figure 3e). These observations clearly demonstrate that suppressing dynamic strain aging (by strongly reducing thermal energy and thus retarding thermal activation), leads to homogenously deformed microstructures.
The fracture surfaces of the tensile specimens were also investigated in greater detail by SEM, figure 4. The fracture surface of the tensile specimens that were tested at RT indicates a shear fracture behavior (figure 4a and b, taken at different magnifications). A smooth fracture surface was formed by deformation along a PLC band. At -60 °C (figures 4c and d), dynamic strain aging was reduced and a partial transition from brittle to ductile fracture was observed. Typical dimple structures can be observed on the fracture surfaces. At -196 °C, the fracture surface contains fine dimples and indicates a completely ductile fracture (figure 4e and f), which agrees well with the strongly increased ductility observed in the corresponding stress-strain curves.

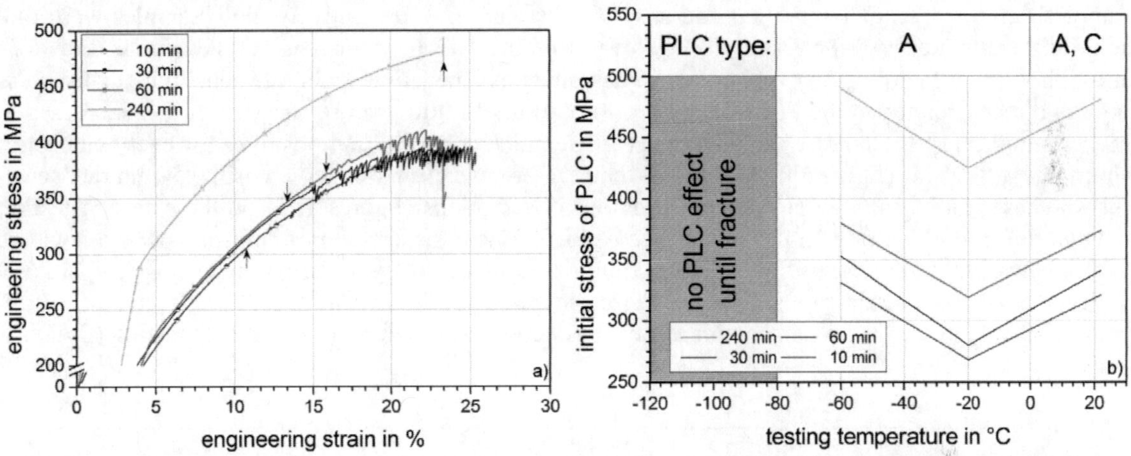


**Figure 5.** a) Influence of aging time (10 min to 240 min) on the onset of serrated flow at RT. The first occurrence of PLC bands is indicated by arrows. Longer aging times shift these to higher strains. The flow stress can be increased by increasing the aging time. b) Critical stresses for the onset of PLC effects as a function of testing temperature and for different natural aging times.

Besides the deformation temperature, one further important parameter that has an effect on the occurrence of PLC effects is the formation of precipitates after solutionizing. Depending on the aging time, these precipitates can already be formed at RT. The early formation of Guinier-Preston (GP) zones is strongly promoted by the high amount of copper in AA7075, which can lead to an immediate nucleation of GP zones after the solution heat treatment at RT. The influence of natural aging on the occurrence of PLC effects can be analyzed by considering tensile tests (performed at RT) of differently aged samples (figure 5a): All samples were naturally aged at RT for various times (10, 30, 60 and 240 min). With increasing aging time, strength increases with a simultaneous decrease of ductility; both can be related to the formation of GP zones. In addition it is obvious that natural aging has a strong influence on the occurrence of the PLC effect: The initial points of serrated flow (i.e., the first points where serrated flow could be clearly identified in the stress-strain curves; these points are marked by arrows in figure 5a) are shifted to higher stress/strain values with increasing aging time, which is in a good agreement with literature [20]. Natural aging for 10, 30 and 60 min first results in the presence of PLC type A. Afterwards, the curves exhibit a PLC type C character, including a subsequent crack initiation and failure of the specimens. After natural aging for 240 min, the fracture was exactly initiated when the first PLC bands occurred. A further increase of aging time resulted in a sufficient formation of GP zones, which suppress strain localization because a higher amount of GP zones leads to a decreased number of point defects [20].

Thermal activation also plays an important role in determining the occurrence of PLC bands; this is obvious from figure 5b, which summarizes the initial stress values of serrated flow as a function of testing temperature and for samples aged naturally for different times. As known from the literature the maximum susceptibility to PLC effects is about one third of the melting temperature [21]. Especially for the AA7075 alloy our investigations indicate a critical value of about -20 °C, which is in good agreement with the literature. Below RT, all samples showed the PLC type A. Even without natural aging, tensile testing below -60 °C was not affected by PLC effect. At this temperature and strain rate ($1 \cdot 10^{-3}$ $s^{-1}$), diffusive processes are slowed down enough to fully prevent dynamic strain aging.
Dynamic strain aging, as a thermally activated process, depends on the deformation temperature as well as on the strain rate. In order to further characterize the influence of PLC effects on the strain rate sensitivity, we also consider the results of the jump-tests. Figure 6 shows the stress–strain behavior of samples deformed at RT (figure 6a) and at -196 °C (figure 6b), respectively; both samples were first solution heat-treated and then naturally aged for 10 min. At RT, an increase of strain rate leads to a stress drop (by a factor of $\Delta\sigma$), which can be described as a negative strain rate sensitivity. This stress drop is directly related to the PLC effect (i.e., formation of a PLC band), and it is further increased at higher strain rates. In contrast, the amount of $\Delta\sigma$ remains constant (and positive) for every change in strain rate at -196 °C (figure 6b). At this low temperature, we clearly observe positive strain rate sensitivity. These results fully agree with our previous observations (figures 1, 5) on the effect of testing temperature and strain rate on the occurrence of PLC bands: localization of deformation is associated with insufficient hardening and/or negative strain rate sensitivity. By lowering the testing temperature, both hardening and strain rate sensitivity can be increased and therefore the PLC effect can be suppressed, which in turn is favorable with respect to ductility and failure during tensile testing (figures 3 and 4).

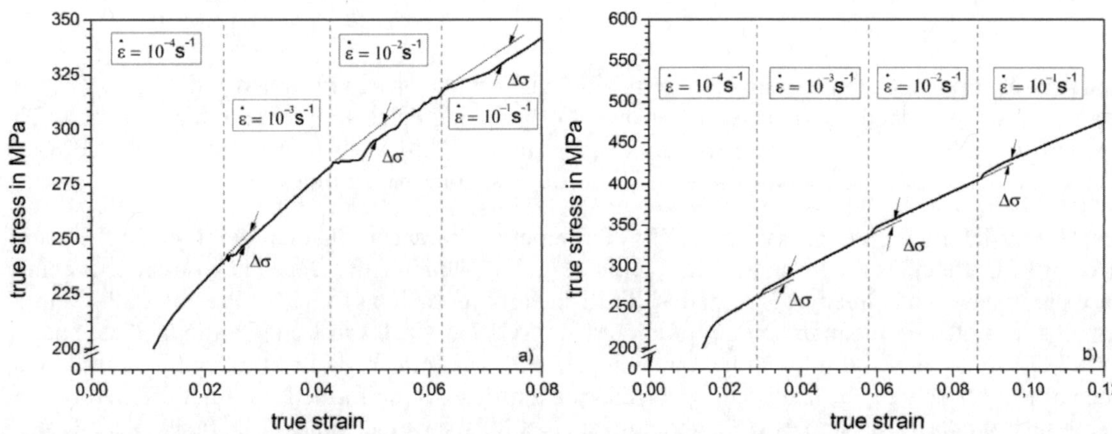


**Figure 6.** Influence of testing temperature and strain rate on the mechanical behavior under tensile loading at a) RT and b) -196 °C. The change of strain rate (by one decade in each jump) is marked by dotted lines. The yield stress decreases (by $\boldsymbol{\Delta\sigma}$) with increasing strain rate at RT. Because of the suppression of dynamic strain aging, this effect is reversed at low temperatures.

Based on our fundamental analysis of the effect of testing temperature on mechanical behavior and, particularly, on the occurrence of localized deformation related to PLC bands, we now discuss, from an engineering point of view, the influence of thermally activated processes on the possibility to manufacture homogenously deformed billets by ECAP at low temperatures. In figure 7, we present pressing forces that were recorded during ECAP at -60 °C as a function of the displacement of the plunger (and of the billet while passing through the ECAP die, respectively) with backpressure (80 bar, as controlled in the hydraulic cylinder; note that the actual mechanical compressive stress at the billet is different from this value: 176 MPa), or without backpressure, and with different pressing speeds ranging from 10 to 50 mm/min. The data in figure 7a correspond to measurements performed while ECAP processing the AA7075 alloy in solution heat-treated condition, while the data in figure 7b were measured, as a reference material, using an AA6082 alloy in peak aged condition. Because of the chemical composition and because we used the peak aged condition of the 6xxx alloy, no PLC effects could occur during deformation. A comparison of the pressing force data for both alloys demonstrates that fluctuations in the pressing forces (as observed for the AA7075 alloy) cannot be simply assigned to stick–slip effects (and hence friction within the ECAP die), but clearly must be related to shear localization in the ECAP shear plane when deforming this alloy. The strain rate in the ECAP shear plane for our setup (cross section: 15 x 15 $mm^2$) is in the range from $10^{-2}$ $s^{-1}$ to $10^{-1}$ $s^{-1}$ (for the pressing speeds considered here), which allows a careful comparison with the data from our jump-tests. Because of the multi-axial stress state associated with the ECAP process, the results given in figure 6 cannot be directly compared. However, the principal trends in terms of strain rate and testing temperature can be correlated with the processing data to estimate limitations and useful temperature/ strain rate regimes for low temperature ECAP.

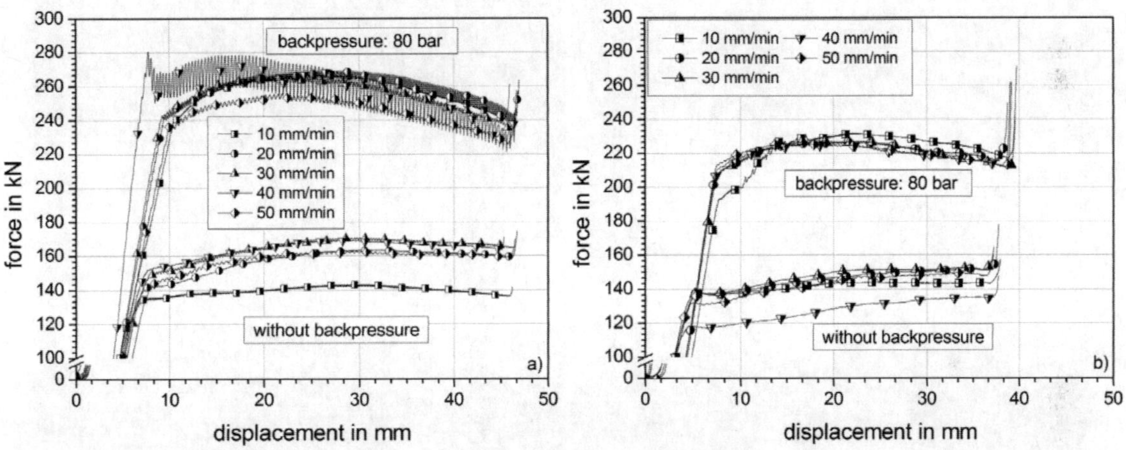


**Figure 7.** Pressing forces during the ECAP process (first ECAP pass) with and without backpressure depend on the deformation speed: a) AA7075 in solution heat-treated condition and b) AA6082 in peak aged condition (T6). Fluctuations of the pressing force can be related to dynamic strain aging in AA7075.

Figure 7 shows that, for both materials, the pressing forces are higher when an active backpressure is applied and lower when no backpressure is used. In the ECAP experiments on AA6082, the pressing forces were hardly affected by the deformation velocity (figure 7b). Because of the coarse grained condition deformed during this first ECAP pass, the material was not prone to shear localization. In contrast, the pressing forces for the deformation of AA7075 (figure 7a) did depend on the deformation velocity, as is expected when dynamic strain aging dominates the deformation behavior. When (without backpressure) the deformation velocity was increased from 10 mm/min to 30 mm/min (which is equivalent to a shear zone strain rate of about $10^{-2}$ $s^{-1}$), pressing forces continuously increased. At a velocity of 40 mm/min, the pressing forces were nearly the same as at 30 mm/min, but the amount of serrated flow (as evidenced by the fluctuations in the pressing force curves) was considerably decreased. At 50 mm/min (corresponding to a strain rate in the shear zone of about $10^{-1}$ $s^{-1}$), both the amount of serrated flow and the pressing forces decreased, which agrees well with the reduction of the flow stress and the reduction of dynamic strain aging observed at the corresponding strain rates in the jump-tests (figure 6a).

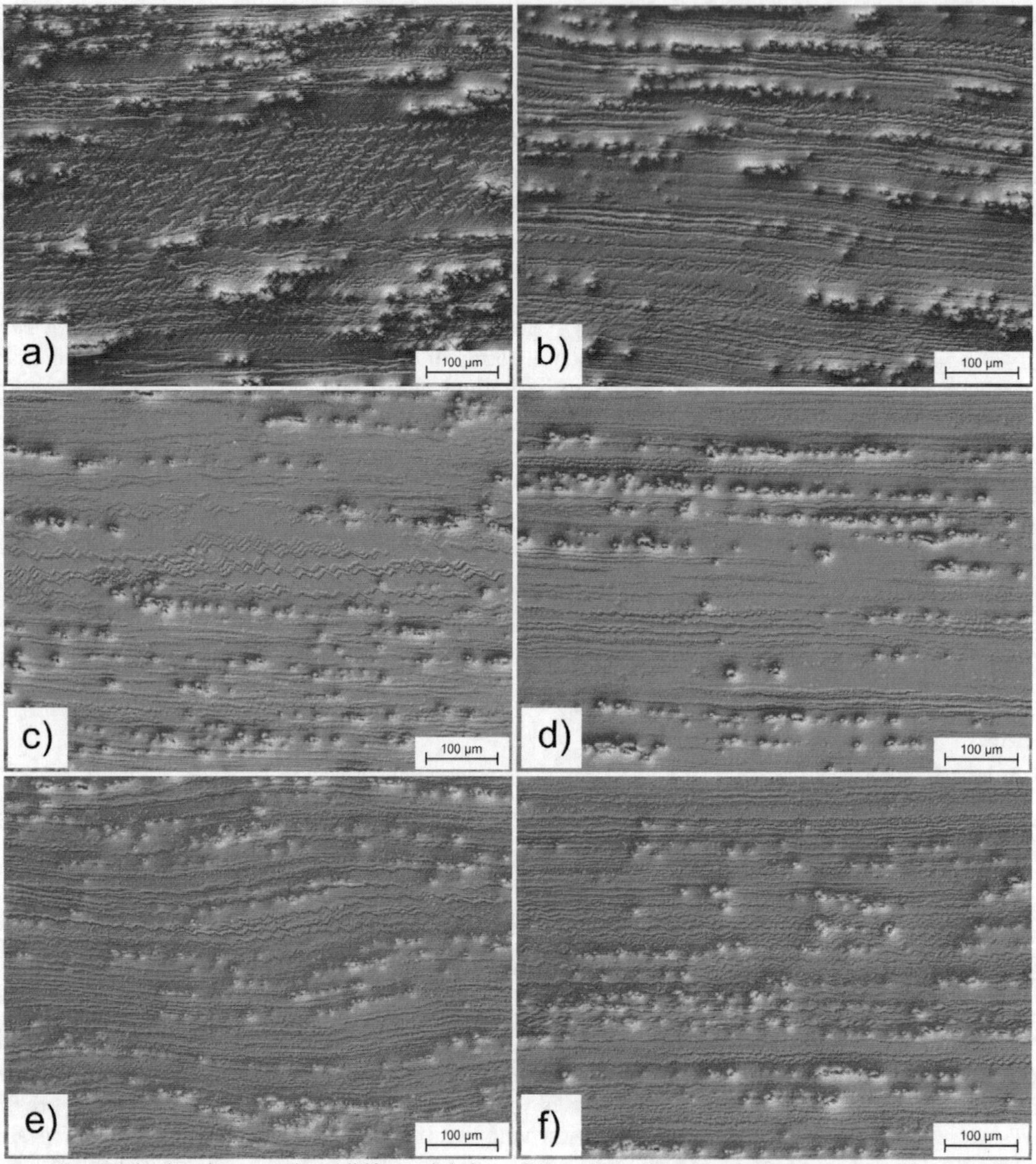


**Figure 8:** Optical micrographs (differential interference contrast) of the ECAP billets deformed at -60 °C at a) 1 mm/min, c) 20 mm/min, e) 50 mm/min, and with an additional backpressure (80 bar) at the similar velocities b), d) and f). The displacements of the coarse impurity phases indicate shear localization.

Similar observations hold when one considers the measurements during ECAP with an additional backpressure; because of the multi-axial stress-state in the shear zone, the maximum susceptibility for the PLC effect was shifted to a slightly lower deformation velocity (40 mm/min). This analysis of pressing forces as a function of deformation velocity allowed to define useful deformation parameters for successful ECAP processing at low temperatures: a relatively low temperature (-60 °C) with a medium strain rate ($10^{-2}$ $s^{-1}$) provides optimum conditions for a suppression of the PLC effect.

Finally, the occurrence of shear bands – as directly related to the deformation velocity – can be confirmed from optical micrographs of the ECAP billets (figure 8; all billets were deformed at -60 °C). After ECAP processing at 1 mm/min, a high amount of (relatively broad) shear bands was observed (figure 8a). By applying an additional backpressure, the number of these bands can be clearly reduced (figure 8b). Increasing the deformation velocity to 20 mm/min resulted in a strongly decreased amount of shear bands (figure 8c) and backpressure leads to smaller widths (figure 8d). The strongly increased amount of serrated flow at 50 mm/min documented in figure 7a fully agrees with the corresponding microstructure (figure 8e). This velocity also has a negative effect on the production of homogeneously deformed billets, which can be inferred from the wavy displacement of the coarse impurities. While this phenomenon can be compensated by using an active backpressure, the number of shear bands cannot be significantly decreased (figure 8f). The key result of these first microstructural observations on ECAP billets deformed at low temperatures is that billet homogeneity and the material's tendency to form shear bands can be directly related to the uniaxial mechanical properties studied in greater depth in this paper.

## 4. Summary and conclusions

We have studied the influence of thermally activated processes, i.e., the role of strain rate and temperature, on the workability and mechanical properties of the high strength aluminum alloy 7075. Due to the increased strain hardening capability of this alloy, low deformation temperatures result in an increased ductility and improved workability. Moreover, we studied how these improvements benefit from the reduction of dynamic strain aging by suppressing the PLC effect at low temperatures. A deformation temperature of -60 °C leads to a strongly decreased amount of shear bands (resulting from PLC effects) and to a sufficient ductility. Our results also show that natural aging at room temperature has a strong influence on the occurrence of dynamic strain aging. By increasing the natural aging time, PLC effects can be reduced. This effect is related to the formation of GP zones, which shift the strength of the material to higher values and lead to increased pressing forces during the ECAP process. To verify our observations from uniaxial testing at different temperatures and strain rates, ECAP was performed with varied deformation velocities at -60 °C. Processing at 20 mm/min, with the use of an active backpressure, was found to be appropriate to produce homogenously deformed billets.

## Acknowledgements

The authors gratefully acknowledge funding by the German Research Foundation (Deutsche Forschungsgemeinschaft, DFG) within the framework of SFB 692. Furthermore we would like to thank S. Seipp, S. Hahn and T. Petschowsky (all from the Chair of Materials Science, TUC) for supporting the SEM investigations and ECAP processing, respectively.

## References

[1] Zhao Y H, Liao X Z, Cheng S, Ma E and Zhu Y T 2006 *Adv. Mater.* **18** 2280
[2] Magalhães D C C, Hupalo M F, Cintho O M 2014 *Mater. Sci. Eng., A* **593** 1
[3] Panigrahi S K, Jayaganthan R 2011 *Mater. Des.* **32** 3150
[4] Das P, Jayaganthan R, Chowdhury T, Singh I V 2011 *Mater. Sci. Eng., A* **528** 7124
[5] Panigrahi S K, Jayaganthan R 2011 *Metall. Mater. Trans. A.* **42** 3208
[6] Rangaraju N, Raghuram T, Krishna B V, Rao K P, Venugopal P 2005 *Mater. Sci. Eng., A.* **398** 246
[7] Fritsch S, Hockauf M, Schönherr R, Hunger S, Meyer L W, Wagner M F-X *Materialwiss. Weekstofftechn.* 2010 **41** 697
[8] Fritsch S, Scholze M, Wagner M F-X *Materialwiss. Weekstofftechn.* 2012 **43** 561
[9] Rao P N, Singh D, Jayaganthan R *Mater. Des.* 2014 **56** 97
[10] He L Z, Cao Y H, Liu X T, Zhang H T, Wang P, Lu C, Guo Y P, Cui J Z 2013 *Mater. Character.* **84** 188
[11] Chen Y, Li Y, He L, Lu C, Ding H, Li Q *Mater. Lett.* 2008 **62** 2821
[12] Chatterjee A, Sharma G, Sarkar A, Singh J B, Chakravartty J K 2012 *Mater. Sci. Eng., A.* **556** 653
[13] Podolskiy A V, Ng H P, Psaruk I A, Tabachnikova E D, Lapovok R 2014 *J. Mater. Sci.* **49** 6803
[14] Iwahashi Y, Wang J T, Horita Z, Nemoto M, Langdon T G 1996 *Scr. Mater.* **35** 143
[15] Fritsch S, Hunger S, Scholze M, Hockauf M, Wagner M F-X 2011 *Materialwiss. Weekstofftechn.* **42** 573
[16] Saito Y, Tsuji N, Utsunomiya H, Sakai T, Hong R G 1998 *Scr. Mater.* **39** 1221

[17] Seeger A, Haasen P 1958 *Philos. Mag.* **3** 470
[18] Hart E W 1967 *Acta Metall.* **15** 351
[19] Chung N, Embury J D, Evensen J D, Hoagland R G, Sargent C M 1977 *Acta Metall.* **25** 377
[20] Leacock A G, Howe C, Brown D, Lademo O G, Deering A 2013 *Mater. Des.* **49** 160
[21] Robinson J M, Shaw M P 1994 *Int. Mate. Rev.* **39** 113

IOP Conf. Series: Materials Science and Engineering **118** (2016) 012031 doi:10.1088/1757-899X/118/1/012031

# Plasma electrolytic oxidation of AMCs

**R Morgenstern[1], M Sieber[1] and T Lampke[1]**

[1] Materials and Surface Engineering Group, Institute of Materials Science and Engineering, Technische Universität Chemnitz, 09125 Chemnitz, Germany

roy.morgenstern@mb.tu-chemnitz.de

**Abstract.** Aluminum Matrix Composites (AMCs) consisting of high-strength alloys and ceramic reinforcement phases exhibit a high potential for security relevant lightweight components due to their high specific mechanical properties. However, their application as tribologically stressed components is limited because of their susceptibility against fatigue wear and delamination wear. Oxide ceramic protective coatings produced by plasma electrolytic oxidation (PEO) can solve these problems and extend the possible applications of AMCs. The substrate material was powder metallurgically processed using alloy EN AW 2017 and SiC or $Al_2O_3$ particles. The influence of material properties like particle type, size and volume fraction on coating characteristics is clarified within this work. An alkaline silicate electrolyte was used to produce PEO coatings with technically relevant thicknesses under bipolar-pulsed current conditions. Coating properties were evaluated with regard to morphology, chemical composition, hardness and wear resistance. The particle type proved to have the most significant effect on the coating properties. Whereas compactness and thickness are not deteriorated by the incorporation of thermodynamically stable alumina particles, the decomposition of silica particles during the PEO processes causes an increase of the porosity. The higher silica particle content decreases also the coating thickness and hardness, which leads in particular to reduction of the wear resistance of the PEO coatings. Finally, different approaches for the reduction of the coating porosity of silica reinforced AMCs are discussed.

## 1. Introduction

High strength aluminum alloys are particularly suitable as lightweight materials in vehicle and aircraft engineering. Because of their high specific strength, accelerated masses can be reduced, which enables savings of energy costs and the reduction of $CO_2$ emissions. A further improvement of material properties can be achieved by the creation of aluminum matrix composites (AMCs). An enhancement of strength, hardness, E modulus and creep resistance can be achieved by ceramic particle reinforcement of aluminum alloys as small particles (below 1 µm) contribute to dispersion hardening and grain refinement during material processing. The resistance of materials against abrasion and adhesive wear generally increases with increasing hardness. Thus, AMCs exhibit superior wear resistance compared with aluminum alloys and cast iron under sliding wear conditions. This enables the implementation of AMCs in automotive applications such as brake discs, drum brakes, calipers and cylinder liners [1] and weight reduction due to the substitution of iron base materials. However, as reviewed by Deuis et al. [2], AMCs consisting of high-strength aluminum alloys and micron or submicron scale particles are susceptible to fatigue wear at higher normal loads. Crack initiation takes place at pores, which arise close to particles due to plastic deformation of the matrix alloy at the critical depth beneath the surface. Finally, plate-shaped material volumes are delaminated from the surface.

The wear resistance of materials can be improved by surface and coating technology. Especially ceramic coatings provide efficient wear protection due to high abrasion resistance and low adhesion against metallic counter bodies. Compact oxide ceramic coatings can be generated by plasma electrolytic oxidation (PEO) on aluminum alloys. The conversion coatings exhibit excellent adhesion to the substrate and a maximum thickness of more than 100 µm. Due to the nature of the process, PEO coatings consist of an inner compact and an outer porous layer. The inner layer is compacted by several remelting processes in the immediate surroundings of micro-arc discharges and consists predominantly of α alumina (corundum). Hardness values comparable to monolithic alumina ceramics exceeding 10 GPa (about 1500 to 2000 HV) can be measured there [3]. Particularly for tribological applications, the process parameters (e.g. electrolyte composition, pulsed current) are optimized with respect to the proportion of the compact coating. If necessary, the outer porous layer has to be removed by machining.

The incorporation or dissolution of ceramic particles influences microstructure and properties of PEO coatings. Wang et al. [4] describe that the growth rate of PEO coatings on AMCs is generally lower compared with the unreinforced matrix alloy due to the high electrical resistivity of the particles. Thermodynamically stable and electrochemically inert alumina ($Al_2O_3$) particles are incorporated in the coating without restraining the coating hardness, whereas higher porosity is caused by the presence of silicon carbide (SiC) particles. According to Wang et al. [4], a higher number of cracks and pores develops due to internal stress and the formation of $Al_4C_3$ plates at the substrate-coating interface. However, this correlation is not clear as the authors simultaneously varied the matrix alloys and observed that alloy composition also strongly affects coating thickness and porosity. In contrast, according to the model of Xue [5], SiC particles are dissolved in several steps in the vicinity of microarc discharges without causing pores or cracks. Furthermore, no comparable studies dealing with the influence of particle properties on the wear behavior of PEO coatings exist. However, separate studies on SiC and $Al_2O_3$ reinforced materials give evidence that sliding wear resistance of AMCs against steel counter bodies can be improved by PEO [6, 7].

The aim of the work is to investigate the influence of particle type, size and volume fraction on coating thickness, microstructure, hardness and sliding wear resistance and to give some indications of the particle incorporation or dissolution mechanism.

## 2. Material and methods

The AMC materials were produced by a powder metallurgical process. A gas-atomized alloy AlCu4MgMn powder was used as metallic matrix. The chemical composition was determined by inductively coupled plasma optical emission spectroscopy (ICPOES) in wt%: 3.8 % Cu, 1.2 % Mg, 0.6 % Mn, 0.4 % Pb, 0.2 % Fe, 0.2 % Zn. Cubic silicon carbide (SiC, Plasmachem GmbH) and α alumina ($Al_2O_3$, Plasmachem GmbH) each with different particle sizes 300 nm and 3 µm at maximum were used as ceramic reinforcement. Alloy powder and different particle volume fractions (5, 10, 15 vol%) were mixed by high-energy ball-milling (HEM, Zoz GmbH Simoloyer). The composite powder was encapsulated in a cylinder of commercially pure aluminum, hot degassed, hot isostatically pressed (HIP, 450 °C, 1100 bar, 3 h) and extruded (370 °C) to a cross section of 15 x 15 cm². Subsequently, a T4 temper (solution annealing, natural aging) of the composite was carried out. The process conditions, the microstructure formation of the powder after HEM processing and the material after the extrusion are described in detail in [8].

Samples with a thickness of 5 mm were cut perpendicular to the extrusion direction. The aluminum encapsulation was removed and the front surfaces were grinded until grain size of 600. Samples were electrically bonded by an aluminum wire. All rectangular surfaces were exposed to the electrolyte. The plasma anodized coatings were produced in the alkaline electrolyte consisting of 5 g/l KOH, 3 g/l $Na_2SiO_3 \cdot 5H_2O$ and 0.5 g/l $Na_2HPO_4$. The coating process was carried out under a symmetrical rectangular bipolar-pulsed current regime with pulse amplitude of 50 A/dm² and a pulse length of 10

ms in both, anodic and cathodic half-cycle. The temperature of the electrolyte was kept constant at 20 °C during the process. The principal arrangement of the laboratory plant for the plasma anodizing process is shown elsewhere [3]. The time-dependent behavior of pulsed current and voltage was recorded with a transient recorder Yokogawa DL850 at a sampling rate of 200 kHz.
The coating thickness was measured at the coating cross section by optical microscopy. Cross sections for optical and scanning electron microscopy (SEM) were prepared by cutting, hot mounting in electrically conductive resin, diamond grinding and polishing. SEM samples were finally polished with a silica oxide polishing suspension and carbon coating was used to avoid sample charging. Backscattered electron (BSE) images were obtained at SEM LEO 1455VP using 25 kV excitation voltage. Secondary electron (SE) images as well as Energy and angle selective BSE (ESB) images were obtained at FE-SEM Zeiss NEON40EsB. Hardness values were obtained by hardness measurement at 100 mN testing force with Fischerscope HM2000 XYm at the cross-section of the coatings. A Vickers diamond was used as the indenter. Oscillating dry sliding wear behavior was evaluated by a ball-on-disc arrangement (Tetra) at room temperature using a rotational speed of 90 $min^{-1}$, turning radius of 3 mm, 15 N normal force and a 100Cr6 ball with a diameter of 6 mm as the counter body. The cross section profiles of wear traces were measured by contact stylus instrument Hommelwerke T 4000.

## 3. Results and discussion

### *3.1 Energy efficiency*

For many tribological applications, a PEO coating thickness of about 100 µm is pursued in order to provide sufficient supporting effect and long-term wear protection. From an economical point of view, this should be achieved in a minimum time and with the lowest possible energy consumption. As the PEO process is conducted with constant current density and process time, the total energy consumption grows with increasing process voltage. It is known from conventional electrolytic anodizing of AMCs that the process voltage increases especially with the presence of electrically insulating and chemically inert $Al_2O_3$ particles [9]. However, as evident from figure 1, the height of voltage amplitudes is not significantly influenced by particle reinforcement during PEO process with the pulsed current. The positive voltage amplitude increase at the beginning of the process, which can be observed for $Al_2O_3$ reinforced and unreinforced material, can be explained by the increasing electrical resistance during the growth of a continuous compact oxide layer. After approximately 1200 s, the curves converge to a similar voltage height for all investigated materials. At this stadium, multiple microarc discharges remelt the continuous primary oxide layer at points of lower coating thickness. This leads to a macroscopic uniform coating growth and the compaction of the inner layer. It can be concluded that the electrical energy requirement is similar for matrix alloy and AMCs.
Despite the similar energy consumption, the coating thickness depends strongly on the particle type as can be seen from figure 2. In general, the total coating thickness is not deteriorated by the incorporation of $Al_2O_3$ particles except from slightly lower thickness values at 10 vol% particle content. Furthermore, there is a tendency that the thickness of the porous layer (empty bars in figure 2) grows with increasing particle size. In contrast to this, significantly lower coating thicknesses can be observed on SiC reinforced AMCs. Moreover, a higher portion of the total coating thickness can be attributed to the outer porous layer and there are higher standard deviations with regard to the coating thickness. The SiC particle volume content does not influence the coating thickness significantly. However, total coating thickness and the thickness of the inner compact layer are slightly reduced by increasing particle size.

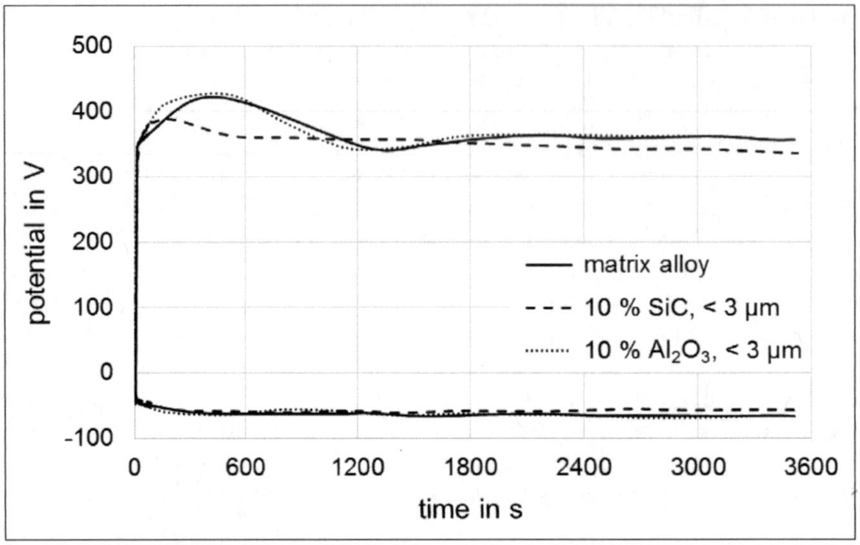


**Figure 1.** Envelopes of positive and negative potential amplitudes as a function of time during PEO process with pulsed current.

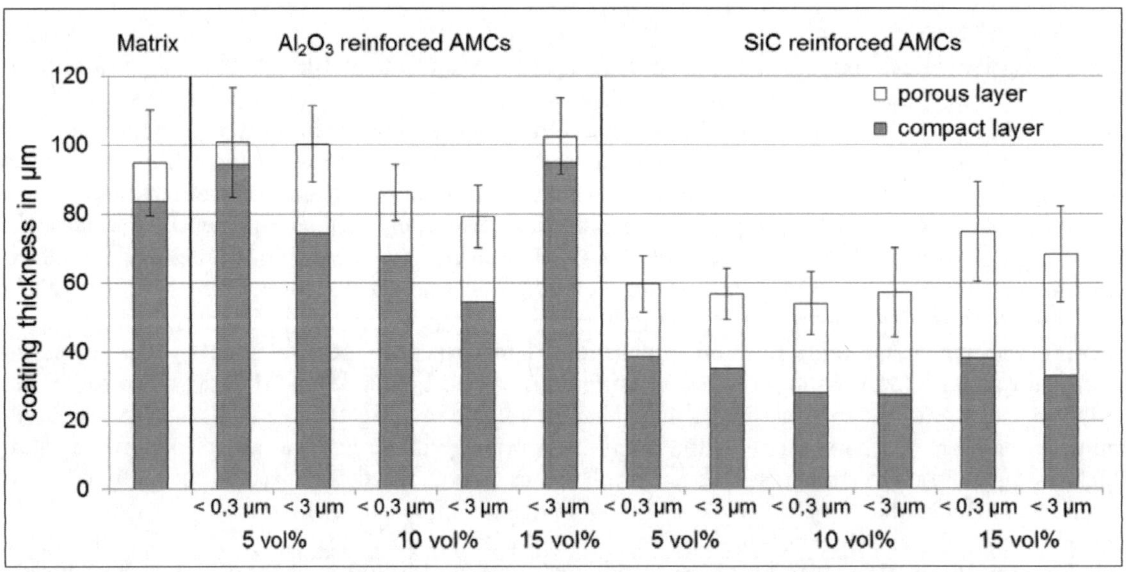


**Figure 2.** Comparison of total thickness, thickness of porous layer and compact layer for PEO coatings on matrix alloy as well as SiC and $Al_2O_3$ reinforced AMCs with varying particle sizes and particle volume contents

*3.2 Coating properties*

The observation of Wang et al. [4] that PEO coatings on SiC reinforced AMCs exhibit higher porosity than those on $Al_2O_3$ reinforced material can be approved by SEM investigations (figures 3a and 4a). It can be seen from high magnification images that both materials show microscale porosity in close vicinity to the substrate coating interface (figures 3b and 4b) which is a typical phenomenon in PEO processes. Fine submicron scale porosity (highlighted in figure 3b) with a similar size as copper containing precipitates (highlighted in figure 4b) can be observed in the PEO coatings on both AMC substrates. It might be concluded that some of those pores evolved due to the dissolution of the intermetallic phases (IMP). In accordance with the literature, it can be assumed that $Al_2O_3$ particles are incorporated into the PEO coating without any chemical interactions. However, they can hardly be differentiated from the $Al_2O_3$ matrix and, therefore, are not detectable within the PEO coating by SEM investigations. PEO coatings on SiC reinforced AMCs exhibit larger pores additionally to the submicron scale porosity. As indicated in figure 3b, silicon rich particles can be detected qualitatively within these pores by EDS spot measurements. In accordance with the theory of Xue et al. [5], SiC particles are gradually transformed into oxide phases and finally dissolved. However, it has to be emphasized that conversion and dissolution of SiC particles cause the increasing coating porosity. As SiC particles are not electrically insulating like $Al_2O_3$ particles, the particle dissolution might be accompanied by oxygen evolution and, therefore, pore formation. The side process of gas evolution consumes the electrical charge, which means that less electrical energy is available for coating growth. This would be an explanation for the lower thickness of PEO coatings on SiC reinforced AMCs.

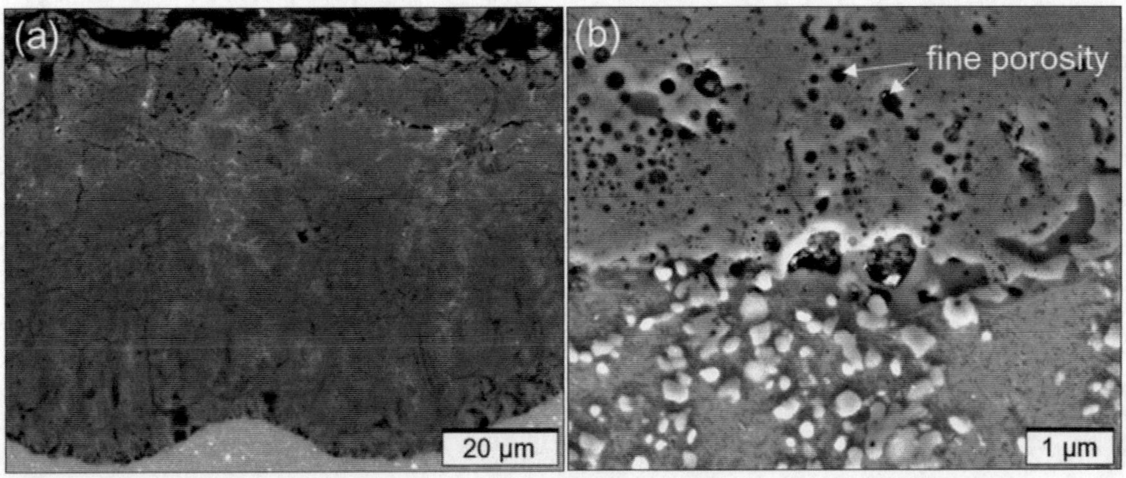


**Figure 3.** Microstructure of PEO coatings on AMCs with 10 vol% $Al_2O_3$ particles < 3 µm: a) BSD overview image, b) SE close-up image of the substrate-coating interface

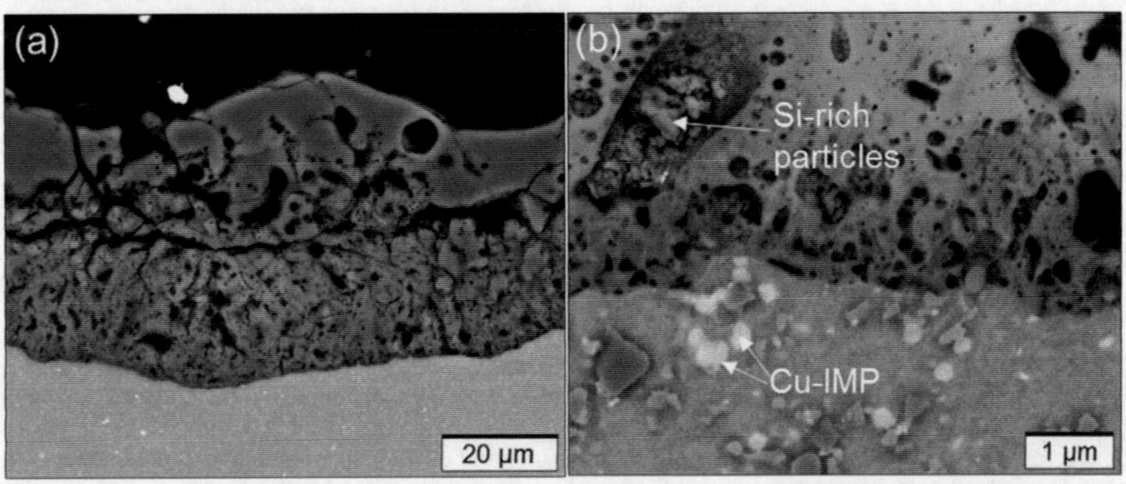


**Figure 4.** Microstructure of PEO coatings on AMCs with 10 vol% SiC particles < 3 µm: a) BSD overview image, b) ESB close-up of the substrate-coating interface

IOP Conf. Series: Materials Science and Engineering **118** (2016) 012031 doi:10.1088/1757-899X/118/1/012031

Generally, the coating hardness decreases with increasing distance to the substrate/coating interface within the inner compact layer due to the growing amount of amorphous phases and slightly increasing porosity. It can be seen from figure 5a, that PEO coatings on SiC reinforced AMCs exhibit significantly lower hardness values in comparison to those on the matrix alloy. More precisely, the hardness of PEO coatings decreases with increasing SiC particle content as a higher particle density in the AMC enhances the formation of pores. Furthermore, coating hardness is decreasing with increasing particle size at constant particle content because large particles leave large pores after dissolution. In contrast to this, the coating hardness is slightly improved by the presence of small $Al_2O_3$ particles (filled symbols in figure 5b) compared with PEO coatings on the unreinforced matrix alloy (dotted line in figure 5b). Larger $Al_2O_3$ particles (unfilled symbols in figure 5b) do not influence the coating hardness significantly. Furthermore, there is no considerable correlation between coating hardness and particle volume fraction.

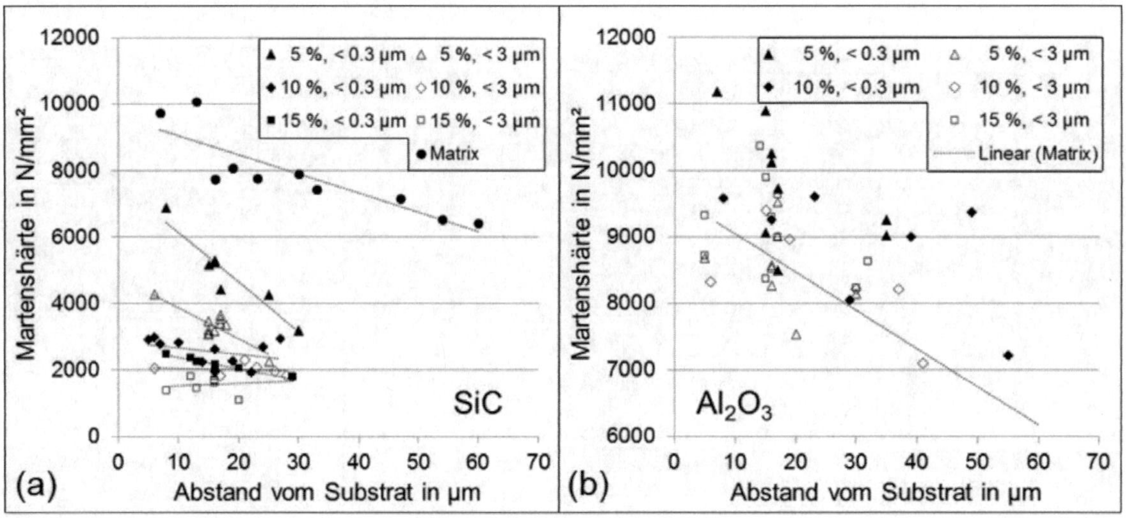


**Figure 5.** Hardness profiles of PEO coatings on a) SiC and b) $Al_2O_3$ reinforced AMCs with different particle volume fractions and particle sizes

Sliding wear behavior of uncoated SiC and $Al_2O_3$ reinforced AMCs with 10 vol% particle content and particle size < 3 µm was investigated as a reference for the wear resistance of PEO coatings. Both types of AMCs exhibit a comparable wear behavior. As numerous cracks and flat pittings can be observed within the wear tracks by SEM investigations, it can be verified that fatigue and delamination wear are the dominating wear mechanisms. In many cases the edges of pittings were overlapped by plastic deformation of the surrounding material. Wear depths and cross section profiles of wear traces are also similar for both uncoated AMCs, thus, only one profile for $Al_2O_3$ reinforced material is displayed infigure 6a. It can be seen from this profile that wear depth was about 10 µm and material accumulation took place at the edges of the wear trace due to plastic deformation. However, the porous PEO coatings on SiC reinforced AMCs exhibit distinct wear traces with depths of 20 to 30 µm (figure 6a) and significantly higher wear volumes compared with the untreated material (figure 6b). Furthermore, it can be seen from the BSD cross section image (figure 7a) and the cross section profile (figure 6a) that there is a sharp transition between unworn surface and wear trace. The outer porous layer is removed quickly; however, the inner compact layer withstands sliding wear. Fatigue cracks are stopped by fine pores within the compact layer and do not reach the substrate surface. Moreover, the steel counter body is worn when it gets in contact with the hard inner layer. The adherent layer of compacted fine-grained debris from PEO coating and counter body consisting of iron rich and alumina particles with sizes below 2 µm can occasionally be observed at the worn PEO surface. In contrast to this, no significant wear of PEO coatings on $Al_2O_3$ reinforced AMCs can be observed. However, due to the high hardness of the PEO coating, the steel counter body is severely worn during sliding wear tests. Thus, wear traces are almost completely covered by a layer of compacted wear debris, which also consists of fine iron-rich and alumina particles. This accumulated debris can be suggested by the cross sectional profile of the wear trace in figure 6a and can be seen as a bright layer on top of the PEO coating in BSE image figure 7b.

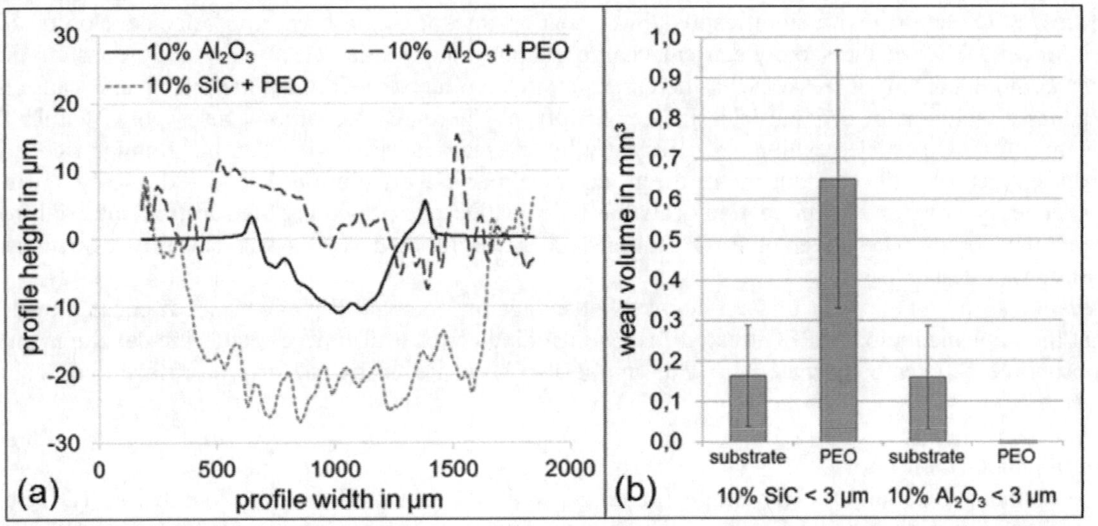


**Figure 6.** a) Elevation profile and b) wear volume of wear tracks on uncoated and PEO treated AMCs s

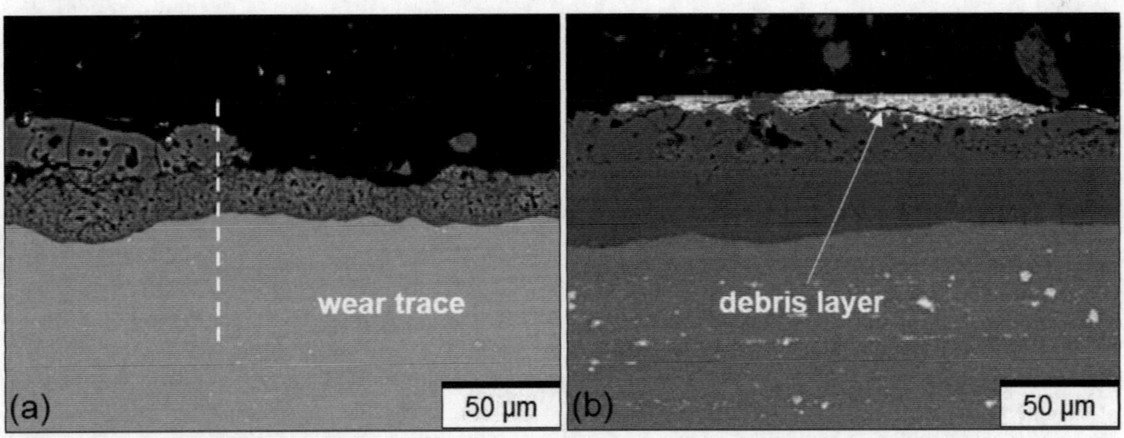


**Figure 7.** Cross sections of wear traces of PEO coatings on AMCs reinforced with 10 vol%, < 3 µm a) SiC and b) $Al_2O_3$ particles (BSD images)

One possibility of improving the wear resistance of porous PEO layers on SiC reinforced AMCs could be the infiltration of pores by sol-gels or nanoparticle reinforced polymers. Similar process combinations were recently applied to porous PEO coatings on magnesium alloys in order to improve their corrosion and wear resistance [10]. Furthermore, the application of an aluminum layer by thermal spraying enables the formation of compact wear resistant coatings on numerous substrates that are not well suited for PEO treatment [11].

## 4. Conclusions

The influence of different particle types, sizes and volume contents on the PEO process as well as on the microstructure evolution and the properties of PEO coatings on AMCs were investigated. It is apparent that the coating thickness and hardness are not significantly deteriorated by the incorporation of $Al_2O_3$ particles. A slight increase in coating hardness can be observed when small $Al_2O_3$ particles (< 0.3 μm) are present in the AMC. As a consequence of high hardness and low porosity, PEO coatings on $Al_2O_3$ reinforced AMCs do not show any wear damage after sliding wear tests. In contrast to this, PEO coatings with significantly lower coating thickness and higher porosity develop on SiC reinforced AMCs at the same electrical charge quantity and similar electrical energy consumption. Furthermore, porosity increases with increasing particle volume fraction and particle size, because the gradual dissolution of SiC particles in the vicinity of microarc discharges causes pores within the coating. As SiC particles exhibit a significantly lower electrical resistivity than the alumina rich PEO layer, oxygen might evolve during the electrochemical particle dissolution. This could also be a reason for higher porosity and lower coating growth efficiency. Despite protecting the substrate material from wear, the porous outer layer of PEO coatings on SiC reinforced AMCs was severely worn during sliding wear tests.

Because of the severe wear of steel counter bodies against compact PEO coatings, it became evident that the implementation of PEO treated AMCs in sliding contacts also necessitates the development of appropriate counter body materials. This might also be achieved by the use of surface and coating technologies.

**Acknowledgement**

The authors gratefully acknowledge funding by the German Research Foundation (Deutsche Forschungsgemeinschaft, DFG) within the framework of the Collaborative Research Centre 692 (SFB HALS 692). The technical assistance of C. Cai, D. Spieler, Dr. D. Dietrich, K. Muhr and C. Gläser is also gratefully acknowledged.

## 5. References

[1] Prasad S V and Asthana R 2004 *Tribol. Lett.* **17** 445
[2] Deuis R L, Subramanian C, Yellup J M 1997 *Compos. Sci. Technol.* **57** 415
[3] Sieber M, Mehner T, Dietrich D, Alisch G, Nickel D, Meyer D, Scharf I and Lampke T 2014 *Surf. Coat. Technol.* **240** 96
[4] Wang Y K, Sheng L and Xiong R Z, Li B S 1999 *Surf. Eng.* **15** 112
[5] Xue W 2006 *Appl. Surf. Sci.* **252** 6195
[6] Cui S, Han J, Du Y and Li W 2007 *Surf. Coat. Technol.* **201** 5306
[7] Ceschini L, Martini L, Sambogna G and Tarterini F 2011 *Mater. Sci. Forum* **678** 61
[8] Podlesak H, Siebeck S, Mücklich S, Hockauf M, Meyer L W, Wielage B and Weber D 2009 *Materialwiss. Werkstofftech.* **40** 500
[9] Morgenstern R, Nickel D, Dietrich D, Scharf I and Lampke T 2015 *Mater. Sci. Forum* **825-826** 636
[10] Castellanos A, Altube A, Vega J M, García-Lecina E, Díez J A and Grande H J 2015 *Surf. Coat. Technol.* **278** 99
[11] Wielage B, Meyer D, Pokhmurska H, Alisch, G and Grund T 2008 *Materialwiss. Werkstofftech.* **39** 920

IOP Conf. Series: Materials Science and Engineering **118** (2016) 012032 doi:10.1088/1757-899X/118/1/012032

# Forging of eccentric co-extruded Al-Mg compounds and analysis of the interface strength

**Part I W Förster[1], C Binotsch[1] and B Awiszus[1]**
**Part II T Lehmann[2], J Müller[2], C Kirbach[2], M Stockmann[2] and J Ihlemann[2]**

[1] Professorship of Virtual Production Engineering, Technische Universität Chemnitz, Germany
[2] Professorship of Solid Mechanics, Technische Universität Chemnitz, Germany

**Abstract.** Within the subproject B3 of the Collaborative Research Center 692 it has been shown that Al-Mg compounds with a good bonding quality can be produced by hydrostatic co-extrusion. During processing by forging, the aluminum sleeve is thinned in areas of high strains depending on the component geometry. To solve this problem an eccentric core arrangement during co-extrusion was investigated. Based on the results of FE-simulations, the experimental validation is presented in this work. Rods with an offset of 0.25, 0.5 and 0.75 mm were produced by eccentric hydrostatic co-extrusion. Ultrasonic testing was used to evaluate the bonding quality across the entire rods. For the forging investigations the basic process Rising was chosen. The still good bonding quality after forging was examined by dye penetrant testing and optical microscopy.
For an optimal stress transfer between the materials across the entire component, a sufficient bonding between the materials is essential. To evaluate the interface strength, a special bending test was developed. For the conception of the bending specimens it was required to analyze the Rising specimens geometry. These analyses were performed using a reconstruction of the geometrical data based on computer tomography (CT) investigations. The comparison with the numerically deter-mined Rising specimen geometry shows good correlation. Parametric Finite Element Analyses of the bending test were used to develop the load case and the specimen geometry. By means of iterative adaption of load application, bearing and specimen geometry parameters, an advantageous stress state and experimentally applicable configuration were found. Based on this conception, the experimental setup was configured and bending tests were performed. The interface strength was deter-mined by the calculation of the maximum interlaminar interfacial tension stress using the experimental interface failure force and the bending FE model.

## 1. Motivation

The challenge of using composites and compounds is in creating a sufficient bonding between the materials. However, not only an existing bonding is important but also a reliable transfer of stresses in operation, especially for components for industrial or safety applications. By using aluminum and magnesium as compound materials, both goals could be reached. The production by hydrostatic co-extrusion – with AZ31 as core material and AA-6082 as sleeve material – results in a complete bonding [1]. The bonding could be maintained during further processing by forging [2]. The interface strength was investigated with different methods and high interface strengths were found, but with a high variation [3, 4]. Furthermore, another effect occurred at components with asymmetric cross sections which were forged by Rising as shown in figure 1. In areas of high strains – increased by an opposing material flow – the flank thickness of the sleeve material is reduced [2]. This could become a problem, if the sleeve material breaks up. Then, the stresses would be applied to just one material which means the loss of the compound. Additionally, corrosion could start damaging the magnesium core. To increase the flank thickness and to enhance the field of application of the compounds, an eccentric core arrangement during co-extrusion was investigated.

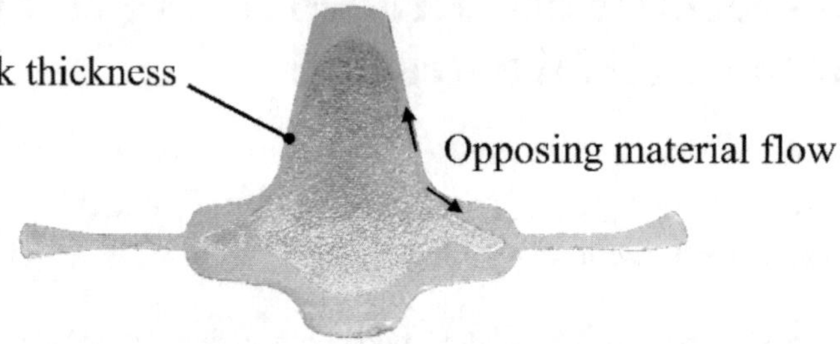


**Figure 1.** Transversal cross section of a component with centric core arrangement forged by Rising with reduced flank thickness in the region of high strains.

## 2. Part I – Production and Processing

### *2.1. Introduction*

The idea of an eccentric co-extrusion was developed by FEUERHACK [2] and a schematic presentation of the process is given in figure 2a. Thereby, the eccentric arrangement of the magnesium core is given by an offset of center axes $\Omega$ of the aluminum and the magnesium billet. It can be calculated with equation 1 and the extrusion ratio $\psi$, if $d$ are the diameters of the pressing billet respectively rod.

$$\psi = \frac{d^2_{\text{Billet}}}{d^2_{\text{Rod}}} = \frac{\Omega^2_{\text{Billet}}}{\Omega^2_{\text{Rod}}} \qquad (1)$$

The offset leads to a minimum ($s_{\text{min}}$) respectively maximum ($s_{\text{max}}$) wall thickness of the aluminum sleeve as shown in figure 2b. FEUERHACK investigated eccentric co-extrusion with five different offsets within the pressing billet in the range of 1 to 5 mm numerically. The further processing of the eccentric rods was also numerically investigated. Therefore, FEUERHACK chose the basic forging process Rising to compare the results with the former used centric core arrangement. He showed, that with increasing the offset, the critical flank thickness also increases [2].

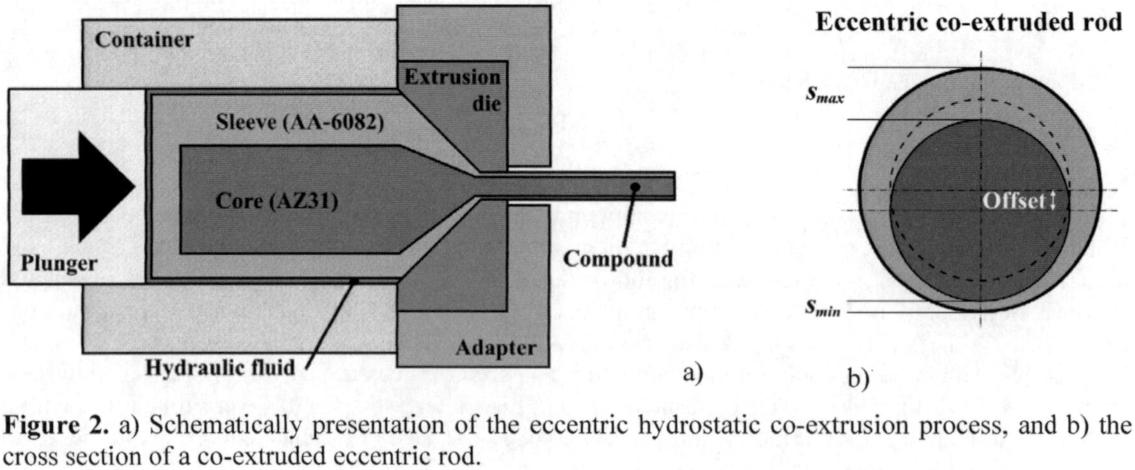


**Figure 2.** a) Schematically presentation of the eccentric hydrostatic co-extrusion process, and b) the cross section of a co-extruded eccentric rod.

Based on the numerical findings of FEUERHACK the experimental validation is presented in this article. Therefore, rods with an offset of center axes of 0.25, 0.5 and 0.75 mm were produced.

### *2.2. Experiments*

*2.2.1. Eccentric hydrostatic co-extrusion.* The extrusion experiments were performed at the Compound Extrusion Products GmbH in Freiberg, Germany. The production of eccentric pressing billets and rods are based on the experimental experience in co-extrusion of Al-Mg compounds with a centric core arrangement. In opposite to the assumption of FEUERHACK, the core diameter is 58 mm instead of 60 mm due to technical limitations of processing a loose fit between the materials. Thus, the comparability of the numerical and experimental results is limited. Table 1 gives a summary of the calculated values of the experimental setup used in this work.

**Table 1.** Calculated dimensions of the pressing billets and the co-extruded rods, based on the extrusion ratio.

| All dimensions in mm | Offset | $s_{min}$ | $s_{max}$ |
|---|---|---|---|
| Pressing billet | 0 | 11 | 11 |
| | 1 | 10 | 12 |
| | 2 | 9 | 13 |
| | 3 | 8 | 14 |
| Co-extruded rod | 0 | 2.75 | 2.75 |
| | 0.25 | 2.5 | 3.0 |
| | 0.50 | 2.25 | 3.25 |
| | 0.75 | 2.0 | 3.5 |

The parameters – with an extrusion temperature of about 300 °C and a plunger velocity of 3.6 mm/s – for eccentric co-extrusion were the same as used for a centric co-extrusion with a tapered extrusion die. As sleeve material the aluminum alloy AlMgSi1 (AA-6082) and as core material MgAl3Zn (AZ31) were used. A detailed description of the co-extrusion process of Al-Mg compounds can be found at KITTNER [1]. The bonding quality of the rods has been examined by ultrasonic testing. With this nondestructive method the rod could be verified entirely. In addition, the interface formation was evaluated by optical microscopy.

*2.2.2. Forging.* The forging experiments were performed on a 1000 kN eccentric press (Raster-Zeulenroda) with the same forging die and process parameters as used by FEUERHACK [2]. Specimens with a length of 40 mm were cut off from the co-extruded rod and heated at 350 °C for 30 minutes prior to forging. For significant results, the correct positioning of the specimens into the forging die is important. The specimens have to be rotated that $s_{min}$ and $s_{max}$ are in a six o`clock position (figure 2b). The lower the offset, the more difficult the positioning, due to an optical identification of $s_{min}$ and $s_{max}$. For lubrication an aerosol with a composition of $MoS_2$ and graphite was applied to the upper and lower die prior to forging. The specimen temperature was about 300 °C and the die temperature 200 °C. After forging, some components were cut in longitudinal or transversal direction for an investigation by dye penetrant testing and preparation for optical microscopy.
In addition, a comparison of CT data of the component geometry and simulation results was performed. Therefore, new simulations with the adapted core diameter of 58 mm were necessary. The experimental CT investigations are described in part II of this paper, but the results of the comparison between simulation and experiment are presented under paragraph 2.3.2.

IOP Conf. Series: Materials Science and Engineering **118** (2016) 012032 doi:10.1088/1757-899X/118/1/012032

*2.3 Results and discussion*

*2.3.1. Eccentric hydrostatic co-extrusion.* By using ultrasonic testing, the bonding quality across the entire rod could be evaluated. This method was used for co-extruded Al-Mg rods by LEHMANN [3] at first. The ultrasonic emitter/receiver is positioned on the lateral area of the rod. Thus the ultrasonic waves are spread across its cross section. In case of a non-bonding condition, the emitted ultrasonic waves are reflected at the interface resulting in a strong peak after a short runtime. If there is a sufficient bonding most of the ultrasonic waves pass the interface and only a small part is reflected. This leads to a smaller, but still significant peak – when the interface is passed – and to one larger peak indicating the bottom echo. By this characteristic echo signals the corresponding areas could be qualitatively identified and were marked on the rods, as can be seen in figure 3.

**Figure 3.** Developed view of a rod with a qualitatively indication of the bonding quality investigated by ultrasonic testing; red areas indicate no bonding and green areas a good bonding.

To give an overview of the experiments, the results of the specimens with the largest offset (0.75 mm within the rod) are presented. The investigated rods had a completely formed bonding with exception of the beginning of the rods. A poor bonding quality in this area was also shown by KITTNER [1] due to an unsteady material flow in the beginning of the co-extrusion process. To characterize the interface formation, specimens for optical microscopy were prepared in longitudinal and transversal direction. The good bonding quality – detected by ultrasonic testing – could be confirmed by this investigation. A faultless interface with a thickness of about 1 µm has formed completely, as the macro- and microstructures in figure 4 show. This is in accordance to the results that have been found during co-extrusion with centric core arrangement [1] and have been expected.

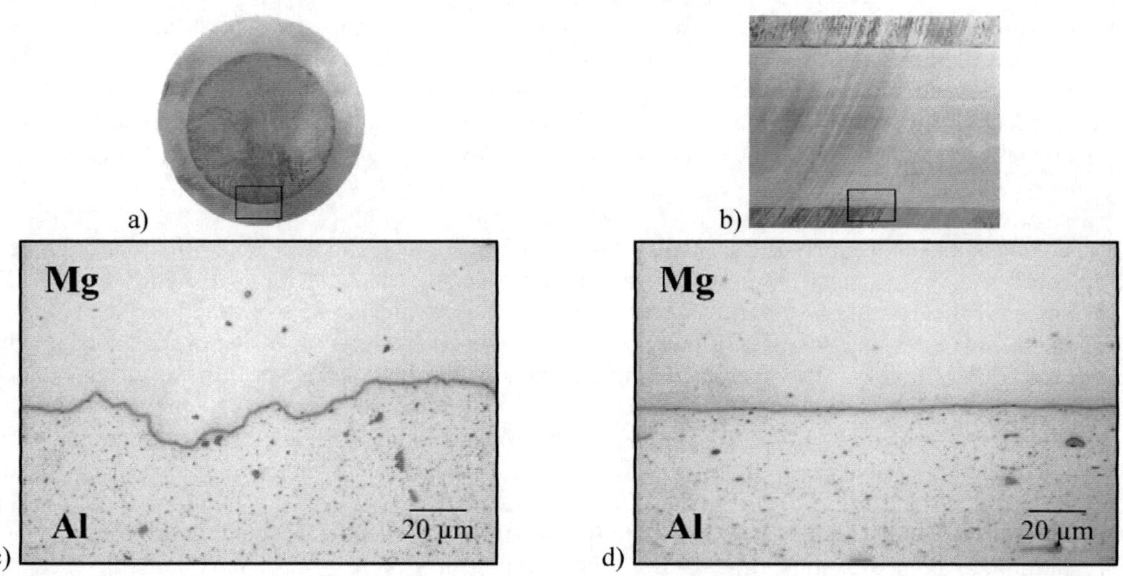


**Figure 4.** Microstructural investigations confirmed the good bonding quality: a) transversal and b) longitudinal cross section, c) and d) corresponding microstructures.

*2.3.2. Forging.* During the forging process no peculiarities could be observed compared to forging with a centric core arrangement. After cutting the forged compounds, a dye penetrant test was performed on the longitudinal and transversal cross section. The transversal cross section shows no macroscopic damage of the interface at the observed area. But in the longitudinal cross section a damage of the interface is indicated in the $s_{min}$ region (figure 5b). Due to the material flow in longitudinal direction the interface is fragmented as it has been found by FEUERHACK [2]. He described the fragmentation process of compounds with a centric core arrangement by four levels. The indicated damage in figure 5b corresponds to level three in his classification. In this level, the previously fragmented interface is elongated due to the material flow in longitudinal direction. Thereby, the fragments drift away from each other and voids develop between them. If the forming pressure and elongation are high enough, the materials are pressed into the voids again [2]. In the present case, the forming process is completed before the materials can be pressed into this voids as shown in figure 5d and 5e. Thus, they are detected by the dye penetrant test.

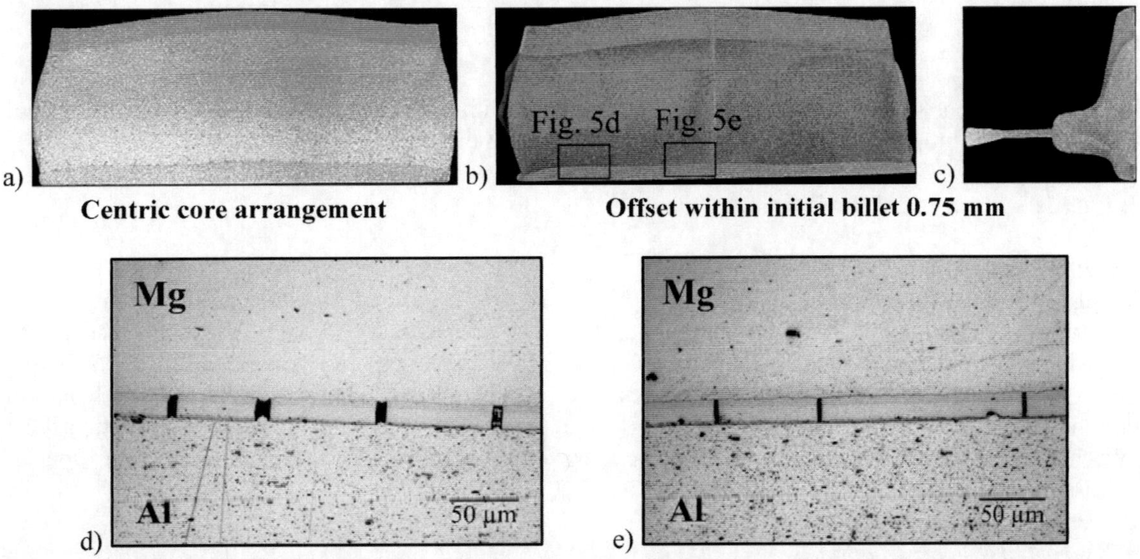


**Figure 5.** Comparison between the cross sections of a Rising specimen with a) centric core arrangement and b-c) offset of 0.75 mm within initial billet show a shifted Mg-core, d-e) Microstructures of fragmented interface areas detected by dye penetrant testing.

Comparing the longitudinal cross sections of a component with no offset (figure 5a) and with an offset within the initial billet of 0.75 mm (figure 5b), it can be seen, that the Mg-core is shifted. The critical flank thickness in the component with centric core arrangement is in range of 160 to 170 µm. For the component with an offset of 0.75 mm within the initial billet the flank thickness is in range of 400 to 450 µm (figure 5c). Thus, the flank thickness could be increased by the eccentric core arrangement.
The results of the comparison between simulation and experimental results are in a good conformity. For a better comparison a model based on STL (Standard Tessellation Language) was derived from the CT-data, as well as from the geometry of the simulation results. Figure 6 shows the outline of the component cross sections in transversal (figure 6a) and longitudinal (figure 6b) direction. The blue lines indicate the real component identified by CT. The red lines indicate the component based on the simulation results. At the transversal cross section it can be seen, that the outer outlines do not match exactly. This is due to information losses during data transformation and a limited precision in the positioning of both datasets. The differences are about 0.1 mm, which is very small. The modelled material behavior is based on flow curves identified by compression tests at three different temperatures and strain rates. Nevertheless, it cannot reproduce the real material flow exactly. Thus, a

higher deviation of both datasets occurs in the upper region and in the flash region. Additionally, in FE-simulations, the material behavior is supposed to be more rigid than in reality. The deviation can also be seen at the material flow in longitudinal direction. Due to this problem the critical flank thickness in the simulation is thinner than in the experiment. The deviation of the resulting geometries of simulation and experiment at this point is about 0.1 mm. In general the deviations are in range of 0.1 to 0.6 mm depending on the position of the measurement.

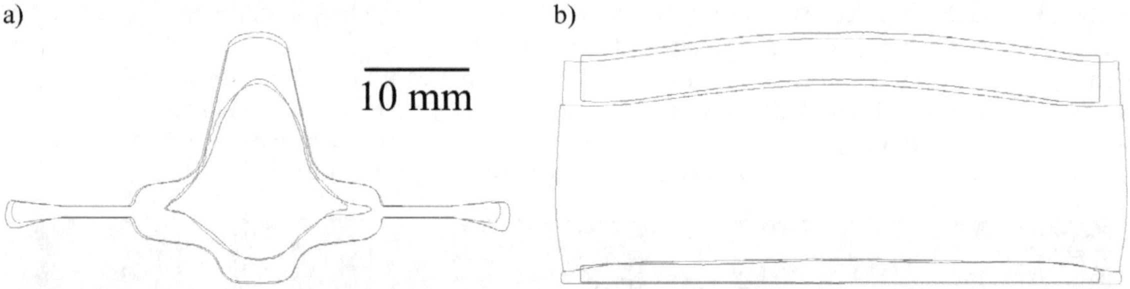


**Figure 6.** Comparison of the datasets of a component with an offset of 0.75 mm within the initial billet, blue lines indicate the CT-data, and red lines indicate the simulation model.

## 3. Part II – Strength of the interface

### *3.1. Introduction*
For the further manufacturing of the forged billet and application of the structural component it is essential to achieve a sufficient bonding. Therefore, it was necessary to analyze the material in the forged state. In the presented investigations, Rising specimens described in paragraphs 2.2.2. and 2.3.2. were analyzed. Interlaminar tensile stresses are particularly critical for the interface. Hence, in previous investigations special bending tests were used to characterize the tensile strength of the interface [3, 4]. For that reason, a new bending configuration which is adapted to the special requirements was developed and applied.

### *3.2. Geometry reconstruction*
For the conception of the load case and the specimen geometry it was required to know the geometry of the Rising specimen. Thus, investigations using computer tomography (CT) were performed. In these analyses Rising specimen with an offset of 0.75 mm of the initial billet were examined. An example of the primary results of the CT investigation is given in figure 7.

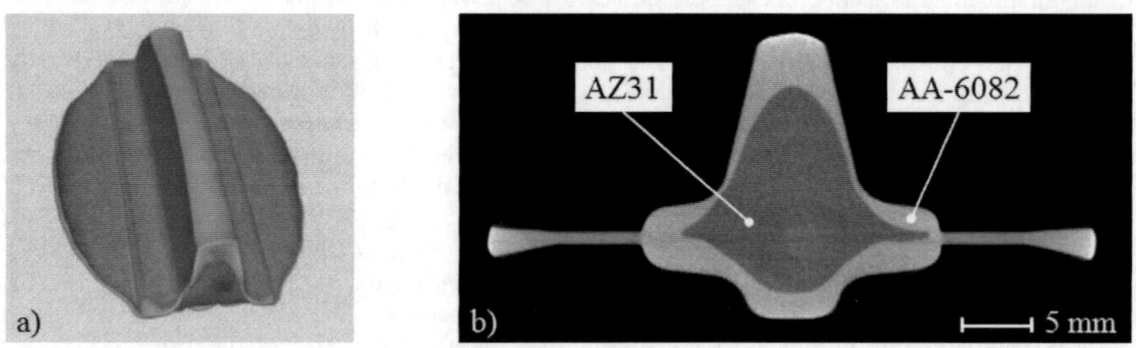


**Figure 7.** Example of CT data of a Rising specimen, offset of the initial billet: 0.75 mm: a) 3D model, b) Transversal cross section in the middle part (CT analyses performed by CWM GmbH, Chemnitz, CT scanning resolution: voxel edge length ≈ 55 µm).

As final result, the geometrical data were transformed into a 3D STL model of the core and sleeve. Hence, beside the outer contour, the interface geometry can be determined. The STL file was imported into the software Paraview 4.3.1 for analysis and editing to compare the geometry with the results of the simulation and to develop the bending specimen geometry. The comparison of the CT and numerical geometry is given in paragraph 2.3.2. An example of the analyzed transversal cross section of the Rising specimen and the position of specimen extraction is shown in figure 8. The general design and proportions of the bending specimen were developed using numerical simulations, see paragraph 3.3. Based on one exemplary reconstructed geometrical data, a parametrical finite element model of the bending specimen was configured. For the determination of strength values the measured transversal geometry of the specimens after manufacturing was used.

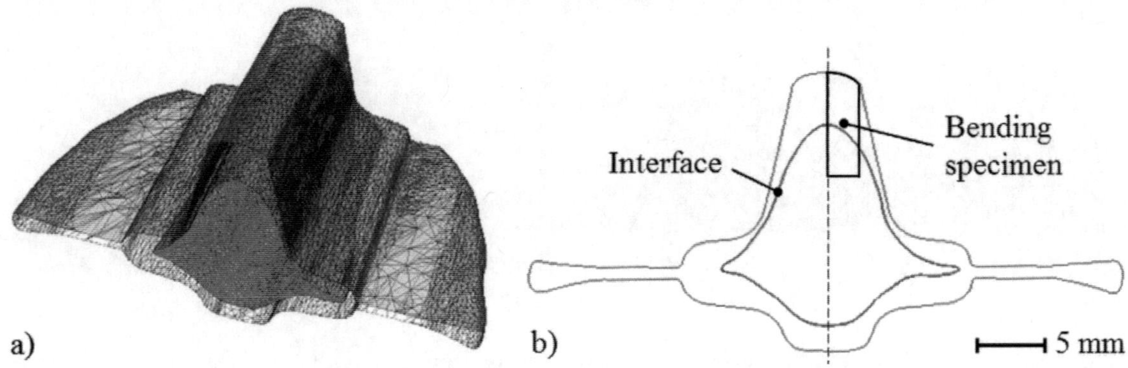


**Figure 8.** Example of reconstructed STL data of a Rising specimen in Paraview, offset of the initial billet: 0.75 mm: a) Separated 3D model, b) Transversal cross section in the middle part, position of the bending specimen.

### *3.3. Development of load case and specimen geometry*

Generally, many configurations of load cases are conceivable for the determination of the interface strength. In previous investigations different specimen geometries and loading devices were applied, see [3, 4]. Every version of bending configurations considered the requirements of the respective geometry of the semi-finished product or the further processed component. In case of the Rising specimen presented in this paper, the interface geometry is complex. The wall thickness of the sleeve is strongly varying within the transversal cross section. Therefore, it was required to find a part for an applicable extraction of bending specimen. For all analyzed configurations, the section adjacent to the middle section (in longitudinal direction) in the upper part of the Rising specimen was chosen, see figure 8b. A small specimen width enables a separation of three specimens from the limited middle section. Due to the small cross section of the bending specimen geometry it was reasonable to use a punch loading with supporting bearings without pre-stressing instead of the clamping system which was used in [3]. For that reason, a symmetrical three point bending loading device was developed.

The conception of the load case and the specimen geometry was performed using parametric finite element analyses (FEA) with ANSYS 14.0. The 3D model includes a monolithic specimen based on the reconstructed geometrical data with linear elastic material behavior (one material: elastic parameters of Al). This approach provides a stress value which occurs in the direct adjacency of the interface. The complicated stress distribution due to stress singularities at the interface edge is not included. Analytical investigations to the Al-Mg interface edge resulted in very weak stress singularities [3]. Thus, it can be concluded that this approach is legitimate. The criteria for developing the load case and dimensioning the specimens which were considered in the FEA are: high interlaminar tensile stress at the interface, limited v. Mises equivalent stress in the adjacent region of the load application and bearings and a controllable handling of the specimens while manufacturing and test preparation. The resulting FE model including the symbolic boundary conditions as well as an example of the longitudinal normal stress distribution are given in figure 9. Although the model includes only one material (Al), the interface contour is considered. Beside the calculation of the interfacial stress at the bottom of the specimen, the modeling of the interface contour enables further stress evaluations (not presented in this paper). For the variation of the specimen geometry the parameters of the model can be adapted. The maximum tensile stress is located at the interface at the bottom, due to the symmetrical configuration. As a result of the deviation from the elementary bending theory it is shown that the position of the neutral axis is varying depending on the x-coordinate (not constant in the middle of the specimen height). Furthermore, the parametric numerical model is the basis for the interface strength determination presented in paragraph 3.5.

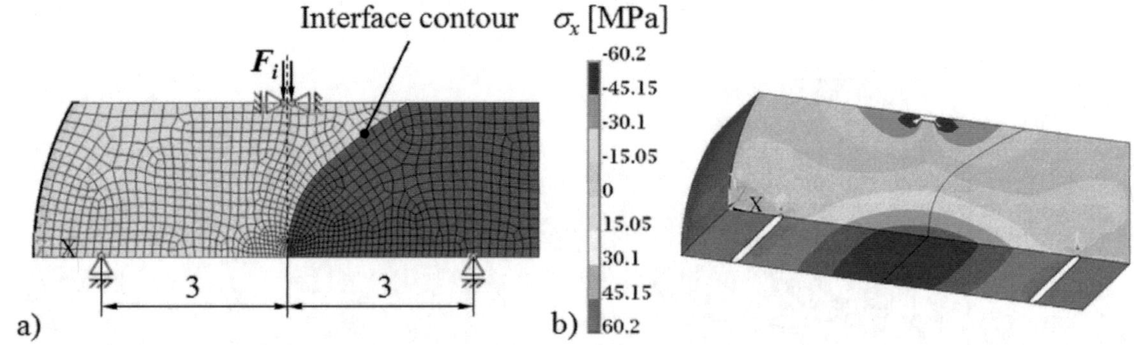


**Figure 9.** a) 3D FE model, loadcase, b) Stress distribution σx on the surface of the specimen (minimum stress is limited to -60.2 MPa, white areas).

*3.4. Experimental setup, preparing and performing of the bending tests*

The tests were performed using Rising specimens with an offset of 0.25, 0.50 and 0.75 mm of the initial billets. Three bending specimens were separated from the middle section of each analyzed Rising specimen by wire eroding with subsequent grinding (example, see figure 10a). Furthermore, a maximum of three Rising specimens per variation of the offset value were investigated. To detect possible pre-existing defects at the interface, dye penetrant tests were performed. By means of damage detection some specimens had to be separated out which could not be mechanical tested. The material with the offset of 0.5 mm showed a particularly high damage ratio.

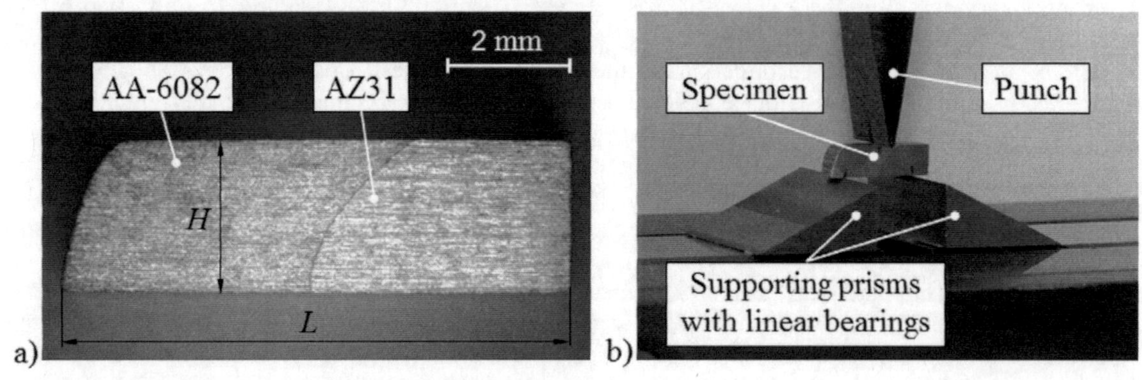


**Figure 10.** a) Example of a bending specimen W x H x L= 2.5 x 2.5 x 8.5 mm (W...width) – nominal geometry, b) Experimental setup of the bending test.

The bending device which is shown in figure 10b was assembled in a 100 kN testing machine of Zwick/Roell, using a 5 kN load cell. Two linear bearings were used to provide the displacement degree of freedom in longitudinal direction (x-direction) of the specimen. To support the specimen at defined sections, two prisms are bonded with the top of the linear bearings. The loading punch is connected to the traverse of the testing machine. To achieve a sufficient accuracy, the position of load application (punch) and bearings (supporting prisms) is marked on the surface of the small specimen. The bending tests were performed at room temperature with a punch velocity of 3 mm/min until interface failure occurs. As a consequence of different offset values, the sleeve thickness at the bottom of the specimens is in a range of 3.25 to 4.15 mm.

### *3.5. Results and discussion*

After reaching the maximum force value at interface failure, an abrupt force decrease occurred, which is an indication of brittle failure behavior. It correlates with the previous results of interface strength investigations with similar Al-Mg compounds at room temperature [3, 4]. The experimental determined maximum forces are used as input value for strength determination. Furthermore, the specific transversal geometry of the specimens which is varying because of the individual grinding process was considered. For the strength calculation the FE model of the bending test was used. In table 2, the results of the interface strength (including the average value $\sigma_{IF\ av}$) for the different offsets 0.2, 0.50 and 0.75 mm of the initial billets are given. The average strength of the specimens with offset 0.75 mm shows a higher level compared to the material with an offset of 0.5 and 0.25 mm. Generally, high variation of the interface strength between specimens of the same offset configuration is observed. The lower strength values (e.g. 11 MPa, see table 2) could indicate that these specimens are already damaged at the inner section of the interface which could not be detected by dye penetrant tests. Furthermore, the average interface strength is in the range of the values determined in [4]. Additional to the fundamental investigations in [3], by means of the presented Rising tests it was demonstrated that the interface of the co-extruded rods can be successfully deformed in further manufacturing steps.

**Table 2.** Determined interface strength $\sigma_{IF}$ of Al-Mg specimens forged by Rising with different offsets of the initial billets.

| Offset [mm] | Number of bending specimens | $\sigma_{IF\ min}$ [MPa] | $\sigma_{IF\ max}$ [MPa] | $\sigma_{IF\ av}$ [MPa] |
|---|---|---|---|---|
| 0.25 | 7 | 11 | 66 | 39 |
| 0.5 | 2 | 37 | 49 | 43 |
| 0.75 | 9 | 36 | 89 | 61 |

## 4. Summary and outlook

With the new process eccentric hydrostatic co-extrusion, rods with an offset of center axes up to 0.75 mm could be produced. Investigations on the bonding quality by ultrasonic testing and optical microscopy showed a complete bonding as expected. Further processing by the forging process Rising revealed a good formability of these compounds. In addition, the critical flank thickness could be increased from about 170 µm to 450 µm (compounds with offset 0.75 mm). A comparison of FE-simulation an experimental results show a very good conformity. Based on this results, the offset will be increased up to 1.25 mm in future investigations.

The interface strength of material forged by Rising was analyzed using a new developed bending test which was configured by means of FEA and geometrical investigations. The results of the bending tests show a wide range of strength values up to 89 MPa. Compared to compounds with an offset of 0.25 and 0.5 mm the material with offset 0.75 mm shows higher strength values. Subject of further investigations could be the analysis of strength variations at different positions in the Rising specimens. Furthermore, possible reasons of the strength variations should be examined.

**Acknowledgments**
The authors thank the German Research Foundation (DFG) for its funding within the framework of the Collaborative Research Centre 692 (SFB HALS 692).

**References**

[1] Kittner K, Feuerhack A, Förster W, Binotsch C and Graf M 2015 Recent developments for the production of Al-Mg compounds *Mater. Today:. Proc.* 2S pp 225–232
[2] Feuerhack A 2014 Experimentelle und numerische Untersuchungen von Al-Mg Verbunden mittels Verbundschmieden *PhD-Thesis TU Chemnitz* (online: http://nbn-resolving.de/urn:nbn:de:bsz:ch1-qucosa-146654)
[3] Lehmann T 2012 Experimentell-numerische Analyse mechanischer Eigenschaften von Aluminium/Magnesium-Werkstoffverbunden *PhD-Thesis TU Chemnitz* (online: http://nbn-resolving.de/urn:nbn:de:bsz:ch1-qucosa-100576)
[4] Kirbach C, Stockmann M, Ihlemann J 2014 Experimental and numerical investigations of Al/Mg-compound specimens under bending after mass forming processes *Proceedings of the 31st Danubia-Adria Symposium on Advances in Experimental Mechanics* pp 32–33

IOP Conf. Series: Materials Science and Engineering **118** (2016) 012033 doi:10.1088/1757-899X/118/1/012033

# The effect of plasma electrolytic oxidation on the mean stress sensitivity of the fatigue life of the 6082 aluminum alloy

**L Winter[1], R Morgenstern[1], K Hockauf[1] and T Lampke[1]**

[1] Materials and Surface Engineering Group, Institute of Materials Science and Engineering, Technische Universität Chemnitz, 09125 Chemnitz, Germany

E-Mail: lisa.winter@mb.tu-chemnitz.de

**Abstract.** In this work the mean stress influence on the high cycle fatigue behavior of the plasma electrolytic oxidized (PEO) 6082 aluminum alloy (AlSi1MgMn) is investigated. The present study is focused on the fatigue life time and the susceptibility of fatigue-induced cracking of the oxide coating and their dependence on the applied mean stress. Systematic work is done comparing conditions with and without PEO treatment, which have been tested using three different load ratios. For the uncoated substrate the cycles to failure show a significant dependence on the mean stress, which is typical for aluminum alloys. With increased load ratio and therefore increased mean stress, the fatigue strength decreases. The investigation confirms the well-known effect of PEO treatment on the fatigue life: The fatigue strength is significantly reduced by the PEO process, compared to the uncoated substrate. However, also the mean stress sensitivity of the fatigue performance is reduced. The fatigue limit is not influenced by an increasing mean stress for the PEO treated conditions. This effect is firstly shown in these findings and no explanation for this effect can be found in literature. Supposedly the internal compressive stresses and the micro-cracks in the oxide film have a direct influence on the crack initiation and growth from the oxide film through the interface and in the substrate. Contrary to these findings, the susceptibility of fatigue-induced cracking of the oxide coating is influenced by the load ratio. At tension-tension loading a large number of cracks, which grow partially just in the aluminum substrate, are present. With decreasing load ratio to alternating tension-compression stresses, the crack number and length increases and shattering of the oxide film is more pronounced due to the additional effective compressive part of the load cycle.

## 1. Introduction

Aluminum alloys are widely used for lightweight structures in the automotive and aircraft industries due to their high specific strength and fatigue resistance. To improve their poor tribological behavior, anodic oxidation is a well known method to meet the requirements. By plasma electrolytic oxidation (PEO) treatment excellent wear and high corrosion resistance can be achieved. Improved mechanical properties, i.e. hardness and E-modulus, a good adhesion on the substrate and also good biocompatibility are positive characteristics of PEO coatings [1–3].

Literature data to fatigue behavior of PEO treated aluminum alloys are limited, but the existing investigations show an inferior fatigue performance in comparison to the uncoated substrate [4–7]. With increased coating thickness this effect is more pronounced [6,7]. The loss in fatigue strength is attributed to the compression residual stresses in the coating which initiate tensile residual stresses in the substrate and lead to early crack initiation and fatigue fracture [4,6].

The high cycle fatigue performance usually depends on the applied load ratio and therefore the attached mean stress [8]. With increasing mean stress the fatigue strength is reduced. This effect can be presented graphically in Haigh or Smith diagrams. As numeric method, the mean stress influence factor M can be calculated according to Schütz [9]:

$$M = \frac{\sigma_{ad}(R=-1)}{\sigma_{ad}(R=0)} - 1 \quad (1)$$

With the fatigue strength amplitudes at the load ratios R = -1 and R = 0 a factor M between 0 (fatigue strength is not sensitive to mean stress) and 1 (major influence of mean stress on fatigue strength) is calculated.

The fatigue strength of uncoated aluminum alloys shows a significant dependence on the load ratio R [10,11]. To the best of our knowledge, the effect of plasma electrolytic oxidation on the mean stress sensitivity of fatigue life has not been investigated, yet. However, Shazad et al. [12] examined the influence of load ratio on the fatigue performance for an electrolytic anodic oxidized and sealed 2214 aluminum alloy. In their study, the presence of the anodic oxide film did not significantly modify the effect of the mean stress in comparison to the uncoated substrate.

The purpose of the present study is to investigate, which effect PEO treatment has on the mean stress sensitivity of the fatigue strength. Further, the mean stress influence on the susceptibility for cracking of the anodic film is examined. As substrate material the 6082 aluminum alloy was chosen, as it is widely used in structural applications and anodic oxidation processes can be conducted without complications.

## 2. Experimental Methods

For this study the age-hardening 6082 aluminum alloy (AlSi1MgMn) of commercial purity was used. The chemical composition is shown in table 1. The material was solid-solution treated at 530 °C for 1 h, quenched in water and subsequently aged at 170 °C for 65 h to achieve peak strength. Mechanical properties determined by tensile testing are given in table 2.

**Table 1.** Chemical composition of the 6082 aluminum alloy (AlSi1MgMn).

| Element | Si | Fe | Cu | Mn | Mg | Zn | Al |
|---|---|---|---|---|---|---|---|
| wt.-% | 1.00 | 0.21 | 0.05 | 0.76 | 0.90 | 0.02 | balance |

**Table 2.** Tensile properties of the uncoated peak-aged 6082 aluminum alloy.

| Yield strength in MPa | Ultimate tensile strength in MPa | Uniform elongation in % | Elongation to failure in % |
|---|---|---|---|
| 297 ± 1 | 307 ± 4 | 5.2 ± 0.8 | 23.7 ± 0.9 |

Prior to PEO, fatigue testing specimens with a geometry shown in figure 1 were machined from the peak-aged substrate.

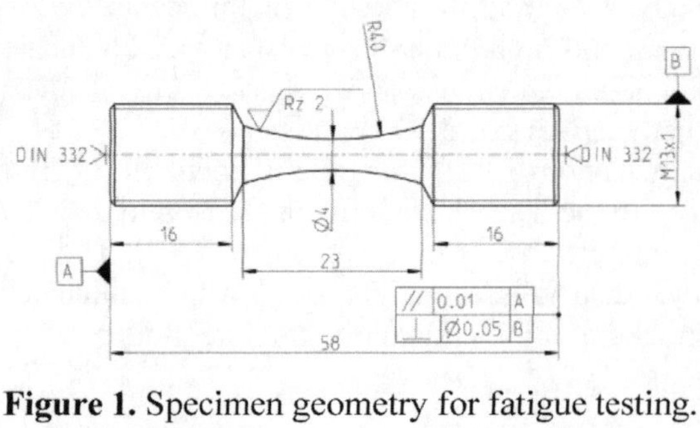


**Figure 1.** Specimen geometry for fatigue testing.

In preparation for the PEO treatment, the peak-aged aluminum alloy was pickled using 3 vol.-% aqueous sodium hydroxide solution (NaOH) for 1 min at 50 °C and subsequent 10 vol.-% aqueous nitric acid ($HNO_3$) for 30 s at room temperature. For the anodic oxidation process, a silicate-hydroxide-electrolyte, which contained 5 g/l sodium metasilicate ($Na_2SiO_3 \cdot 5H_2O$) and 5 g/l potassium hydroxide (KOH) in aqueous solution, was used. The PEO treatment was carried out under a symmetrical rectangular bipolar current regime with a current density of 30 A/dm² and a frequency of 50 Hz. The temperature of the electrolyte was kept constant at 20 °C and the duration of the process was 45 min. Afterwards the coated specimens were rinsed in water and dried in air.

The fatigue testing was performed in a RUMUL Testronic resonant testing machine (Russenberger Prüfmaschinen AG, Switzerland) under alternating compression-tension and tension-tension loading with load ratios of R = -1, R = 0 and R = 0.1. The fatigue tests were carried out until the endurance limit of $N_D = 10^7$ cycles was reached or until a crack occurred, which was detected by a drop in the resonant frequency of 2 Hz or more.

After fatigue testing from each specimen, longitudinal- and cross-sections were cut from the gauge length and the coating thickness and hardness was determined at the metallographically polished samples. Measurement of the coating thickness was done via optical microscopy with an Olympus GX51 (Olympus Deutschland GmbH, Germany). Hardness measurements were performed on a Fischerscope HM 2000XYm (Helmut Fischer GmbH + Co. KG, Germany). Afterwards, the polished samples were analyzed by scanning electron microscopy (SEM) on a LEO 1455VP (LEO Elektronenmikroskopie GmbH, Germany) to examine coating damage through fatigue loading. The quadrant backscattering detector (QBSD) was used for coating characterization and visualization of coating irregularities and cracks.

## 3. Results and Discussion

Table 3 lists the measured coating thickness and hardness. A coating thickness of about 45 µm was achieved. The non-uniformity in oxide film thickness, which shows in minimum and maximum values of 14.1 and 103.9 µm, respectively, was also observed in [4]. The large deviation of the coating thickness is owed to the PEO process and the short oxidation time, which was chosen in order to generate a thin oxide film and therefore to reduce the negative effect on fatigue strength. With longer PEO process time and therefore a thicker oxide film, the waviness of the PEO film can be reduced but the negative influence on fatigue strength would be enhanced. The coating exhibits a hardness of about 7880 MPa.

**Table 3.** Properties of the PEO coating on the 6082 aluminum alloy.

| | Average | Minimum | Maximum |
|---|---|---|---|
| **Coating thickness in µm** | 47.3 | 14.1 | 103.9 |
| **Martens hardness in MPa** | 7881.3 | 5840.3 | 9661.1 |

The fatigue behavior of the aluminum alloy in uncoated and plasma electrolytic oxidized condition and its dependence on the load ratio is shown in figure 2. In table 4 the fatigue limit for $N_D = 10^7$ cycles is listed. The uncoated aluminum substrate exhibits a fatigue strength amplitude of 120 MPa at alternating tension-compression stresses. With increasing load ratio to R = 0.1 and therefore increased mean stress, the stress amplitude is reduced to 90 MPa. This significant influence of the mean stress on the fatigue life time for the aluminum substrate is in good accordance to the investigations from Zhao et al. [10] and Mayer et al. [11].

Plasma electrolytic oxidation leads in general to a decreased fatigue life time, which is explained by the internal compressive stresses in the PEO film, which, in turn, results in tensile stresses in the adjacent substrate [4–7]. Our investigation confirms this effect. For R = -1 the fatigue limit is decreased by 50 %, as compared to the uncoated substrate. Contrary to the uncoated substrate, the load ratio and therefore the applied mean stress does not influence the fatigue life time of the PEO coated substrate. The fatigue strength amplitudes at the load ratios R = 0 and R = 0.1 are just as high as for alternating compression-tension stresses.

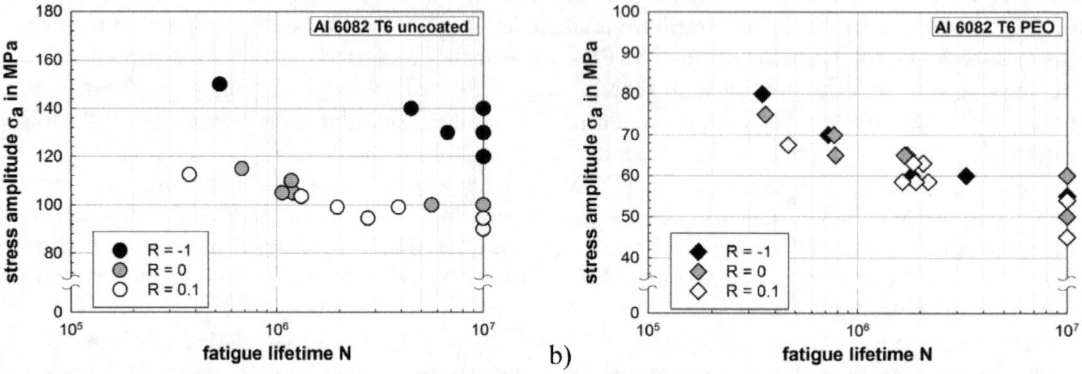

a) b)

**Figure 2.** Fatigue behavior of the 6082 aluminum alloy depending on the load ratio: a) uncoated, b) plasma electrolytic oxidized (PEO).

**Table 4.** Fatigue limit at $N_D = 10^7$ cycles of the uncoated and PEO coated 6082 aluminum alloy at the tested load ratios.

| | load ratio R | Maximum stress $\sigma_o$ in MPa | Minimum stress $\sigma_u$ in MPa | Stress amplitude $\sigma_a$ in MPa | Mean stress $\sigma_m$ in MPa |
|---|---|---|---|---|---|
| **uncoated** | -1 | 120 | -120 | 120 | 0 |
| | 0 | 180 | 0 | 90 | 90 |
| | 0.1 | 200 | 20 | 90 | 110 |
| **plasma electrolytic oxidized** | -1 | 55 | -55 | 55 | 0 |
| | 0 | 120 | 0 | 60 | 60 |
| | 0.1 | 120 | 12 | 54 | 66 |

The Haigh diagram depicted in figure 3 presents the fatigue limit at $N_D = 10^7$ cycles depending on the mean stress for the uncoated substrate and the PEO treated condition. The mean stress sensitivity factor M is indicated by the slope of the secant of the data points for the tested conditions. It is clearly visible, that the mean stress influence on the fatigue strength is diminished by PEO. The uncoated substrate exhibits a mean stress sensitivity of M = 0.2, calculated according to equation 1, which is typically for aluminum alloys [8,13].

By the PEO treatment the mean stress influence is neutralized to M = 0. For this condition our results show a fatigue stress amplitude which is unattached by the load ratio and therefore independent from the applied mean stress. This decrease in mean stress sensitivity through PEO treatment is firstly shown in these findings and for this effect no explanation can be found in literature. We cannot conclude with certainty, but supposedly the neutralized mean stress sensitivity is linked to the oxide film characteristic, especially the present micro-cracks and the residual stresses in the PEO film. It is well-known, that PEO films exhibit internal compressive stresses [4,14,15] and we assume these have direct influence on crack initiation and growth from the oxide film through the interface and in the substrate.

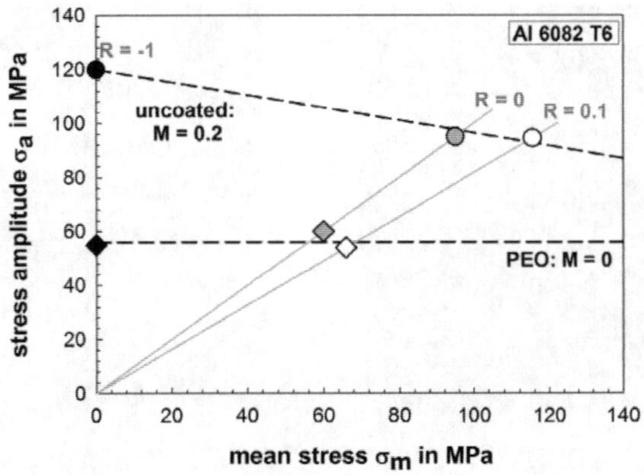


**Figure 3.** Haigh diagram representing the fatigue strength amplitudes at $N_D = 10^7$ cycles for the un-coated and the PEO coated condition at the tested load ratios. The slope of the secant indicates the mean stress sensitivity factor M.

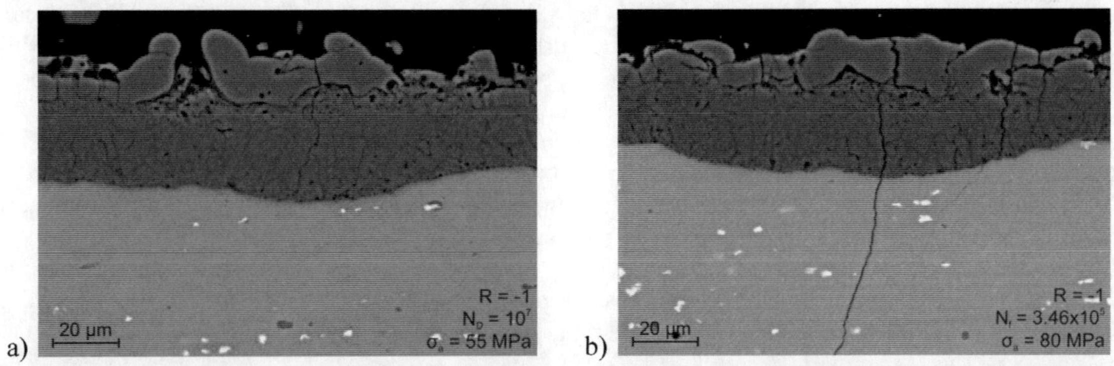


**Figure 4.** SEM micrographs (QBSD) of the PEO treated 6082 aluminum alloy after fatigue testing at load ratio R = -1: a) runout (non-failure), b) fatigued specimen.

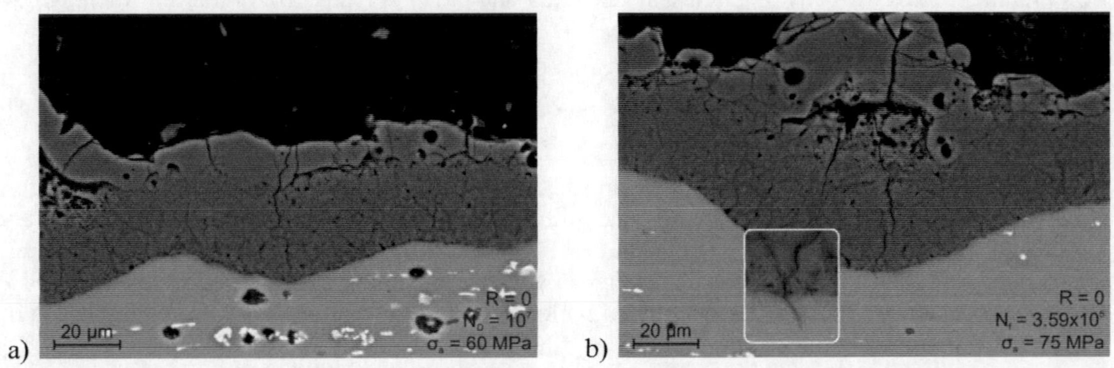


**Figure 5.** SEM micrographs (QBSD) of the PEO treated 6082 aluminum alloy after fatigue testing at load ratio R = 0: a) runout (non-failure), b) fatigued specimen.

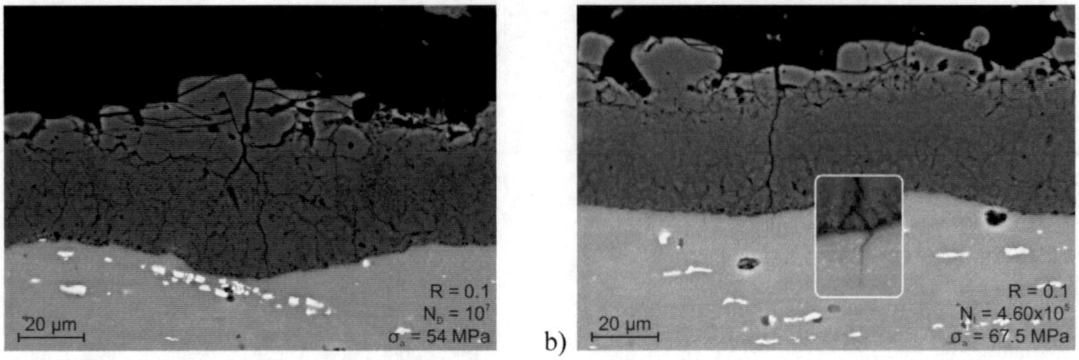


a) b)

**Figure 6.** SEM micrographs (QBSD) of the PEO treated 6082 aluminum alloy after fatigue testing at load ratio R = 0.1: a) runout (non-failure), b) fatigued specimen.

SEM micrographs (QBSD) in the figures 4-6 show the PEO film after fatigue testing at three different load ratios. The oxide film has a porous nature and branched micro-cracks are clearly visible in all tested specimens, even despite failure through fatigue did not occur (see figures 4-6 a)). The appearance of the oxide film for the fatigued specimen differs, dependent on the load ratio. At alternating tension-compression stresses a large amount of fatigue-induced radial cracks are present, which grow from the oxide film just into the substrate. A few major cracks grow long-ranging in the substrate and lead to failure (shown in figure 4 b)). With increased load ratio and therefore higher applied tension stresses, the number of cracks is still high, but, contrary to the fatigued PEO treated specimens at R = -1, only short-ranging cracks are present. It cannot be concluded with certainty if this crack length effect is fatigue-induced or originated by preparation and the considered sample planes. But under fully reversed tension-compression loading at R = -1, the oxide film exhibits major shattering due to additional fatigue damage through the compression part of the load cycle. That is why we expect that the majority of these cracks originates from the fatigue testing.

## 4. Summary and Conclusions

The effect of plasma electrolytic oxidation on the mean stress sensitivity of the fatigue life time of the 6082 aluminum alloy was investigated. An uncoated and a PEO treated condition and their fatigue strength at three different load ratios was compared. The following conclusions can be drawn:

(1) The fatigue strength is drastically reduced by PEO treatment at all tested load ratios. The loss in fatigue performance can be attributed to the porous nature of the oxide film, which promotes early crack initiation.

(2) The PEO treatment neutralizes the mean stress influence on the fatigue life time and therefore the fatigue strength is unattached by the load ratio. These results are firstly shown in this paper and no explanation for this effect can be found in literature. Supposedly the neutralized mean stress sensitivity is linked to the present micro-cracks and the internal compressive stresses in the PEO film which have direct influence on crack initiation and growth from the oxide film through the interface and in the substrate.

(3) The susceptibility for fatigue induced cracking of the oxide film is influenced by the load ratio. Fatigue loading at R = 0 and R = 0.1 causes a high amount of radial cracks in the coating, which grow partially just in the substrate. Under alternating tension-compression stresses, the number of cracks and their length increases. Shattering of the oxide film is more pronounced then, due to the fatigue damage through the tension and additionally through the compression part of the load cycle.

**Acknowledgements**
This work was financially supported by the German Research Foundation (Deutsche Forschungsgemeinschaft, DFG) in the framework of Collaborative Research Centre (Sonderforschungsbereich, SFB) 692 "High-strength aluminium-based lightweight materials for safety components". The assistance of Dagmar Dietrich, Christel Gläser, Juliane Mehnert, Karla Muhr and Gabriele Tauchmann is gratefully acknowledged.

**References**

[1] Godja N, Kiss N, Löcker C, Schindel A, Gavrilovic A, Wosik J, Mann R, Wendrinsky J, Merstallinger A and Nauer G 2010 *Tribol. Int.* **43** 1253
[2] Khan R H U, Yerokhin A Li X, Dong H and Matthews A 2010 *Surf. Coatings Technol.* **205** 1679
[3] Malayoglu U, Tekin K C, Malayoglu U and Shrestha S 2010 *Mater. Sci. Eng. A* **528** 7451
[4] Lonyuk B, Apachitei I and Duszczyk J 2007 *Surf. Coatings Technol.* **201** 8688
[5] Rajasekaran B, Ganesh Sundara Raman S, Rama Krishna L, Joshi S V and Sundararajan G 2008 *Surf. Coatings Technol.* **202** 1462
[6] Wasekar N P, Ravi N, Suresh Babu P, Rama Krishna L and Sundararajan G 2010 *Metall. Mater. Trans. A* **41** 255
[7] Dejun K, Hao L and Jinchun W 2015 *J. Alloys Compd.* **650** 393
[8] Haibach E 2006 *Betriebsfestigkeit* (Berlin Heidelberg: Springer)
[9] Schütz W 1967 *Zeitschrift für Flugwissenschaften* **15** 407
[10] Zhao T and Jiang Y 2008 *Int. J. Fatigue* **30** 834
[11] Mayer H, Schuller R and Fitzka M 2013 *Int. J. Fatigue* **57** 113
[12] Shahzad M, Chaussumier M, Chieragatti R, Mabru C and Rezai-Aria F 2012 *Surf. Coatings Technol.* **206** 2733
[13] Ostermann F 2014 *Anwendungstechnologie Aluminium* (Berlin Heidelberg: Springer)
[14] Khan R H U, Yerokhin A L, Pilkington T, Leyland A and Matthews A 2005 *Surf. CoatingsTechnol.* **200** 1580
[15] Asquith D T, Yerokhin A L, Yates J R and Matthews A 2006 *Thin Solid Films* **515** 1187

IOP Conf. Series: Materials Science and Engineering **118** (2016) 012034 doi:10.1088/1757-899X/118/1/012034

# Kinematic assumptions and their consequences on the structure of field equations in continuum dislocation theory

**C B Silbermann and J Ihlemann**

Professorship of Solid Mechanics, Technische Universität Chemnitz, Reichenhainer Straße 70, 09126 Chemnitz, Germany

E-mail: christian.silbermann@mb.tu-chemnitz.de

**Abstract.** Continuum Dislocation Theory (CDT) relates gradients of plastic deformation in crystals with the presence of geometrically necessary dislocations. Therefore, the dislocation tensor is introduced as an additional thermodynamic state variable which reflects tensorial properties of dislocation ensembles. Moreover, the CDT captures both the strain energy from the macroscopic deformation of the crystal and the elastic energy of the dislocation network, as well as the dissipation of energy due to dislocation motion. The present contribution deals with the geometrically linear CDT. More precise, the focus is on the role of dislocation kinematics for single and multi-slip and its consequences on the field equations. Thereby, the number of active slip systems plays a crucial role since it restricts the degrees of freedom of plastic deformation. Special attention is put on the definition of proper, well-defined invariants of the dislocation tensor in order to avoid any spurious dependence of the resulting field equations on the coordinate system. It is shown how a slip system based approach can be in accordance with the tensor nature of the involved quantities. At first, only dislocation glide in one active slip system of the crystal is allowed. Then, the special case of two orthogonal (interacting) slip systems is considered and the governing field equations are presented. In addition, the structure and symmetry of the backstress tensor is investigated from the viewpoint of thermodynamical consistency. The results will again be used in order to facilitate the set of field equations and to prepare for a robust numerical implementation.

## 1. Introduction

The plastic behavior of metallic crystals strongly depends on the underlying defect structure. For a broad class of metals it is the motion of dislocations that carries plastic deformation and new macroscopic properties (e.g. hardening mechanisms) emerge from collective self-organization of dislocations. Among the resulting dislocation structures, cell patterns are often observed and seem to be of special relevance. In order to simulate the mechanical behavior of metals with dislocation cells, a model was developed which operates with scalar dislocation densities. This micro model was incorporated into a macroscopic viscoplasticity framework that allows studying the effects of complex load path changes [1]. Still, a desired validation of model assumptions as well as a reasonable estimation of introduced microstructural parameters requires a lower scale theory.

An appropriate means is the Continuum Dislocation Theory (CDT) (cf. e.g. [2-6]). In contrast to approaches such as [1], measures of dislocation density and energy are now related to gradients of the plastic distortion. Furthermore, CDT incorporates the size effect, i.e. the material response depends on the size of the body due to some inherent length scale. Hence, a better understanding of physical phenomena may be expected. Despite the success of the CDT, some basic questions remain still open, e.g. if the theory should be based on the dislocation density tensor or on individual slip system gradients. The present study makes a contribution towards answering this question.

## 2. Kinematics of continuously dislocated crystals

### *2.1. Basic assumptions*

The kinematics of continuously distributed dislocations in the geometrically linear theory is based on the additive split of the displacement gradient into elastic and plastic parts: $\mathrm{grad}(\underline{u}) =: \underline{\underline{\beta}} = \underline{\underline{\beta}}_{\mathrm{e}} + \underline{\underline{\beta}}_{\mathrm{p}}$. They are named elastic and plastic distortion tensor, respectively. Typical for the theory, the additional primary field is considered a result of the collective motion of continuously distributed dislocations. Since crystals have a discrete lattice with different slip systems, the plastic distortion is restricted. In order to incorporate this into the continuum theory, the following expansion is used:

$$\underline{\underline{\beta}}_{\mathrm{p}} = \sum_{n=1}^{N} \beta_n(\underline{r})\ \underline{s}_n \otimes \underline{m}_n = \sum_{n=1}^{N} \beta_n(\underline{r})\ \underline{\underline{G}}_n\ , \tag{1}$$

where $N$ is the number of slip systems and the unit vectors $\underline{s}_n$ and $\underline{m}_n$ are the slip direction and slip plane normal, respectively (figure 1). From Eq. (1) it follows that $\underline{\underline{\beta}}_{\mathrm{p}}$ is isochoric and thus has only 8 independent coordinates. This assumption is well supported by many experiments that demonstrate the incompressibility of plastic flow. However, in real crystals much more than

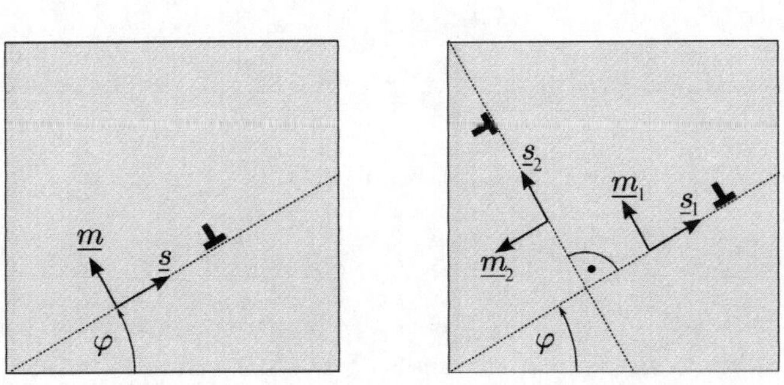


**Figure 1.** Left: Single-slip with one active slip system & straight edge dislocations (symbol **T**). Right: Multi-slip, special case of two orthogonal slip systems with straight edge dislocations

8 slip systems are present and can be active at the same time. Furthermore, the slip system tensors $\underline{\underline{G}}_n$ need not be and cannot be linearly independent in case of $N \leq 8$ and $N > 8$ slip systems, respectively. At first this might seem as an unbearable ambiguity, but a remedy becomes obvious by introducing dissipation (cf. Subsection 4.3). The resulting slip (shear) with respect to a certain shear plane is then a linear combination

$$\begin{aligned} \beta_{\mathrm{p}1} &= \underline{s}_1 \cdot \underline{\underline{\beta}}_{\mathrm{p}} \cdot \underline{m}_1 = \beta_1 + \beta_2(\underline{s}_1 \cdot \underline{s}_2)(\underline{m}_1 \cdot \underline{m}_2) + \beta_3(\underline{s}_1 \cdot \underline{s}_3)(\underline{m}_1 \cdot \underline{m}_3) + \dots\ , \\ \beta_{\mathrm{p}2} &= \underline{s}_2 \cdot \underline{\underline{\beta}}_{\mathrm{p}} \cdot \underline{m}_2 = \beta_2 + \beta_1(\underline{s}_1 \cdot \underline{s}_2)(\underline{m}_1 \cdot \underline{m}_2) + \beta_3(\underline{s}_1 \cdot \underline{s}_3)(\underline{m}_1 \cdot \underline{m}_3) + \dots\ . \end{aligned} \tag{2}$$

Separating the individual contributions $\beta_n$ is thus only possible for orthogonal slip systems where the scalar products on the right hand side of Eq. (2) vanish.

IOP Conf. Series: Materials Science and Engineering **118** (2016) 012034 doi:10.1088/1757-899X/118/1/012034

### *2.2. Preliminary consideration for a single dislocation segment*

For the sake of physical interpretation it is useful to first consider the dislocation density tensor of a single dislocation line segment, defined as

$$\underline{\underline{\alpha}} = \tfrac{\alpha(\underline{r})}{b}\,\underline{b} \otimes \underline{t} = \rho(\underline{r})\ \underline{b} \otimes \underline{t}\ , \tag{3}$$

with the crystal's Burgers Vector $\underline{b}$, its norm $b = |\underline{b}|$, the line tangent $\underline{t}$ (unit vector) and the (signed) scalar dislocation density $\rho(\underline{r})$ as a function of space. Since $\underline{\underline{\alpha}}$ is an unsymmetric rank-one tensor, the second and third principal invariant vanish, i.e. $I_2(\underline{\underline{\alpha}}) = I_3(\underline{\underline{\alpha}}) = 0$. In fact, among all six independent invariants only three remain:

$$\begin{aligned} I_1(\underline{\underline{\alpha}}) &= \underline{\underline{I}} \cdot\cdot\, \underline{\underline{\alpha}} = \rho\ (\underline{b}\cdot\underline{t}) \quad , \\ \bar{I}_2(\underline{\underline{\alpha}}) &= \tfrac{1}{2}\left\{I_1^2(\underline{\underline{\alpha}}) - \underline{\underline{\alpha}} \cdot\cdot\, \underline{\underline{\alpha}}^{\mathrm{T}}\right\} = \tfrac{1}{2}\rho^2\left\{(\underline{b}\cdot\underline{t})^2 - (\underline{b}\cdot\underline{b})\right\} \quad , \\ \bar{I}_3(\underline{\underline{\alpha}}) &= \tfrac{1}{6}\left\{I_1^3(\underline{\underline{\alpha}}) - 3I_1(\underline{\underline{\alpha}})(\underline{\underline{\alpha}} \cdot\cdot\, \underline{\underline{\alpha}}^{\mathrm{T}}) + 2\underline{\underline{\alpha}}^2 \cdot\cdot\, \underline{\underline{\alpha}}^{\mathrm{T}}\right\} = \tfrac{1}{6}\rho^3\left\{(\underline{b}\cdot\underline{t})^3 - (\underline{b}\cdot\underline{b})(\underline{b}\cdot\underline{t})\right\} = \tfrac{1}{3}I_1\bar{I}_2. \end{aligned} \tag{4}$$

Now it is very illustrative to consider both pure dislocation characters. If the dislocation line segment has screw character (denoted by $\odot$), $\underline{b}$ and $\underline{t}$ are parallel, hence:

$$I_1(\overset{\odot}{\underline{\underline{\alpha}}}) = \overset{\odot}{\rho}\ b \quad , \quad \bar{I}_2(\overset{\odot}{\underline{\underline{\alpha}}}) = 0 \quad , \quad \bar{I}_3(\overset{\odot}{\underline{\underline{\alpha}}}) = 0 \quad . \tag{5}$$

If the dislocation line segment has edge character (denoted by $\perp$), $\underline{b}$ and $\underline{t}$ are perpendicular, hence:

$$I_1(\overset{\perp}{\underline{\underline{\alpha}}}) = 0 \quad , \quad \bar{I}_2(\overset{\perp}{\underline{\underline{\alpha}}}) = -\tfrac{1}{2}\overset{\perp}{\rho}{}^2\ b^2 \quad , \quad \bar{I}_3(\overset{\perp}{\underline{\underline{\alpha}}}) = 0 \quad . \tag{6}$$

A mixed dislocation segment can always be decomposed into edge and screw part. Consequently, the corresponding densities can be calculated from the invariants:

$$\overset{\odot}{\rho} = \tfrac{1}{b}I_1(\underline{\underline{\alpha}}) \quad , \quad \overset{\perp}{\rho} = \tfrac{1}{b}\sqrt{-2\bar{I}_2(\underline{\underline{\alpha}})} \quad . \tag{7}$$

Thus, a physical significance/interpretation of the invariants of tensor $\underline{\underline{\alpha}}$ is found.

### *2.3. Properties and invariants of the dislocation (density) tensor*

The Nye-Bilby-Kröner dislocation (density) tensor $\underline{\underline{\alpha}} := \mathrm{curl}(\underline{\underline{\beta}}_{\mathrm{p}})$ reflects tensorial dislocation properties of geometrically necessary dislocations as a result of the incompatibility of plastic (or elastic) deformation. It follows from the definition of the resultant Burgers vector using Stokes' theorem:

$$\begin{aligned} \underline{b}_{\mathrm{r}} &= \oint_C \underline{\underline{\beta}}_{\mathrm{p}} \cdot \mathrm{d}\underline{r} = \int_A \mathrm{rot}\left(\underline{\underline{\beta}}_{\mathrm{p}}\right) \cdot \underline{n}\ \mathrm{d}A\ , \\ \mathrm{d}\underline{b}_{\mathrm{r}} &= \underline{\underline{\alpha}} \cdot \underline{n}\ \mathrm{d}A\ \Rightarrow\ \underline{\underline{\alpha}} := \mathrm{rot}(\underline{\underline{\beta}}_{\mathrm{p}}) \quad . \end{aligned} \tag{8}$$

From Eq. (8) it follows immediately that $\underline{\underline{\alpha}}$ is unsymmetric. Furthermore, the resulting Burgers vector is an integral quantity. However, multiplying the dislocation tensor $\underline{\underline{\alpha}}$ with some cut plane normal $\underline{n}$, a corresponding local measure $\underline{\alpha}$ is obtained:

$$\underline{\alpha}(\underline{n}) = \frac{\mathrm{d}\underline{b}_{\mathrm{r}}}{\mathrm{d}A} = \underline{\underline{\alpha}} \cdot \underline{n} = \overset{\odot}{\alpha}\ \underline{n} + \overset{\perp}{\alpha}_1\ \underline{\tau}_1 + \overset{\perp}{\alpha}_2\ \underline{\tau}_2\ . \tag{9}$$

As shown, this vector can be decomposed into normal ($\underline{n}$) and tangential parts ($\underline{\tau}_1$ , $\underline{\tau}_2$), which may be attributed to screw and edge character dislocations, respectively [4]. Nevertheless, it

IOP Conf. Series: Materials Science and Engineering **118** (2016) 012034 doi:10.1088/1757-899X/118/1/012034

should be noted that procedure (8) is in fact a homogenization leaving only geometrically necessary dislocations over. All information about statistically stored dislocations is lost. If slip system expansion (1) is applied, the dislocation density tensor has the form:

$$\underline{\underline{\alpha}} = \mathrm{curl}(\underline{\underline{\beta}}_{\mathrm{p}}) = \sum_{n=1}^{N} \underline{s}_n \otimes \underline{\nabla}\beta_n \times \underline{m}_n \; . \tag{10}$$

Note that in this study, a summation over the slip systems is always indicated by a sigma sign. Else, there is no summation convention for the slip system indices.

In order to capture the energy of a continuously dislocated crystal (cf. Subsection 3.1), physically meaningful invariants of $\underline{\underline{\alpha}}$ have to be found. As an unsymmetric second-order tensor, six independent invariants may exist. As shown in the previous subsection, the first principal invariant $I_1(\underline{\underline{\alpha}})$ is related to the dislocation density of screw segments. However, in the following only straight edge dislocations shall be considered. Since $\underline{s}_n$ and $\underline{m}_n$ are perpendicular, Cartesian basis systems $\{\underline{s}_n, \underline{m}_n, \underline{t}_n\}$ can be defined [4], where $\underline{t}_n := \underline{s}_n \times \underline{m}_n$ is the tangent vector on the edge dislocation bundle. The assumption of straight edge dislocation lines implies zero plastic slip gradients in the tangent direction, i.e $\underline{\nabla}\beta_n \cdot \underline{t}_n = 0$. This has further consequences on the properties of $\underline{\underline{\alpha}}$. Calculating the first principal invariant

$$I_1(\underline{\underline{\alpha}}) = \underline{\underline{I}} \cdot\cdot \sum_{n=1}^{N} \underline{s}_n \otimes \underline{\nabla}\beta_n \times \underline{m}_n = \sum_{n=1}^{N} \underline{s}_n \cdot (\underline{\nabla}\beta_n \times \underline{m}_n) = \sum_{n=1}^{N} \underline{\nabla}\beta_n \cdot (\underline{m}_n \times \underline{s}_n) = 0 \; , \tag{11}$$

it is found that the tensor is now indeed deviatoric (traceless). Hence, $\underline{\underline{\alpha}}$ can have only five independent invariants. Imposing further the restriction of single slip ($N = 1$), it is found that $\underline{\underline{\alpha}}$ is also second-order nil-potent, i.e. $\underline{\underline{\alpha}}^n = \underline{\underline{0}} \; \forall \; n \geq 2$, e.g.:

$$\underline{\underline{\alpha}}^2 = \underline{\underline{\alpha}} \cdot \underline{\underline{\alpha}} = (\underline{s} \otimes \underline{\nabla}\beta \times \underline{m}) \cdot (\underline{s} \otimes \underline{\nabla}\beta \times \underline{m}) = \underline{s} \otimes \{(\underline{\nabla}\beta \times \underline{m}) \cdot \underline{s}\} \otimes \underline{\nabla}\beta \times \underline{m} = \underline{\underline{0}} \; . \tag{12}$$

Thus, all basic invariants $J_n(\underline{\underline{\alpha}}) := \underline{\underline{I}} \cdot\cdot \underline{\underline{\alpha}}^n = 0$ for a crystal with only one active slip system and straight edge dislocation lines. Then, $I_2(\underline{\underline{\alpha}}) = I_3(\underline{\underline{\alpha}}) = 0$ follows immediately. Only one non-vanishing invariant remains:

$$\bar{I}_2(\underline{\underline{\alpha}}) = \frac{1}{2}\left\{I_1(\underline{\underline{\alpha}}) - \underline{\underline{\alpha}} \cdot\cdot \underline{\underline{\alpha}}^{\mathrm{T}}\right\} = \frac{1}{2}\left\{0 - \|\underline{\underline{\alpha}}\|^2\right\} \; . \tag{13}$$

Consequently, the Frobenius norm $\|\underline{\underline{\alpha}}\|$ carries all scalar information of $\underline{\underline{\alpha}}$ in this special case. Thus, the (total) dislocation density may be defined as

$$\rho = \tfrac{1}{b}\|\underline{\underline{\alpha}}\| = \tfrac{1}{b}|\underline{\nabla}\beta \times \underline{m}| = \tfrac{1}{b}|\underline{\nabla}\beta \cdot \underline{s}| \quad . \tag{14}$$

However, in the general case (10) without any restrictions, the three principal invariants do not vanish. Moreover, they contain interaction terms as can be seen e.g. in the case of two slip systems with $\underline{\underline{\alpha}} = \underline{s}_1 \otimes \underline{\nabla}\beta_1 \times \underline{m}_1 + \underline{s}_2 \otimes \underline{\nabla}\beta_2 \times \underline{m}_2$ from Eq. (10):

$$\begin{aligned} \|\underline{\underline{\alpha}}\|^2 = \underline{\underline{\alpha}} \cdot\cdot \underline{\underline{\alpha}}^{\mathrm{T}} &= \underline{\nabla}\beta_1 \otimes \underline{\nabla}\beta_1 \cdot\cdot (\underline{\underline{I}} - \underline{m}_1 \otimes \underline{m}_1) + \underline{\nabla}\beta_2 \otimes \underline{\nabla}\beta_2 \cdot\cdot (\underline{\underline{I}} - \underline{m}_2 \otimes \underline{m}_2) \\ &\quad + 2\,(\underline{s}_1 \cdot \underline{s}_2)\; \underline{\nabla}\beta_1 \otimes \underline{\nabla}\beta_2 \cdot\cdot \left(\underline{\underline{I}}(\underline{m}_1 \cdot \underline{m}_2) - \underline{m}_1 \otimes \underline{m}_2\right) \\ &= (\underline{\nabla}\beta_1 \cdot \underline{s}_1)^2 + (\underline{\nabla}\beta_2 \cdot \underline{s}_2)^2 + 2\,(\underline{s}_1 \cdot \underline{s}_2)(\underline{\nabla}\beta_1 \cdot \underline{s}_1)(\underline{\nabla}\beta_2 \cdot \underline{s}_2) \; . \end{aligned} \tag{15}$$

For this purpose, the Identity tensor was represented by $\underline{\underline{I}} = \underline{m}_n \otimes \underline{m}_n + \underline{s}_n \otimes \underline{s}_n + \underline{t}_n \otimes \underline{t}_n$ and $\underline{\nabla}\beta_n \cdot \underline{t}_n = 0$ was exploited according to the assumptions above. Since the physical significance of these interactions is not clear, a definition of scalar dislocation densities from principal invariants or from the norm $\|\underline{\underline{\alpha}}\|$ is not favorable. Instead, another slip-system-based set of invariants is preferred. First, the corresponding local dislocation vector of all dislocations piercing a plane with $\underline{n} = \underline{t}_n$ is calculated, i.e. the dislocation tangents $\underline{t}_n$ are cut perpendicularly. Now the resulting vector is decomposed according to Eq. (9). This is demonstrated on the example of two slip systems again:

$$\begin{aligned} \alpha_1 &= \underline{s}_1 \cdot \underline{\underline{\alpha}} \cdot \underline{t}_1 = (\underline{\nabla}\beta_1 \cdot \underline{s}_1) + (\underline{\nabla}\beta_2 \cdot \underline{s}_2)(\underline{s}_1 \cdot \underline{s}_2)\ , \\ \alpha_2 &= \underline{s}_2 \cdot \underline{\underline{\alpha}} \cdot \underline{t}_2 = (\underline{\nabla}\beta_2 \cdot \underline{s}_2) + (\underline{\nabla}\beta_1 \cdot \underline{s}_1)(\underline{s}_1 \cdot \underline{s}_2)\ . \end{aligned} \tag{16}$$

Recalling Eq. (3) it becomes clear that these invariants measure the (signed) dislocation density multiplied by $b$ from all segments with Burgers vector $\underline{b} = b\,\underline{s}_n$ at some position $\underline{r}$ . Consequently, unsigned dislocation densities can be defined as $\rho_n = |\alpha_n|/b$ . In case of orthogonal slip systems (here: $\underline{m}_1 \cdot \underline{m}_2 = \underline{s}_1 \cdot \underline{s}_2 = 0$), there are no contributions from one slip system to the other. Hence, the dislocation densities are proportional to the gradients of plastic slip in the slip direction only:

$$\rho_1 = \tfrac{1}{b}|\alpha_1| = \tfrac{1}{b}|\underline{\nabla}\beta_1 \cdot \underline{s}_1|, \quad \rho_2 = \tfrac{1}{b}|\alpha_2| = \tfrac{1}{b}|\underline{\nabla}\beta_2 \cdot \underline{s}_2|\ . \tag{17}$$

The total dislocation density is then simply a sum: $\rho_{\text{tot}} = \rho_1 + \rho_2$. However, in the general case of non-orthogonal slip systems (as in most real crystals), the mere gradients $\underline{\nabla}\beta_n \cdot \underline{s}_n$ are no invariants of $\underline{\underline{\alpha}}$. In other words, it is not possible to only extract them from $\underline{\underline{\alpha}}$ in a way that does not depend on the chosen coordinate system. Instead, additional terms result from the tensor nature of the involved quantities (as can be seen in Eq. (16)). Taking them into account, the resultant field equations for $\beta_1, \beta_2, \ldots, \beta_N$ will be in fact coupled, but invariant with respect to a change of the coordinate system.

From the physical point of view the CDT can neither determine the dislocation configuration which has led to the plastic distortion, nor can the CDT uniquely determine the character of the dislocations that are geometrically necessary at the moment. Rather, the only information which is accessible is the resultant Burgers vector with respect to any cut plane as a measure of the crystal's closure failure.

## 3. Thermodynamics of continuously dislocated crystals

### *3.1. State variables, stored energy and external power*

The free energy density of the dislocated crystal depends on the thermodynamical state variables $\underline{\underline{\alpha}}, \underline{\underline{\varepsilon}}_{\text{e}} = \text{sym}(\underline{\underline{\beta}}_{\text{e}})$. An additive decomposition $\phi = \phi_{\text{e}}(\underline{\underline{\varepsilon}}_{\text{e}}) + \phi_{\text{p}}(\underline{\underline{\alpha}})$ is assumed. The (macroscopic) elastic strain energy is given by Hooke's law:

$$\phi_{\text{e}}(\underline{\underline{\varepsilon}}_{\text{e}}) = \mu\|\underline{\underline{\varepsilon}}_{\text{e}}\|^2 + \tfrac{1}{2}\lambda I_1^2(\underline{\underline{\varepsilon}}_{\text{e}})\ , \tag{18}$$

where $\mu$ and $\lambda$ denote Lam's constants. The dislocation energy $\phi_{\text{p}}$ depends on the number of active slip systems as well as the slip mechanisms (e.g. glide or cross slip). However, in this study only two special cases are considered. For single slip conditions, motivated by statistical mechanics [2], the dislocation energy can be specified as

$$\phi_{\text{p}}(\underline{\underline{\alpha}}) = k\mu \ln[(1 - \rho/\rho_{\text{s}})^{-1}]\ , \tag{19}$$

with a scaling parameter $k > 0$ and the dislocation density $\rho$ according to definition (14). This ansatz features both the linear dependence of the dislocation energy for small densities and the steep upturn when $\rho$ approaches the saturation density $\rho_{\text{s}}$. In case of two active slip systems many forms are possible depending on the choice of invariants of the dislocation tensor. For the reasons discussed in Subsection 2.2, invariants $\rho_n = |\alpha_n|/b$ from Eq. (16) should be preferred. Consider an example from [3]:

$$\phi_\mathrm{p}(\underline{\underline{\alpha}}) = k\mu \left( \ln\left[\left(1-\frac{\rho_1}{\rho_\mathrm{s}}\right)^{-1}\right] + \ln\left[\left(1-\frac{\rho_2}{\rho_\mathrm{s}}\right)^{-1}\right] + \chi\frac{\rho_1}{\rho_\mathrm{s}}\frac{\rho_2}{\rho_\mathrm{s}} \right) . \tag{20}$$

Here, the interaction between the two slip systems is modeled by the quadratic term with some interaction parameter $\chi > 0$ (figure 2). For two orthogonal slip systems ( $\underline{m}_1 \cdot \underline{m}_2 = \underline{s}_1 \cdot \underline{s}_2 = 0$, cf. Eq. (17)) this is the only interaction. In the general case ($\underline{m}_1 \cdot \underline{m}_2 \neq 0$), the first two summands of Eq. (20) include some inherent interaction as well. Consequently, the numerical simulations for $\underline{m}_1 \cdot \underline{m}_2 = 0$ and $\underline{m}_1 \cdot \underline{m}_2 \neq 0$ are expected to give insights into the effects of these differences.

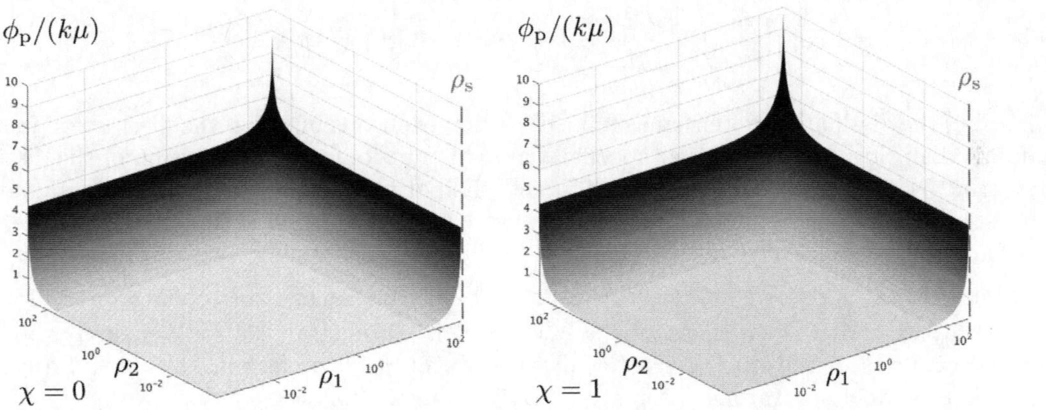


**Figure 2.** Surface plot of the dislocation energy according to Eq. (20). As indicated by the dashed lines, the dislocation densities cannot exceed the saturation value (here $\rho_\mathrm{s} = 4{\cdot}10^2 \mu\mathrm{m}^{-2}$). In addition, a minor effect of parameter $\chi$ is visible.

Next, the internal power (external power minus the rate of kinetic energy) has to be defined in a suitable way. Postulating the existence of a hyper stress tensor conjugate to the plastic distortion field, the internal power density is finally obtained as

$$p_\mathrm{in} = \mathrm{div}(\underline{\dot{u}} \cdot \underline{\underline{\sigma}}) + \underline{\dot{u}} \cdot \rho_\mathrm{m}(\underline{f} - \underline{\ddot{u}}) + \mathrm{div}(\underline{\underline{\dot{\beta}}}_\mathrm{p} \cdot\cdot \underline{\underline{\underline{\Sigma}}}) , \tag{21}$$

with $\rho_\mathrm{m}$ denoting the mass density, $\underline{\underline{\sigma}}$ the Cauchy stress tensor and $\underline{f}$ the body forces. The first two terms in Eq. (21) are classical while the third one arises from the incorporation of higher gradients in the CDT.

*3.2. Evaluation of the Clausius-Duhem inequality*

The field equations of the CDT are obtained exploiting the laws of thermodynamics in the form of the Clausius-Duhem inequality. In the isothermal case it reads $p_\mathrm{in} - \dot{\phi} \geq 0$. With the previous assumptions the dissipation (dissipated power $p_\mathrm{in} - \dot{\phi}$) follows as:

$$\left(\mathrm{div}(\underline{\underline{\sigma}}) + \rho_\mathrm{m}(\underline{f}\text{-}\underline{\ddot{u}})\right) \cdot \underline{\dot{u}} + \left(\underline{\underline{\sigma}}\text{-}\frac{\partial\phi}{\partial\underline{\underline{\beta}}}\right) \cdot\cdot\, \underline{\underline{\dot{\beta}}}^\mathrm{T} + \left(\mathrm{div}\left(\underline{\underline{\underline{\Sigma}}}\right) - \frac{\partial\phi}{\partial\underline{\underline{\beta}}_\mathrm{p}}\right) \cdot\cdot\, \underline{\underline{\dot{\beta}}}_\mathrm{p}^\mathrm{T} + \left(\underline{\underline{\underline{\Sigma}}} + \frac{\partial\phi}{\partial\underline{\underline{\alpha}}} \cdot \underline{\underline{\underline{\epsilon}}}\right) \cdot\cdot\cdot\, \mathrm{grad}(\underline{\underline{\dot{\beta}}}_\mathrm{p})^\mathrm{T} \geq 0.$$

where $\underline{\underline{\underline{\epsilon}}}$ denotes the Ricci tensor and $\underline{\underline{\beta}}$ stands for the displacement gradient. Now, Galilean invariance of the dissipation implies $\mathrm{div}(\underline{\underline{\sigma}}) + \rho_\mathrm{m}(\underline{f}\text{-}\underline{\ddot{u}}) = \underline{0}$ (local balance of forces). In this contribution, a special form of the theory is considered employing the following potential relations for the stress tensors:

$$\underline{\underline{\sigma}} = \frac{\partial\phi(\varepsilon_{\mathrm{e}}, \underline{\underline{\alpha}})}{\partial\underline{\underline{\beta}}} = \frac{\partial\phi_{\mathrm{e}}}{\partial\underline{\underline{\varepsilon}}_{\mathrm{e}}} \quad , \quad \underline{\underline{\underline{\Sigma}}} = \text{-}\frac{\partial\phi}{\partial\underline{\underline{\alpha}}}\cdot\underline{\underline{\underline{\epsilon}}} \quad . \tag{22}$$

Consequently, the only remaining part of the dissipation inequality to be fulfilled is

$$\left(\operatorname{div}\left(\underline{\underline{\underline{\Sigma}}}\right) - \frac{\partial\phi}{\partial\underline{\underline{\beta}}_{\mathrm{p}}}\right)\cdot\cdot\,\dot{\underline{\underline{\beta}}}_{\mathrm{p}}^{\mathrm{T}} = \left(\text{-}\operatorname{curl}\left(\frac{\partial\phi}{\partial\underline{\underline{\alpha}}}\right) - \frac{\partial\phi}{\partial\underline{\underline{\beta}}_{\mathrm{p}}}\right)\cdot\cdot\,\dot{\underline{\underline{\beta}}}_{\mathrm{p}}^{\mathrm{T}} \geq 0. \tag{23}$$

It can either be set zero (zero dissipation case, cf. also [6]) or equal to some corresponding partial derivative of a so-called dissipation potential $\mathcal{D}$ :

$$\text{i)}\ \text{-}\operatorname{curl}\left(\frac{\partial\phi}{\partial\underline{\underline{\alpha}}}\right) - \frac{\partial\phi}{\partial\underline{\underline{\beta}}_{\mathrm{p}}} = \underline{\underline{0}} \quad \text{or} \quad \text{ii)}\ \text{-}\operatorname{curl}\left(\frac{\partial\phi}{\partial\underline{\underline{\alpha}}}\right) - \frac{\partial\phi}{\partial\underline{\underline{\beta}}_{\mathrm{p}}} = \frac{\partial\mathcal{D}}{\partial\dot{\underline{\underline{\beta}}}_{\mathrm{p}}} \ . \tag{24}$$

Choosing the dissipation potential as a positive, homogeneous function the dissipation inequality is fulfilled a priori without further specifications. In addition, the introduction of the dissipation potential facilitates the modeling of different dissipative phenomena in a thermodynamically consistent way: It is experimentally confirmed that dislocation motion requires exceeding a critical resolved shear stress $\tau_{\mathrm{cr}}$. This can be taken into account by $\mathcal{D}(\dot{\underline{\underline{\beta}}}_{\mathrm{p}}) = \tau_{\mathrm{cr}}\|\dot{\underline{\underline{\beta}}}_{\mathrm{p}}\|$. The quadratic form $\mathcal{D}(\dot{\underline{\underline{\beta}}}_{\mathrm{p}}) = \frac{1}{2}\eta_{\mathrm{p}}\|\dot{\underline{\underline{\beta}}}_{\mathrm{p}}\|^2$ captures viscous interaction of moving dislocations and crystal lattice. For the same reasons as discussed in Subsection 2.2 for tensor $\underline{\underline{\alpha}}$, slip system based invariants of the rate of plastic distortion should be preferred. Recalling Eq. (2), it is possible to define an invariant dissipation potential in the following way:

$$\mathcal{D}(\dot{\underline{\underline{\beta}}}_{\mathrm{p}}) = \tau_{\mathrm{cr}}\left(|\dot{\beta}_{\mathrm{p2}}| + |\dot{\beta}_{\mathrm{p2}}|\right) + \tfrac{1}{2}\eta_{\mathrm{p}}\left(\dot{\beta}_{\mathrm{p2}}^2 + \dot{\beta}_{\mathrm{p2}}^2\right) \ . \tag{25}$$

Note that taking merely the slip rates $\dot{\beta}_n$ would, in general, imply that the dissipation is not invariant with respect to a change of coordinate system anymore.

### *3.3. Thermodynamic consistency of the case with zero dissipation*

If decomposition (1) is applied, i.e. the crystal nature is taken into account using slip systems, further considerations are necessary. This stems from the fact that the plastic distortion is then not arbitrary anymore. Let us introduce some abbreviations:

$$\operatorname{curl}\left(\frac{\partial\phi}{\partial\underline{\underline{\alpha}}}\right) =: \underline{\underline{X}} \quad , \quad \frac{\partial\phi}{\partial\underline{\underline{\beta}}_{\mathrm{p}}} = \frac{\partial\phi_{\mathrm{e}}(\underline{\underline{\varepsilon}} - \underline{\underline{\varepsilon}}_{\mathrm{p}})}{\partial\underline{\underline{\beta}}_{\mathrm{p}}} = \text{-}\frac{\partial\phi_{\mathrm{e}}}{\partial\underline{\underline{\varepsilon}}_{\mathrm{e}}} = \text{-}\underline{\underline{\sigma}} \quad . \tag{26}$$

The second-order tensor $\underline{\underline{X}}$ is the well-known backstress tensor, which is obviously unsymmetric in general. The Cauchy stress tensor however is symmetric for the chosen strain energy (18). Rewriting Eq. (24) for zero dissipation in the form of an effective stress $\underline{\underline{\sigma}} - \underline{\underline{X}}$ reveals a contradiction: the difference of a symmetric and an unsymmetric tensor can never vanish. However, this discrepancy can be solved realizing that the problem originates going from Eq. (23) to (24). Implicitly, it was assumed that inequality (23) must hold for arbitrary plastic distortions, which is now not the case anymore. Instead, expansion (1) must be introduced into inequality (23). In case of zero dissipation this yields

$$(\underline{\underline{\sigma}} - \underline{\underline{X}})\cdot\cdot\sum_{n=1}^{N}\dot{\beta}_n\,\underline{s}_n\otimes\underline{m}_n = \sum_{n=1}^{N}\underline{s}_n\cdot(\underline{\underline{\sigma}}\text{-}\underline{\underline{X}})\cdot\underline{m}_n\,\dot{\beta}_n = 0\ . \tag{27}$$

Now demanding this equation to be fulfilled for arbitrary plastic slip rates $\dot{\beta}_n$ yields a much weaker condition.

Let us first consider single slip ($N = 1$). In order to obtain zero dissipation, only one coefficient $\underline{s} \cdot (\underline{\underline{\sigma}}\text{-}\underline{\underline{X}}) \cdot \underline{m}$ must vanish. In other words, only this coefficient is physically significant. All other components are simply undefined. The next interesting scenario is $1 < N < 8$. Now $N$ coefficients of $\underline{\underline{\sigma}}\text{-}\underline{\underline{X}}$ (or linear combinations of them) have to vanish. However, the backstress tensor $\underline{\underline{X}}$ may still be unsymmetric. Nevertheless, special cases can occur where $\underline{\underline{X}}$ is symmetric, e.g. for two orthogonal slip systems (see Subsection 4.2). In the case of $N > 8$ *independent* slip systems the plastic distortion tensor has a sufficient degree of freedom such that it can take arbitrary values (though it still is a sum of slip system contributions). As a consequence, $\underline{\underline{X}}$ now must be and will be symmetric in order to fulfill the partial differential equation $(24)_\mathrm{i}$, in short $\underline{\underline{\sigma}}\text{-}\underline{\underline{X}} = \underline{\underline{0}}$ (cf. [5]). This is in accordance with symmetric backstress tensors known from purely phenomenological models (cf. e.g. [7]). The results are summarized below.

**Table 1.** Restrictions imposed on the effective stress $\underline{\underline{\sigma}}\text{-}\underline{\underline{X}}$ for zero dissipation

| Case | Number of active slip systems | Restrictions | Symmetry of $\underline{\underline{X}}$ |
|---|---|---|---|
| a) | $N = 1$ | $\underline{s} \cdot (\underline{\underline{\sigma}}\text{-}\underline{\underline{X}}) \cdot \underline{m} = 0$ | unsymmetric |
| b) | $1 < N < 8$ | $\underline{s}_n \cdot (\underline{\underline{\sigma}}\text{-}\underline{\underline{X}}) \cdot \underline{m}_n = 0 \ \forall\, n \in [1, N]$ | unsymmetric |
| c) | $N \geq 8$ | $\underline{\underline{\sigma}}\text{-}\underline{\underline{X}} = \underline{\underline{0}}$ | symmetric |

There is another important consequence: Obviously, for $N < 8$ slip systems some components of $\underline{\underline{X}}$ are neither defined nor physically significant. This also holds for non-zero dissipation. Hence, the field equations cannot be formulated for the entire plastic distortion tensor in this case. Instead, scalar field equations operating on the slip system level have to be used. Still, for $N \geq 8$ independent slip systems, this approach is depleted and a tensor field equation has to be solved.

## 4. Field equations for special slip conditions

### *4.1. One active slip system with straight dislocation lines*

In this case (cf. figure 1, left), the plastic distortion tensor has the simple form $\underline{\underline{\beta}}_\mathrm{p} = \beta\ \underline{s} \otimes \underline{m}$. It is interesting to note that thus plastic slip results both in plastic strain and plastic spin, as it can be seen from:

$$\underline{\underline{\varepsilon}}_\mathrm{p} = \mathrm{sym}(\underline{\underline{\beta}}_\mathrm{p}) = \tfrac{\beta}{2}(\underline{s} \otimes \underline{m} + \underline{m} \otimes \underline{s}) \quad , \quad \underline{\underline{w}}_\mathrm{p} = \mathrm{skw}(\underline{\underline{\beta}}_\mathrm{p}) = \tfrac{\beta}{2}(\underline{s} \otimes \underline{m} - \underline{m} \otimes \underline{s}) \ . \tag{28}$$

Hence, it is not possible to obtain an irrotational plastic distortion field. Furthermore, case a) from table 1 applies. Consequently, the field equations for the plastic distortion tensor are simplified to a scalar field equation for the plastic slip $\beta(\underline{r})$ :

$$-\underline{s} \cdot \left( \frac{\partial \phi}{\partial \underline{\underline{\beta}}_\mathrm{p}} + \mathrm{curl}\left( \frac{\partial \phi}{\partial \underline{\underline{\alpha}}} \right) \right) \cdot \underline{m} = \underline{s} \cdot \frac{\partial \mathcal{D}}{\partial \dot{\underline{\underline{\beta}}}_\mathrm{p}} \cdot \underline{m} \quad \Rightarrow \quad -\frac{\partial \phi}{\partial \beta} + \mathrm{div}\left( \frac{\partial \phi}{\partial \underline{\nabla} \beta} \right) = \frac{\partial \mathcal{D}}{\partial \dot{\beta}} \ . \tag{29}$$

Together with the other field equation, which reduces to $\mathrm{div}(\underline{\underline{\sigma}}) = \underline{0}$ in absence of body and inertia forces, the set of field equations is complete. Defining proper initial and boundary conditions, the evolution of the fields $\underline{u}(\underline{r})$ and $\beta(\underline{r})$ can be predicted. Numerical solutions of the corresponding initial boundary value problems were obtained for the case of plane deformations using an in-house finite difference simulation code. The results will be presented in a forthcoming article.

### 4.2. Two orthogonal slip systems with straight dislocation lines

In this case (cf. figure 1, right) $\underline{\underline{\beta}}_{\rm p} = \beta_1\ \underline{s}_1 \otimes \underline{m}_1 + \beta_2\ \underline{s}_2 \otimes \underline{m}_2$. Consider an example:

$$\underline{s}_1 = \underline{e}_x,\ \underline{m}_1 = \underline{e}_y \quad , \quad \underline{s}_2 = \underline{e}_y,\ \underline{m}_2 = \text{-}\underline{e}_x \quad , \tag{30}$$

for the basis $\underline{e}_a = \{\underline{e}_x, \underline{e}_y, \underline{e}_z\}$. The plastic distortion has then the following coefficient matrix:

$$[\underline{e}_a \cdot \underline{\underline{\beta}}_{\rm p} \cdot \underline{e}_b] = \begin{bmatrix} 0 & \beta_1 \\ \text{-}\beta_2 & 0 \end{bmatrix} \quad . \tag{31}$$

In contrast to the previous subsection, $\underline{\underline{\beta}}_{\rm p}$ may be symmetric or skew-symmetric now. The dislocation density tensor $\underline{\underline{\alpha}} = \text{curl}(\underline{\underline{\beta}}_{\rm p}) = \underline{s}_1 \otimes \underline{\nabla}\beta_1 \times \underline{m}_1 + \underline{s}_2 \otimes \underline{\nabla}\beta_2 \times \underline{m}_2$ follows straightforward. Without loss of generality, let us for now assume a quadratic form of the dislocation energy $\phi_{\rm p}(\underline{\underline{\alpha}}) = C/2\ \|\underline{\underline{\alpha}}\|^2$ [4] such that the backstress tensor follows as

$$\underline{\underline{X}} = \text{curl}\left(\frac{\partial\phi}{\partial\underline{\underline{\alpha}}}\right) = C\left(\underline{s}_1 \otimes \underline{m}_1 \cdot (\Delta\beta_1 \underline{\underline{I}} - \underline{\nabla}\otimes\underline{\nabla}\beta_1) + \underline{s}_2 \otimes \underline{m}_2 \cdot (\Delta\beta_2 \underline{\underline{I}} - \underline{\nabla}\otimes\underline{\nabla}\beta_2)\right), \tag{32}$$

with the Laplace operator defined as $\Delta = \underline{\nabla}\cdot\underline{\nabla}$. Introducing the relations (30) yields

$$[\underline{e}_a \cdot \underline{\underline{X}} \cdot \underline{e}_b] = C\begin{bmatrix} \nabla_y\nabla_x\beta_1 & \text{-}\Delta\beta_1 + \nabla_y\nabla_y\beta_1 \\ \Delta\beta_2 - \nabla_x\nabla_x\beta_2 & \text{-}\nabla_x\nabla_y\beta_2 \end{bmatrix} = C\begin{bmatrix} \nabla_y\nabla_x\beta_1 & \text{-}\nabla_x\nabla_x\beta_1 \\ \nabla_y\nabla_y\beta_2 & \text{-}\nabla_x\nabla_y\beta_2 \end{bmatrix} \quad . \tag{33}$$

As this corresponds to case b) from table 1, the following restrictions must hold for zero dissipation:

$$\underline{s}_1 \cdot (\underline{\underline{\sigma}}\text{-}\underline{\underline{X}}) \cdot \underline{m}_1 = 0 \quad , \quad \underline{s}_2 \cdot (\underline{\underline{\sigma}}\text{-}\underline{\underline{X}}) \cdot \underline{m}_2 = 0 \quad . \tag{34}$$

Inserting relations (30) again yields $\sigma_{xy} = X_{xy}$ and $\sigma_{yx} = X_{yx}$. Considering the symmetry of the Cauchy stress tensor, it is inferred that $X_{yx} = X_{xy}$. Hence, the backstress tensor must be symmetric, which implies

$$\text{-}\nabla_x\nabla_x\beta_1 = \nabla_y\nabla_y\beta_2 \quad \Leftrightarrow \quad \text{-}\underline{s}_1\cdot\underline{\nabla}\otimes\underline{\nabla}\beta_1\cdot\underline{s}_1 = \underline{s}_2\cdot\underline{\nabla}\otimes\underline{\nabla}\beta_2\cdot\underline{s}_2 \quad . \tag{35}$$

This theoretical statement is of high importance for the numerical solution of the field equations in this case. They are obtained employing now the dislocation energy in the form (20) and the dissipation according to Eq. (25) (with $\tau_{\rm cr} = \chi = 0$ for simplicity):

$$\begin{aligned} (\lambda+\mu)\underline{\nabla}\otimes\underline{\nabla}\cdot\underline{u} + \mu\underline{\nabla}\cdot\underline{\nabla}\otimes\underline{u} - \mu\,\text{sym}(\underline{\underline{G}})\cdot(\underline{\nabla}\beta_1 - \underline{\nabla}\beta_2) &= \underline{0} \quad , \\ 2\mu\,\text{sym}(\underline{\underline{G}}) \cdot\cdot\ \underline{\nabla}\otimes\underline{u} + \mu(\beta_2 - \beta_1) + C\ \underline{s}_1\cdot\underline{\nabla}\otimes\underline{\nabla}\beta_1\cdot\underline{s}_1\ \left(1 - \frac{|\underline{\nabla}\beta_1\cdot\underline{s}_1|}{b\rho_{\rm s}}\right)^{\text{-}2} &= \eta_{\rm p}\dot{\beta}_1 \ , \\ \text{-}2\mu\,\text{sym}(\underline{\underline{G}}) \cdot\cdot\ \underline{\nabla}\otimes\underline{u} - \mu(\beta_2 - \beta_1) + C\ \underline{s}_2\cdot\underline{\nabla}\otimes\underline{\nabla}\beta_2\cdot\underline{s}_2\ \left(1 - \frac{|\underline{\nabla}\beta_2\cdot\underline{s}_2|}{b\rho_{\rm s}}\right)^{\text{-}2} &= \eta_{\rm p}\dot{\beta}_2 \ . \end{aligned} \tag{36}$$

Here, the abbreviations $\underline{\underline{G}} = \underline{s}_1 \otimes \underline{m}_1 = \text{-}\underline{m}_2 \otimes \underline{s}_2$ and $C = k\mu/(b\rho_{\rm s})^2$ were used. System (36) constitutes a set of coupled partial differential equations. Let us first consider its stationary points. To this end, suppose a constant displacement gradient such that the total shear $\gamma = 2\,\text{sym}(\underline{\underline{G}}) \cdot\cdot\ \underline{\nabla} \otimes \underline{u}$ is constant as well. Further, assume homogeneous plastic slip fields. In equilibrium the temporal change of the plastic slip in system 1 and 2 comes to an end. Hence, the left hand side of system (36) becomes zero, i.e.:

$$\begin{aligned} \text{sym}(\underline{\underline{G}})\cdot(\underline{\nabla}\beta_1 - \underline{\nabla}\beta_2) &= \underline{0} \ , \\ \mu\gamma + \mu(\beta_2 - \beta_1) &= 0 \ , \\ \text{-}\mu\gamma - \mu(\beta_2 - \beta_1) &= 0 \ . \end{aligned} \tag{37}$$

The solution is a fixed line $\beta_2^0 = \beta_1^0 - \gamma$. Consider some small perturbations $\hat{\beta} \ll 1$ around these stationary points:

$$\begin{bmatrix}\beta_1\\ \beta_2\end{bmatrix} = \begin{bmatrix}\beta_1^0\\ \beta_2^0\end{bmatrix} + \hat{\beta}\begin{bmatrix}v_1\\ v_2\end{bmatrix}\exp(S\ t\mu/\eta_\mathrm{p}) = \begin{bmatrix}\beta_1^0\\ \beta_1^0-\gamma\end{bmatrix} + \hat{\beta}\begin{bmatrix}v_1\\ v_2\end{bmatrix}\exp(S\ t\mu/\eta_\mathrm{p})\ . \tag{38}$$

Inserting the perturbations into system (36) yields an eigenvalue problem in the form

$$([S] - S_K[I])\,[v_K] = [0] \quad \rightarrow \quad S_I = 0, [v_I] = \begin{bmatrix}1\\ 1\end{bmatrix}, \quad S_{II} = -2, [v_{II}] = \begin{bmatrix}1\\ \text{-}1\end{bmatrix}, \tag{39}$$

with the matrix $[S] = \left[\begin{smallmatrix}-1 & +1\\ +1 & -1\end{smallmatrix}\right]$ and the identity matrix $[I] = \left[\begin{smallmatrix}1 & 0\\ 0 & 1\end{smallmatrix}\right]$. Looking at the eigenvectors $[v_K]$ and at the eigenvalues $S_K$ as growth rates of the perturbations, the following conclusion is drawn: Perturbations with $\beta_2 = \text{-}\beta_1$ will decay very quickly ($S_{II} < 0$), other than perturbations with $\beta_2 = \beta_1$ that persist ($S_I = 0$). Numerical simulations showed this type of behavior which is considered unphysical. Fortunately, a remedy is already given by condition (35). Inserting it into system (36) reveals that the evolution of both plastic slips is identical. Hence, only one such equation must be solved. This is also evident from the physical point of view: as the conditions in both slip systems are exactly the same considering (critical) resolved shear stress and viscosity, there is no reason why the evolution should be different. The consequence is a symmetric plastic distortion (no plastic spin) and a symmetric backstress tensor. As shown in [5], the symmetry of $\underline{\underline{X}}$ can also contribute to the well-posedness of the numerical problem.

### *4.3. Multiple slip systems with straight dislocation lines – an outlook*

The extension towards multi-slip is a non-trivial task due to the proper definition of the scalar dislocation densities and the corresponding free energy of the dislocation network. If merely the plastic slips $\beta_n$ and their gradients are used without any further considerations, a major problem arises: the field equations could depend on the chosen coordinate system. This is in contradiction to general principles of physics and should therefore be avoided. Instead, invariants of the tensor quantities need to be used. However, this might yield strongly coupled field equations with respect to the plastic slips (only for linearly independent slip systems a weaker coupling seems feasible, cf. Subsection 4.2). From the physical point of view, this stems from the multitude of active slip system combinations that result in the same overall plastic distortion. Considering the critical resolved shear stress as threshold for dislocation motion in a slip system offers a first possibility to reduce this ambiguity: Depending on the corresponding resolved shear stress only a subset of all slip systems is activated. Moreover, in real crystals other involved defects will always lead to a preference of some slip system, even though it is crystallographically equivalent to others. If this happens, the symmetry of the setting is broken and there should be no ambiguity anymore. The major task seems to consistently incorporate this behavior into the CDT.

## 5. Conclusions

For a single dislocation segment, the physical significance of the invariants of the dislocation tensor was shown. However, in case of continuously distributed dislocations it is more difficult to interpret the (principal) invariants due to interaction terms of different slip systems. Hence, slip system based invariants were suggested that facilitate the definition of invariant energies and dissipation potentials. Further, it was demonstrated that the structure of the plastic distortion field equation depends on the number of slip systems. For less than eight independent slip systems, only scalar field equations (flow rules) for a subset of coefficients need to be solved. Else, a tensor field equation has to be solved. This entails further consequences on the symmetry of the backstress tensor. The results of this study promote the idea that a physically meaningful backstress tensor ought to be symmetric in general.

IOP Conf. Series: Materials Science and Engineering **118** (2016) 012034 doi:10.1088/1757-899X/118/1/012034

## Acknowledgments

This research was supported by the German National Science Foundation (DFG) within collaborative research center SFB 692 High-strength aluminum-based light-weight materials for safety components. In addition, the help of Professor Khanh C. Le, Hatice Tavli, Robert Scherzer and Hendrik Donner is gratefully acknowledged.

## References

[1] Silbermann C B, Shutov A V, Ihlemann J 2014 Modeling the evolution of dislocation populations under non-proportional loading *Int. J. Plasticity* **55** 58–79

[2] Berdichevsky V 2006 Continuum theory of dislocations revisited *Continuum Mech. Thermodyn.* **18** 195–222

[3] Le K C, Sembiring P 2008 Plane constrained shear of single crystal strip with two active slip systems *J. Mech. Phys. Solids* **56** 2541–54

[4] Gurtin M E 2002 A gradient theory of single-crystal viscoplasticity that accounts for geometrically necessary dislocations *J. Mech. Phys. Solids* **50** 5–32

[5] Ebobisse F, Neff P, Reddy D 2015 Existence results in dislocation based rate-independent isotropic gradient plasticity with kinematical hardening and plastic spin: The case with symmetric local backstress *arXiv:*1504.01973

[6] Wulfinghoff S, Forest S, Böhlke T, 2015 Strain gradient plasticity modeling of the cyclic behavior of laminate microstructures *J. Mech. Phys. Solids* **79** 1–20

[7] Shutov A V, Kuprin C, Ihlemann J, Wagner M F-X, Silbermann C, 2010 Experimental investigation and numerical simulation of the incremental deformation of a 42CrMo4 steel *Materialwiss. Werkstofftech.* **41.9** 765–75

# Evaluation of the Technical-Economic Potential of Particle-Reinforced Aluminum Matrix Composites and Electrochemical Machining

**A Schubert[1,2], U Götze[3], M Hackert-Oschätzchen[1], N Lehnert[1], F Herold[3], G Meichsner[2], A Schmidt[3]**

[1]Technische Universität Chemnitz, Professorship Micromanufacturing Technology, Chemnitz, Germany
[2]Fraunhofer Institute for Machine Tools and Forming Technology, Chemnitz, Germany
[3]Technische Universität Chemnitz, Professorship of Management Accounting and Control, Chemnitz, Germany

E-mail: franziska.herold@wirtschaft.tu-chemnitz.de

**Abstract.** Compared to conventional cutting, the processing of materials by electrochemical machining offers some technical advantages like high surface quality, no thermal or mechanical impact on the work piece and preservation of the microstructure of the work piece material. From the economic point of view, the possibility of process parallelization and the absence of any process-related tool wear are mentionable advantages of electrochemical machining. In this study, based on experimental results, it will be evaluated to what extent the electrochemical machining is technically and economically suitable for the finish-machining of particle-reinforced aluminum matrix composites (AMCs). Initial studies showed that electrochemical machining – in contrast to other machining processes – has the potential to fulfil demanding requirements regarding precision and surface quality of products or components especially when applied to AMCs. In addition, the investigations show that processing of AMCs by electrochemical machining requires less energy than the electrochemical machining of stainless steel. Therefore, an evaluation of electrochemically machined AMCs – compared to stainless steel – from a technical and an economic perspective will be presented in this paper. The results show the potential of electro-chemically machined AMCs and contribute to the enhancement of instruments for technical-economic evaluations as well as a comprehensive innovation control.

## 1. Introduction

The combination of two or more physically and chemically different materials is a composite material. In composite materials, the different materials exist in separated phases. The major phase is the matrix material and the minor phase is the reinforcement. The properties of the composites, such as E-modulus or density, are defined by the matrix material as metal, ceramic or polymer, and by the material and the type, e.g. particle or fiber, of the reinforcement [1].

In the Collaborative Research Centre SFB 692 HALS at Technische Universität Chemnitz inter alia particle-reinforced aluminum matrix composites (AMCs) are investigated. In SFB 692 several academic institutions analyze and evaluate the potential of AMCs referring to material properties, different methods of machining and economic aspects.

This study is focused on the electrochemical machining (ECM) as a method for finish-machining of AMCs and the evaluation of its technical and economic potential. Electrochemical machining is already industrially established for the machining of tools but also for series production [2]. In this study the technical and economic benefits and drawbacks of a material replacement from stainless steel 1.4301 by an AMC will be evaluated.

## 2. Evaluation Approach

As this paper deals with issues of two different scientific disciplines, the usage of a methodology contemplating technical as well as economic outcomes and integrating them into a comprehensive evaluation is worthwhile. Amongst others, it is important to have a consistent data base of technical and economic evaluation with regard to process parameters, resource consumption, output quantities and material properties etc. in order to avoid double work and to achieve a high significance of results. The methodology of integrated evaluation applied here [3], offers the option of a structured and integrated technical-economic evaluation of production factors (such as materials), process chains and processes, products (or work pieces) as well as technologies by using three modules: the technical evaluation, the economic evaluation and, finally, the integrated evaluation (figure 1).

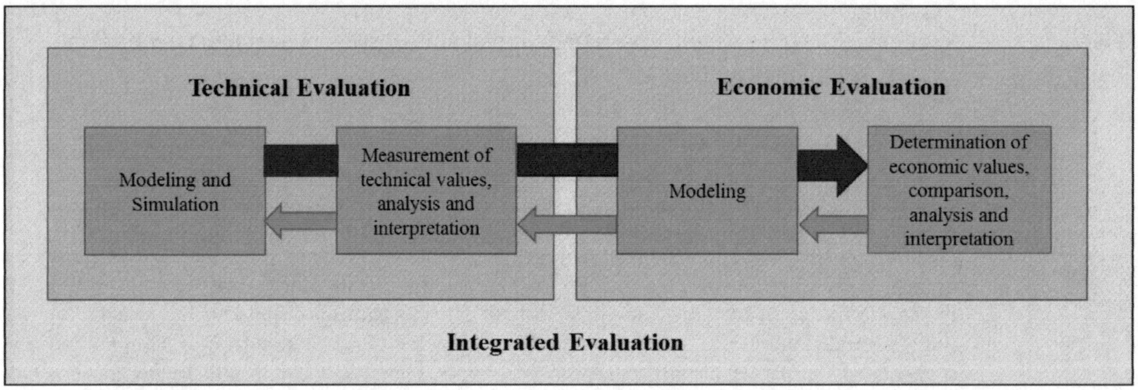


**Figure 1.** Approach of integrated technical and economic evaluation according to [3].

The technical evaluation considers technical criteria and applies evaluation methods from engineering sciences. The results of this evaluation are the basis of the subsequent economic as well as integrated evaluation. For the economic evaluation, methods originating from the scientific field of management accounting can be used for calculating economic target figures and analyzing profitability of alternatives. Its results may initiate reconsiderations and further development of technical alternatives in terms of resource efficiency or other targets. Finally, based on technical as well as economic outcomes an integrated evaluation is performed intending to reveal the overall performance of alternatives. In the following, this integrated evaluation approach is applied to the assessment of electrochemically machined AMCs (compared to stainless steel).

## 3. Technical Evaluation

The electrochemical machining is based on the local anodic dissolution of metallic work pieces. Figure 2 shows a scheme of the dissolution during electrochemical machining.

The work piece is the anode and the tool is the cathode. Between these two electrodes is the working gap, flushed by electrolyte. By applying an electric potential to the electrodes complex electrochemical reactions occur in the electrolyte. As a result, the work piece dissolves by emitting ions into the electrolyte [2, 4, 5].

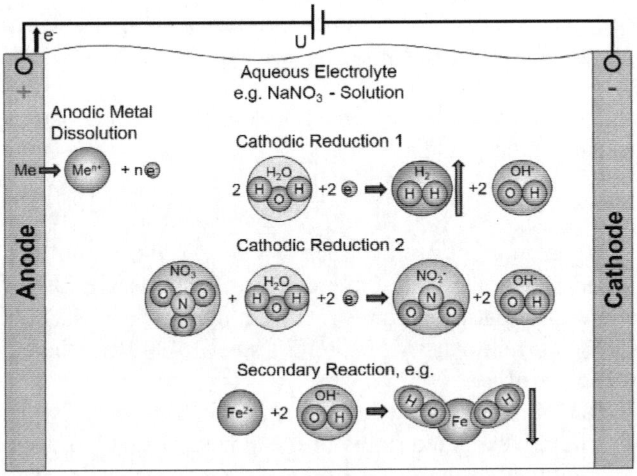


**Figure 2.** Scheme of the electrochemical dissolution process [4].

During the electrochemical machining process temperatures are below 80 °C. Furthermore, no mechanical contact of work piece and tool exists, so no process-related tool wear occurs. Because of the low temperature and the lack of contact, the micro-structure of the work piece is barely influenced [2, 4, 5].

The investigations on finish-machining of AMCs by ECM were performed inter alia by the pulsed electrochemical machining (PECM) with oscillating cathode. The experimental analysis in this study was performed with a PEMCenter 8000 from PEMTec SNC, France. The PEMCenter 8000 is a system containing an electrolyte preparation system, a power supply, a control unit and the process chamber.

For the process design and the economic evaluation, the electrochemical dissolution characteristic of the work piece material is needed. This characteristic is hard to predict and should be determined experimentally. For this experimental characterization, an experimental setup was developed which is shown in figure 3.

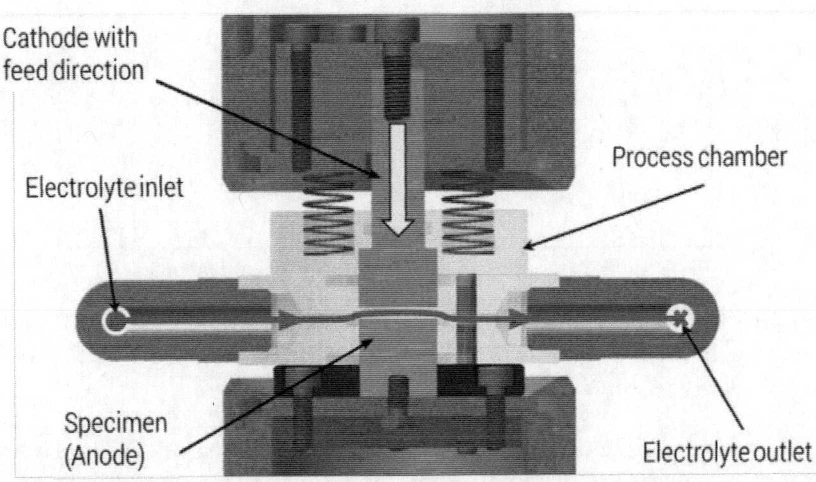


**Figure 3.** Section view of the experimental setup [6].

The cathode is fed vertically down towards the AMC specimen. Both have cylindrical shapes with a diameter of 12 mm. The electrolyte is supplied in horizontal direction through the working gap which leads to a good electrolyte replacement. With this experimental setup, the electrochemical dissolution characteristic of stainless steel 1.4301 and the AMC consisting of EN AW 2017 + 10 % SiC-particles were determined [6, 7].

One possibility for describing the electrochemical dissolution characteristic is the current efficiency $\eta$, which is calculated by using equation (1).

$$\eta(J) = \frac{V_{eff}(J)}{V_{Sp}} \tag{1}$$

The current efficiency is the quotient of the effectively removed volume $V_{eff}(J)$ and the theoretically removable volume $V_{Sp}$ based on Faraday's law. Since the effectively removed volume depends on the current density $J$, the current efficiency depends on current density as well. Figure 4 shows the current efficiency as a function of the current density of stainless steel 1.4301 and the analyzed AMC which are machined by using $NaNO_3$ electrolyte.

It can be seen that the functions for the two materials differ significantly. The current efficiency of the AMC is above 90 % in the investigated range of the current density. The maximal value is 110 %. This is characteristic for an active dissolution. In contrast, the current efficiency of the stainless steel is 0 % for small current densities and rises rapidly between 10 A/cm$^2$ and 30 A/cm$^2$ up to a maximal value of 70 %. The stainless steel shows a trans passive dissolution characteristic. Due to the active dissolution the removal rate of the AMC is higher than that of the stainless steel. By the experimental results it can be derived additionally, that for a current density of 50 A/cm$^2$ the removable volume per coulomb of the AMC has a value of approximately 0.042 mm$^3$/C while the value of the stainless steel is approximately 0.019 mm$^3$/C.

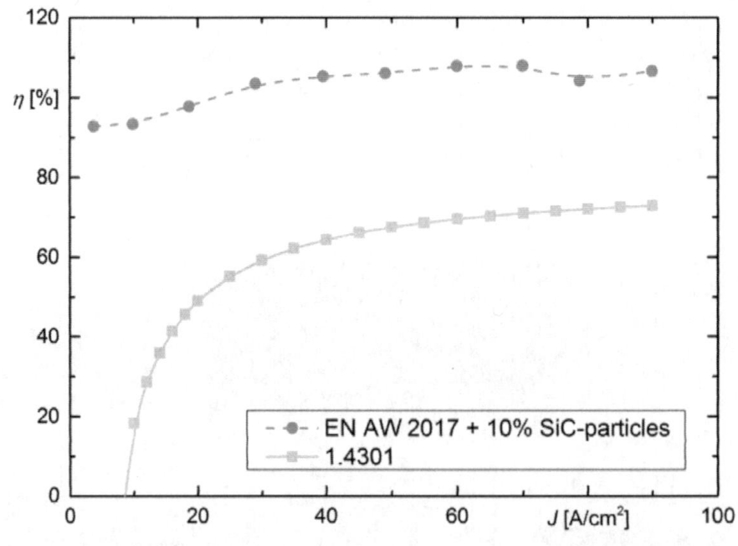


**Figure 4.** Current efficiency $\eta$ as a function of the current density $J$ of 1.4301 and the EN AW 2017 + 10 % SiC-particles in NaNO3 electrolyte with a conductivity of 70 mS/cm.

As a result, ECM shows a considerable potential for finish-machining of AMCs. Thereby, the current efficiency of the investigated AMC is significantly higher than the current efficiency of the stainless steel. This means that a defined volume of AMC can be removed electrochemically in distinctly less time than the same volume of 1.4301.

For the economic evaluation, a work piece with an initial volume of 3800 mm$^3$ and an end volume of 2600 mm$^3$ is taken as basis. This work piece is formerly made of stainless steel 1.4301 and should be made of AMC in future. Based on technical evaluation, the work piece made of AMC can be machined faster. Besides, the density of the AMC (2.9 g/cm$^3$) is about a third of the density of stainless steel 1.4301 (7.9 g/cm$^3$). When switching from steel to AMC, however, the lower E-modulus must be regarded (200 GPa for 1.4301 and 115 GPa for the AMC). Additionally, the economic evaluation has to consider that the process chain consists of the electrochemical machining process in a strict sense and the cleaning of the work pieces afterwards in the case of AMCs as well as steel.

## 4. Economic Evaluation

For a structured economic and integrated evaluation of innovative (and often complex) products, processes, production factors (such as materials) and combinations of these object types, a procedure model has proven to be worthwhile [8–11]. This model is characterized by a hierarchical breakdown of the evaluation task as well as a decision theory-based sequence of six steps at each level:

1. Determination of evaluation goal(s), requirements on the evaluation task, evaluation scope, and system boundaries
2. Determination of relevant target figure(s)
3. Identification, preselection, modelling and analysis of alternatives
4. Identification, analysis and forecast of relevant influencing factors (scenarios)
5. Determination of target figure(s) outcomes
6. Comparison, sensitivity analysis, interpretation

In the following, the procedure model will be applied to the economic evaluation of electrochemical machining of AMCs and the resulting AMCs. Correspondingly, the evaluation goal is the economic assessment of the process chain of electrochemical machining as well as the machined AMCs compared to stainless steel. The requirements on the evaluation task comprise a preferably high transparency as well as significance. The evaluation scope includes the process chains of AMCs and stainless steel with the relevant processes of electrochemical machining (in a strict sense) and final cleaning as well as the properties of the work piece (part). The evaluation is based on a laboratory scale for particle reinforced aluminum matrix composites (AMCs) and on the data ascertained in the SFB 692. In industrial applications, cost-saving effects like volume discounts for materials may appear that cannot be included here. The related system boundaries are the start and the end of the manufacturing process; at these points, amongst others, the properties of the raw material and of the output (work piece with machined and cleaned AMCs/steel) have to be determined.

Within the second step of the procedure model, one or more target figures have to be determined. Here, it is assumed that the properties of the output of electrochemically machined AMCs/steel enable the applicability of both in a couple of fields and effects of these properties on revenues can be neglected. Furthermore, for means of simplification, the monetary effects of possibly different output quantities will not be considered. Additionally, a dynamic calculation does not seem to be necessary. Then, costs are an appropriate economic target figure for the evaluation of the process chain as well as its output [12]. Corresponding with the target figure, the evaluation methods should be chosen. For forecasting costs of innovative processes, methods of cost accounting (such as calculation with activity units [12] and the related activity based costing [13, 14]) and/or development-concurrent cost-calculations [15, 16] can be used, in general. In the following, a specific variant of calculation with activity units is applied.

In the third step, the alternatives have to be identified, preselected, modelled and analyzed. Here, the alternatives are AMCs and stainless steel 1.4301 each manufactured by electrochemical machining. The results of the technological analysis of the output materials have already been presented in section 3. Each of the manufacturing process chains consists of the processes of electrochemical machining in the strict sense (described for AMCs in section 3 as well) and final cleaning. Main production factors used in these process chains are material (AMCs and stainless steel), energy, machinery and equipment (PEMCenter 8000)), labor (one employee who has to insert and remove the parts into and from the machine and another employee for cleaning), and production space. The economic evaluation has to refer to the consumption of these factors – correspondingly, material costs, energy costs, (other) machine-dependent costs such as depreciation, maintenance costs, imputed interest, labor costs and occupancy costs have to be taken into account. The evaluation also has to consider the main differences between the process chains and the materials: The processing time of steel (for electrochemical machining in a strict sense) is longer due to a smaller removable volume per coulomb resulting in higher energy consumption. This longer processing time results in higher machine-dependent costs and possibly labor costs; in contrast, material costs of AMCs are considerably higher and more maintenance activities are expected to be necessary for them since the filter used in the process of electrochemical machining requires more cleaning/replacement due to the existence of particles.

At the fourth step, relevant (cost) influencing factors are identified, analyzed and forecast. These comprise factors from inside a company: Output volume, processing time and transition time, labor time, needed space as well as other input-, process- and output-related factors such as current density and removable volume. Additionally, external factors are relevant, especially factor prices such as energy prices, hourly wage rates, interest rates, asset costs. Material costs are strongly depending on the output volume. For energy and other machine-dependent costs, the processing time is the decisive influencing factor, labor costs are strongly related with the labor time (and thus, processing as well as transition time), occupancy costs depend on the needed space. Thus, following the logic of calculation with activity units for these cost categories processing time, labor time and space are used as activity units and the corresponding costs are calculated based on them.

In case of different possible outcomes of these factors, several scenarios (as specific constellations of various outcomes of the influencing factors) can be modelled and analyzed. Here, a first differentiation of scenarios is made referring to different output volumes and the corresponding outcomes of other influencing factors, e.g. processing time. For a laboratory scale, the simultaneous processing of 4 parts (= one batch) with the machine PEMCenter 8000 is assumed. This output volume is considered as a starting point for calculation. A higher output volume of 100,000 parts is chosen for means of comparison.

Based on the results of the previous steps, target figure outcomes will be determined in the fifth step. As mentioned above, a specific variant of cost calculation with activity units is conducted for electrochemical machining of AMC as well as stainless steel. Therefore, firstly the total amount of the relevant cost categories has to be determined based on the expected outcomes of the influencing factors and corresponding activity units, then costs per activity unit are calculated and finally the process costs are derived based on the number of activity units needed by the process.

At first, the machining of AMC work pieces is examined. For calculation of the machine-dependent costs, several further assumptions regarding the machine have to be made. Amongst others, a maximum machine operation time (of 365 days and 24 hours per day minus maintenance time) is expected; maintenance time amounts to 48 hours per year. The initial investment outlay for the machine is 500,000 € and the economic life of the machine is assumed to be 25 years with a liquidation value of zero. The depreciation of 20,000 € per year is calculated by using straight-line method [17]. The maintenance costs comprise costs for general maintenance of the machine as well as costs for filter cleaning and amount to 2,100 € per year. The imputed interest is calculated by multiplying the average capital tie-up of the machine (250,000 €) with the assumed interest rate of 4 %. Based on the – debatable – assumption that the operation time is fully utilized, cost rates per minute are calculated for depreciation (0.04 €/min), maintenance costs (0.004 €/min) and imputed interest (0.02 €/min) and then assigned to the process by multiplying with the processing time of electrochemical machining which is expected to be 3 minutes per batch (4 parts). Additionally, a transition time of 0.5 minutes per batch is needed for inserting the work pieces in and removing them from the rack. This operation is done by one employee and can be executed while other work pieces are processed within the machine. Thus, no machine downtimes occur. The handling time for the cleaning amounts to 20 minutes per batch. For calculating the energy costs of the process, an energy price of 0.14 €/kWh as well as a power consumption of 1.4 kWh per work piece are estimated (the latter based on technical calculations). Occupancy costs are determined on the basis of average rent values and heating costs for factories in Chemnitz. Hence, an annual price of 82.88 € per square meter has been taken into account. For determining labor costs, again different scenarios may be realistic. One is that the employee works only in this process chain and cannot do other work (scenario 1). In that case, the relevant labor times are the processing times of electrochemical machining as well as the cleaning process. In the second extreme case (scenario 2) the employee would be capable of doing other work during the total processing of electrochemically machining (including transition). Then, the relevant times are the transition time and the processing time of the cleaning process (the electrochemical machining is an automated process). Of course, scenarios between these extreme cases may also be realistic. In each case, the labor costs are calculated by multiplying the resulting times (in hours) by the assumed hourly wage rate of 50 € (including incidental wage costs). Tab. 1 shows the results of the cost calculation for the process chain of electrochemical machining of AMCs for both scenarios.

The results show that – due to the underlying assumptions – all cost elements show a linear dependency on the output volume. Correspondingly, the costs per unit are constant. In scenario 1, labor costs are almost six times higher because of the inability of using the workforce during the processing time of electrochemical machining. Correspondingly, the total costs are considerably higher.

**Table 1.** Total costs and costs per part for electrochemical machining of AMCs for scenario 1 and scenario 2.

| Process chain | Scenario 1 | | Scenario 2 | |
|---|---|---|---|---|
| | 4 parts | 100,000 parts | 4 parts | 100,000 parts |
| **Electrochemical Machining** | | | | |
| Processing time | 3.00 min | 75,000.00 min | 3.00 min | 75,000.00 min |
| Transition time | 0.50 min | 12,500.00 min | 0.50 min | 12,500.00 min |
| Depreciation | 0.11 € | 2,869.61 € | 0.11 € | 2,869.61 € |
| Maintenance costs | 0.01 € | 301.31 € | 0.01 € | 301.31 € |
| Imputed interest | 0.06 € | 1,434.80 € | 0.06 € | 1,434.80 € |
| Labor costs | 2.50 € | 62,500.00 € | 0.42 € | 10,416.67 € |
| Energy costs | 0.04 € | 980.00 € | 0.04 € | 980.00 € |
| Occupancy costs | 0.01 € | 321.09 € | 0.01 € | 321.09 € |
| **Total process costs** | **2.74 €** | **68,406.81 €** | **0.65 €** | **16,323.47 €** |
| **Cleaning** | | | | |
| Processing time | 20.00 min | 500,000.00 min | 20.00 min | 500,000.00 min |
| Labor costs | 16.67 € | 416,666.67 € | 16.67 € | 416,666.67 € |
| **Total process costs** | **16.67 €** | **416,666.67 €** | **16.67 €** | **416,666.67 €** |
| **Manufacturing costs** | **19.40 €** | **485,073.47 €** | **17.32 €** | **432,990.14 €** |
| + **Material costs** | **4 parts** | **100,000 parts** | **4 parts** | **100,000 parts** |
| Basic raw material (AMC) | 3.771 € | 94,270.00 € | 3.771 € | 94,270.00 € |
| Electrolyte (NaNO3) | 0.001 € | 35.87 € | 0.001 € | 35.87 € |
| **Material costs** | **3.772 €** | **94,305.87 €** | **3.772 €** | **94,305.87 €** |
| = **Total costs** | **23.18 €** | **579,379.34 €** | **21.09 €** | **527,296.01 €** |
| **Costs per part** | **5.79 €** | **5.79 €** | **5.27 €** | **5.27 €** |

In the next step, the process chain costs for AMCs are compared to the costs for processing stainless steel 1.4301. Due to the longer processing time, most of the processing time-depending cost items of the steel variant are higher than those of AMCs: depreciation, imputed interest, maintenance costs, occupancy costs and energy costs. In contrast, the material costs for the basic raw material are lower when producing stainless steel work pieces. Finally, the labor costs of processing AMC as well as steel

IOP Conf. Series: Materials Science and Engineering **118** (2016) 012035 doi:10.1088/1757-899X/118/1/012035

are equal for scenario 2, because the transition time of electrochemical machining and the processing time of the cleaning process are identical. The resulting total costs and costs per part of processing stainless steel 1.4301 by ECM are shown in table 2 for both scenarios. Again, under the assumption of scenario 1 the labor costs and corresponding manufacturing costs of processing stainless steel will be considerably higher than that of scenario 2.

**Table 2.** Total costs and costs per part for electrochemical machining of stainless steel for scenario 1 and scenario 2.

| | Process chain | Scenario 1 | | Scenario 2 | |
|---|---|---|---|---|---|
| | | 4 parts | 100,000 parts | 4 parts | 100,000 parts |
| | **Electrochemical Machining** | | | | |
| | Processing time | 15.00 min | 375,000.00 min | 15.00 min | 375,000.00 min |
| | Transition time | 0.50 min | 12,500.00 min | 0.50 min | 12,500.00 min |
| | Depreciation | 0.57 € | 14,348.03 € | 0.57 € | 14,348.03 € |
| | Maintenance costs | 0.05 € | 1,291.32 € | 0.05 € | 1,291.32 € |
| | Imputed interest | 0.29 € | 7,174.01 € | 0.29 € | 7,174.01 € |
| | Labor costs | 12.50 € | 312,500.00 € | 0.42 € | 10,416.67 € |
| | Energy costs | 0.20 € | 4,900.00 € | 0.20 € | 4,900.00 € |
| | Occupancy costs | 0.06 € | 1,605.45 € | 0.06 € | 1,605.45 € |
| | **Total process costs** | **13.67 €** | **341,818.81 €** | **1.59 €** | **39,735.48 €** |
| | **Cleaning** | | | | |
| | Processing time | 20.00 min | 500,000.00 min | 20.00 min | 500,000.00 min |
| | Labor costs | 16.67 € | 416,666.67 € | 16.67 € | 416,666.67 € |
| | **Total process costs** | **16.67 €** | **416,666.67 €** | **16.67 €** | **416,666.67 €** |
| | **Manufacturing costs** | **30.34 €** | **758,485.48 €** | **18.26 €** | **456,402.14 €** |
| | **Material costs** | **4 parts** | **100,000 parts** | **4 parts** | **100,000 parts** |
| + | Basic raw material (stainless steel) | 1.458 € | 36,450.00 € | 1.458 € | 36,450.00 € |
| | Electrolyte (NaNO3) | 0.007 € | 179.35 € | 0.007 € | 179.35 € |
| | **Material costs** | **1.465 €** | **36,629.35 €** | **1.465 €** | **36,629.35 €** |
| = | **Total costs** | **31.80 €** | **795,114.83 €** | **19.72 €** | **493,031.49 €** |
| | **Costs per part** | **7.95 €** | **7.95 €** | **4.93 €** | **4.93 €** |

The last step of the procedure model consists of the comparison, sensitivity analysis and interpretation of the outcomes. As tables 1 and 2 show, for scenario 2 the total costs (and costs per part) of processing stainless steel are slightly lower than that of AMC for every production volume considered. Under scenario 1, the result is different. Here, the total costs of electrochemically machined AMCs are lower than that of steel for all production volumes due to the stronger influence of the longer processing time of steel (resulting in five times higher labor costs). This shows that the economical advantageousness of materials and their processing will depend on the characteristics of the production system.

Concluding, it has to be admitted that a couple of simplifying assumptions have been made concerning the calculations. One of them is the supposition of linear relationship between machine operation time (transition and processing time) and costs. Thus, learning effects and other economies of scale effects such as declining factor prices are neglected. Some of these assumptions may be cancelled or modified in the integrated technical-economic evaluation, the last element of the evaluation approach.

## 5. Integrated Technical-Economic Evaluation

An integrated technical-economic evaluation is necessary if not all of the decision-relevant technical features and/or their influences have been included in the economic evaluation. It can be conducted in different ways: by building a list of additional arguments for or (dis)advantages of the alternatives, by visualizing measures for the technical and economic attractiveness or profitability in diagrams [19] or by using multi-criteria decision-making methods such as utility value method [17]. Here, for sake of simplification and due to the restricted extent of the paper, the first option is chosen.

A first additional point concerns the lower processing time of AMCs compared to steel. This difference does not only influence costs (as considered in section 4) but also the maximum output quantity of the process chains which is considerably larger for AMCs. Thus, also the achievable revenues and the corresponding margins tend to be higher. Secondly, AMCs are currently at an earlier stage of the life cycle compared to stainless steel. Thus, more improvements and cost reductions have to be expected.

Furthermore, the properties of the materials and their influence on the work pieces have to be considered. The most important differing properties are the density and the E-Modulus (see section 3). As shown, the density of AMC is more than two times lower than that of stainless steel 1.4301. In terms of weight saving, this implies that parts consisting of AMC are more than two times lighter than comparable stainless steel parts. This property is inquired especially for lightweight construction in the automobile or aircraft industry. In these industries, already slight weight reductions result in notable economic benefits that justify higher prices of the material/parts. The economic evaluation indicates that the costs of AMC parts are not extensively higher than that of stainless steel parts (if they are higher at all). For being advantageous, the economic benefit of using AMCs should exceed the additional costs. For example, in aircraft industry, a weight saving of one kilogram results in a monetary benefit between 100 and 500 € [18]. Thus, it can be assumed that this technical property of AMCs generates a considerable economic potential. Indeed, it has also to be taken into account that the E-Modulus of AMC is only half as large as that of stainless steel. This implies a lower ductility. Thus, AMC parts should be used when lightweight is more important than ductility and not for highly stressed parts.

## 6. Summary and Outlook

In this paper, the procedure of electrochemical machining of AMCs as well as the resulting AMCs have been evaluated from a technical as well as economic point of view. After presenting an integrated evaluation approach, the processing of electrochemical machining has been described and a technical evaluation of electrochemical machining of AMC compared to stainless steel has been conducted. Based on this, the process chains of electrochemical machining of AMCs and steel were evaluated by cost calculation. The results show relatively small differences of total costs with the results strongly depending on the scenarios. The integrated evaluation finally reveals additional advantages of AMCs in terms of smaller capacity utilization and above all a lower density generating a high potential for lightweight applications. Further work should be directed to refining the methodology with respect to the modelling of production scenarios and cost functions as well as its verification by applying the methods in industrial practice. Furthermore, the studies should be extended to life cycle assessment in order to include ecological effects and thereby the contribution of the entire process chains and materials to sustainability.

## Acknowledgment

We gratefully thank the Deutsche Forschungsgemeinschaft (DFG) for funding this work within the Collaborative Research Centre SFB 692.

## References

[1] Chawla K K 2012 *Composite Materials 3rd ed.* (New York: Springer)
[2] Hackert-Oschätzchen M 2015 *Gestaltung von elektrochemischen Abtragprozessen durch Multiphysiksimulation gezeigt an der Endformgebung von Mikrobohrungen*

Habilitationsschrift (Verlag Wissenschaftliche Scripten)

[3] Götze U, Koriath H J, Kolesnikov A, Lindner R and Paetzold J 2012 Integrated methodology for the evaluation of the energy- and cost-effectiveness of machine tools *CIRP J. Manuf. Sci. Technol.* **5** 3 151–163

[4] Klocke F and König W 2007 *Fertigungsverfahren 3. Abtragen, Generieren und Lasermaterialbearbeitung 4th ed.* (Berlin Heidelberg: Springer)

[5] Schubert A, Meichsner G, Hackert-Oschätzchen M, Zinecker M and Edelmann J 2011 Pulsed Electrochemical Machining of Powder Metallurgy Steels *Proc. 7th Int. Symp. Electrochem. Mach. Technol* 68–75

[6] Schubert A, Hackert-Oschätzchen M, Lehnert N, Martin A and Meichsner G 2015 Analysis of the Electrochemical Dissolution Characteristics of High-Strength Particle Reinforced Aluminium Matrix Composites *ICSMA-17*

[7] Schubert A, Meichsner G, Boenig L, Hackert-Oschätzchen M, Krönert M and Edelmann J 2013 Resource-Efficient Design of Precise Electrochemical Machining *Proc. 9th INSECT* 53–64

[8] VDI-Richtlinie 2221 1993 *Methodik zum Entwerfen und Konstruieren technischer Systeme und Produkte* 5th ed. (Düsseldorf: VDI)

[9] Götze U, Hertel A, Schmidt A, Päßler E and Kaufmann J 2014 Integrated Framework for Life Cycle Oriented Evaluation of Product and Process Technologies: Conceptual Design and Case Study *Technology and Manufacturing Process Selection* ed E Henriques et al (London: Springer) 193–215

[10] Götze U, Schmidt A and Weber T 2010 Vorgehensmodell zur Abbildung und Analyse des Lebenszykluserfolges von Werkstoffen – Konzeption und beispielhafte Veranschaulichung *Materialwiss. Werkstofftech.* **41** 6 464–475

[11] Laux H, Gillenkirch R M and Schenk-Mathes H Y 2014 *Entscheidungstheorie* 9th ed. (Berlin Heidelberg: Springer)

[12] Götze U, Hache B, Schmidt A and Weber T 2011 Methodik zur kostenorientierten Bewertung von Prozessketten der Werkstoffverarbeitung *Materialwiss. Werkstofftech.* **42** 7 647–657

[13] Kaplan R S and Anderson S R 2007 *Time-driven activity-based costing: A simpler and more powerful path to higher profits* (Boston, Mass.: Havard Business School Publishing)

[14] Bhimani A, Horngren C T, Datar S M and Rajan M V 2012 *Management and Cost Accounting* 5th ed. (Harlow et al: Financial Times/Prentice Hall)

[15] Ehrlenspiel K, Kiewert A und Lindemann U 2007 *Cost-Efficient Design* ed M S Hundal (Berlin Heidelberg: Springer)

[16] Konarsky M, Götze U and Leidich E 2014 An IT-based cost information system for cost planning and monitoring in customized single-unit production. Re-Engineering Total Cost Management. *Proc. of ICEC 2014 - IX World Congress* 20.-22.10.2014

[17] Götze U, Northcott D and Schuster P 2015 *Investment Appraisal Methods and Models* 2nd ed. (Berlin Heidelberg: Springer)

[18] Reuter M 2014 *Methodik der Werkstoffauswahl. Der systematische Weg zum richtigen Material* 2nd ed. (München: Hanser)

[19] Peças P, Ribeiro I and Henriques E 2014 Life Cycle Engineering for Materials and Technology Selection: Two Models, One Approach *Procedia CIRP15* 543–548

IOP Conf. Series: Materials Science and Engineering **118** (2016) 012036 doi:10.1088/1757-899X/118/1/012036

# Jet Electrochemical Machining of Particle Reinforced Aluminum Matrix Composites with Different Neutral Electrolytes

**M Hackert-Oschätzchen, N Lehnert, A Martin and A Schubert**

Technische Universität Chemnitz, Professorship Micromanufacturing Technology, Chemnitz, Germany

E-mail: norbert.lehnert@mb.tu-chemnitz.de

**Abstract.** Conventional mechanical machining of particle reinforced aluminum matrix composites (AMCs) is challenging because the hard ceramic particles in the soft aluminum matrix lead to an increased tool wear. Furthermore, the mechanical and thermal impact during conventional machining affects the microstructure of the AMCs. Electrochemical machining (ECM) is an alternative method to machine AMCs. Based on anodic dissolution, ECM has a slight influence on the work piece material structure and is independent of material strength and hardness. So the microstructure of the work piece remains unaffected. One method of ECM is electrochemical machining with continuous electrolytic free jet (Jet-ECM). Hereby the electrochemical removal is localized by the geometry of the electrolyte jet. By moving the electrolyte jet micro-structures and microgeometries can be generated quickly and flexibly in metallic parts [1]. Another advantage of Jet-ECM is the low consumption of electrolyte which allows an easy and inexpensive change of electrolyte for investigations with different types of electrolyte. In this study AMCs reinforced with different amounts of SiC-particles are machined with two pH-neutral electrolytes using Jet-ECM. The results provide information about the suitability of the selected electrolytes for the machining of AMCs. In addition, the influence of the particle content on the electrochemical removal result will be evaluated.

## 1. Introduction

The requirements concerning material properties are rising steadily, especially for safety-related components. If a material should have a low density to reduce mass, but also should have a high stiffness, conventional materials like metals, ceramics or polymers, can hardly meet the demands. Composite materials, which are a combination of two or more physically and chemically different materials, can fulfill such requirements. The different materials exist as separated phases in the composite. The major phase is the matrix material and the minor phase is the reinforcement, which is shown schematically in figure 1 for a particle reinforced metal matrix composite. The properties of the composite material, such as E-modulus or thermal expansion, are defined by the used materials and their proportions. So the new material may have properties similar to steel, though with lower density, which leads to a mass reduction. Such materials are used e.g. in aeronautics and automotive parts [2].

In the Collaborative Research Center SFB 692 HALS at the Technische Universität Chemnitz several academic institutions work together in the investigation of aluminum matrix composites

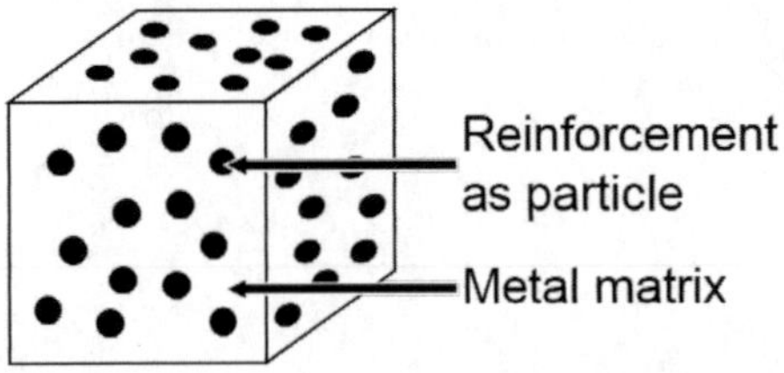


**Figure 1.** Scheme of the microstructure of a particle reinforced metal [3]

(AMCs). One main task is the finishing machining of AMCs by electrochemical machining (ECM).

In this study jet electrochemical machining (Jet-ECM) of AMCs using neutral electrolytes is investigated. The AMCs are the aluminum alloy EN AW 2017 reinforced with SiC-particles with diameters less or equal to 1 µm. The applied electrolytes are aqueous solutions of $NaNO_3$ or NaCl with a conductivity of approximately 185 $\frac{mS}{cm}$ at a temperature of 25 °C.

## 2. Jet Electrochemical Machining

Electrochemical machining is based on the anodic dissolution of work piece material through electric charge transport. The dissolution takes place at the interface between the work piece surface and the electrolyte. The special characteristic of Jet-ECM is the supply of fresh electrolyte through a micro nozzle with a mean jet velocity of approximately 20 $\frac{m}{s}$ [1, 3, 4]. Figure 2 shows the principle of Jet-ECM.

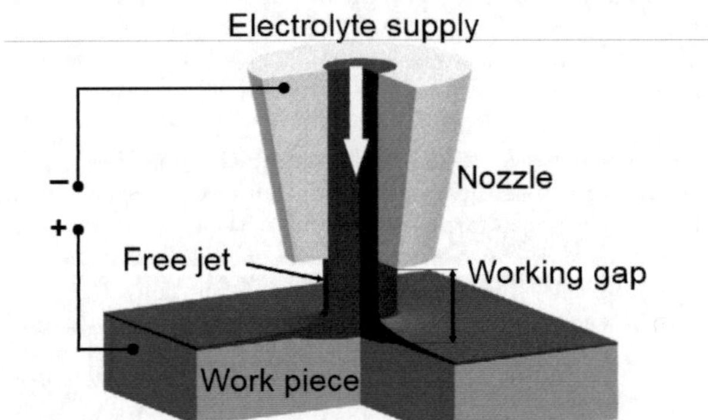


**Figure 2.** Principle of Jet-ECM [1, 3, 4]

The electrolyte is ejected vertically towards the work piece surface. Due to the high velocity, a closed free jet is formed in the surrounding atmospheric air. The current density is locally confined by the electrolyte jet leading to a highly localized machining area [1, 4].

In the jet current densities up to 2000 $\frac{A}{cm^2}$ can be realized [5]. As a result, excellent surface qualities can be achieved. The high velocity of the electrolyte jet leads to a very good supply with fresh electrolyte [1, 4].

The experiments of Jet-ECM of particle reinforced AMCs with different neutral electrolytes were performed on a modular prototype system. Figure 3 shows a scheme of the setup.

The experimental setup is mounted on a table and a portal made of granite to guarantee the required mechanical and thermal stiffness. The relative movement between nozzle and work piece is carried out by a three-axis positioning system. A pulsation-free pump transports the electrolyte to the nozzle. The electrolyte is ejected in z direction towards the work piece. Used electrolyte is collected in a disposal tank. A generator system supplies electric voltage between the nozzle and the work piece, providing the process current. A personal computer serves as control system for all electrical and kinematic operations of the system [1, 3, 4].

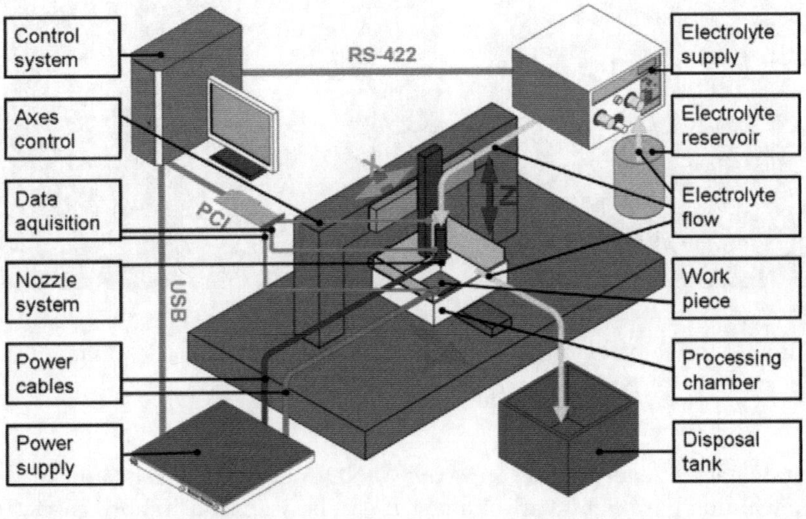


**Figure 3.** Scheme of the experimental setup of Jet-ECM [1, 4]

## 3. Experimental Parameters

The work piece materials are AMCs with a matrix of EN AW 2017 and a reinforcement content of 5% or else 10% SiC-particles. The particles have a diameter of less or equal 1 µm. Both AMCs were produced within the Collaborative Research Center SFB 692 HALS. As mentioned above, two pH-neutral electrolytes were used to perform the experiments. For comparability the aqueous solutions of $NaNO_3$ or NaCl have the same electric conductivity of approximately 185 $\frac{mS}{cm}$ at a temperature of 25 °C. The used nozzle has an inner diameter of 100 µm.

For the experiments a standardized machining strategy, point machining, was used. Point machining generates calottes on the work piece surface. Thereby the machining time and the applied voltage were varied. Table 1 summarizes selected parameters of the experiments.

**Table 1.** Selected experimental parameters

| Parameter | Values |
|---|---|
| Work piece materials | EN AW 2017 + 5% SiC-particles |
| | EN AW 2017 + 10% SiC-particles |
| Electrolytes | $NaNO_3$, NaCl |
| Electrolyte conductivity | 185 $\frac{mS}{cm}$ |
| Nozzle diameter | 100 µm |
| Working gap | 100 µm |
| Machining time for point machining | 0.5 s, 1 s, 1.5 s, 2 s |
| Applied voltage | 10 V, 15 V, 20 V, 25 V |

## 4. Experimental Results

The generated geometries were evaluated qualitatively by scanning electron microscopy and quantitatively by confocal microscopy. Figure 4 shows SEM images of calottes generated on EN AW 2017 + 10% SiC at 20 V and 2 s machining time.

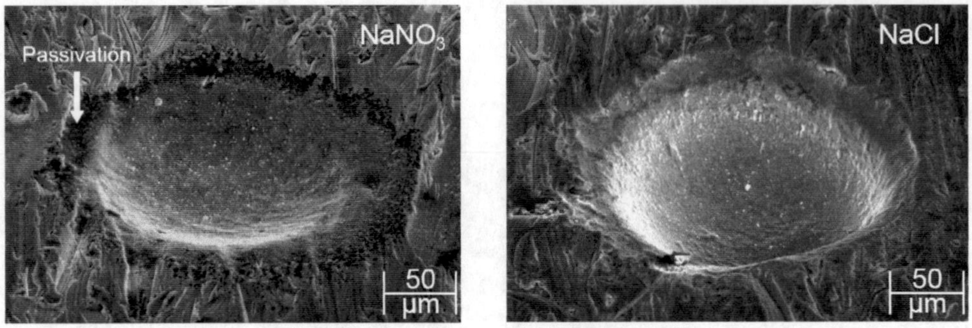


**Figure 4.** SEM images of calottes on EN AW 2017 + 10% SiC-particles machined with 20 V, 2 s and aqueous electrolyte of $NaNO_3$ and NaCl

The images show calottes generated by applying $NaNO_3$ electrolyte (left) and NaCl electrolyte (right). Further machining parameters are similar. It can be seen that in both cases the machined surface is much smoother than the initial work piece surface. At the edge of the calotte machined with $NaNO_3$, a region can be spotted that appears black in the SEM image. This region indicates passivation effects [3]. The SEM image of the calotte generated by applying NaCl electrolyte indicates no passivation effects. The edge appears more rounded than with $NaNO_3$. The calotte generated with NaCl appears deeper than the one generated by applying $NaNO_3$ electrolyte. This observation is confirmed by confocal microscopy. Figure 5 shows section views of calottes according to the SEM-images in figure 4.

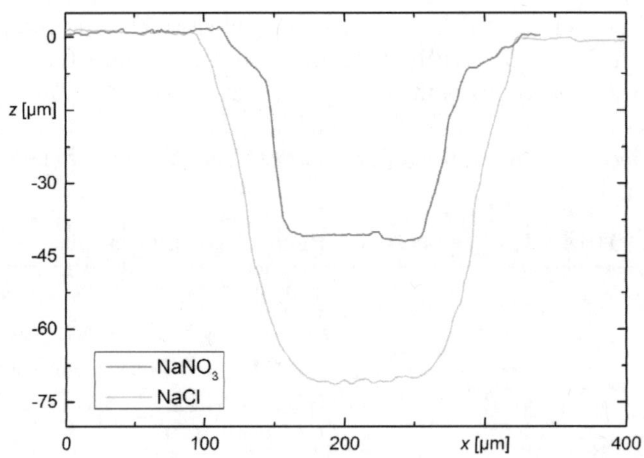


**Figure 5.** Section views of calottes on EN AW 2017 + 10% SiC-particles machined with 20 V, 2 s and aqueous electrolyte of $NaNO_3$ and NaCl

Obviously, the calotte generated by applying NaCl is significantly wider and deeper than the other calotte, which implies that the AMC dissolves much faster by using NaCl electrolyte. Figure 6 shows diagrams of diameters d and depths z of the calottes on EN AW 2017 + 10% SiC-particles as a function of time with variation of the applied voltage for each electrolyte.

In both cases it can be seen that diameter and depth increase with machining time and voltage. The calottes have equal diameters of approximately twice the nozzle diameter,

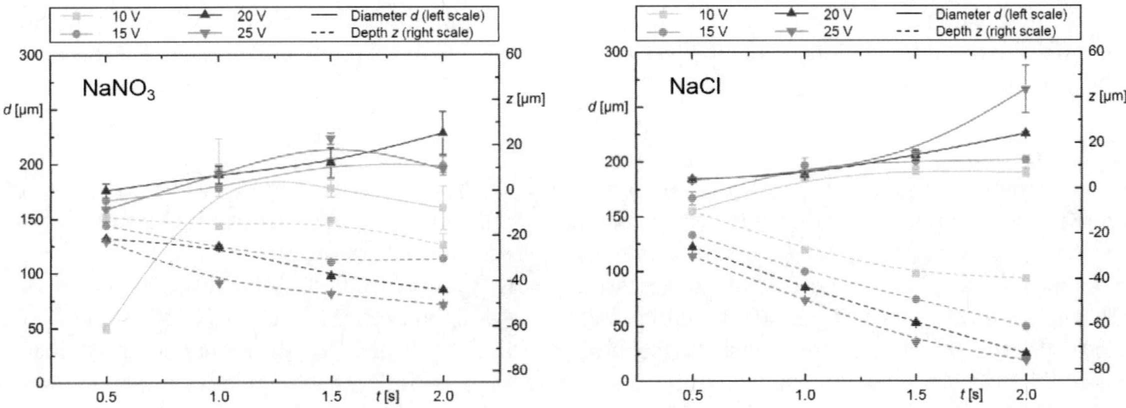


**Figure 6.** Diameter $d$ and depth $z$ of calottes on EN AW 2017 + 10% SiC-particles machined with aqueous electrolyte of $NaNO_3$ and NaCl as a function of machining time $t$ and applied voltage

regardless of the used electrolyte. This is a result of the high localization of the current density in Jet-ECM. The function of the diameter of the calottes machined applying $NaNO_3$ electrolyte at 10 V is significantly different. This implies that, by using $NaNO_3$ as electrolyte, a voltage of 10 V is too low to establish a reproducible dissolution process. As mentioned above, the calottes machined with the NaCl electrolyte are deeper for most of the experiments. It can be estimated that the calottes machined using NaCl are up to twice as deep as the calottes machined using $NaNO_3$. Figure 7 shows diagrams similar to figure 6 for the calottes on EN AW 2017 + 5% SiC-particles.

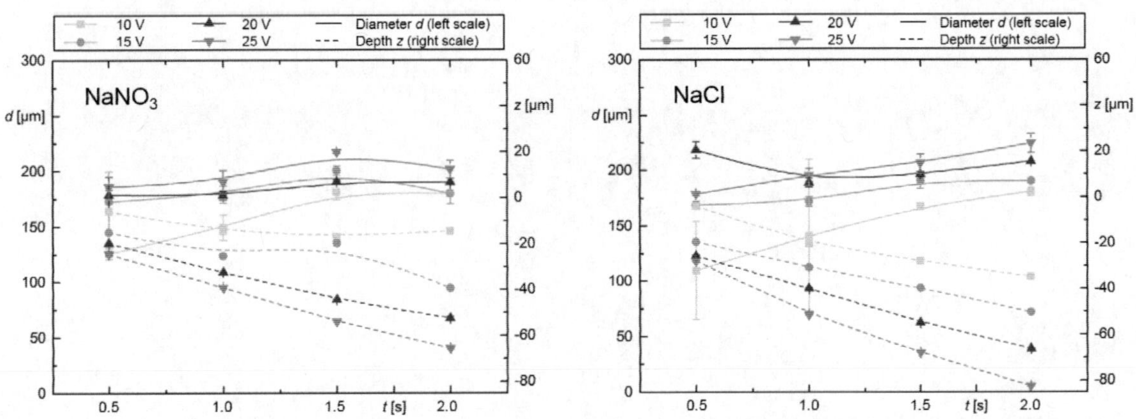


**Figure 7.** Diameter $d$ and depth $z$ of calottes on EN AW 2017 + 5% SiC-particles machined with aqueous electrolyte of $NaNO_3$ and NaCl as a function of machining time $t$ and applied voltage

Similar to the AMC with 10% SiC-particles, diameter and depth increase with machining time and voltage. Also, the calottes machined by applying NaCl are deeper than the calottes machined by applying $NaNO_3$ in most cases. Comparing the results of the two work piece materials, as shown in figures 6 and 7, the different amounts of SiC-particles have no significant influence on the dimensions of the generated calottes. This can be attributed to the small difference of their percentage.

## 5. Conclusion

Jet electrochemical machining of particle reinforced aluminum matrix composites with different neutral electrolytes was evaluated in this study. The suitability of aqueous electrolyte of $NaNO_3$ and NaCl for machining AMCs and the machining results were analyzed.

Generally speaking, Jet-ECM of the investigated AMCs was applied successfully using the two pH-neutral electrolytes. It could be shown, that the aqueous electrolytes of $NaNO_3$ and NaCl cause different electrochemical dissolution characteristics. While the diameters of the calottes created with both electrolytes are similar, the usage of NaCl electrolyte results in significantly deeper calottes for machining times of about 1.5 to 2 seconds. By using $NaNO_3$ passivation effects can be spotted. To achieve reproducible dissolution behavior when applying $NaNO_3$ electrolyte the voltage should be adjusted above 10 V. The slight variation of the amount of SiC-particles, from 10% to 5%, has no significant influence on the machining results.

## Acknowledgment

The authors acknowledge the DFG (German Research Foundation) for supporting this work carried out within the framework of Collaborative Research Center SFB 692 HALS.

## References

[1] Hackert-Oschätzchen M, Meichsner G, Zinecker M, Martin A and Schubert, A 2012 *Precis. Eng.* **36 4** 612–9
[2] Chawla K K 2012 *Composite Materials 3rd ed.* (New York: Springer)
[3] Hackert-Oschätzchen M, Morgenstern R, Lehnert N, Martin A, Lampke T and Schubert A 2015 *Proc. of 11th International Symposium on Electrochemical Machining Technology* 35–42
[4] Hackert-Oschätzchen M, Martin A, Meichsner G, Zinecker M and Schubert A 2013 *Precis. Eng.* **37 3** 621–34
[5] Schubert A, Hackert-Oschätzchen M, Meichsner G, Zinecker M and Martin A 2011 *Proc. of 7th International Symposium on Electrochemical Machining Technology* 47–54.

IOP Conf. Series: Materials Science and Engineering **118** (2016) 012037 doi:10.1088/1757-899X/118/1/012037

# Microstructure of arc brazed and diffusion bonded joints of stainless steel and SiC reinforced aluminum matrix composite

**M. Elßner[1], S. Weis[1], T. Grund[1], G. Wagner[1], S. Habisch[2] and P. Mayr[2]**

[1] Chair of Composite Materials, Institute of Materials Science and Engineering, Technische Universitaet Chemnitz, 09125 Chemnitz, Germany
[2] Institute of Joining and Assembly, Technische Universitaet Chemnitz, 09107 Chemnitz, Germany

michael.elssner@mb.tu-chemnitz.de
stefan.habisch@mb.tu-chemnitz.de

**Abstract.** Joint interfaces of aluminum and stainless steel often exhibit intermetallics of Al-Fe, which limit the joint strength. In order to reduce these brittle phases in joints of aluminum matrix composites (AMC) and stainless steel, diffusion bonding and arc brazing are used. Due to the absence of a liquid phase, diffusion welding can reduce the formation of these critical intermetallics. For this joining technique, the influence of surface treatments and adjusted time-temperature-surface-pressure-regimes is investigated. On the other hand, arc brazing offers the advantage to combine a localized heat input with the application of a low melting filler and was conducted using the system Al-Ag-Cu. Results of the joining tests using both approaches are described and discussed with regard to the microstructure of the joints and the interfaces.

## 1 Introduction

In the field of light-weight materials AMC are the subject of current research work. The advantages of AMC in comparison to aluminum alloys are a high specific strength, an increased wear resistance and a low coefficient of thermal expansion (CTE) [1-3]. Numerous applications in aerospace and automotive industry as well as plant manufacturing can be found. Thermal expansion matched heat transfer components can be mentioned for example. To realize this application, an appropriate joining technology for dissimilar joints of AMC and stainless steel is required. In recent works, the suitability of different joining technologies and treatments was investigated. For example, joints of aluminum and steel are realizable by using the TIG (Tungsten Inert Gas) welding process. The TIG welding of AMC leads to an inhomogeneous welding area. Furthermore an unequal distribution of the reinforcing particles and the formation of intermetallics can occur. The properties of these joints are insufficient [4]. Brittle intermetallics like $Al_{13}Fe_4$, $Al_5Fe_2$, $Al_2Fe$ and $Al_5Fe_4$ downgrade the properties of the joints. To reduce the formation of the intermetallics, a zinc coating is used [5]. Another approach is the decrease of the heat input while joining. This is also important because of the thermal induced damage of the AMC that leads to ineligible porosities and inhomogeneities [6]. This can be avoided by another process to join steel and AMC: Friction stir welding [7]. During the process the tool, however, is affected by the reinforcement particles, which leads to a considerable wear. Therefore, the process conditions change continuously, which affect the reproducibility of the process results [6]. Transient liquid phase bonding (TLP bonding) is another suitable process for joining AMC. One advantage is that the surface condition is not important due to the liquid state of the interlayer at joining temperature [8, 9]. Askew et al. investigated TLP bonding of AMC with nickel interlayer [9]. The results indicate, that the shear strength increases by limiting the bond line width to 1 µm, but the joints were not free of defects and failures.

A high potential has arc brazing. The melting temperature of the filler is lower than the base materials one, which avoids thermal damage of the AMC. Due to the focused arc, the heat input is reduced in comparison to TIG welding. Another advantage is the degradation of the oxide layer on the AMC induced by the arc, which improves the wetting. The formation of intermetallics can be reduced by an adapted filler. For brazing of aluminum alloys, Al-Si based [10-12] and Zn-Al based [13, 14] fillers are commonly used. The disadvantages are the limited number of aluminum alloys brazable by Al-Si fillers and the poor corrosion resistance of Zn-based fillers. In this work, a filler based on the ternary system Al-Ag-Cu, designed for arc brazing, is investigated. The eutectic composition of 40 wt% Al, 40 wt% Ag and 20 wt% Cu (named Al40Ag20Cu) has a melting temperature of 506 °C, measured by differential scanning calorimetry (DSC) [14].

Diffusion bonding is another suitable joining technique to avoid the formation of intermetallics within joints of aluminum and stainless steel [15, 16]. Zhang et al. investigated diffusion joints of AMC, with different SiCp volume fractions and interlayer materials [17]. The tensile strength of these joints decreases with an increasing volume fraction [17, 18]. It is also limited by the surface preparation or surface roughness of the samples [17-19]. Furthermore, an interlayer was used to compensate the difference in surface roughness of the two samples and to improve the joint properties [17, 20, 21]. By using an active interlayer material, e.g. lithium, the surface oxides can be reduced. Also the formation of intermetallics could be avoided [21]. However, there is still a need for research for joints of AMC and stainless steel.

## 2 Experimental Procedure

In the present study the used base materials are a stainless steel and a powder-metallurgically produced aluminum matrix composite (AMC). The stainless steel is grade X2CrNi18-10 (material no.: 1.4301 / AISI 304). The matrix of the AMC is similar to EN AW-2017, reinforced by 10 % SiC particles with an average particle size of 1.5 µm. The feedstock of the AMC is mechanically alloyed in a high energy ball mill. A rod is produced by hot isostatic pressing, extruding is carried out afterwards. Different sample geometries are necessary for arc brazing and diffusion bonding experiments, figure 1. The generated rods with a cross-section of 15 x 15 mm² are cut in half lengthwise and divided into 50 mm sections for arc brazing. The cut surfaces are ground and polished (Rz < 1 µm) prior joining tests. The stainless steel sheets with a thickness of 1.5 mm have dimensions of 20 x 50 mm². The surface is polished using a hand grinder and cleaned by ethanol afterwards. For diffusion bonding the same materials in a smaller sample size were used. The dimensions of the AMC samples are 14 x 14 x 10 mm³, that of stainless steel samples are 14 x 14 x 1,5 mm³. Important parameters for diffusion bonding are in addition to the temperature-time-load-regime the surface condition or the surface roughness. Therefore, the joining surfaces of the samples were analysed in accordance with ISO 25178 and in dependence of the machining process. The stainless steel surface was used in the as-delivered condition with a surface roughness Sz = 15.4 µm. The AMC joining surface was turned to a surface roughness of Sz = 14.2 µm and additionally sand-blasted to Sz = 32.3 µm. The higher surface roughness of the sand-blasted samples causes micro surface deformations during the application of the load, which increases the number of lattice defects at the interface and thus increases the diffusibility. The plane parallelism of each sample was measured to ensure a deviation of less than 10 µm, so that a tight interface contact and hence defect-free joint can be reached. To remove the oxide layer or impurities on the joining surfaces, which limit the diffusivity, all samples were chemically treated in a glove box in inert atmosphere. The AMCs were etched in 5% NaOH for 45 s and the stainless steel samples were cleaned with acetone. After the chemical surface treatment, both materials were batched and moved under inert atmosphere protection into the joining chamber.

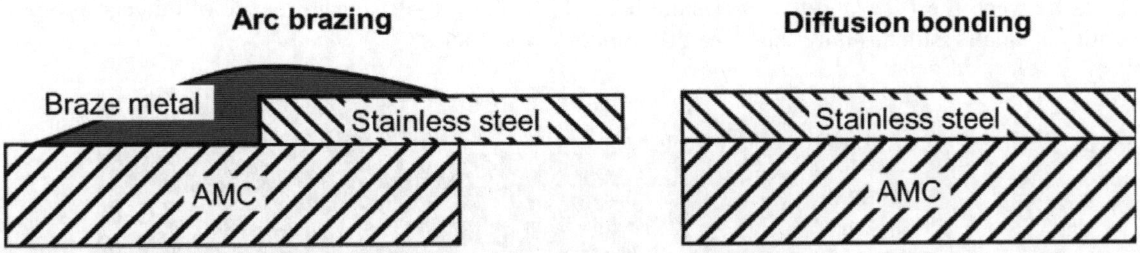


**Figure 1.** Different joining set ups for arc brazing and diffusion bonding

### 2.1 *Arc brazing*

For arc brazing a new developed filler alloy of the system Al-Ag-Cu-Si is used [22]. The base composition is the eutectic with 40 wt% Al, 40 wt% Ag and 20 wt% Cu, named Al40Ag20Cu. In a previous work the wettability of stainless steel was improved by additional alloying with 1.5 wt% Si [23]. Al40Ag20Cu + 1.5 wt% Si has a melting temperature of 498 °C, measured by DSC. All filler alloys are prepared by melting and continuous casting. Therefor B-Ag72Cu28 as master alloy is molten in the continuous caster with additions of Al (purity 99.99 vol%), Cu (purity 99.9 vol%) and Si (purity 99.9999 vol%). The induction melting under Ar atmosphere (purity 99.996 vol%) takes place in a graphite crucible of the quality GM-125i. A wire of 2 mm diameter is produced by continuous casting. The cast specimens are analyzed by X-ray fluorescence analysis (voltage 30 keV) to ensure a good homogeneity. All brazing tests are carried out in an inert atmosphere (Ar). A TIG welding source with a maximum current of 250 A which enable alternating current (AC) is employed for this purpose. The applied current intensity of 40 A is a result of previous wetting tests [23].

### 2.2 *Diffusion bonding*

All experiments are carried out in a diffusion bonding machine with a maximum joining temperature of 1000 °C, 80 t max. load and a chamber size of 300 x 300 220 mm³. Additionally, a glove box is directly connected with the diffusion bonding machine, so that it is possible to treat the joining surfaces mechanically or chemically in defined atmosphere and move the samples into the chamber under inert atmosphere. The joining tests are carried out under inert atmosphere 700 hPa Argon (99,999% Ar) at different temperatures (380-560 °C) over a time of 60 min. A constant load of 10 N/mm² is used in the present study. Only the joining temperature is varied to examine the microstructural changes of the stainless steel-AMC interface as a function of the temperature. The parameter sets were selected to avoid the formation of the critical Al-Fe intermetallics by a systematic temperature reduction. 560 °C as highest joining temperature is chosen, since the intermetallic compound CuAl and Cu11,5Al9 form at higher temperatures and the possible interaction of the Cu-Al and Al-Fe intermetallics can be investigated. The characterization of the joined samples after arc brazing and diffusion bonding is performed by light microscopy (LM) and scanning electron microscopy (SEM). The composition of the phases and the diffusion zones are examined by energy dispersive X-ray spectroscopy (EDXS) and X-ray diffraction analysis (XRD). The hardness is determined as Martens hardness (HM) profiles across the joint with a measuring force of 50 mN. The values in HM are transferred in Vickers hardness (HV) to compare the hardness results with typical literature values.

## 3 Results and Discussion

### 3.1 *Arc brazing*

For arc brazing the AMC/braze metal interface and the stainless steel/braze metal interface are considered. The first one shows a diffusion zone in the direction of braze metal near to the AMC and a zone of alloying afterwards, figure 2. The alloying results in an increase in hardness at the interface and the alloying zone, 280 HV0.005 to about 350 HV0.005 at the interface. The hardness of the pure filler was measured at 306 HV1, previously. The difference in the hardness values is a result of the alloying processes between the filler and the base material. Additionally, the rising influence of defects like porosities at small test loads may cause the differences in the values.

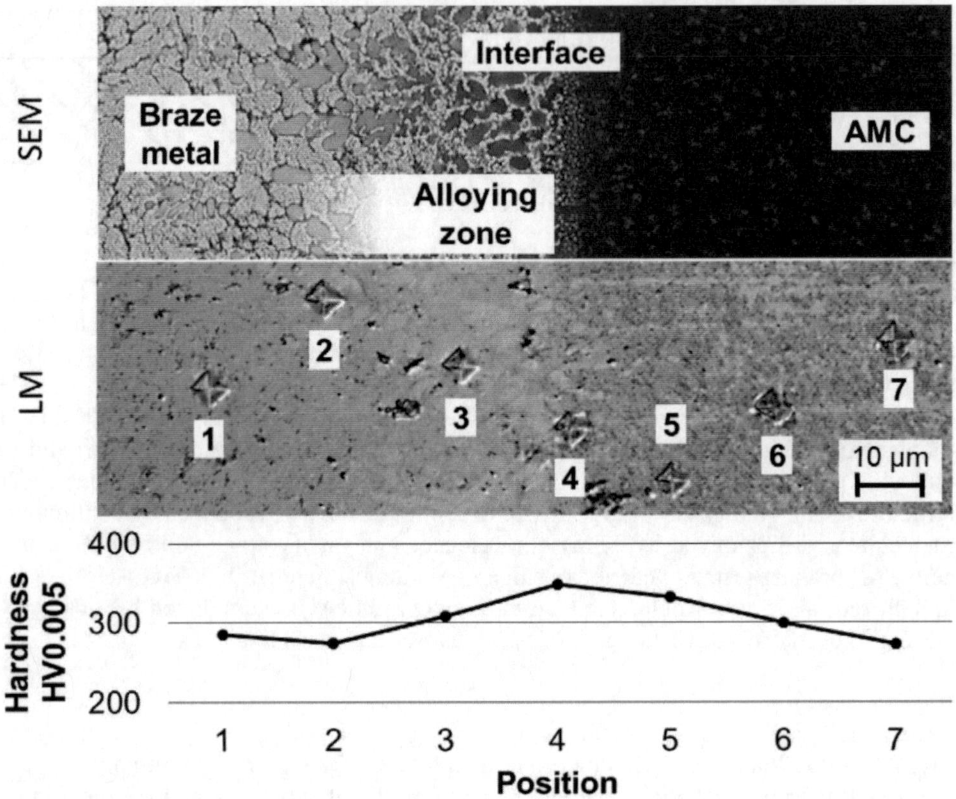


**Figure 2.** Microstructure (SEM) of AMC/braze metal interface, positon of indents (LM) and hardness

Before brazing the eutectic filler consists of the phases $Ag_2Al$, $Al_2Cu$, a solid solution of Al and additional Si, analyzed by XRD. The peak in hardness is a result of alloying processes between the filler and the base material and the development of a solid solution of aluminum and intermetallics. After brazing the phases $Ag_2Al$, $Al_2Cu$ and a solid solution of Al are detected using XRD analysis. The EDXS shows an increased content of Ag (up to 5 at%) in the solid solution of Al. This amount is higher than the solubility in Al in equilibrium state (< 2 at%). Therefore, Ag rich intermetallics are precipitated and lead to an increased hardness. Subsequently the hardness slowly decreases to about 280 HV0.005 towards the AMC. The hardness of the AMC doesn't drop to lower values, which indicates a diffusion zone or a section of heat treatment near the joint. The EDXS mapping in this section shows that no element of the filler enriches the AMC near the interface. A diffusion zone is not the reason for the hardness profile in AMC. It can be seen that the distribution of Cu within the AMC matrix metal is finer as in the untreated base material. Due to the heat treatment during brazing, a precipitation hardening effect is one possibility to explain the increase in the hardness of the aluminum alloys. No cracks can be detected at the interface to AMC.

At the interface to stainless steel the diffusion and reaction zones appear differently in comparison to the AMC, figure 3. The morphology of the braze metal and the filler is dissimilar. The reaction zone is large in its areal extent. The hardness of the braze metal (550 HV0.005) increases after brazing due to the formation of intermetallics. The hardness values reach a maximum of about 870 HV0.005 at the interface between the phase seam and the braze metal. In the direction of the braze metal the hardness decreases faster as in the direction of the stainless steel. The hardness value of the braze metal at the interface to the reaction zone near the stainless steel (570 HV0.005) is approx. two times higher than at the interface to the AMC, figure 2.

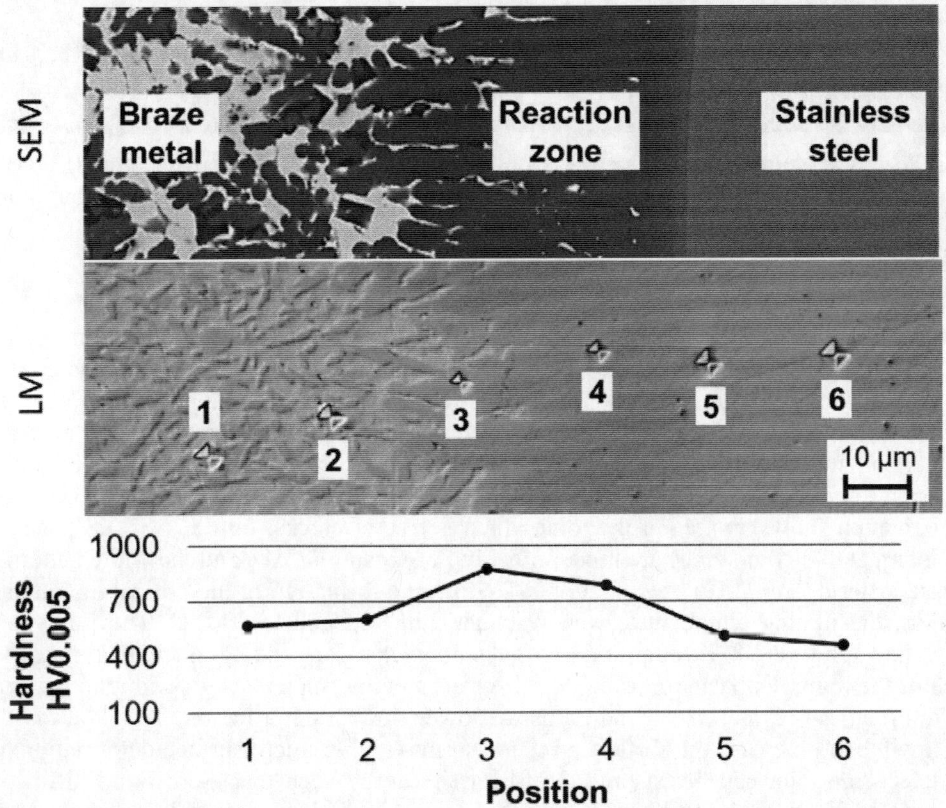


**Figure 3.** Microstructure (SEM) of stainless steel/braze metal interface, positon of indents (LM) and hardness

The EDXS shows a reduction of the content of Al in the reaction zone from braze metal to the stainless steel. The increased hardness is probably a result of the formation of intermetallics of the system Fe-Al. Using XRD analysis, the different phases AlFe, Al-Cr-Fe and Al-Fe-Ni are determined. The correlation of the pattern to the phases is difficult due a marked texture. The highest hardness (870 HV0.005) indicates the phase FeAl3 (892 HV1) [24]. In some places, cracks can be observed in the joining zone, figure 4 and figure 5. Cracks in joints of aluminum and stainless steel mainly occur in the brittle intermetallics. The zone of reaction and diffusion consists of different layers, figure 5. Cracks occur at the interface between the Al rich phases of the system Al-Fe and the Fe rich phases. The reason is the difference in hardness, FeAl3 (892 HV1) and AlFe (470 HV1) [24].

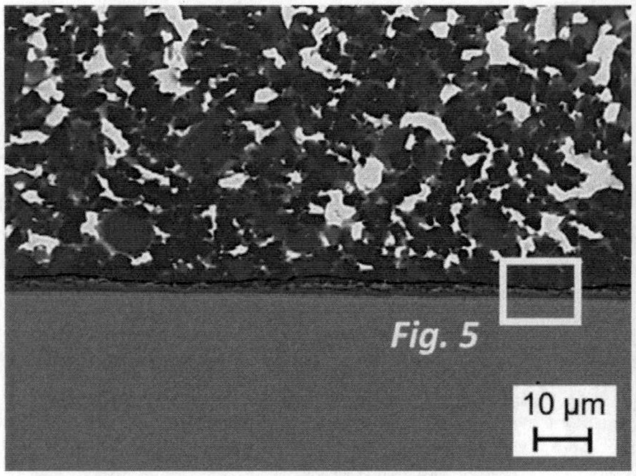


**Figure 4.** Microstructure (SEM) of stainless steel/braze metal interface

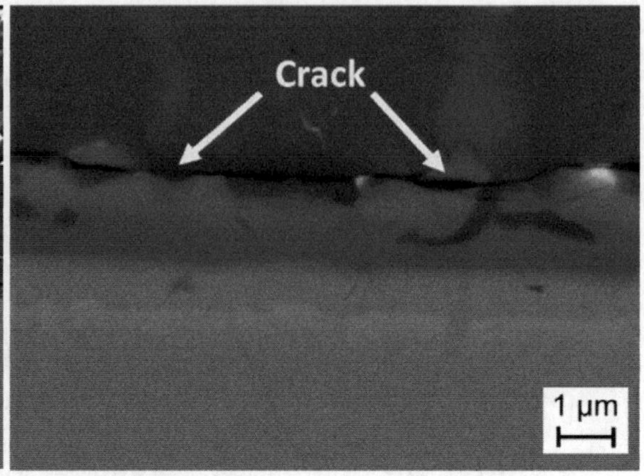


**Figure 5.** Detailed microstructure (SEM) stainless steel/braze metal interface with crack in the reaction zone

*3.2 Diffusion welding*

Figure 6 shows a cross-section of the stainless steel-AMC joint with an almost 20 µm wide diffusion zone across the interface. No pores have been detected in this zone, as the AMC-material at 560 °C gets so soft that both surfaces align very well to each other. Furthermore, no differences between turned and sand-blasted AMC joining surfaces were detected. In some joints, cracks, probably caused by the thermal stresses during cooling, were formed in-between the two intermetallic compounds of the diffusion zone. Areas of potential lower thermal stresses showed no cracks. The backscattered SEM-micrograph shows an area of the joint which is free of cracks, figure 7. EDXS have shown that the two intermetallic compounds are most probably $Fe_2Al_5$ and $Fe_4Al_{13}$ and the white pattern inside the AMC base material are Al-Cu precipitations. To limit the formation of the critical intermetallic Al-Fe compounds, the joining temperature was systematically reduced to 380 °C. The joints with turned surfaces, shown in figure 8, exhibit no pores, gaps or cracks along the interface. In contrast to that, the interface of the joints with sand-blasted AMC surfaces is interrupted by gaps, due to the limited load of 10 N/mm² at 480 °C and 380 °C and thus the lower surface deformation. To avoid these gaps one approach will be to increase the surface pressure, so that higher micro surface deformation occurs and thus a defect-free joint can be ensured. However, it can be seen that there is no diffusion zone at 480 °C and 380 °C. Also the SEM analysis of the joints showed no formation of the intermetallic Al-Fe compounds.

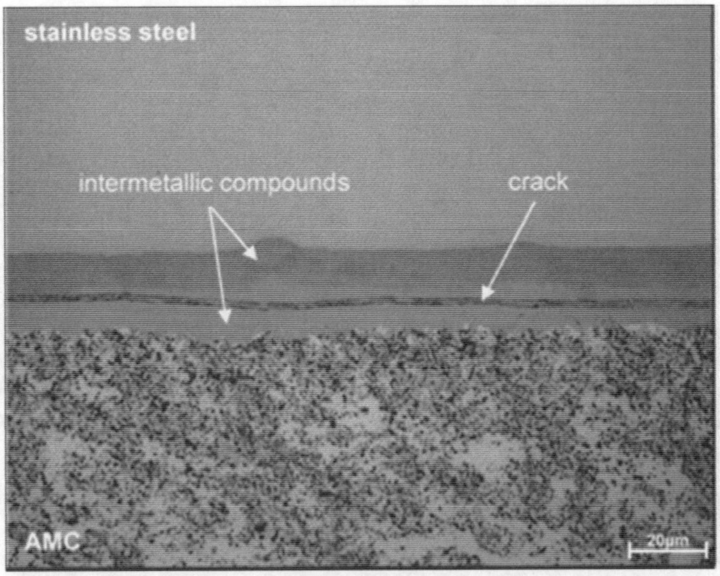


**Figure 6.** Interface of a diffusion bonded AMC-stainless steel joint formed at 560 °C with approx. 20 µm wide diffusion zone along the interface; turned joining surface

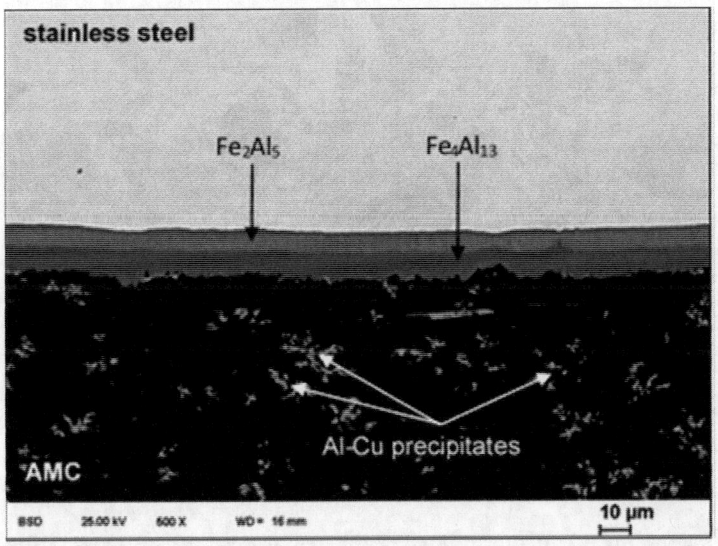


**Figure 7.** SEM micrograph of the AMC-stainless steel joint without cracks inside the diffusion zone; turned joining surface, joining temperature 560 °C

Only an increase of preparation artefacts, caused by the reaction of the base materials and the polishing suspension, along the interface and inside the AMC base material at 380 °C could be observed. Figure 9 shows, that these artefacts contain just some cracks with a dimension of less than 4 µm in length and 1 µm in width. The experiments with sand-blasted joining surfaces and a joining temperature of 380 °C resulted in weak joints and broke apart during cutting.

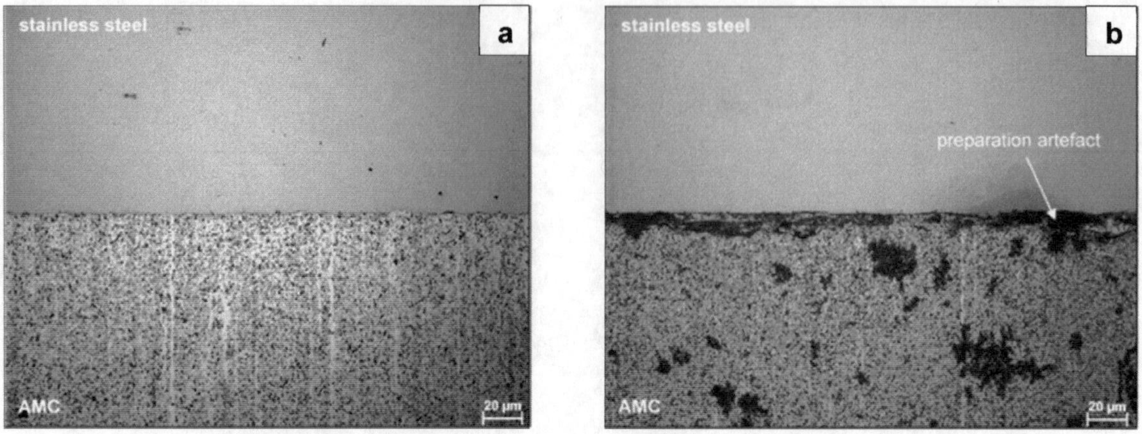


**Figure 8.** Influence of joining temperature on interface formation of diffusion bonded stainless steel-AMC joints with turned joining surfaces at a) 480 °C and b) 380 °C

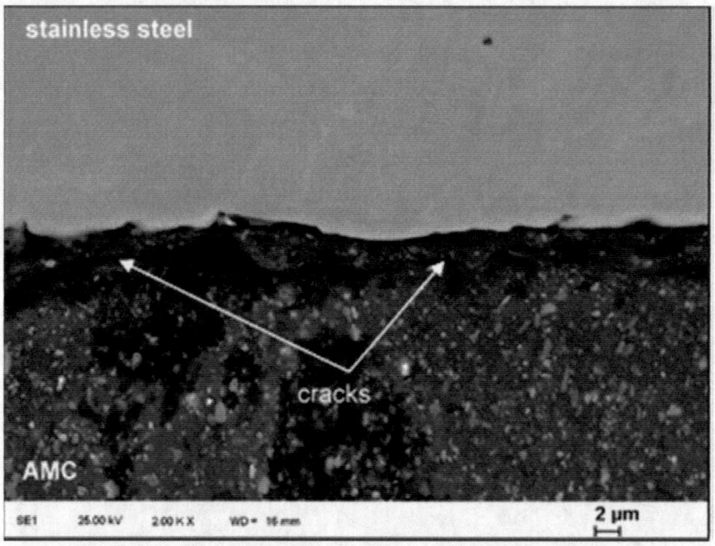


**Figure 9.** SEM micrograph of the stainless steel to AMC diffusion bonded joint with some cracks on the AMC side along the interface

Figure 10 summarizes the results of the micro-hardness measurements. Especially the hardness profile across the diffusion zone of the joint at 560 °C confirms the expected critical behavior: an increase of hardness of approx. 800 HV0.01 in the $Fe_4Al_{13}$ intermetallic compound compared with the AMC base material. Due to its small dimension, the hardness of the $Fe_2Al_5$ phase could not be measured directly. The approximation of the hardness profile shows, that the $Fe_2Al_5$ phase has a hardness of around 580 HV0.01. Thus, both intermetallic compounds form a continuous metallurgical notch at the interface. This is the reason why cracks will be formed along the interface, promoted by thermal stresses or other (external) loads, figure 7. However, the diffusion bonding experiments at 480 °C and 380 °C did not show such a metallurgical notch with an increase of hardness at the interface. From figure 10, a much more remote change in hardness between the AMC material and the stainless steel, based on the lower joining temperatures and thus the limited formation of brittle intermetallic compounds, can be seen.

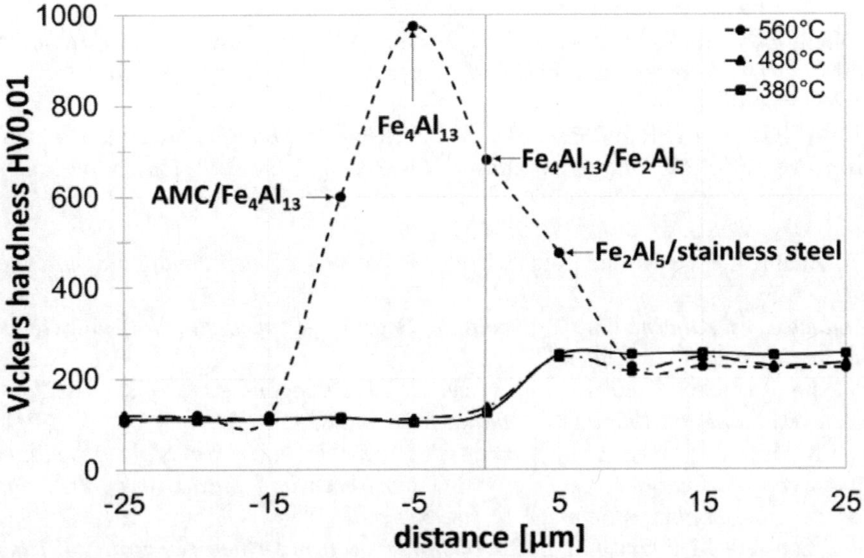


**Figure 10.** Micro-hardness profile in HV0.01 of the stainless steel to AMC diffusion bonded joints at 560 °C, 480 °C and 380 °C

## 4 Conclusion

Both processes show new promising approaches for joining particle reinforced alu-minum matrix composites (AMC) to stainless steel. The investigation of the microstructures of the joints leads to the following results:

1) During arc brazing of AMC and stainless steel, alloying of the filler material occurs, and a diffusion/reaction zone is recognizable. The slope of the hardness profile at the interfaces of braze metal to the AMC is more remote than that to the stainless steel. That is a result of the formation of Al-Fe intermetallics at the interface to stainless steel. The hardness values and XRD analyses indicate the phases $FeAl_3$ and AlFe. The reaction/diffusion zone at the stainless steel side obviously consists of different phases.
2) Diffusion bonding of AMC to stainless steel at 560 °C results in brittle intermetallic phases in the joint area that comprise Al and Fe ($Fe_2Al_5$ and $Fe_4Al_{13}$), so that cracks will form along the interface most likely. Al-Cu intermetallics were not detected along the interface. Micro hardness measurements across the diffusion zone showed that the $Fe_2Al_5$ phase exhibits an approx. 6 times higher and the $Fe_4Al_{13}$ phase an approx. 10 times higher hardness than the AMC.
3) By reducing the joining temperature to a range of 380 °C to 480 °C in combination with surface treatments, Al-Fe intermetallics are avoided and thus a continuous change in hardness can be attained.

Arc brazing and diffusion bonding are suitable to join AMC and stainless steel. The presented results of both approaches show the potential and differences of the investigated processes. The determination of the mechanical properties with respect to process parameters and the resulting microstructure are the purpose of the future work.

**Acknowledgements**
The authors gratefully acknowledge the funding by the German Research Foundation (Deutsche Forschungsgemeinschaft, DFG) within the framework of the Collaborative Research Centre 692 (SFB HALS 692).

**References**

[1] Xiu Z, Yang W, Chen G, Jiang L, Mac K, Wu G, 2012. *Microstructure and tensile properties of Si3N4p/2024Al composite fabricated by pressure infiltration method. Mat. a. Designs 33, 350-358.*

[2] Sajjadi S A, Ezatpour H R and Parizi M T, 2012.*Comparison of microstructure and mechanical properties of A356 aluminium alloy/Al2O3 composites fabricated by stir and compo-casting processes. Mat. a. Designs 34,106-111.*

[3] Qu X-h, Zhang L, Wu M and Ren S-b, 2011. *Review of metal matrix composites with high thermal conductivity for thermal management applications. Progress in Natural Science: Materials International. Vol. 21, 189-197.*

[4] Steiners M and Höcker F, 2007. *Einfluss der Beschichtungen beim stoffschlüssigen Lichtbogen-fügen von Stahl mit Aluminium. Mat. u. Werkst. 38, 559-564.*

[5] Staubach M, Jüttner S, Füssel U and Dietrich M, 2007. *Fügen von Stahl-Aluminium-Mischver-bindungen mit energiereduzierten MSG-Verfahren und Zusatzwerkstoffen auf Aluminium und Zinkbasis. Schw. u. Schneid. 6, 302-313.*

[6] Ureña A, Escalera M D and Gil L, 2000. *Influence of interface reactions on fracture mecha-nisms in TIG arc-welded aluminium matrix composites. Comp. Scien. a. Tech. 60, 613-622.*

[7] Lean P P, Gil L and Ureña A, 2003. *Dissimilar welds between unreinforced AA6082 and AA6092/SiC/25p composite by pulsed-MIG arc welding using reinforced filler alloys (Al-5Mg and Al-Si). J. of Mat. Proc. Tech. 143-144, 846-850.*

[8] Nami H, Halvaee A and Adgi H, 2011. *Transient liquid phase diffusion bonding of Al/Mg2Si metal matrix composite. Mat. and Design 32, 3957-3965.*

[9] Askew J R, Wilde J F and Khan T I, 1998. *Transient liquid phase bonding of 2124 aluminium metal matrix composite. Mat. Sci. and Technol. 14, 920-924.*

[10] Qin G, Lei Z, Su Y, Fu B, Meng X and Lin S, 2014. *Large spot laser assisted GMA brazing-fusion welding of aluminium alloy to galvanized steel. J. of Mat. Proc. Tech. 214, 2684-2692.*

[11] Song J L, Lin S B, Yang C L, Ma G C and Liu H, 2009. *Spreading behavior and microstructure characteristics of dissimilar metals TIG welding-brazing of aluminum alloy to stainless steel. Mat. Scien. a. Eng.: A 509, 31-40.*

[12] Wielage B and Klose H, 1995. *Das Aluminiumlöten von Wärmetauschern. DVS-Berichte 166, 88-90.*

[13] Wielage B and Martinez L, 2001. *Aluminiumlöten bei 550°C – Eigenschaften von ZnAl-Verbindungen. DVS-Berichte 212, 214-217.*

[14] Wielage B and Trommer F, 2003. *Löten von Aluminium mit Zinkbasisloten. Schweiß. u. Schneid. 5, 273-275.*

[15] Howlader M M R, Kaga T and Suga T, 2010. *Investigation of bonding strength and sealing behaviour of aluminium/stainless steel bonded at room temperature. Vacuum 84, 1334-1340.*

[16] Bhanumurthy K, Fotdar R K, Joyson D, Kale G B, Pappachan A L, Grover A K and Krishnan J, 2006. *Development of tubular transition joints of aluminium/stainless steel by deformation diffusion bonding. Mat. Sci. and Tech. 22, 321-329.*

[17] Zhang X P, Ye L, Mai Y W, Quan G F and Wei W, 1999. *Investigation on diffusion bonding characteristics of SiC particulate reinforced aluminium metal matrix composites (Al/SiCp-MMC). Composites Part A 30, 1415-1421.*

[18] Muratoglu M, Yilmaz O and Aksoy M, 2006. *Investigation on diffusion bonding characteristics of aluminium metal matrix composites (Al/SiCp) with pure aluminium for different heat treatments. J. for Mat. Process. and Tech. 178, 211-217.*

[19] Huang J, Dong Y, Wan Y, Zhao X and Zhang H, 2007. *Investigation on re-active diffusion bonding of SiCp/6063 MMC by using mixed powders as interlayers. J. of Mat. Process. and Tech. 190, 312-316.*

[20] Feng T, Chen X, Wu L and Lou S, 2006. *Diffusion welding of SiCp/2014Al composites using Ni interlayer. J. of University Scien. and Tech. Beijing 13, 267-271.*

[21] Urena A, Gomez de Salazar J M, Escalera M D, 1996. *Diffusion bonding of an alloy reinforced with silicon carbide particles (AA2014/SiC/13p) using metallic interlayers. Scripta materialia 35, 1285-1293.*

[22] Elßner M, Weis S, Grund T, Hausner S, Wielage B and Wagner G, 2014. *Lichtbogenlöten von Aluminiummatrix-Verbundwerkstoffen mit AlAgCu-Loten, Werkstoffe und werkstoff-technische Anwendungen 52, ISBN 978-3-00-046877-3.*

[23] Elßner M, Weis S and Wagner G, 2015. *Joining of Aluminum Matrix Composites and Stainless Steel by Arc Brazing. Materials Science Forum. Vol. 825-826, 393-400.*

[24] Potesser M, Schoeberl T, Antrekowitsch H and Bruckner J, 2006. *The characterization of the intermetallic Fe-Al layer of steel-aluminum weldings. Conference proceedings EPD Congress, 167-176.*

IOP Conf. Series: Materials Science and Engineering **118** (2016) 012038 doi:10.1088/1757-899X/118/1/012038

# Anodisation with dynamic current control for tailored alumina coatings

**M Sieber[1], I Althöfer[1], D Höhlich[1], I Scharf[1], D Böttger[2], S Böttger[2], E Böttger[2] and T Lampke[1]**

[1] Materials and Surface Engineering Group, Institute of Materials Science and Engineering, Technische Universität Chemnitz, D-09125 Chemnitz, Germany
[2] Eckhard Böttger oberflächenveredlung, D-09465 Sehmatal-Neudorf, Germany

maximilian.sieber@mb.tu-chemnitz.de

**Abstract.** The anodic oxidation process is commonly used to refine the surface of aluminium and its alloys. Compared to the substrate, the alumina layers produced by anodising exhibit an increased hardness and chemical resistance. Thus, the corrosion and wear resistance are generally improved. The coatings are also electrically isolating and may serve decorative purposes. Applying a time-variant, dynamic electrical process control by pulse-current or current-steps is a promising approach to improve the coating properties, which is partially deployed in an industrial scale. In the present work, the influence of dynamic electrical process control on the coating properties is examined by means of a design of experiments (DOE). The effects of various electrolyte compositions and temperatures as well as processing time are considered with regard to coating thickness, hardness, wear resistance and the electrical energy consumption during the formation of the coatings. Information about the statistical significance of the effects of the parameters on the considered properties is obtained by an analysis of variance (ANOVA).

## 1. Introduction

The surface refinement of aluminium and its alloys by anodic oxidation in sulfuric acid electrolytes generally improves both corrosion and wear resistance. Therefore, the formed oxide-ceramic layers offer the possibility to access a wider range of application for these lightweight materials. However, the anodic oxidation process is rather costly with regard to the required energy, especially electrical energy for the process itself and the cooling of the electrolyte. The latter is necessary to produce harder coatings with a reduced volume of the characteristic pores, which form during the process because of a complex interrelation between oxide formation and the subsequent dissolution of the oxide material in the electrolyte.

Several approaches have been established to improve the energy efficiency of the anodic oxidation process. This includes the use of electrolyte additives [1–4] as well as the optimisation of the electric parameters [5–8]. The application of time-variable voltage- and/or current-regulation to tailor process and coating properties has been investigated [9–13] and discussed controversially [14] by many researchers.

The aim of the present work is to clarify the effect of time-variable electric regimes for the anodic oxidation on the properties of the formed oxide coatings. The influence of pulsed and stepwise variable current regimes is compared to the influence of electrolyte additives, temperature and process time by means of a DOE (design of experiments). Thereby, the total charge transfer quantity is kept constant within the process to allow for the comparison of both coating properties and the necessary amount of electrical energy to produce the coating. To avoid burning of the samples in the later process stages, the voltage is limited to 70 V. For the sake of comparability between direct current and dynamic current control, high mean current densities (above 10 A/dm²) are avoided.

## 2. Experimental methods

### *2.1. Coating formation – materials and processing*

The aluminium alloy AlMgSi1 (nominal composition 0.7–1.3 % Si, ≤0.5 % Fe, ≤0.1 % Cu, 0.4–1.0 % Mn, 0.6–1.2 % Mg, ≤0.25 % Cr, ≤0.2 % Zn, ≤0.1 % Ti, Al balance) was used as substrate (dimensions: 6 mm x 12.7 mm x 25.4 mm). The pretreatment of the samples included etching in 3 wt-% sodium hydroxide at 50 °C for 1 min and pickling in 30 vol-% nitric acid at room temperature for 30 s followed by rinsing under deionised water.

The electrolyte volume for the anodic oxidation was 2 l. The electrolytes were prepared using deionised water, sulfuric acid (Merck, Germany), glycolic acid (Sigma-Aldrich, Germany) and the commercial electrolyte additive Anodal EE (Clariant, Germany). All chemicals were of analytical grade. Three different electrolyte compositions were used: (A1) 20 vol-% sulfuric acid + 20 g/l glycolic acid; (A2) 20 vol-% sulfuric acid; (A3) 20 vol-% sulfuric acid + 20 ml/l Anodal EE. The electrolyte temperature was maintained at (B1) 5 °C, (B2) 15 °C and (B3) 25 °C with the help of a thermostat. Galvanostatic process control was applied with a pE1028 power source (Plating Electronic, Germany). Three different types of galvanostatic control were used: (C1) direct current; (C2) pulsed current with a pulse width of 2 min at the higher current and 30 s at the lower current, the ratio of the higher to the lower current being 5/1; (C3) stepwise time-variant current with 3 equally spaced steps over the process time using 2/3, 3/3 and 4/3 of the mean current density. The process time was set at (D1) 30 min, (D2) 45 min and (D3) 120 min resulting in a mean current density of 6.3 A/dm², 4.2 A/dm² and 1.575 A/dm², respectively, to ensure a constant turnover of electric charge in the process. The parameters A–D were varied according to a DOE using a modified Box-Behnken-Design [15] with 25 different combinations of the four parameters. Four samples were produced and tested for every set of parameters. Figure 1 shows a graphical representation of the DOE. After the coating formation process, the specimens were rinsed under water and dried under cool air.

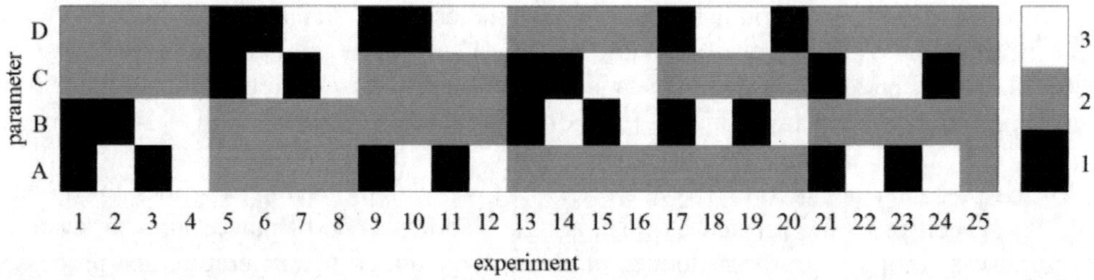


**Figure 1.** Graphical representation of the DOE used in this work showing the assignment of parameter levels (1–3) to 25 experiments with 4 repetitions, each, with parameters A – electrolyte composition, B – temperature, C – current regime, D – process time.

### *2.2. Process and coating characterization*

During the anodic oxidation process, current and voltage values were logged with a sampling rate of 1 Hz using the internal logging function of the power device. Afterwards, the electrical energy $W_{el}$ used for the process was calculated by integrating the product of current density and voltage depending on the process time.

The coating thickness $s$ and the hardness $H$ were determined in the cross sections of the coatings (Martens hardness HM 0.05/30/30 by Fischerscope HM2000 XYm, Fischer, Germany). It is well known form the literature, that the oxide coatings produced by anodising declines with increasing distance to the substrate. Therefore, the distance of the hardness indents to the substrate/coating interface was determined to assess the gradient of the hardness decline (hardness gradient $H'$ in N/mm³). For each sample, three hardness profiles with five indents were measured.

As an integral property, the scratch energy density $w$ of the coatings was determined. With a Revetest-RST (CSM Instruments, Switzerland), a Rockwell diamond (radius 200 µm) was scratched across the sample surface with a constant load of 15 N for 5 mm. The scratch energy density $w$ was calculated according to *Eq. 1* using the mean tangential force during the scratch test $F$, the scratch length $l$ and the scratch volume $v$, which was determined by measuring a linear surface profile (tactile measurement, T4000, Hommelwerke, Germany) normal to the scratch direction at five different positions:

$$w = F \cdot l \cdot v^{-1} \quad (1)$$

*2.3. DOE exploitation*

For each of the properties $W_{el}$, $s$, $H'$ and $w$, the mean values and the standard deviation for each level of each parameter were determined. All values for the respective property obtained at a specific parameter level were taken into account. For example, the standard deviation of parameter A at level 1 includes all results with varying parameters B, C, D. Further, an error probability $p_{err}$ for considering the parameters as significant with respect to a specific property was obtained by performing an analysis of variance (ANOVA) with the Matlab built-in function "anovan" [16]. The error probability $p_{err}$ characterises the probability of committing an error in considering the effect of a parameter on a specific property as significant. It provides information, whether the results obtained for at least one level of the parameter are statistically significantly different from the results obtained at the other levels.

## 3. Results and discussion

A typical voltage/current density characteristic during the anodic oxidation process is shown in figure 2 (experiment no. 25, compare figure 1). It is apparent, that the voltage increases with the process time. This behaviour is due to the growing of the oxide layer, which results in an increased electrical resistance. Figure 3 shows the influence of the process parameters on the consumption of electrical energy during the anodic oxidation $W_{el}$. With the exception of the current regime, every parameter wields a statistically significant influence over $W_{el}$. The addition of organic components (A1, A3) into the electrolyte as well as the reduction of the temperature attenuate the dissolution of formed oxide material and therefore increase the process voltage and hence $W_{el}$. The difference between the two used electrolyte additives is insignificant according to the ANOVA, though there is a tendency towards lower $W_{el}$ for the glycolic acid additive compared to the commercial additive. However, the difference to the sulphuric acid electrolyte without additives is significant. Interestingly, the longest process time is attended by the lowest $W_{el}$ value. Obviously, the low current density used in the longer processes involves (1) a low voltage and (2) a reduced voltage increase over time. Hence, the amount of consumed electrical energy decreases. There is no significant difference in $W_{el}$ between 30 min and 45 min process time.

The effect of the parameters on the coating thickness $s$ is shown in figure 4. Although the influences of electrolyte, temperature and process time are rather small, they are statistically significant according to the ANOVA with error probabilities of less than 3 % in any case. Considering the electrolyte composition, the glycolic acid additive increases the mean coating thickness by approximately 10 % while the commercial additive shows no significant effect compared to the sulfuric acid electrolyte without additions. The relatively lower coating thickness in the electrolyte with the commercial additive is caused by the limiting of the voltage at 70 V, which was sometimes the case in electrolyte A3 because of the strong increase of the process voltage caused by the additive. Consequently, the amount of charge transferred in the process decreases and thus the coating thickness is reduced. With regard to the electrolyte temperature, the coating thickness is at the same level for 5 °C and 15 °C. Only an increase to 25 °C causes a reduction of the thickness. However, the current regime does not wield a significant influence on the coating thickness, which is apparently related to the constant amount of charge transferred in the process. Hence, the same amount of oxide is formed. With reference to the process time, the results for 30 min and 120 min are not statistically significantly different from the results obtained for 45 min, but they are statistically significantly different compared to each other. Therefore, the tendency of a lower coating thickness with increasing process time is small, yet statistically significant. This and the effect of the temperature can be explained with an increased dissolution of formed oxide material because of the longer exposition to the electrolyte with increasing process time and because of the stronger chemical attack on the oxide with increasing temperature.

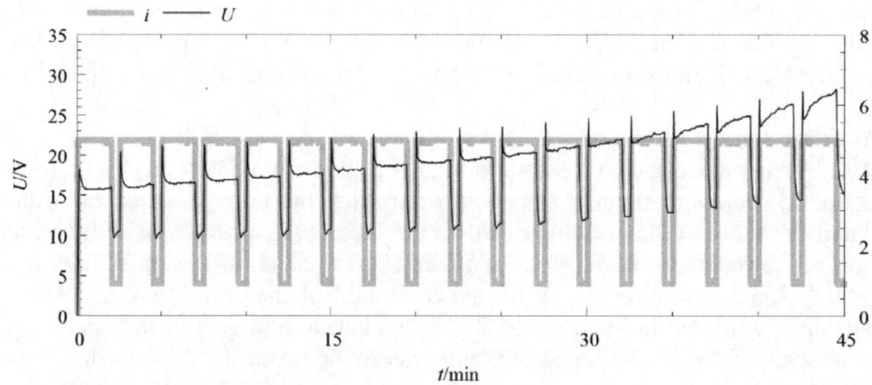


**Figure 2.** Voltage and current density in dependence of the processing time during experiment no. 25.

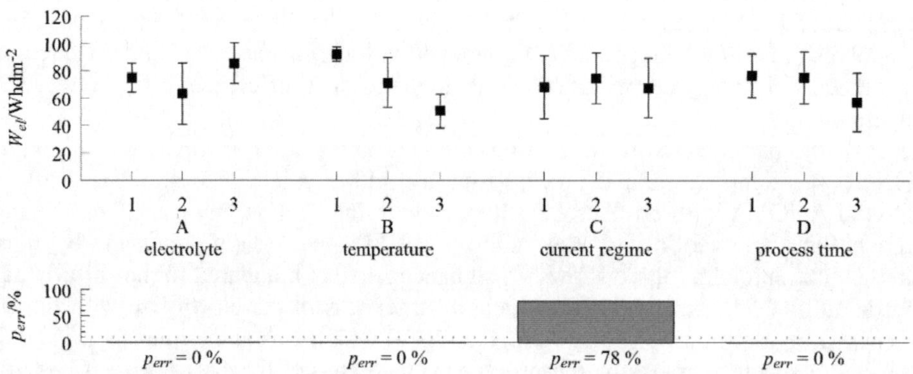


**Figure 3.** Effect of electrolyte, temperature, current regime and process time on the electric energy consumption during the anodic oxidation and corresponding error probability obtained from the ANOVA.

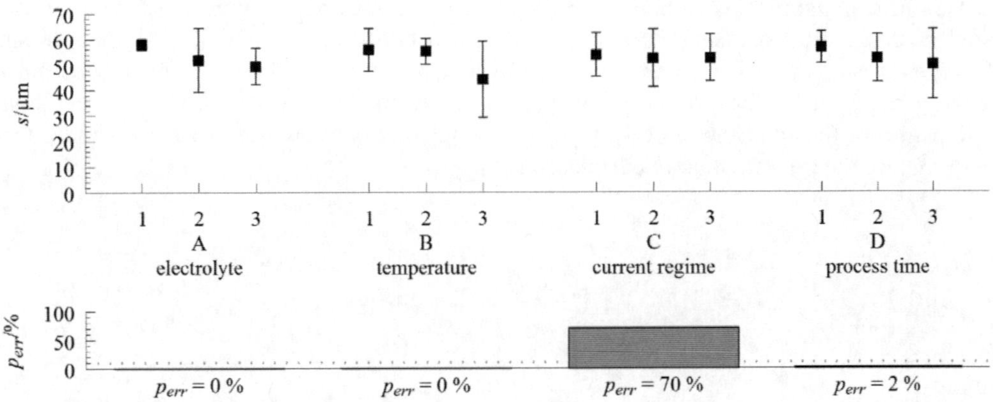


**Figure 4.** Effect of electrolyte, temperature, current regime and process time on the coating thickness and corresponding error probability obtained from the ANOVA.

With regard to the hardness gradient *H'*, i.e. the gradient with which the hardness is reduced with increasing distance to the substrate, the temperature and the electrolyte composition are the factors with the biggest influence (figure 5). Reduced chemical dissolution caused by a decreased electrolyte temperature or the addition of both the tested organic components to the electrolyte leads to a less pronounced widening of the pores because of chemical attack on the formed oxide. Thereby, the results for the two electrolyte additives do not differ significantly from each other but the hardness gradient is significantly decreased compared to the sulphuric acid electrolyte without additives according to the ANOVA. The same results are obtained with regard to the electrolyte temperature: there is no significant difference between the hardness gradient of coatings obtained at 15 °C and 5 °C, while the hardness gradient is significantly increased at 25 °C. The correlation between pore volume and hardness has been established in [17]. As for the other properties, the current regime does not exert a statistically significant effect.

The scratch energy density *w* is strongly dependent of the electrolyte composition and the temperature (figure 6). Interestingly, the mean value of all experiments at an electrolyte temperature of 25 °C (B3) is very small compared to that of the other parameters. This indicates that the variation of neither the electrolyte composition, current regime nor the process duration is suitable to improve the scratch energy density of samples anodized at 25 °C. Both the results obtained for coatings formed at 15 °C and 5 °C show a higher scratch energy density. Thereby, a temperature decrease from 15 °C to 5 °C does not lead to a higher scratch energy density. The influence of the additives on *w* reflects the results of the hardness gradient measurements. The organic additives significantly improve *w* compared to the pure sulfuric acid electrolyte. Again, the ANOVA shows that there is no difference between the two electrolyte additives. Process time and current regime do not exert an influence on the scratch energy density according to the analysis of variance.

The use of the dynamic current control in the chosen parameter field does not exert an influence on the tested coating and process properties. Further, in accordance with [18], any improvement with regard to the coating´s thickness, hardness or wear resistance is attended by a higher amount of electrical energy required for the process. Nonetheless, there might be conditions, e.g. very high mean current densities or the anodisation of other alloys, which do not allow for a formation in direct current mode and require the use of dynamic current control.

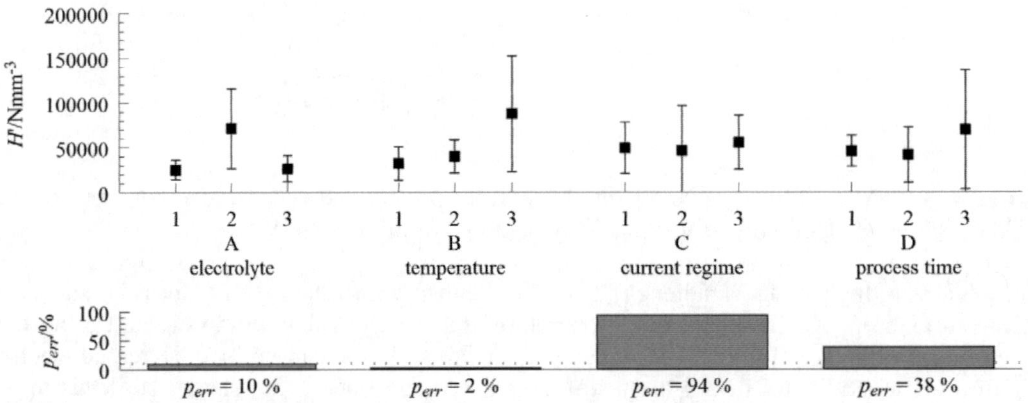


**Figure 5.** Effect of electrolyte, temperature, current regime and process time on the hardness gradient and corresponding error probability obtained from the ANOVA.

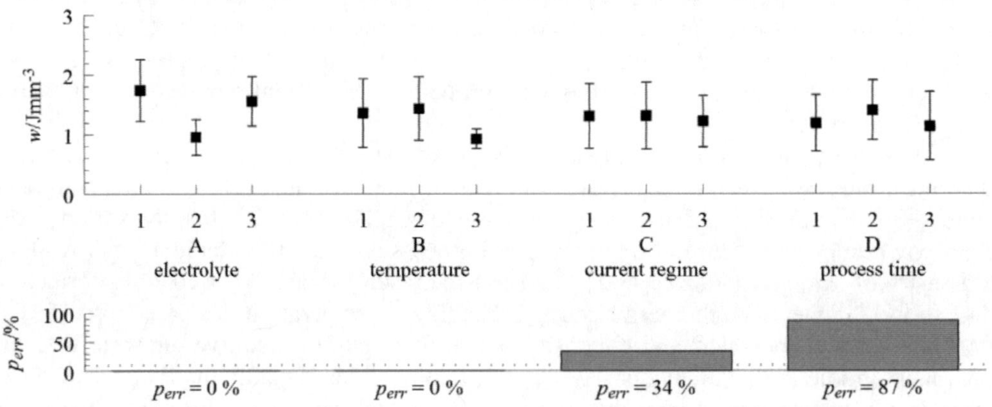


**Figure 6.** Effect of electrolyte, temperature, current regime and process time on the scratch energy density and corresponding error probability obtained from the ANOVA.

## 4. Conclusion

The influence of the electrolyte composition, temperature, current regime and process time on selected properties of the anodic oxidation process and the formed oxide layers were investigated. The used three-level DOE was exploited with regard to electric energy consumption during the coating formation, thickness, hardness gradient and scratch energy density of the coatings. Thereby, the amount of charge turnover in the process was kept constant, given that the process voltage did not exceed 70 V. The obtained results can be summed up with regard to the considered parameters as follows. Both the commercial and the non-commercial organic additive generally increased the energy consumption and improved the coating properties. Hardly a difference was found between the two additives. The reduction of the electrolyte temperature generally increased the electrical energy consumption. With regard to the considered coating properties, however, the reduction of the temperature from 15 °C to 5 °C showed no significant improvement. The choice of the current regime (direct current vs. pulse-current vs. current-steps) affected neither the electrical energy consumption nor the coating properties in the considered range of parameters. With regard to the process time, there was no difference in electrical energy consumption and coating properties between 30 min and 45 min. While an increase to 120 min decreased the electrical energy consumption, it also deteriorated the coating properties in terms of thickness and hardness gradient.

## Acknowledgement

The authors gratefully acknowledge funding by the German Research Foundation (Deutsche Forschungsgemeinschaft, DFG) within the framework of SFB 692. The work of Julia Wendler, Gabriele Tauchmann and Sigrid Kaufmann (all from the Institute of Materials Science and Engineering, TUC) is gratefully acknowledged.

## References

[1] Bensalah W, Elleuch K, Feki M, Wery M and Ayedi H F 2007 *Surf. Coat. Technol.* **201** 7855–64
[2] Yoshimoto M, Morizino Y, Tsurekawa S and Baba T 2012 *J. Ceram. Soc. Jpn.* **120** 276–9
[3] Giovanardi R, Fontanesi C and Dallabarba W 2011 *Electrochim. Acta* **56** 3128–38
[4] Raj V and Mumjitha M 2014 *Mater. Sci. Eng. B.* **179** 25–35
[5] Zhou F, Baron-Wiechec A, Garcia-Vergara S J, Curioni M, Habazaki H, Skeldon P and Thompson G E 2012 *Electrochim. Acta* **59** 186–95
[6] Garcia-Vergara S J, Habazaki H, Skeldon P and Thompson G E 2007 *Nanotechnol.* **18** 415605
[7] Voon C H, Derman M N, Hashim U and Ahmad K R 2013 *Adv. Mat. Res.* **795** 56–9
[8] Curioni M, Skeldon P, Ferguson J and Thompson G E 2011 *J. Appl. Electrochem.* **41** 773–85
[9] Gabe D R and Dowty I H 1987 *Surf. Coat. Technol.* **30** 309–16
[10] Colombini C 1992 *Met. Finish.* **90** 31–6
[11] Schraedel F C 2006 *Prod. Finish.* **71** 30–3
[12] Rasmussen J 1994 *Galvanotechnik* **85** 1841–7
[13] Lee W, Scholz R and Gösele U 2008 *Nano Lett.* **8** 2155–60
[14] Abele M, Burkhardtsdmaier U and Pfeifer H 1994 *Galvanotechnik* **85** 2505–8

[15] Box G E P and Behnken D W 1960 *Technometrics* **2** 455–75
[16] The Mathworks 2014 *Matlab Version R2014b* (Natick USA)
[17] Aerts T, Dimogerontakis T, De Graeve I, Fransaer J and Terryn H 2007 *Surf. Coat. Technol.* **201** 7310–7
[18] Sieber M, Scharf I, Herold F, Schmidt A, Böttger D, Böttger S, Böttger E, Götze U and Lampke T 2016 *Mater. Des.* **89** 1259–69

# FE-simulation of the Presta joining process for assembled camshafts – local widening of shafts through rolling

**R Scherzer, C B Silbermann and J Ihlemann**

Professorship of Solid Mechanics, Technische Universität Chemnitz, Reichenhainer Straße 70, 09126 Chemnitz, Germany

robert.scherzer@mb.tu-chemnitz.de

**Abstract.** Considerable weight benefits and the option to combine various steel alloys of the single parts are the major advantages of assembled over conventional camshafts. The Presta joining process is the leading manufacturing method of assembled camshafts in the global market. The process is divided into two substeps. At first, the outer diameter of the shaft is widened with a profile oriented orthogonal to the shaft axis at the intended cam seat. At this position the shaft is subsequently joined with a cam with an internal profile oriented parallel to the shaft axis. As a result, these perpendicular profiles form a tight fit due to plastic deformations. Consequently the simulation of the manufacturing process has to start with the simulation of the rolling of the shaft. The resulting profile requested in this step is axisymmetric, but the arrangement of tools is not. Thus a three-dimensional model is required, which is presented in this work. Furthermore, the infeed of the rolling tool is unknown and controlled by the stiffness of the holders of the rolling tool. This work shows the modeling of this behavior. To predict realistic results for the underlying process, the use of precise material models is essential in order to take several hardening mechanisms into account. However, the use of complex material models implies additional effort, which is shown in this work.

## 1. Introduction

To this day many manufacturing processes are primarily based on in-house experiences and large experimental efforts. But this approach becomes unsustainable when today's requirements of lightweight and highest efficiency should be met. Hence, the use of computer simulations and more precise numerical methods in development processes is almost inevitable. Consequently, there is a big potential of saving costs and time by reducing the production of expensive prototypes. However, the quality of numerical simulations is based on the quality of the employed boundary conditions, among others. Hence, a deep understanding of the underlying production process is required to obtain reliable predictions. This work discusses the Presta joining process, which is explained in detail below.

The Presta joining process is a shaft-hub-connection creating a combination of press fit and form fit. It is primarily used as the world's leading manufacturing method for assembled camshafts and allows up to 30% weight reduction compared to conventional cast or forged camshafts. This in turn implies a significant inertia reduction of the rotating camshafts [1, 2].

The manufacturing process is divided into two substeps. First, the shaft is formed by rolling the intended cam seat to create a local widening at this position. This diameter enlargement is enabled by rollers with an approximately sinusoidal cross sectional surface and results in a profile oriented orthogonal to the shaft axis (figure 1a).

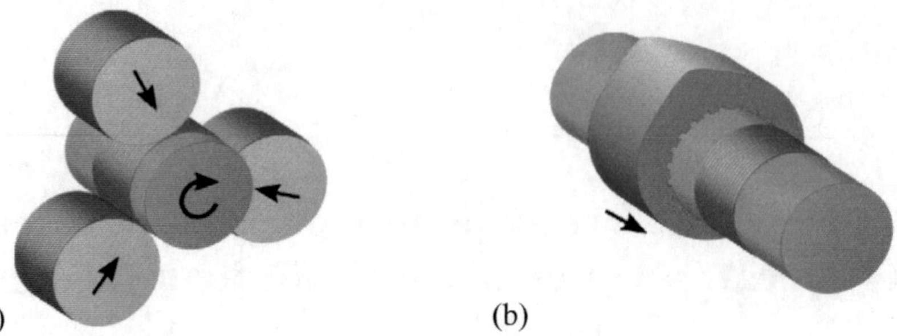

(a) (b)

**Figure 1.** Schematic representation of the Presta joining process: (a) rolling of the shaft, (b) forcing the cam onto the previously rolled cam seat of the shaft

Subsequently, the rolling step is followed by the actual joining of the cam with the shaft. Contrary to the rolled profile of the shaft the cam's inner profile is oriented parallel to the shaft axis. When forced onto each other these perpendicular profiles generate the tight fit that is characteristic for Presta processed camshafts (figure 1b). In Section 2, a computational model is presented, which enables the simulation of the rolling process using the finite element method (FEM).

One has to take into account that the forming during both substeps leads to large local plastic deformations. Hence, the applied material formulation has to be capable of reproducing the real material response and to be applicable in a wide strain range. Within the FE-model of the rolling process, a phenomenological material formulation of finite strain viscoplasticity is used [3, 4]. In Section 3, a short overview of the model is given and the results of the parameter identification of the steel alloy used for the shaft are shown.

Finally, Section 4 presents the results of the forming simulations.

## 2. Three-dimensional FE-model of the rolling process

### *2.1. Geometric model*

The rolling step of the Presta joining process ideally leads to an axisymmetric profile consisting of several grooves oriented orthogonal to the shaft axis. In contrast, the assembly of the rolling tool is not axisymmetric. Thus, a two-dimensional modeling approach of the process is not realizable. However, in respect of a reduction of calculation costs, two other symmetries can be applied. First, the profile of the roller is symmetric and therefore the model size can be reduced to one half. Second, the three rollers are oriented by an angular distance of about 120°. Thus, the remaining model can further be reduced to one third (figure 2a).

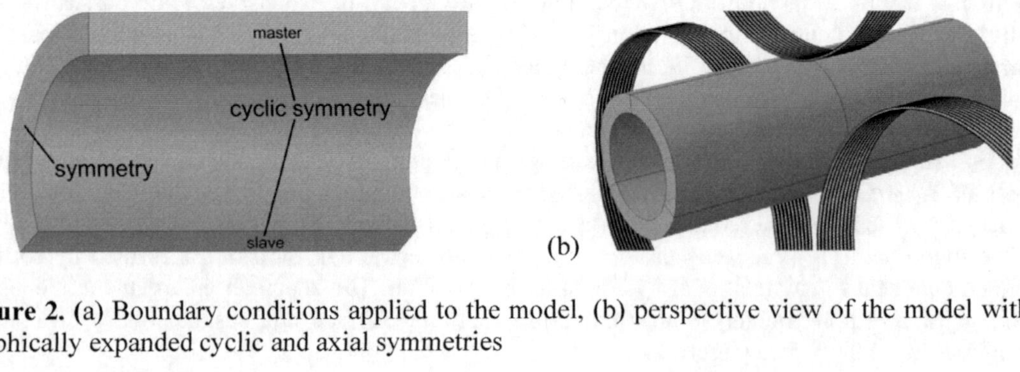


(a) (b)

**Figure 2.** (a) Boundary conditions applied to the model, (b) perspective view of the model with graphically expanded cyclic and axial symmetries

Finally, the modeled shaft geometry is extended beyond the width of the rollers (figure 2b). It is necessary to consider this additional region because it supports the forming zone and without that, the geometrical stiffness of the shaft would be too low. This would lead to a non-representative shaping of the rolling profile. The length of this supportive zone approximately corresponds to the distance between two cam seats of the camshaft.

### 2.2. *FE-mesh of the shaft*

Due to the repetitive cross sectional profile of the rollers and the ideally axisymmetric forming of the shaft, an approximate prediction of the deformation of every single groove is possible. Hence, the FE-mesh is designed to follow the estimated material flow (figure 3).

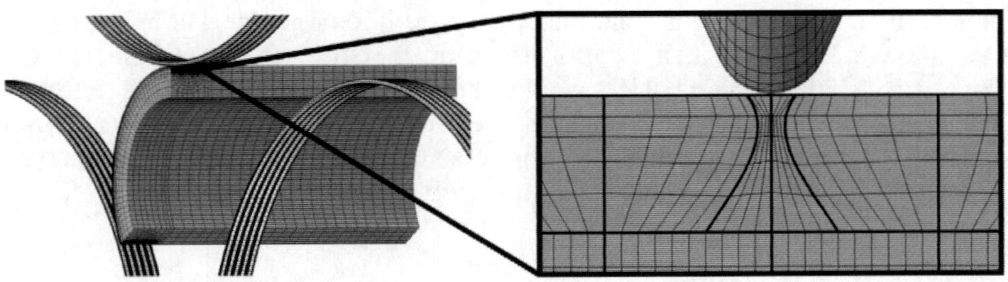

**Figure 3.** FE-mesh: Perspective view of the assembly and a detailed view of the cross sectional mesh topology and its underlying partitioning (thick lines)

Therefore, the geometric part of the shaft has been partitioned before meshing. According to the roller's profile this partitioning is repeated for all grooves. With increasing distance from the forming zone a coarser mesh is applied to save computational cost. Similar to thread rolling processes this plastic deformation zone is limited to the area close to the contact surface [5, 6].

### 2.3. *Loading boundary conditions*

The simulation assembly mentioned above consists of the shaft and the three profiled rollers, which already implies a simplification of the rolling tool used in the manufacturing process (figure 4).

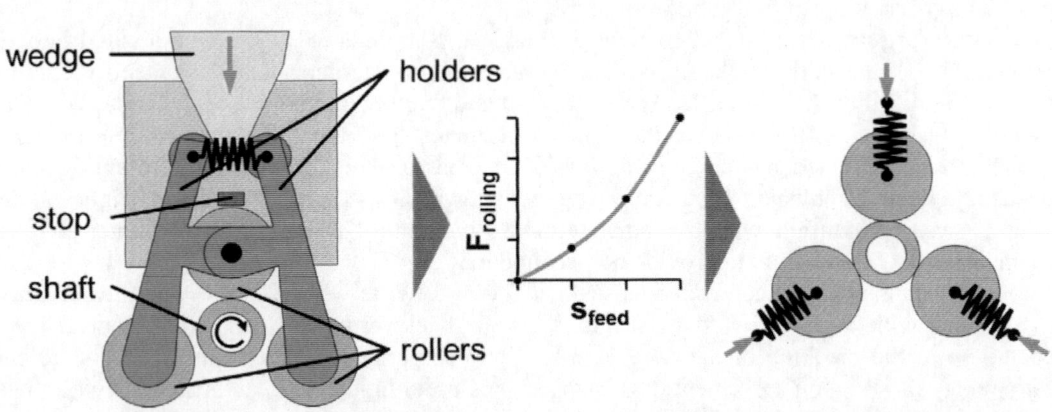


**Figure 4.** Assembly of the rolling tool [1] and the schematic representation of the transfer of its stiffness into the model's boundary conditions

Here, the time-dependent feed moves a wedge to press the upper part of two holders in opposite directions. The displacement of the wedge is limited by a stop. Consequently, the rollers attached to the holders as well as the roller mounted to the base frame squeeze the shaft to initiate forming. Additionally, the rollers start to follow the rotation of the already rotating shaft. As a result, the holders are preloaded and the forming remains in full progress even after the wedge feed is stopped. Hence, every section of the shaft is incrementally profiled by the rollers due to the prestressed holders. During the forming process the preload gradually reduces with increasing penetration of the shaft surface until the pressure is too low for further forming. Once this state is achieved, the wedge will be returned in its initial position.

The Presta joining process is completed by forcing the cam onto the rolled cam seat. However, the simulation presented in this work only relates to the rolling step. The joining step will be part of our future research.

In summary, the rolling process is controlled by the feed of the wedge and by the time period the rolling is performed. Beyond that, it is also driven by the stiffness of the holders. However, an independent FE-modelling of these holders would lead to additional computational effort. Hence, the transfer of this stiffness into the simulation has been realized by a force-displacement relationship (figure 4). This relation has been implemented in a tabular form and determined experimentally by loading the holders in a tensile testing machine.

## 3. Application of viscoplastic material behavior

In the Presta joining process, the rolling of the shaft is an incremental forming process with a characteristic small plastic deformation zone where large local deformations occur [4, 6]. Hence, appropriate material model has to be applied in the simulation in order to reproduce the material behavior. Furthermore, the underlying numerical algorithms have to be stable and robust. Moreover, a reliable material parameter identification needs to be feasible as well [3, 7]. The material model used in this work is a phenomenological model of finite strain viscoplasticity introduced by Shutov and Kreißig [3]. This formulation includes nonlinear kinematic hardening of Armstrong-Frederick type as well as isotropic hardening. In further research the model has been extended to two back-stress tensors and two isotropic hardening mechanisms in order to reproduce the material behavior during incremental forming processes [4].

After the material model is linked to the FE-tool through a material interface, the most important part is the identification of the related material parameters. Therefore, torsion tests on tube samples made from the shaft material have been conducted to provide the required experimental data. This approach has been adopted from Shutov et al. [8].

The viscoplastic material model contains 13 material parameters. They can be divided into elastic constants (K, G), the initial yield stress ($\sigma_F$), parameters for rate-dependency ($\eta$, m), two hardening parameters for each of the two back-stresses ($c_{kin}$, $\kappa$) and two parameters for each of the two isotropic hardening mechanisms ($\beta$, $\gamma$). Here, the elastic parameters and the initial yield stress of the shaft material were determined by the use of material databases. The parameters representing the rate-dependency were established the same way as shown in [4]. The remaining eight hardening parameters were identified with a mathematical software tool using the same material interface implementation as the FE-tool to provide non-redundancy.

The torsion tests have been conducted as straight-forward twists with different velocities and as zig-zag functions with a 10° forward twist and 5° or 8° backward twist. All experiments have been executed up to the breaking of the tube sample. The comparison between measured and calculated shear stresses shows good agreement over many cycles up to high plastic deformations with the same set of parameters (figure 5).

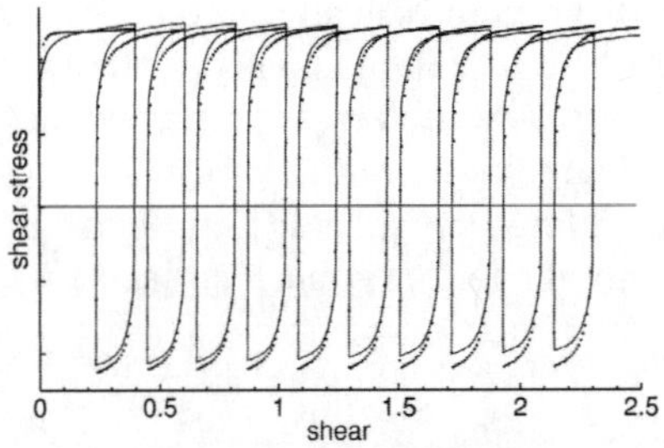


(a)

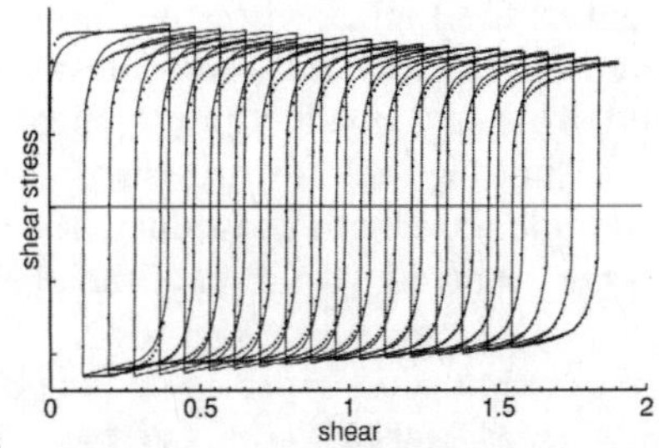


(b)

**Figure 5.** Comparison between experimentally measured (dots) and calculated (lines) shear stresses in a torsion test with (a) 50% and (b) 80% reversal load

## 4. Simulation results and discussion

As already mentioned above, the rolling of the shaft produces an axisymmetric profile in an ideal case. The FE-simulation supports this assumption and shows the expected axial symmetry in plots of the radial displacements as well as in the von-Mises stresses (figure 6).

(a) 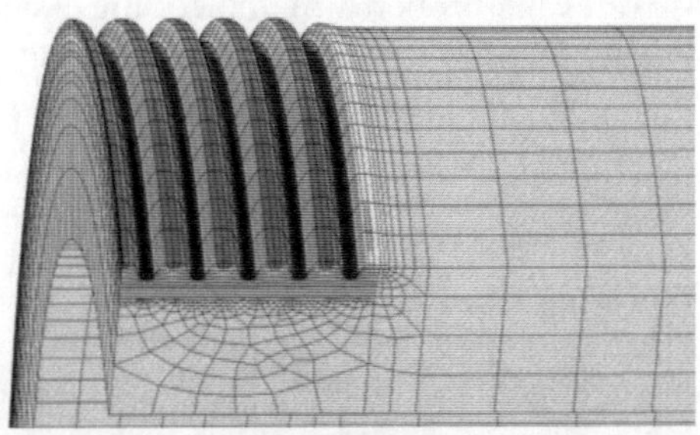

(b) 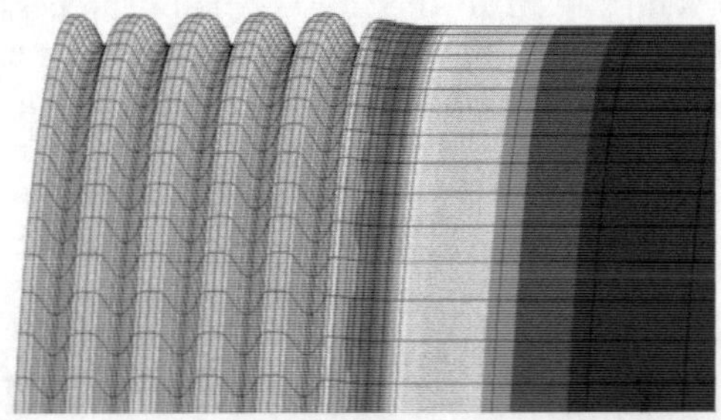

**Figure 6.** Approximately axisymmetric results of a) the radial displacement (sectional view) and b) the von-Mises stresses; tangential gradients in both plots are nearly zero, what proves the estimation of axisymmetry.

The axial symmetry develops over the time of the rolling process. Especially during the first cycles, larger displacement increments occur and axial symmetry does not yet exist. For this purpose, a look at the progression of the holders reaction force shows an evolution towards a constant value (figure 7). Approaching this final value means no further forming occurs.

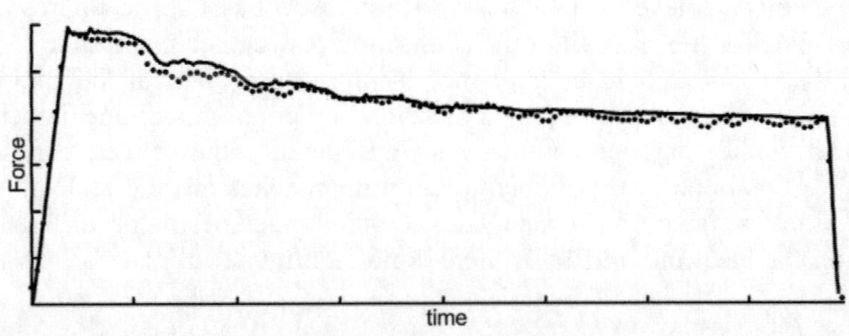


**Figure 7.** Excellent agreement between the holder's reaction force measured in manufacturing (dots) compared with the calculated values of the simulation (lines)

Moreover, the limitation of the forming zone to a small volume beneath the contact surface was considered. To examine this, the inelastic arc length calculated in the implemented material model was evaluated. This scalar value monotonically increases due to plastic deformation independent of their direction. Thus, the inelastic arc length can be seen as the degree of deformation [4]. The contour plot of the inelastic arc length shows patterns similar to thread rolling processes [9]. Furthermore, the plot supports the expected limitation of the plastic zone to the region near the contact surface (figure 8).

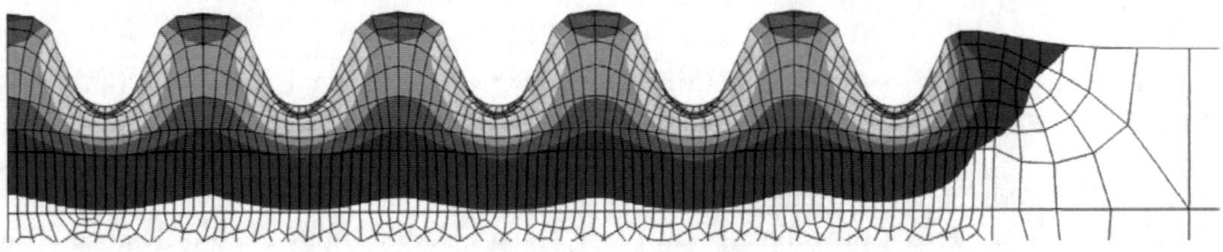

**Figure 8.** The contour plot of the inelastic arc length shows the limitation of plastic deformations to the forming zone with the largest deformations located at the bottom of the grooves.

With regard to the Presta joining process, the most relevant property of the rolling step is the shaping quality of the shaft. This can be measured by analyzing the cross sectional geometry of the resulting profile (figure 9).

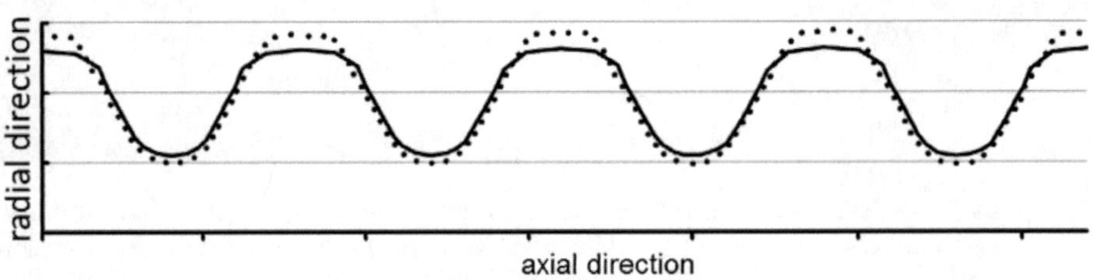


**Figure 9.** Comparison between the measured (dots) and calculated (lines) profile of the shaft after rolling. Reasons for these visible differences may be found once the understanding of the material's viscous behavior is more elaborate.

Although the simulation shows a good general accordance with the measured profile, the calculated forming is not as sharp as the manufactured profile. This indicates that the requested widening of the shaft has not been fully achieved. In future work this aspect has to be improved since the shaft's widening is responsible for the press fit of the connection between cam and shaft [1].

The observed discrepancy could be explained by the viscous part of the applied material model. The deformation velocity in the rolling step is of several orders of magnitude higher than in the torsion tests used to identify the material parameters. Consequently the related parameters (η, m) only represent a rough estimation. Two possible solution approaches are the realization of material tests with higher deformation rates or the analysis of the impact of major differences of the shaft's rotational speed on the resulting profile. If there is no significant influence, the viscous effects could be neglected [8].

## 5. Conclusions

This work introduces a model to perform FE-analyses of the rolling step of the Presta joining process. Based on the technical characteristics of the manufacturing process, the applied geometric and load boundary conditions have been explained. Hence, the model is reduced to one third in circumferential direction and cut in half in axial direction. Furthermore, it was shown that the manufacturing process is driven by the preload and the stiffness of the holders of the rolling tool. A computationally efficient approach to simulate the stiffness of the holders was shown. Instead of using independent FE-models, a force-displacement relationship was applied, which adopts the holder's stiffness.

During the rolling process, large local deformations occur within the shaft. Hence, to ensure a good representation of the material behavior, a phenomenological viscoplastic material model was applied. Here, the procedure of identifying the required material parameters and the result of this identification has been presented.

Finally, the results of the simulation were compared to experimental measurements. It was shown that the simulation supports the expected axial symmetry of the resulting profile. Additionally, as anticipated, the plastic deformation zone is limited to a small area close to the contact zone between roller and shaft.

**Acknowledgements**
The research shown in this work originates from the collaborative research center SFB 692 "High-strength aluminum based lightweight materials for safety components" which is supported by German Research Foundation (DFG). Furthermore, this research was carried out in close cooperation with ThyssenKrupp Presta GmbH in Chemnitz.

**References**
[1] Lengwiler A 2011 *Fehlerfortpflanzung, Simulation und Optimierung von Prozessketten anhand der gebauten Nockenwelle* (Eidgenössische Technische Hochschule ETH Zürich)
[2] Meusburger P 2006 Lightweight design in engine construction by use of assembled camshafts *MTZ worldwide* **67.3** p 10–2
[3] Shutov A V and Kreißig R 2008 Finite strain viscoplasticity with nonlinear kinematic hardening: Phenomenological modeling and time integration *Comp. Methods Appl. Mech. Eng.* **197.21** p 2015–29
[4] Shutov A V, Kuprin C, Ihlemann J, Wagner M F X and Silbermann C 2010 Experimental investigation and numerical simulation of the incremental deformation of a 42CrMo4 steel *Materialwiss. Werkstofftech.* **41.9** p 765–75
[5] Domblesky J P and Feng F 2002 Two-dimensional and three-dimensional finite element models of external thread rolling *Proc. Inst. Mech. Eng., Part B: J. Eng. Manuf.* **216.4** p 507–17
[6] Zhang D W and Zhao S D 2014 New method for forming shaft having thread and spline by rolling with round dies *Int. J. Adv. Manuf. Technol.* **70.5-8** p 1455–62
[7] Shutov A V and Kreißig R 2010 Regularized strategies for material parameter identification in the context of finite strain plasticity *Tech. Mech.* **30.1-3** p 280–95
[8] Shutov A V, Hockauf K, Halle T, Kreißig R and Meyer L W 2009 Experimentelle und phänomenologische Beschreibung des mechanischen Verhaltens der Aluminiumlegierung EN AW-7075 nach großer plastischer Vorverformung *Materialwiss. Werkstofftech.* **40.7** p 551–8
[9] Kukielka K and Kukielka L 2012 Modelling and numerical analysis of the thread rolling process used in food mechanical engineering *KSTU news* **25** p 44–52

IOP Conf. Series: Materials Science and Engineering **118** (2016) 012040 doi:10.1088/1757-899X/118/1/012040

# Deformation behavior of FRP-metal composites locally reinforced with carbon fibers

**M Scholze[1], A Kolonko[2], T Lindner[1], T Lampke[1] and F Helbig[2]**

[1] Chair of Materials and Surface Engineering, Institute of Materials Science and Engineering, Technische Universitaet Chemnitz, Erfenschlager Str. 73, 09125 Chemnitz, Germany
[2] Endowed Chair Textile Plastic Composites, Institute of Lightweight Structures, Technische Universitaet Chemnitz, Reichenhainer Str. 31/33, 09126 Chemnitz, Germany

E-mail: mario.scholze@mb.tu-chemnitz.de, angelika.kolonko@mb.tu-chemnitz.de

**Abstract.** This study investigates variations of hybrid laminates, consisting of one aluminum sheet and a unidirectional glass fiber (GF) reinforced polyamide 6 (PA6) basic structure with partial carbon fiber (CF) reinforcement. To create these heterogeneous FRP laminates, it is necessary to design and produce semi-finished textile-based products. Moreover, a warp knitting machine in conjunction with a warp thread offset unit was used to generate bionic inspired compounds. By the variation of stacking prior to the consolidation process of the hybrid laminate, an oriented CF reinforcement at the top and middle layer of the FRP is realized. In both cases the GFRP layer prevents contact between the aluminum and carbon fibers. In so doing, the high strength of carbon fibers can be transferred to the hybrid laminate in load directions with an active prevention of contact corrosion. The interface strength between thermoplastic and metal component was improved by a thermal spray coating on the aluminum sheet. Because of the high surface roughness and porosity, mechanical interlock was used to provide high interface strength without bonding agents between both components. The resulting mechanical properties of the hybrid laminates are evaluated by three point bending tests in different load directions. The effect of local fiber orientation and layer positioning on failure and deformation mechanism is additionally investigated by digital image correlation (DIC).

## 1. Introduction

The method “The right material at the right place” calls for an interdisciplinary cooperation of science branches to make the potential of clever combinations of materials viable for industrial applications. The Cluster of Excellence MERGE supports this research and development approach. The aim is to design and characterize a new class of effective and efficiently produced hybrid materials. Due to the increasing level of lightweight constructions in the automotive and aviation industry, multi-material concepts become more and more complex comparable with natural structures and solutions of planar and shaped formations. The created bionic inspired materials should be producible in large-scale manufacturing processes. While most metallic materials fit this requirement, work is still needed on the transfer of a complex thermoplastic FRP. Textile-based production systems are suitable for large-scale production. Especially warp knitting machines promise a high productivity of technical textiles for oriented fibrous reinforcements of thermoplastics [1]. To create load path suitable FRP structures, local additional fibrous reinforcement can be integrated into or applied to global reinforced textile-based structures. The research institute Cetex, Chemnitz, developed a unit, which enables the fixation of additional partial continuous fiber reinforcements during the production of warp knitted non-crimp fabrics. The warp thread offset allows for a continuous positioning of defined patterns according to the local requirements of the fiber orientation and fiber volume content (FVC) [2]. Transferred to composites, also the combination of a partially reinforced FRP with metallic sheets is possible. Furthermore, hybrid laminates with load paths at the FRP part and a class A surface at the metallic part can be produced.

A suitable interface and composite design requires an inhibition of contact between elements, which would cause a corrosive behavior of one joining partner. The negative influence of aluminum-carbon fiber interfaces on the corrosion behavior was already investigated in the past [3]. A general way of inhibition is the separation of fiber layer structures in a composite, for example, by a GFRP layer between the CFRP and aluminum. When combining various classes of materials, in addition to corrosion protection also high interface strength is necessary. Adhesive bonding is state of the art in joining metals and plastics [4, 5]. Several disadvantages such as susceptibility to environmental conditions or the required precise adjustment to the joining conditions (temperature and pressure route) can be observed. Beside the influence of the joining conditions and process, the surface preparation has significant influence on the properties of the interface. The pretreatment can be performed by chemical, mechanical or physical processes [6, 7]. Lucchetta et al. studied the influence of the metal surface topography in hybrid joints and concluded that high roughness increases interlocking of the polymer [8]. Furthermore, a general approach for high interface strength is a structured surface which leads to a higher microscopic surface area and undercut grooves for mechanical interlocking. Mechanical pretreatments such as grit blasting can leave a certain amount of residual stresses in the material, promoting the risk of deformation [9]. An adequate way to realize high surface roughness without deformation was shown by Lindner et al., who used a thermal spray coating for an increase of surface roughness [10]. In agreement with other studies, the bonding strength between the aluminum and the FRP was significantly increased by using a NiAl95/5 thermal spray coating without prior grit blasting [10, 11]. With respect to these results, this study focusses on the influence of local CF reinforcement on the deformation behavior of an aluminum-FRP composite. In order to evaluate the resulting mechanical properties of the hybrid laminate, bending tests in different load directions were performed and additionally investigated by DIC.

## 2. Experimental

### *2.1. Design and production of hybrid components*

The selection of compatible materials and suitable process parameters is of great importance for the implementation of a multi-material design. The dimension and performance of the components were specified under the use of a FE-simulation [12]. In the context of this work, each hybrid component was produced in a single process line prior to joining. The 2 mm thick structure of the thermoplastic FRP component consisted of unidirectional (UD) textile-based semi-finished products made of GF/PA6 (basic reinforcement) and CF/PA6 (bionic inspired reinforcement). To determine the performance of partial reinforcement and the influence of its location, a sample pattern of partial CF reinforcement was chosen and positioned as middle or top layer in the FRP structure, see figure 1. For the further hybrid joining process, the FRP structures were provided as preimpregnated sheets.

As metallic component of the hybrid laminate an EN AW6016-T4 (AlSi1.2Mg0.4) aluminum alloy sheet with a thickness of 1 mm was used. In order to generate an integration zone at the joining process by mechanical interlock between the bonding parts, a surface pretreatment of the metallic component was necessary. A suitable technology to realize a high surface roughness for the mechanical interlock is thermal spraying. By means of thermal spraying, it is possible to apply metallic coatings on different substrates. Further advantages of thermal spray coatings are the low deformation of substrate materials and low heat input. Feedstock materials in thermal spraying can be processed as powders or wires. A combination of aluminum and nickel leads to an exothermic reaction during the spray process. In this way, a metallurgical bonding without prior grit blasting of the surface is possible [13]. In this study, an arc wire spraying system with a cored-wire of NiAl95/5 (1.6 mm thickness) was chosen because of its good processability. For more detailed information on the processing parameters, see table 1.

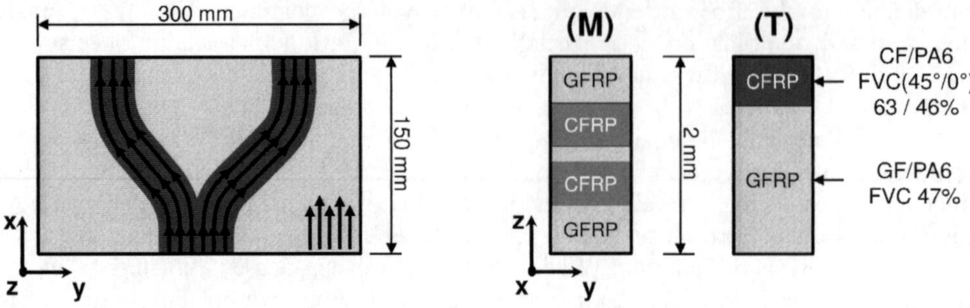


**Figure 1.** Characteristic of FRP components: Basic GF unidirectional reinforcement in the direction of the X-axis (0°), partial CF reinforcement along a simplified bionic inspired path (fiber directions illustrated by arrows). Two types of laminates were produced: CF reinforcement at middle **(M)** and at top layer **(T)**.

**Table 1.** Feedstock material and parameters used for electric arc wire spraying.

| Chemical composition | Current (A) | Voltage (V) | Spraying distance (mm) | Air pressure (bar) | Transverse velocity (m/s) | Line-gap (mm) | Growth by step (µm) |
|---|---|---|---|---|---|---|---|
| Ni – 95 %<br>Al – 5 % | 150 | 30 | 130 | 3.5 | 0.6 | 5 | 50 |

*2.2. Joining process*

The primary objective of the joining process was to combine the hybrid components without a significant deformation of the reinforcement structure inside of the FRP and moreover, to create a defect-free interface. Under variation of the temperature, time and pressure in hot pressing process, fitting parameters were determined. Subsequent to the pretreatment of the FRP and the metal, components were pressed to hybrid laminates with a length of 300 mm and a width of 150 mm by direct heating of the consolidation dies. Temperature and pressure profiles are shown in figure 2. During the joining process, a steady pressure of 5 bar was applied. The temperature measurement was conducted by a thermocouple directly in the interface zone between the metal- and FRP-component. After reaching a temperature of 250 °C in the interface, the tool was cooled down to 80 °C by compressed air. The joining pressure was constant during the air cooling process.

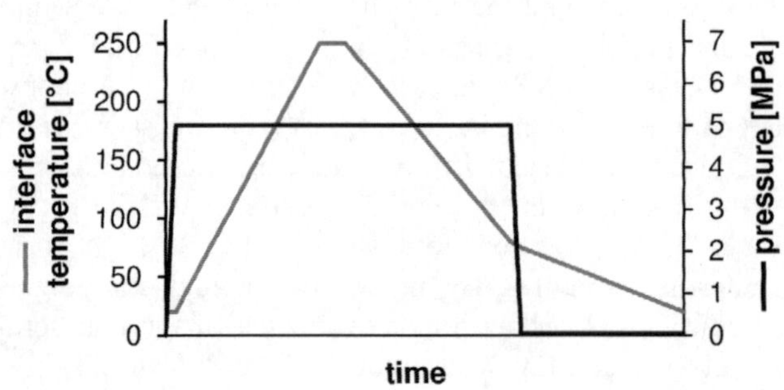


**Figure 2.** Interface temperature and pressure route during the joining process.

### *2.3. Mechanical testing*

In order to evaluate the mechanical properties, three point bending tests according to DIN EN ISO 14125 were conducted [14]. Samples positioned in different fiber directions of the hybrid laminate (see figure 3) with a length of 45 mm and a width of 8 mm were taken by water jet cutting. The bending samples were tested in a universal testing machine (Zwick UPM1475-100 kN) in a three point bending device with a span length of 40 mm at a constant crosshead-speed of 0.6 mm/min. Thereby the FRP-component was loaded with tensile stress. For a precise bending-strain calculation, an inductive displacement measurement was applied. Values shown in this study are based on at least four samples per sample direction and fiber arrangement.

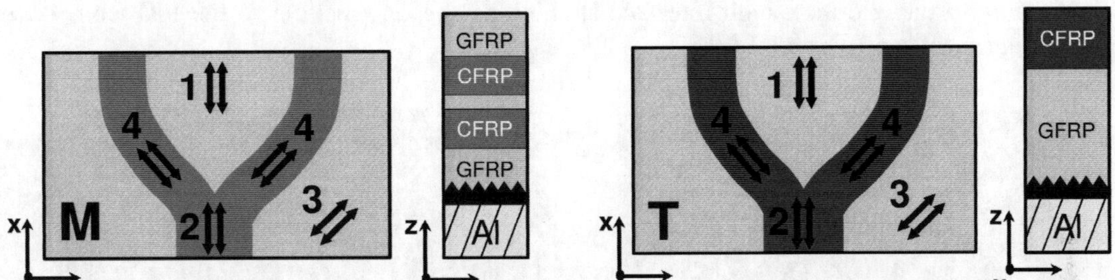


**Figure 3.** Sample positions in hybrid laminates with different layer stacking. Arrows illustrate the position and direction of the cut samples for three point bending tests: **M1:** 0° GFRP; **M2:** 0° GFRP + 0° CFRP at the middle layer; **M3:** 45° GFRP; **M4:** 45° GFRP + 0° CFRP at the middle layer; **T1:** 0° GFRP; **T2:** 0° GFRP + 0° CFRP at the top layer; **T3:** 45° GFRP; **T4:** 45° GFRP + 0° CFRP at the top layer.

In addition to the evaluation of the mechanical properties, the deformation behavior was recorded by an ARAMIS digital image correlation (DIC) system developed by GOM (Braunschweig, Germany). This non-contact optical method provides measurements of surface displacements and calculated strains by tracking random speckle patterns on the surface of the samples. All experiments were performed with a frame rate of 1 Hz using a single CCD camera, which was placed with its optical axis normal to the side surface of the specimens (see figure 4). Subsequent to every single bending test, recorded DIC data were analyzed using the ARAMIS software.

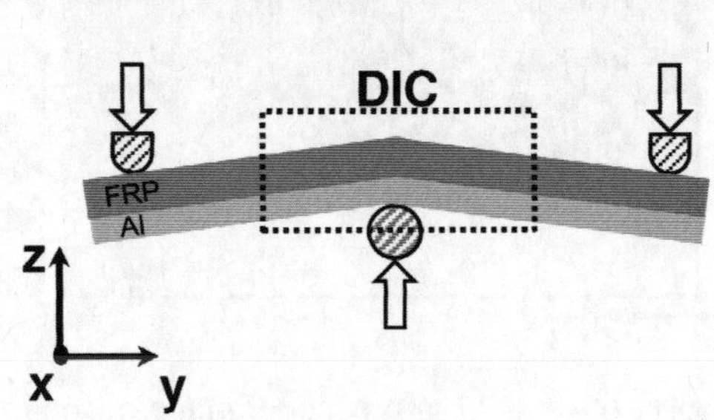


**Figure 4.** Captured DIC area at the bending samples.

## 3. Results and discussion

### *3.1. Microstructure of partially reinforced hybrid laminates*

At first, we discuss the results of the consolidation process of the hybrid laminates by a comparison of different layer stacking with the corresponding microstructures. Figure 5 shows optical microscope images of the hybrid laminates. Different areas of materials are schematically shown next to the images. A clear distinction of the GFRP and CFRP layers and a homogeneous and fully impregnated fibrous structure is visible in the micrographs. Both stacking orders offer an inhibition of contact corrosion by the prevention of contact between the metallic surface and the carbon fibers. The FRP of the composites show in both cross-sections a constant thickness.

To investigate the interface between the FRP and the metal surface, a higher magnification is shown in figure 6. The smooth surface of the aluminum sheet is enlarged by the thermal sprayed NiAl95/5-coating without cavities to aluminum substrate. The rough and porous surface design allows for a mechanical interlock to the FRP-component because of the infiltration of the melted PA6 matrix in the joining process. No glass fibers are interlocked within the coating. This indicates that no fibers are damaged by mechanical stress during processing. The upper GF accumulation (fiber volume content 47%) is homogeneous and mainly fully impregnated with the PA6 matrix. As a result of the microstructure in the interface, high interface strength by the interlocking of the FRP-component to the aluminum sheet can be expected.

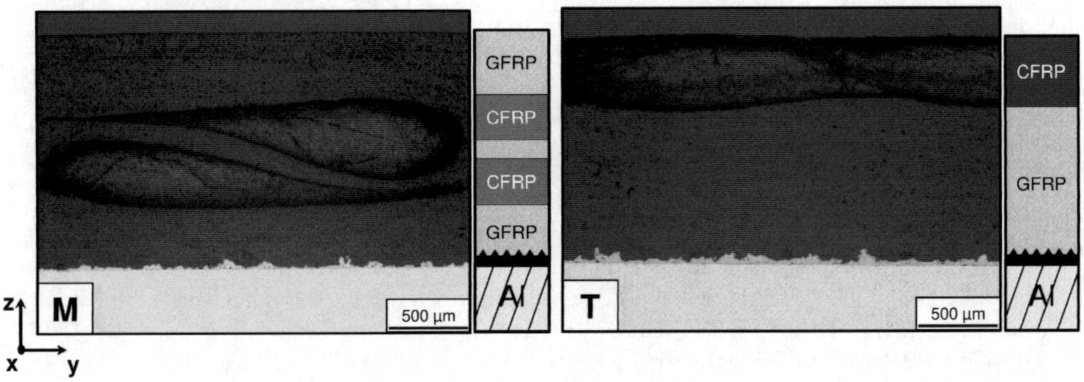


**Figure 5.** Optical microscope images of the hybrid microstructure with different stacking orders: **M:** Local CF reinforcement at the middle layer of the FRP, **T:** Local CF reinforcement at the top layer of the FRP.

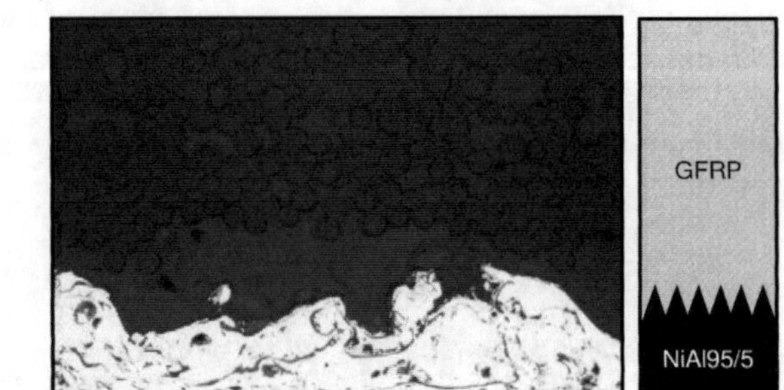


**Figure 6.** Optical microscope image of the interface zone. The porous NiAl95/5-thermal spray coating is fully impregnated with PA6.

### *3.2. Mechanical behavior in different load directions*

Depending on the different load directions of the hybrid laminates, three point bending tests have been performed. Figure 7 shows the maximum bending stresses and bending modules of the different sample orientations (see figure 3 for sample extraction out of hybrid laminates). The mechanical properties of the basic GFRP structure in both laminate types are at the same level (M1 & T1). Also samples with 45° GFRP orientation (M3 & T3) show the same bending strengths and bending modules. This suggests that there is no influence of the local CF reinforcement on the basic GFRP structure. However, resulting mechanical properties show an increase of the bending strength under local CF reinforcement. When CFRP and GFRP are in the same orientation, there is only a slight increase of the bending strength (M1 & T1 compared to M2 & T2). Moreover, no difference in maximum stress is visible between the middle (M2) and the top layer of the CFRP (T2). The maximum bending strength is about 425 MPa. As expected, the local CFRP has a larger influence at 45° GFRP orientation (M3 & T3 compared with M4 & T4). An increase in bending strength is also visible due to the layer variation (M4 & T4). A CFRP layer in tension direction at the top of the hybrid laminate (T4) is characterized by a higher strength than a laminate with CF reinforcement closer to the neutral fiber of the bending sample (M4). In the comparison of maximum stress values in GFRP-direction and local CFRP-direction, we obtained bending strengths of equal values at the same thickness of hybrid laminates when the CFRP is positioned at the top layer (M1 & T1 compared with T4). The highest influence of the partial CF reinforcement is obtained in the enhancement of the local stiffness. Due to the higher elastic modulus of carbon fibers compared to glass fibers, the elastic properties are also transferred to the hybrid laminate. In contrast to the bending strength values, the bending modulus has a significant dependence on the layer position of the local CFRP. With increasing distance of the CFRP to the neutral fiber in bending tests, the bending modulus is increased (M2 & T2, M4 & T4). Under the same orientation of the GF and CF reinforcement at the top layer the bending modulus reached the highest level of 68 GPa (T2). The stiffness of the basic GFRP structure (M1 & T1) is in 45° local reinforcement direction already reached with the middle layer CFRP (M4). In the case of the top layer CFRP (T4), the bending modulus of the samples has even a higher value than the basic GFRP structure in 0°. As a result of that, local CF reinforcement can be used for the enhancement of the local bending stiffness.

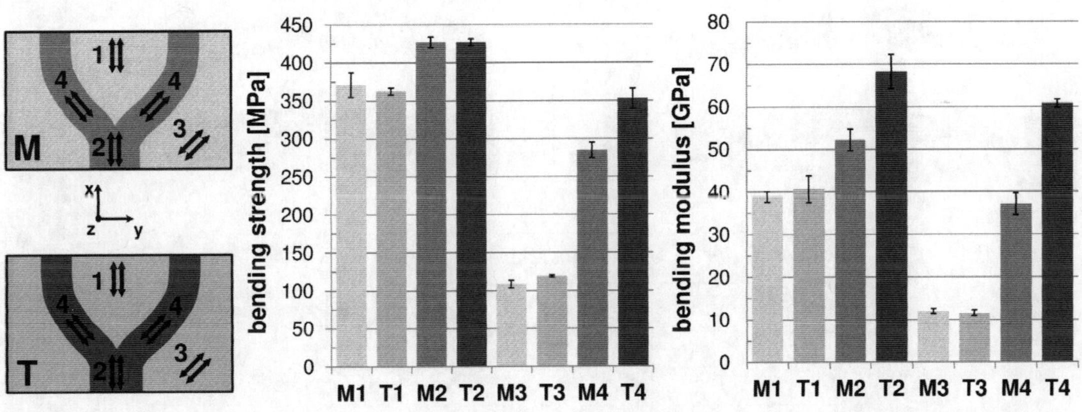


**Figure 7.** Mechanical properties of different layer structures in three point bending tests: **M:** Local CFRP at the middle layer, **T:** Local CFRP at the top layer. Due to the local CF reinforcement, bending strength and bending modulus are increased.

At the top and middle layer, UD fiber orientations of the GF and CF reinforcement are characterized by same bending stress values (see figure 7, M2 & T2), although the layer structure influences the bending modulus. Therefore, the deformation behavior of these layer structures was additionally investigated by using the DIC technique. Figure 8 shows strain field images of M2 and T2 with increasing deformation and the corresponding stress-strain behavior. Strain localizations, resulting from high displacements of local surface patterns, can be monitored by color gradients in DIC images. In the case of a middle layer CFRP, two localization zones parallel to each other can be observed at the bending sample with increasing strain (M2, images 1-5). The same strain levels lead under the top layer CFRP to an early localization in mainly one zone (T2, images 1-5). Furthermore, with increasing strain, the localization leads to delamination along the interface between the CFRP and

GFRP layer. The early localization in the case of the top layer CFRP and the following delamination of the layer structure prevents an increase in bending strength in comparison with the middle layer CFRP. Moreover, only a high bending modulus is reached because at low bending strain levels there is less influence of interlaminar shear stresses between layer structures with different mechanical properties, resulting from layer displacements under bending. The bending modulus is mainly influenced by the extreme upper fiber. As a result of that, a high bending modulus is only reached in the case of a top layer CFRP. In comparison with the microstructures (figure 5) of the two hybrid laminates, the negative influence on delamination by a top layer CFRP can be explained by the distinction of the top layer CFRP and the GFRP. In the case of the middle layer CFRP, a good formation of the CFRP in the surrounding GFRP leads to a more uniform stress and strain distribution in the FRP.

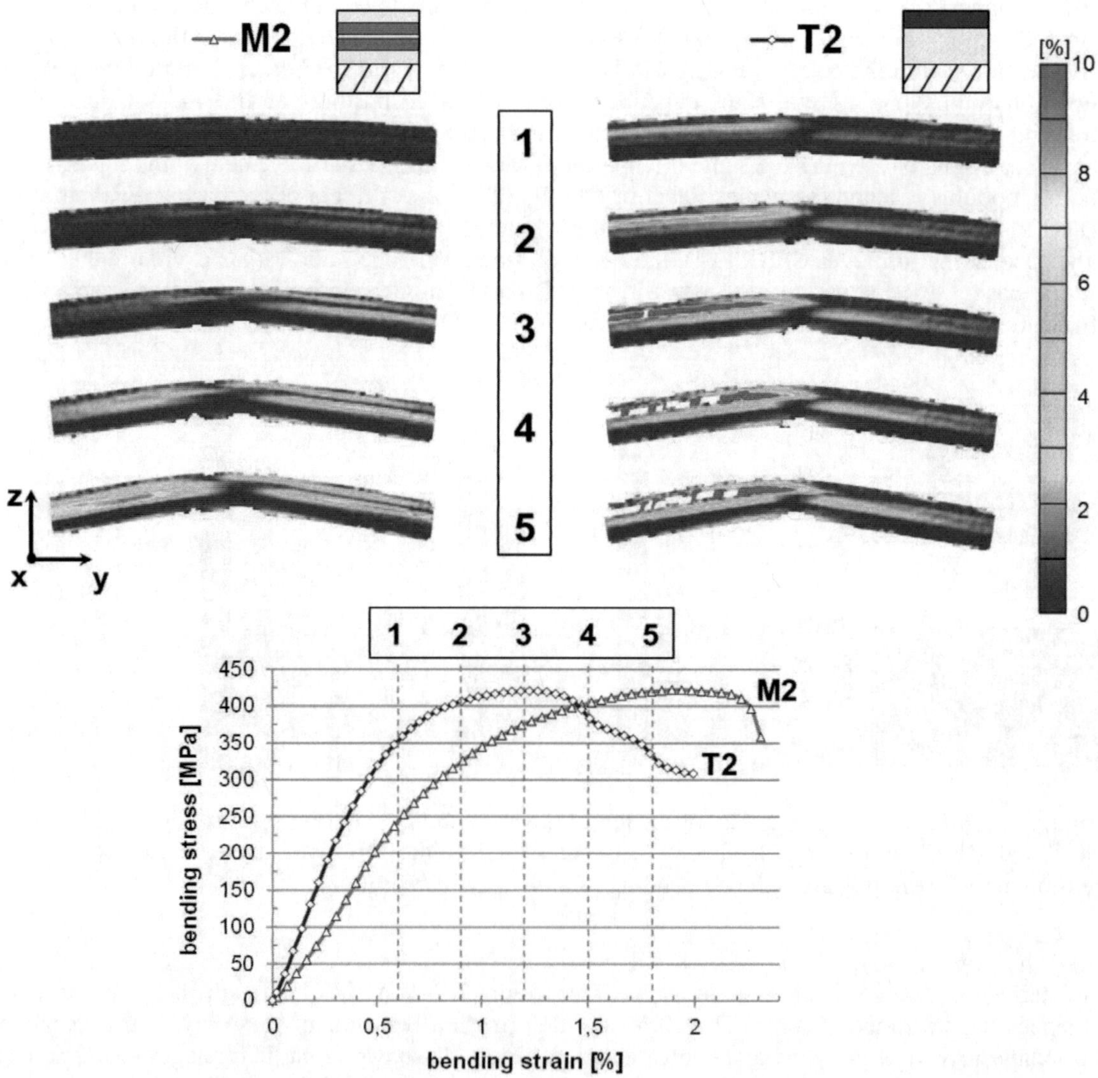


**Figure 8.** Strain field images from DIC: A top layer of the CFRP leads to early localization and delamination between the GFRP and the CFRP. **M2:** 0° GFRP + 0° CFRP at the middle layer; **T2:** 0° GFRP + 0° CFRP at the top layer.

## 4. Summary and conclusions

In this paper, we have investigated the effect of local carbon fiber reinforcement on the deformation behavior of a FRP-metal composite. A warp knitting machine with a warp thread offset unit was used to generate bionic inspired compounds with various directions of a CF in basic GF reinforced PA6. As metallic component an aluminum sheet with a thermal spray coating was used. By the variation of the stacking order prior to the consolidation process, different layer structures in hybrid laminates were produced. The deformation behavior was investigated in different load directions with three point bending tests. Additionally, the digital image correlation technique was used during the deformation to observe localizations in two different layer structures. Several noteworthy results are summarized as follows:

1. The manufacturing of textile-based semi-finished products with bionic inspired endless fiber reinforcement is viable under the usage of a warp knitting machine with a warp thread offset. The local pattern can be transferred into a hybrid laminate structure. After the joining process with the metal component, microstructural investigations show a homogeneous and fully impregnated fibrous structure regardless of the different stacking orders. An active prevention of corrosion is obtained by a GFRP layer, which impedes contact of carbon fibers and the aluminum.

2. The thermal spraying process offers a suitable pretreatment of aluminum surfaces to generate an interface between the metal and the FRP. By using NiAl95/5 as feedstock material, a high bonding strength to the aluminum substrate is achieved. Furthermore, the rough and porous surface of the spray coating with undercuts allows for a mechanical interlock to the FRP-component because of the infiltration of melted PA6 matrix in joining process.

3. A load-optimized structure can be reached at a constant thickness of the laminate. The basic GF-structure is not influenced by the local reinforcement. A comparatively low content of the load-oriented local CF reinforcement in hybrid laminates (0° CFRP/45° GFRP) leads to bending strengths of the same level com-pared to hybrid laminates with an only unidirectional GF reinforcement. Higher bending modules are possible in the case of the CFRP top layer structures, which is due to the highest influence of the extreme upper fiber under bending load.

4. DIC investigations showed an early localization between the layer structures. In the case of the top layer CFRP, load-uniform orientations of the glass and carbon fibers lead to an early delamination compared with the middle layer CFRP. As a result of that, the bending strength is not increased in contrast to the enhancement of the bending modulus with a higher distance of the CFRP from the neutral fiber in the bending samples. According to the delamination in the FRP, further investigations should focus on an improvement in accumulation as well as on the arrangement of carbon and glass fibers in the PA6 matrix system. To increase the hybrid performance, an optimization of the fiber-matrix adhesion and a reduction of internal stresses in the FRP are planned in future research activities.

**Acknowledgement**

This study was supported by the Federal Cluster of Excellence EXC 1075 "MERGE Technologies for Multifunctional Lightweight Structures". Funding by the German Research Foundation (Deutsche Forschungsgemeinschaft, DFG), Germany is gratefully acknowledged.

## References

[1] Helbig F and Scheika M, 2014. *Dünnschichtlaminate aus textilverstärktem FKV. MM MaschinenMarkt 50/51, 28, ISSN 0341-5775.*

[2] Heinrich H-J and Vettermann F, 2009. *Durch neues Gelegedesign Fortschritte beim Ringen um Stärke. Kettenwirk-Praxis 4, 27-29.*

[3] Payan S, Le Petitcorps Y, Olive J-M and Saadaoui H, 2001. *Experimental procedure to analyse the corrosion mechanisms at the carbon/aluminium interface in composite materials. Composites: Part A, 32(3), 585-589.*

[4] Huang Z, Sugiyama S and Yanagimoto J, 2014. *Applicability of adhesive–embossing hybrid joining process to glass-fiber-reinforced plastic and metallic thin sheets. Journal of Materials Processing Technology 214(10), 2018-2028.*

[5] Picard L, Phalip P, Fleury E and Ganachaud F, 2015. *Chemical adhesion of silicone elastomers on primed metal surfaces: A comprehensive survey of open and patent literatures. Progress in Organic Coatings 80, 120-141.*

[6] Velthuis R, Kötter M P, Geiß P L, Mitschang P and Schlarb A K, 2007. *Lightweight Structures Made of Metal and Fiber-Reinforced Polymers. Kunststoffe International 11, 22-24.*

[7] Mitschang P, Velthuis R and Didi M, 2013. *Induction Spot Welding of Metal/CFRPC Hybrid Joints. Advanced Engineering Materials 15(9), 804-813.*

[8] Lucchetta G, Marinello F and Bariani P F, 2011. *Aluminum sheet surface roughness correlation with adhesion in polymer metal hybrid overmolding. CIRP Annals – Manufacturing Technology 60(1), 559-562.*

[9] Tosha K and Iida K, 1989. *Residual stress on the grit blastet surface, International Conference on Residual Stress, Mechanical Behaviour & Surface Engineering, IITT International, Los Angeles, 323-328.*

[10] Lindner T, Friederichs C, Zillmann B, Hockauf K, Wielage B and Lampke L, 2014. *Development of an Integration Zone for Joining Polymer-Metal Hybrid Structures, Proceedings of Euro Hybrid Materials and Structures 2014, Stade, 13-16.*

[11] Nestler D, Arnold S, Jung H, Wielage B and Kroll L, 2014. *Untersuchung geeigneter Oberflächenbehandlungsverfahren der Metallkomponente thermoplastbasierter hybrider Laminate, Tagungsband zum 17. Werkstofftechnischen Kolloquium, 208-216, Chemnitz, 2014, - ISBN: 978-3-00-046877-3.*

[12] Kolonko A, Dura Brisa V J, Helbig F and Kroll L, 2014. *PaFaTherm II – Mehrkomponenten Spritzgießprozesse für strukturvariable textilverstärkte Verbundbauteile – Symposium zum Statusseminar 2014, p. 63, Chemnitz, - ISBN: 978-3-945479-01-8.*

[13] Sheppard J A, 1963. *Sprayed coatings of exothermically formed nickel aluminide. British Welding Journal 10(12), 603-606.*

[14] Norm DIN EN ISO 14125, 2011: *Faserverstärkte Kunststoffe – Bestimmung der Biegeeigenschaften, Berlin: Beuth Verlag.*

IOP Conf. Series: Materials Science and Engineering **118** (2016) 012041 doi:10.1088/1757-899X/118/1/012041

# Effect of new adhesion promoter and mechanical interlocking on bonding strength in metal-polymer composites

**A Schuberth[1], M Göring[2], T Lindner[1], G Töberling[3], M Puschmann[3], F Riedel[3], I Scharf[1], K Schreiter[2], S Spange[2] and T Lampke[1]**

[1]Materials and Surface Engineering Group, Institute of Materials Science and Engineering, Technische Universität Chemnitz, 09125 Chemnitz, Germany
alexandra.schuberth@mb.tu-chemnitz.de
[2]Polymer Chemistry, Technische Universität Chemnitz, 09111 Chemnitz, Germany
stefan.spange@chemie.tu-chemnitz.de
[3]Department of Joining and assembling, Fraunhofer Institute for Machine Tools and Forming Technology IWU, 09126 Chemnitz, Germany
gerd.toeberling@iwu.fraunhofer.de

**Abstract** There are various opportunities to improve the adhesion between polymer and metal in metal-plastic composites. The addition of a bonding agent which reacts with both joining components at the interfaces of the composite can enhance the bonding strength. An alternative method for the adjustment of interfaces in metal-plastic composites is the specific surface structuring of the joining partners in order to exploit the mechanical interlock effect. In this study the potential of using an adhesion promoter based on twin polymerization for metal-plastic composites in combination with different methods of mechanical surface treatment is evaluated by using the tensile shear test. It is shown that the new adhesion promoter has a major effect when applied on smooth metal surfaces. A combination of both mechanical and chemical surface treatment of the metal part is mostly just as effective as the application of only one of these surface treatment methods.

## 1. Introduction

The reduction of the construction weight due to an increasing demand for resource and energy efficiency necessitates further development of metal-plastic composites as substitution for steel for automotive application. Fiber-reinforced plastics (FRPs) with their high stiffness and strength, corrosion resistance and various fabrication technologies [1] are appropriate for the manufacturing of FRP-metal hybrid components. Beside FRPs aluminum alloys can replace steel as metallic part for a further reduction of the construction weight.

For manufacturing metal-plastic composites with high mechanical load capacity, the adjustment of interface properties is necessary. This is because the joined materials exhibit strongly differing chemical and physical properties. Therefore, a chain entanglement effect as observed in polymer joints does not occur [2]. Nevertheless, the final joint strength of such composites can be improved by using a bonding agent. In this case, according to the chemisorption theory, the bonding can be caused by the generation of chemical bonds between the adhesion promoter and both of the joined parts. The mechanical theory of adhesion states that the topography and roughness of adhered surfaces correlate with the measured bonding strength. Depending on the viscosity and surface tension of the adhesive, it can penetrate pores and fill surface hollows of the joined parts. After the curing of the adhesive, a positive connection between adhesive and adherend is created [3]. The main aim of this study is to investigate the interaction of both adhesion mechanisms in connection with a new adhesion promoter.

## 2. Materials and surface treatment methods

### *2.1. Joining components*

A glass-fiber-reinforced polyamide 6 (2-layered, bi-directional reinforcement) with a thickness of 2 mm was used in this study as plastic joining partner. As metallic components, two different bulk materials with a thickness of 1 mm were tested. In a prior investigation [4], the effect of the new bonding agent, produced by twin polymerization, was already approved for the system aluminum alloy–FRP. In this study, an additional aluminum alloy EN AW 6016-T4 (AlSi1.2Mg0.4) and steel DC06 were tested as references.

### *2.2. Adhesion promoter*

The adhesion of a polymer to metals can be significantly improved by applying certain twin monomers (TMs) to the adherent surface [4]. Specific TMs offer the possibility that various functional groups can be introduced by a modular approach. In this work, a primary amino group was used inspired by well-known primer systems such as aminopropyltrimethoxysilane [5,6] or dendrimers. The introduction of amino groups is beneficial in so far as reactions with the carbonic acid end groups of the polyamide component and acid-base interactions with the metallic component surface are possible. Furthermore, the formation of silica will also lead to an improved bonding with the metal surface. The TMs used were synthesized based on commercially available reactants. TM (I) 2.2'-spirobi-[4H-1,3,2-benzodioxasiline] and TM (II) 2-(3-amino-n-propyl)-2-methyl-4H-1,3,2-benzodioxasiline were synthesized according to the literature [7–9]. In a first step, TM (I) and TM (II), with a molar ratio of 15 to 85 [4], were pre-polymerized at 100 °C. The advantage is that the pre-polymerization is a good tool to control the reactivity, sensitivity and solubility of the adhesion promoter system. A 10 wt. % solution of the pre-polymer was prepared in ethanol for coating procedures. The solution was dropped on the metal surface via a syringe, and the volume dropped was approximately 0.15 ml per cm² of the metal surface area. The simultaneous twin polymerization of TM (I) and TM (II), which was performed at 180 °C, leads to nanostructured hybrid materials consisting of SiO2/poly-(3-aminopropylmethyl-siloxane) and phenolic resin (Fig. 1). The adhesion layer is generated in a single process step.

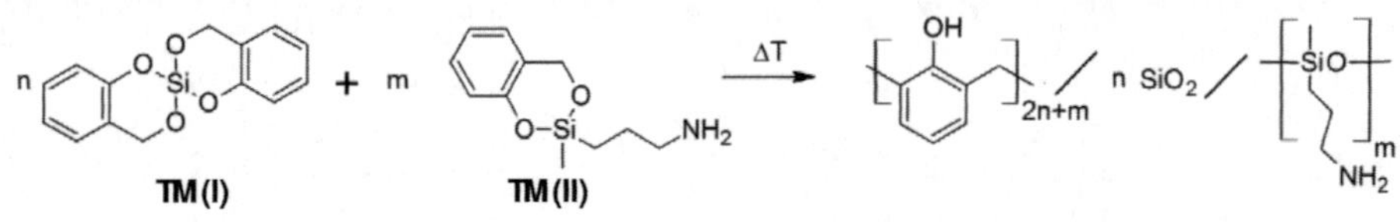


**Figure 1.** Simultaneous twin polymerization of monomer (I) and (II)

### *2.3. Surface treatment*

#### *2.3.1. Grit blasting*

Grit blasting is a widespread method for surface treatment, which provides surface roughening to various extents depending on the grit size and blasting parameters such as pressure, angle, and process time [10]. For this investigation, blasting was conducted with corundum FK-24 with a blasting angle of about 60°. The surface roughness is significantly increased by grit blasting with corundum. The bonding strength of the composite is improved by the mechanical interlock effect between plastic and metal [10].

#### *2.3.2. Laser structuring*

Laser structuring is an innovative surface treatment method based on conventional laser ablation, which can be used to create fine structures at nano- and micro-scale level on the metallic surface. A wide variety of treatable materials and producible structures, no machine wear because of non-tactile machining, short processing times as well as low-level, localized laser energy input to the component are the main advantages of this technology. Due to the low energy input and fast beam guiding, laser structuring is suited for the treatment of localized zones and temperature sensitive component parts [11]. In the course of laser structuring, a laser beam is focused on the surface of the parts being machined. In so doing, high temperature and evaporation pressure are induced, ejecting the molten and partially vaporized material from the surface. As a consequence of the melt solidification at the edge of formed cavities and above the metallic sheet surface, undercut structures can be produced. These undercuts provide a good possibility for the clamping of reinforcing fibers and polymer matrix material of FRPs during the manufacture of metal-plastic composites [11]. The geometrical shape (depth and width) and dimensions of surface structures depend on the material properties and process parameters such as laser power, pulse duration, feed rate, and number of run. For example, excessive energy input during the treatment of materials with low-viscosity melts may affect or reclose already formed structures by increased spattering. Hence, no capacity for mechanical interlocking will be produced. Therefore, the parameter sets for the surface treatment of aluminum and steel have to be adapted to each specific material.

The laser structuring of the test samples was performed at the laser machining center Orca-μ (Acsys Lasertechnik GmbH). As a beam source, a pulsed fiber laser with maximal power of 20 W and wavelength of 1070 nm from IPG was used.

In order to avoid the above-mentioned welding of the formed cavities during the treatment of the aluminum surface, a laser power of about 13–14 W was used and the number of laser runs was reduced to one. When surface structuring the steel samples, it was found that a multiple treatment results in a significant increase in structure depth with a simultaneous increase of the bulging height above the sheet surface. In contrast to aluminum, no welding or reclosures of cavities were observed. For the treatment of steel, a laser power of 18 W was applied and the number of laser runs increased to 10.

*2.3.3. Thermal spray coating*

Thermal spraying is a conventional method for application of coatings on various substrate materials with relatively low deformation. During this process, the coating material is fed to a torch most commonly in form of powder or wire and usually heated close to or above its melting point. The molten or softened particles of the feed-stock material produced this way are accelerated by a gas or plasma stream and applied onto a substrate. During the impact on the substrate, the droplets are deformed and solidified [12]. Depending on the particle size and process parameters, the surface roughness and open porosity can be adjusted in order to generate surface cavities and undercuts, which can be exploited for mechanical interlocking. Particularly arc spraying is capable of generating high surface roughness due to its characteristically broad particle size distribution. The Visu Arc 350 arc spray system was used for applying a coating on aluminum. As feedstock material, a cored wire of NiAl 95/5 (1.6 mm thickness) was chosen for this study, which is typically applied as bonding layer beneath ceramic coatings to enhance adhesion to the substrate. The current and voltage of the arc were 150 A and 30 V respectively. The distance between the spraying gun and aluminum substrate surface was 130 mm.

## 3. Characterization methods

The shear tension test according to DIN EN 1465 was used to evaluate the effectiveness of the twin polymer as well as the various techniques of mechanical surface structuring individually and in combination with each other. The metal and plastic parts were bonded by thermal joining at an elevated temperature (350 °C for aluminum and 400 °C for steel). When the required joining temperature was attained, the bonding force of 250 N was applied and the sample was cooled by compressed air under constant load. The overlapping length was selected to be 5 mm in order to avoid plastic deformation of thin metallic joined parts during the mechanical test. The shear tension test was carried out at room temperature and at a strain rate of 1 mm/min. A universal testing machine (Zwick AllroundLine Z020-20 kN) was used for these tests. The maximum tensile force was measured and the bonding strength was calculated. At least five tests were carried out for each interface design. Prior to the application of the adhesion promoter, the surface topography of the metallic part after the surface structuring treatment was investigated by means of an optical three-dimensional profilometer (3D profilometer, GFMesstechnik GmbH).

## 4. Results and discussion

The images depicted in Table 1 were recorded by a non-contact optical 3D-profilometer. They show the surface topography of metallic samples (4 mm × 3 mm) after different treatments prior to the joining with the plastic part. Due to the different hardness levels of the used metallic materials, the resulting surface roughness Ra of the grit-blasted aluminum samples is higher than that of similarly treated steel

IOP Conf. Series: Materials Science and Engineering **118** (2016) 012041 doi:10.1088/1757-899X/118/1/012041

samples. Thermally sprayed coatings of NiAl 95/5 on both substrates show no significant difference. The laser-structured surface of the steel sheets exhibits much deeper cavities and much higher bulging compared to the aluminum sheets because of multiply repeated laser treatment, as discussed in 2.3.2.

**Table 1.** Surface topography of metallic parts after different treatments.

| | Steel | Aluminum |
|---|---|---|
| Grit-blasted | Ra = 10.2 µm   | Ra = 7.3 µm   |
| Thermally sprayed NiAl 95/5 | Ra = 7.5 µm   | Ra = 8.4 µm   |
| Laser-structured | Ra = 9.5 µm 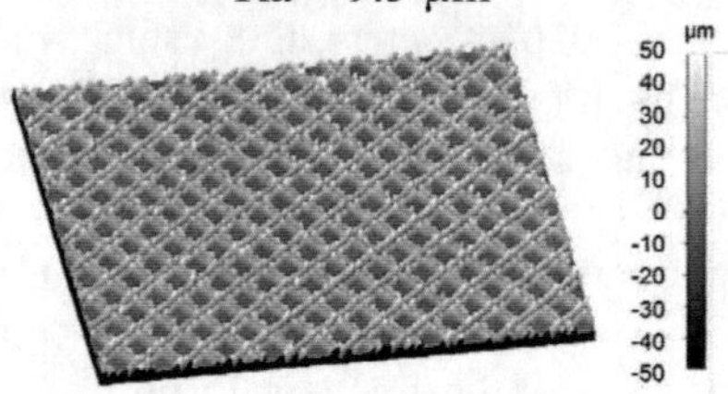  | Ra = 3.4 µm 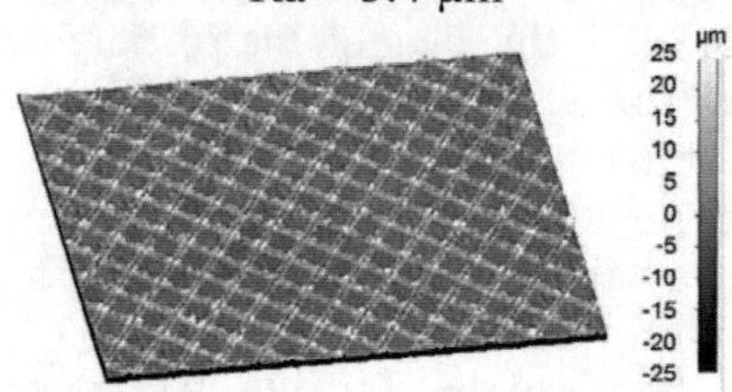  |

The results of the shear tension test confirm the effect of the adhesion promoter produced by twin polymerization, as was already observed in previous investigations with pull-off tests [4]. The developed bonding agent has also a major effect as an adhesion promoter for the joining of steel with FRPs. By applying this adhesion promoter to an unstructured, untreated metallic surface, it is possible to increase the bonding strength of metal-plastic composites by the factor of 7 to 10 (Figs. 2 and 3). Mechanical surface treatments of metallic parts without the application of an adhesion promoter also lead to a significant enhancement of the resulting bonding strength in comparison to untreated metallic parts. Only grit blasting is less effective than the chemical treatment with the novel bonding agent, as it was already discussed in [4].

Not only the surface roughness, but also more complex aspects of the surface topography determine the mechanical interlocking effect when joining different materials [13]. Grit-blasted surfaces offer less anchorage points for mechanical interlocking compared to thermal sprayed surfaces with open porosity or laser-structured surfaces with deep grooves and high bulging. As a consequence of this, the bonding strength of samples with a grit-blasted metallic part is lower than those with a thermally sprayed or laser-structured metallic part. The highest values of the bonding strength, which were determined by testing samples with solely mechanically treated metallic parts, are about 23–25 MPa for both metals investigated.

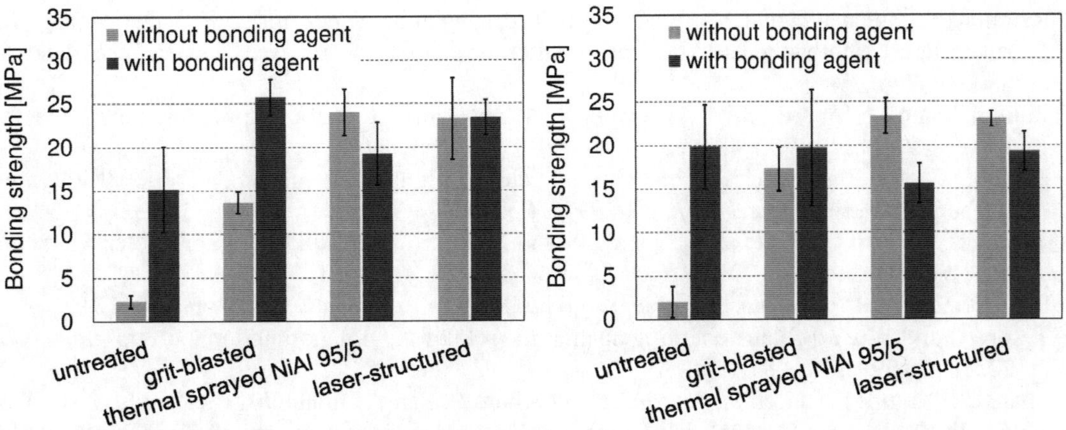


**Figure 2.** Bonding strength of different composites determined by shear tension test: steel-FRP (left) and aluminum-FRP (right)

Generally, the application of an adhesion promoter on the surface-structured samples exhibited no improvement of the results. This indicates that no closed homogeneous film of the adhesion promoter was formed on the structured metallic surface, possibly due to poor wettability of the metallic surfaces or deficient viscosity of the adhesion promoter. The only marked exception to this was observed for grit-blasted steel surfaces. Here, the resulting bonding strength exceeds 25 MPa. The determining mechanism should be investigated in detail. In addition, the influence of prepolymerization parameters, quantity as well as the application method of the applied adhesion promoter are subject to further investigation.

Nestler et al. [13] used chemically and mechanically differently treated aluminum in their study on aluminum–FRP joints. They found that the application of an adhesion promoter based on twin polymerization or grit blasting of metallic part results in similar average bonding strengths of 15–16 MPa. Lower bonding strengths, compared to the results presented here, may be caused by different joining parameters, used materials and the sample geometry. The maximum bonding strength of 17–19 MPa was achieved in [13] by using chemical pretreatments like etching or anodizing or by the application of a thermally sprayed coating. Therefore, it should be investigated whether a chemical pretreatment could be advantageous in combination with the newly developed adhesion promoter.

## 5. Summary and Conclusion

In this study, the separate and combined effect of an adhesion promoter produced by twin polymerization and structured metallic surfaces on the tensile shear strength of metal-polymer composites was investigated. The novel chemical bonding agent caused a significant increase of the bonding strength of metal-plastic composites, particularly when using aluminum as the metallic part. A surface-structuring pretreatment of the metallic part by thermal spraying or laser structuring resulted in higher bonding strength values. A combination of both types of surface treatment was generally not more effective than the individual surface treatments. Hence, the surface treatment method should be adjusted with regard to the practical application of metal-plastic composites, the dimension and shapes of individual composite parts.

## Acknowledgments

The authors gratefully acknowledge funding by the German Research Foundation (Deutsche Forschungsgemeinschaft DFG) through the Cluster of Excellence 1075 "MERGE" (Merge Technologies for Multifunctional Lightweight Structures).

## References

[1] Reinhold B, Blücher D and Korte M Herausforderungen an Füge- und Oberflächentechnik für zukünftige Leichtbaukonstruktionen im Automobilbau *Materialwiss. Werkstofftech.* 2013 **44** 58–63

[2] Ramarathnam G, Libertucci M, Sadowski M M and North T H Joining of Polymers to Metall *Welding Journal* 1992 **7** 483–90

[3] Bischof C ND-Plasmatechnik im Umfeld der Haftungsproblematik bei Metall-Polymer-Verbunden *Matwiss. Werkstofftech.* 1993 **24** 33–41

[4] Yulinova A, Göring M, Nickel D, Spange S and Lampke T Novel adhesion promoter for metal-plastic composites *Adv. Eng. Mater.* 2015 **17** 802–09

[5] Plueddemann E P 1982 *Silane Coupling Agents* (Plenum Press, New York) pp. 153–181

[6] Tesoro G and Wu Y J Silane coupling agents: the role of the organofunctional group *Adhes. Sci. Technol.* 1991 **5** 771–84

[7] Spange S, Kempe P, Seifert A, Auer A A, Ecorchard P, Lang H, Falke M, Hietschold M, Pohlers A, Hoyer W, Cox G, Kockrick E and Kaskel S Nanokomposite mit 0.5 bis 3 nm großen Strukturdomänen durch Polymerisation von Silicium-Spiroverbindungen *Angew. Chem.* 2009 **121** 8403–08

[8] Spange S and Grund S Nanostructured Organic–Inorganic Composite Materials by Twin Polymerization of Hybrid Monomers *Adv. Mater.* 2009 **21** 2111–16

[9] Göring M, Seifert A, Schreiter K, Müller P and Spange S A non-aqueous procedure to synthesize amino group bearing nanostructured organic–inorganic hybrid materials *Chem. Commun.* 2014 **50** 9753–56

[10] Islam M S, Tong L and Falzon P J Influence of metal surface preparation on its surface profile, contact angle, surface energy and adhesion with glass fibre prepreg *Int. J. Adhes. Adhes.* 2014 **51** 32–41

[11] Bobzin K, Bührig-Polazcek A, Haberstroh et al. 2012 *Integrative Production Technology for High-Wage Countries* chapter 5 ed C Brecher (Berlin Heidelberg, Springer-Verlag) pp 480–551

[12] Tucker R C 2013 *Thermal Spray Coatings ASM Handbook* vol 5A ed C M Cotell, Sprague J A et. al (ASM International) pp 497–509

[13] Nestler D, Arnold S, Jung H, Wielage B and Kroll L 2014 *Proc. 17 Werkstofftechnisches Kolloquium (Chemnitz)* vol 52 (TU Chemnitz Press) pp 208–16

IOP Conf. Series: Materials Science and Engineering **118** (2016) 012042 doi:10.1088/1757-899X/118/1/012042

# Process combinations for the manufacturing of metal-plastic hybrid parts

**W-G Drossel, C Lies[1], A Albert, R Haase, R Müller and P Scholz**
Fraunhofer Institute for Machine Tools and Forming Technology IWU, Chemnitz, Germany
[1] Corresponding author

**Abstract.** The usage of innovative lightweight materials and processing technologies gains importance in manifold industrial scopes. Especially for moving parts and mobility products the weight is decisively. The aerospace and automotive industries use light and high-strength materials to reduce weight and energy consumption and thereby improve the performance of their products. Composites with reinforced plastics are of particular importance. They offer a low density in combination with high specific stiffness and strength. A pure material substitution through reinforced plastics is still not economical. The approach of using hybrid metal-plastic structures with the principle of "using the right material at the right place" is a promising solution for the economical realization of lightweight structures with a high achievement potential. The article shows four innovative manufacturing possibilities for the realization of metal-plastic-hybrid parts.

## 1. Introduction

In the automotive industry, there is a longstanding effort to find lightweight construction solutions that on the one hand ensure that car body components have the necessary stability, but on the other hand allow the parts to be produced at low cost. Along with lightweight materials such as aluminium, magnesium and high strength steels, nowadays multi-material design and hybrid components made of plastics and metals can help achieve these objectives. As a result, it is not only possible to further reduce weight [1], but also to implement more complex structures, which for instance can be achieved by injection moulding. Beside the integration of additional functions, the implementation of functional elements for joining or assembly purposes becomes feasible by means of plastic injection moulding. However, the metal guarantees the strength of the component and assembly to the car body by well-established processes. It is essential that the production technology applies these composite structures made of plastics and metal. As a rule, the metal components are made first in separate processes. Afterwards they are joined with the plastic component in the injection moulding process. To guarantee reliable adhesion between metal and plastics, the metal components must also be coated with a bonding agent. The resulting process chains are extensive, labour-intensive and costly. In the following four innovative manufacturing possibilities are presented for the realization of metal-plastic hybrid parts.

## 2. State of the art

### *2.1. Metal/plastic hybrid components*

Metal-plastic hybrid components applied in the car industry can generally be classified into two classes according to the underlying metal component: sheet based (Figure 1) and tube based (Figure 2) parts.

**Figure 1.** Audi TT front end carrier [2]

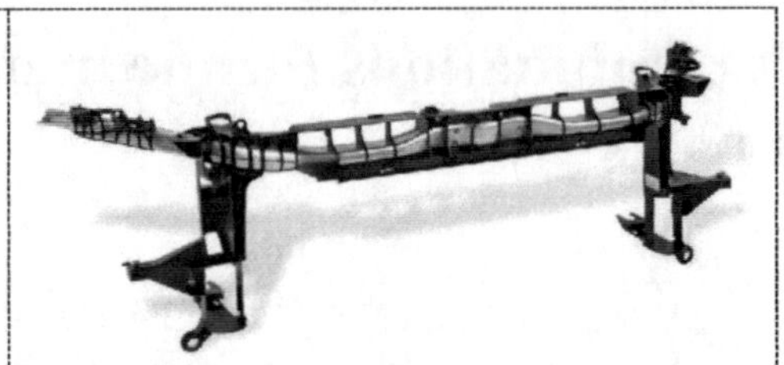

**Figure 2.** Mercedes B class front end carrier [3]

In sheet based parts, one or several – mainly deep drawn – metal component(s) are combined with an injection moulding component. This approach is for instance applied to front end carriers. In tube based metal-plastic hybrid components, a hollow profile made by hydroforming is used as the metal carrier component, which is com-pounded with an injection moulding component. These parts are employed as cockpit cross members. In the state of the art, the tube or sheet parts are produced separately from the plastic parts. The injection moulding process is applied to form the plastic parts and join the components only. The current process chains are accordingly long. As a consequence, new approaches have to be investigated to shorten these process chains.

### *2.2. Components based on hybrid materials*

Within mobility applications, most structures are assemblies of shell-like components. Conventional solutions e.g. for a car body are based on monolithic sheet metal and forming and joining technologies which are able to provide large outputs and high productivity at the same time. Following the demand for lightweight solutions and improved performance, a replacement of the pure metal sheets by hybrid laminates is intended. Aiming at high bending inertia for structural applications, the typical metal-plastic-metal composition arises. Steel and aluminium are the common options for the cover sheets, while the core layer can consist of different plastics, including optional fibre reinforcement. Different hybrid laminates are available on the market as pre-materials. The thickness of the metal skins is commonly within the range of 0.2–0.25 mm. The core layer usually exceeds that value by a factor of 2–5 and is within the range of 0.4–1.0 mm. Materials with thicker cores are also available on the market, but of less importance to forming caused by the decrease in formability.

In parallel, the requirement for high efficiency of the manufacturing processes re-mains. Accordingly, the initial approach was to use hybrid laminates as pre-material for well-established sheet forming technologies. The increased laminate thickness on the one hand allows improved mechanical properties, but on the other hand appears as an obstacle to conventional forming technologies (Figure 3).

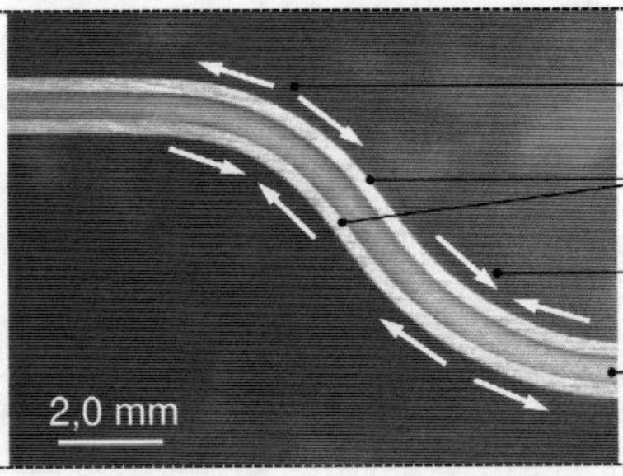


**Figure 3.** Layer composition and stress state in a deep-drawn laminate component.

New failure modes came up such as delamination of the layers due to excessive shear tension and wrinkling of the metallic cover sheets into the less compression resistant core material. As a consequence, the forming limit was significantly decreased for bending and deep drawing processes compared to monolithic blanks. To overcome these limitations, alternative process routes have to be developed. By temperature-supported forming, the fixed bond can be released and allows a shear deformation between the single layers. Another option is to form the single components first and to establish the layer bond subsequently. The solidification of the core material needs to be completed before ejecting the component from the tool. The use of this approach enables to increase the forming limit and the manufacturing of composite parts which are free from forming induced defects.

### *2.3. Lightweight structures with aluminium foam core*

Aluminium foam combines low density with high stability; it shows a good mass/stiffness ratio, and due to its cellular structure it has high energy absorption and damping capacities. Because of this diversity of properties, aluminium foam offers excellent possibilities for the production of statically and dynamically stable lightweight construction [4].

Usually, semi-finished aluminium foam products such as sandwiches and foamed profiles are produced, which can be easily assembled to large assemblies. Today, foaming procedures and manufacturing of semi-finished products such as sandwiches are well mastered. Many prototypes and also serial assemblies show the trend towards an intensive use of aluminium foam as a material for lightweight construction, either the pure material or in composites [5].

Basically, two ways of producing aluminium foam/steel sandwiches are used:

- Direct application of foam and thus bonding of the cover sheets;
- Gluing of the pre-fabricated foam core with the cover sheets.

The sandwich fabrication by direct foam application is realized in several process steps (Figure. 4):

- Production of a mixture of the metal base powder (e.g. Al) and a foaming agent (e.g. TiH2);
- Compacting the powder mixture, e.g. by the process steps of axial pressing and extrusion moulding;
- Positioning and fixing the sandwich cover sheets in the foaming mould;
- depositing the foamable pre-material, and sealing the border areas;
- Foaming of the charged pre-material in the foaming mould at the melting temperature of the aluminium alloy, and subsequent cooling.

IOP Conf. Series: Materials Science and Engineering **118** (2016) 012042 doi:10.1088/1757-899X/118/1/012042

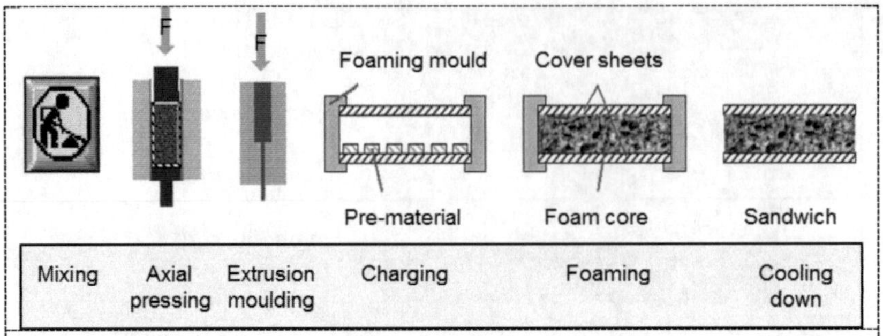


**Figure 4.** Fabrication of aluminium foam/steel sandwiches by the melting powder metallurgical procedure.

## 3. New processes for the in-situ manufacturing of metal-plastic-hybrid parts

### *3.1. Process combination of deep drawing and injection moulding*

Within the framework of the Federal Cluster of Excellence MERGE, in cooperation between the Fraunhofer IWU and the TU Chemnitz, a hybrid process was developed, combining the technologies of deep drawing, injection moulding and forming by means of the molten plastic. This combination of technologies allows in-situ manufacturing of sheet-based metal/plastic hybrid components in one die using only one hybrid process. The designed manufacturing process consists of two stages (see Figure 5). In the first step, the sheet to be formed is fed into the die and deep drawn mechanically. After deep drawing, the plastic is injected. This is used to manufacture the plastic component and furthermore for media-based forming of the metal component. The test part consists of a cup with interior ribbing and undercut, shown in Figure 6).

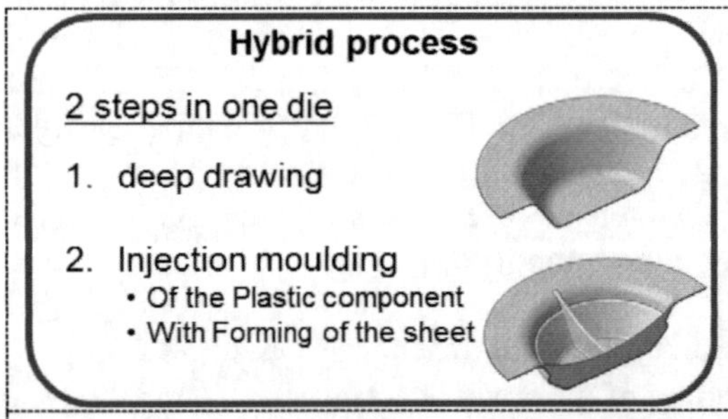


**Figure 5.** basic principle of the combined deep drawing and injection moulding process

**Figure 6.** manufactured demonstrator part

To realize deep drawing, the die was designed for utilization on a deep drawing press. A bolt-on unit was used for the injection moulding procedure (see Figure 7). The adhesion was not only achieved by the form fit that was generated by the manufactured undercut, but also by using a chemical bonding agent previously deposited on the blank surface.

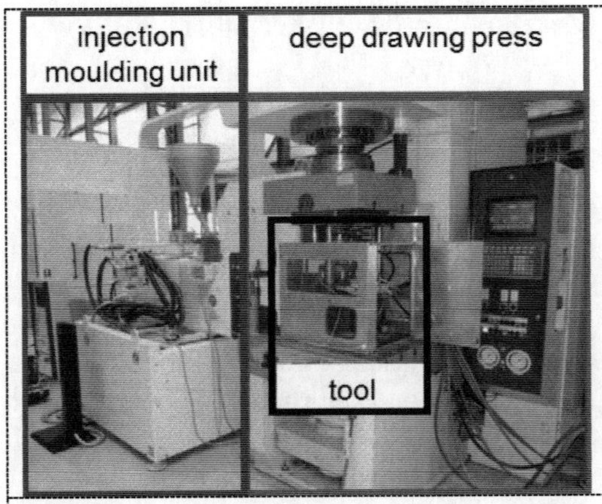


**Figure 7.** deep drawing and injection moulding: experimental setup

| Process parameter | Value |
|---|---|
| Tool temperature | 80 °C |
| Melt temperature | 265 °C |
| Plastic | Durethan BKV 60 |
| Metal | DC04, with and without bonding agent |

**Figure 8.** process parameters

*3.2. Process combination of hydroforming and injection moulding*

The second method is focused on the combination of tube-based hydroforming and injection moulding (see Figure 6). This technology entails a tube that is formed by an inside-operating medium (here: nitrogen), which generates secondary design elements. With the necessary cavities on the outer side, plastic components are applied via injection moulding. The compressed medium inside the tube ensures the necessary counter–pressure, which avoids the collapse of the tube.

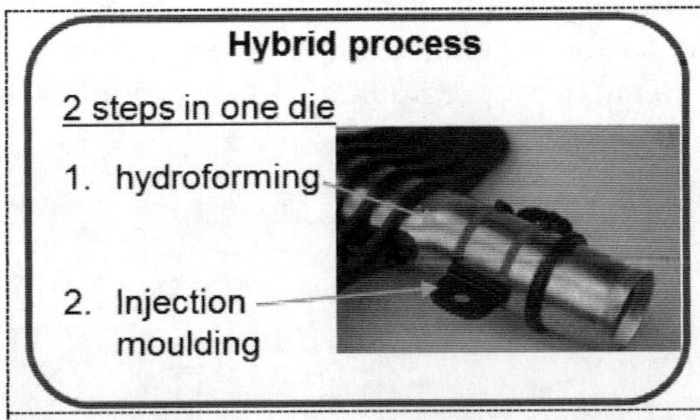


**Figure 9.** basic principle of the combined hydroforming and injection moulding process

**Figure 10.** manufactured demonstrator parts

A focal point is to evaluate the bonding strength (tensile, shear, torsional) between the metal tube and the plastic component. For this purpose, different tool inserts were realized for applying plastic elements with varying parameters. To ensure the bonding strength, mostly chemical bonding agents are used such as Vesta-melt®X1333. However, they are very cost intensive. The aim is to substitute these chemical bonding agents by surface structuring of the metal tubes. The surface structuring methods blasting (several grain sizes), knurling and laser structuring were investigated. For comparison with the preferential state of the art solution, a chemical bonding agent (Vestamelt®X1333) and tubes without any treatment were also examined. The following factors were considered in the experiments:

- Tube material: aluminium EN AW 6060; stainless steel (grade 1.4301)
- Tube temperature: room temperature, 80 °C, 140 °C, 200 °C, 250 °C
- Surface treatment: without, Vestamelt®X1333, corundum blasting, laser
- structuring, knurling

PA6 with 60 % fibre glass reinforcement (Lanxess Durethan® DP BKV 60 H2.0 EF) was used as the plastic material.

In terms of surface treatment, maximum shear strength was achieved at 19.7 N/mm², and thus the best composite bonding was achieved by using the chemical bonding agent Vestamelt®X1333. In case of the laser structured parts, the joint failed only slightly earlier at 18.1 N/mm². Even in case of the knurled tubes, it was possible to achieve good maximum shear strength values of 6.1 N/mm². In contrast, the tubes without coating showed shear strength values below 1 N/mm², and the sandblasted tubes values below 2 N/mm², scarcely providing any potential for good shear strength of the compound as a whole. An influence of the tube material on the shear strength could only be found in the laser structured and knurled test parts. In the case of the laser structured tubes, maximal shear strength of the adhesion is approx. 18.1 N/mm² (steel tubes) and approx. 12.3 N/mm² (aluminium tubes). The reason for these results is the different failure behaviour of the joint as well as the different laser structuring. During laser structuring, the surface of the tubes is melted and the molten material swirled. In the case of the steel/plastic composite components, only the joint breaks. In the case of the aluminium/ plastic joints, the failure is partially represented by the fracture of the filigree aluminium structures. Tempering of tubes does not significantly affect the bonding strength relating to shear strength.

Tensile strength of the bond between the hydroformed metal tube and the injection moulded plastic component could only be realized with the chemical bonding agent or by laser structuring of the surface. The best bonding strength with the use of Ves-tamelt®X1333 could be realized with a tube at room temperature. By using the laser structured tube, the metal components have to be heated up for better tensile strength. The bonding strength with the laser structured steel tubes with a tube temperature of 250 °C is nearly the same bonding strength compared to using Vesta-melt®X1333 with tubes at room temperature. Like the shear strength, also the tensile strength of the parts with laser treated aluminium tubes is lower than using Vesta-melt®X1333 treated tubes. Figures 11 and 12 shows the realized bonding strength values determined for components hydroformed in combination with injection moulding (metal tube and injected PA6/GF, test speed 2 mm/min, 3 repetitions of every test).

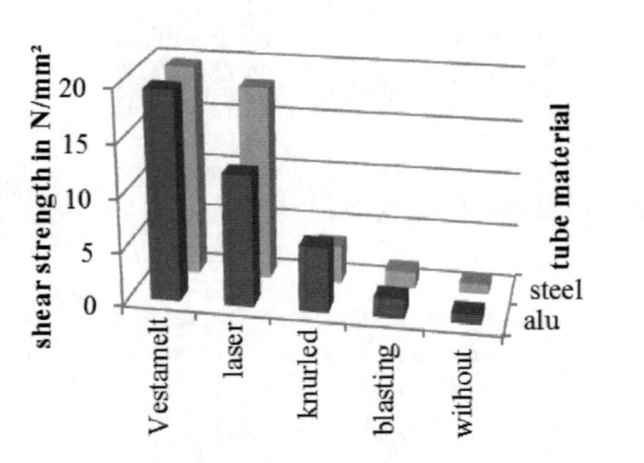


Figure 11. *shear strength*

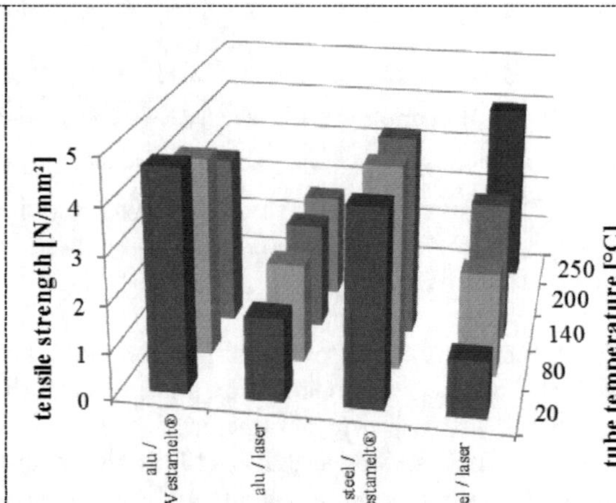


Figure 12. *tensile strength*

In the previous investigations, the authors immediately achieved outstanding shear strength and tensile strength values that are commensurable with those of the chemical bonding agent. These results were obtained by laser structuring without any additional optimizations in the first test loop. Depending on the requirements to be fulfilled by the corresponding part, a knurling treatment may be sufficient as a standalone solution. However, when using knurled tubes, the joint created cannot withstand tensile nor shear stresses. In contrast, laser structured parts can be subjected to tension without any global form fit.

## 4. Production technologies for hybrid laminates

Aiming for an increased formability and to create high-volume production compatible technologies, different process routes (Figure 13) were benchmarked.

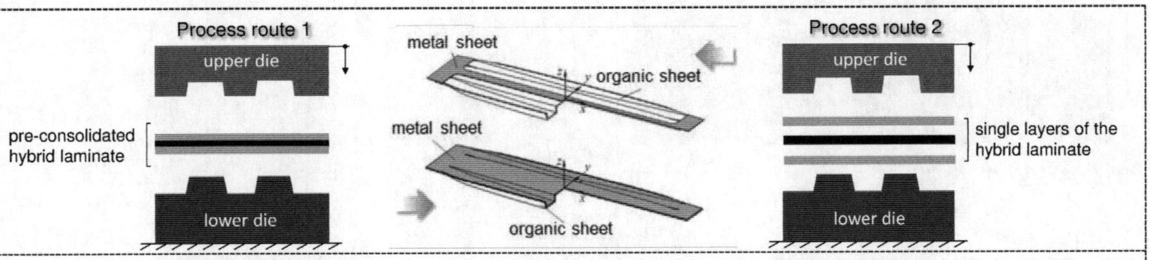


**Figure 13.** Process routes for the manufacturing of hybrid laminates, left: forming of pre-consolidated hybrid laminates, right: consolidation during forming process

Pre-consolidated hybrid laminates are used in forming process route 1 providing the following characteristics:

+ hybrid joint is already established
+ no bonding process needs to be considered
+ processing quite similar to monolithic material
- significantly lower forming limit
- forming induced defects possible

The hybridization of the single components during the forming operation as shown in process route 2 does provide the following properties:

+ sliding of layers during forming is possible
+ increased forming limit
+ prevention of forming induced defects
- bonding needs to be considered as an integrated / subsequent step
- higher cycle time due to tempering of part and tool

Within the DFG-AiF Cluster 'Process chains for complex parts made of hybrid laminates', an actively heated and cooled forming die (Figure 14) was developed with the aim to realize short cycle times. To investigate the feasible cycle times the hybrid part was equipped with thermocouples and heated up in the forming tool to 230 °C (melting temperature polymeric core material) to ensure a complete consolidation. Afterwards the forming tool was cooled down until the ejection temperature of the hybrid part from 180 °C. The heating and cooling took 3.5 min and 2.5 min respectively. The thermal behaviour concerning the heating phase is illustrated in Figure 15).

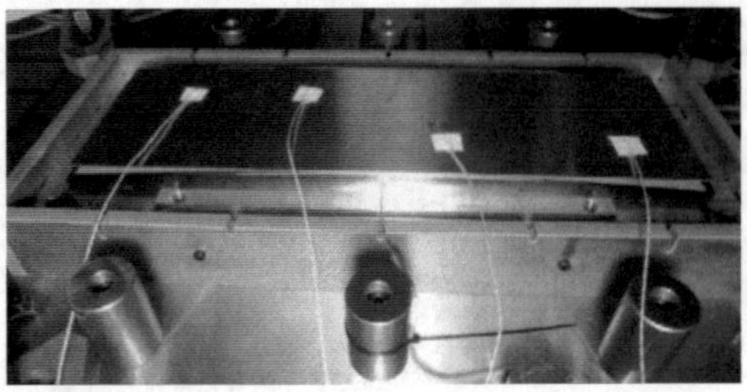

**Figure 14.** *Lower die with stacked laminate and temperature sensors*

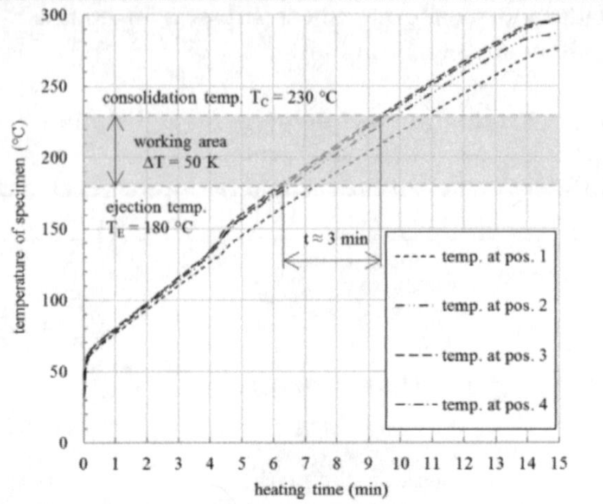


**Figure 15.** *Temperature behaviour of tool and clamped laminate in heating phase*

At a heating rate of 0.25 K/s and a cooling rate of 0.33 K/s, an overall cycle time be-low 6.5 min could be achieved. For potential mass production the cycle time has to be reduced further. For example lower cycle times could be achieved by reducing the tempered mass in combination with an adaption of the design of the temperature conditioning devices. The correlation between the critical cooling rate and thus induced residual stresses is one major focus for future research activities.

## 5. New production technology for hybrid-lightweight-structures with aluminium foam core

One focus of the Federal Cluster of Excellence MERGE is the development of light-weight sandwich composites made of metal foam and fibre-reinforced plastics (FRP). Metal foams are highly suitable for lightweight structures. The combination of metal foam with sheet metal covers as sandwich structures is already used in manifold applications [6].

In order to enhance the level of lightweight construction in metal sandwich structures, the substitution of metal cover sheets with fibre-plastic composites provides an effective approach. By adapting the thickness of core and FRP layers in combination with high performance fibres to the applied load, it is possible to realize specific structural properties with a customized design [7]. The permanent bonding of the sandwich components is realized in a pressing process. Therefore, the sandwich components need to be preheated. To increase the energy efficiency, a technology will be developed using the process heat of the foaming instead of additional preheating. For achieving a high functional density, an injection moulding process can be used to implement thermoplastic elements, e.g. stiffening rips or mounting interfaces, into these sandwich structures based on metal foam cores. One key aspect is the bonding of both components. To find an optimal combination, many experiments have been carried out to investigate the influence of various parameters, e.g. the modification of the metal foam surface and thermoplastics or the use of adhesive agents. To compare different surfaces and surface treatments, tensile tests were executed as part of this investigation. For the tensile tests sample plates of 170 mm x 170 mm were foamed. As an additional step, some of them were pre-treated by the SACO-process (Figure 17). It is a sandblasting process with a chemically modified blasting material to create a certain roughness to increase the bonding surface and additionally coat the surface [8].

As an alternative to additional surface treatment with sand blasting, a structure was implemented on further samples to increase the surface area. Different structures are feasible. The first tested structure is displayed in Figure 16. The structure of the foaming tool is very well reproduced by the foaming process. It is a lattice structure with a distance of approx. 1 mm between 2 peeks and the peaks have a height of approx. 230 µm. Afterwards the samples were compression moulded with the cover plates – polyamide 6 with glass fibre reinforcement. Out of these combinations smaller samples of 50 mm x 50 mm were cut by waterjet. The samples were glued onto adapters for the tensile testing machine. Then the samples were tested with an extension rate of 0.5 mm/min at room temperature based on DIN 53292: "testing of sandwiches; tensile test perpendicular to the faces".

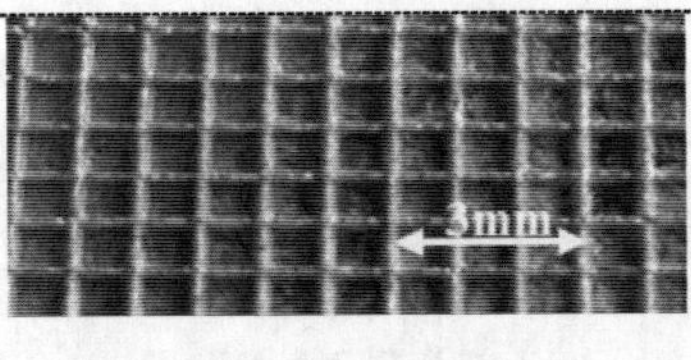


**Figure 16.** *Foam with structured aluminium skin,*

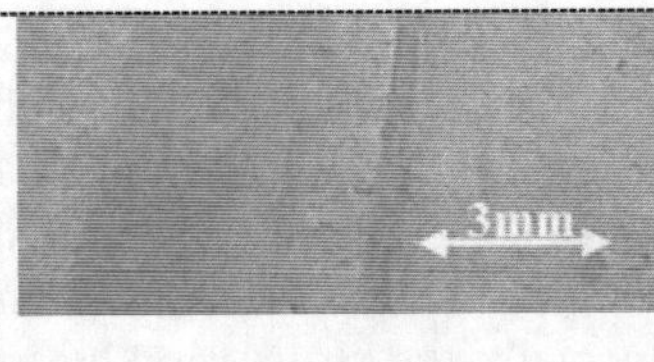


**Figure 17.** *SACO-treated foam*

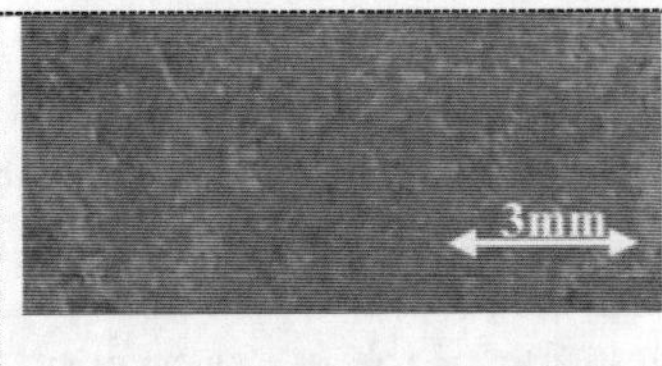


**Figure 18.** *untreated foam*

The test results show that the SACO-treated samples sustain a very good average test force of 10.8 kN. A couple of samples failed in the foam core, which was due to a significantly lower density compared to the other samples. These values were left out of the calculation of the average force.

Looking at the valid samples, the failure occurred not in the sample itself, but the glued test adapters detached. Therefore the mechanical strength of the samples themselves should clearly exceed the joint strength. More experiments with a better adhesive have to be executed to derive a more accurate bonding strength of the samples. Nevertheless the aim is to have a bonding strength higher than the tensile strength of the foam core. For densities up to 1 g/cm³ this was already achieved. For most of the applications, lower foam densities are used. The structured samples break at an average force of 0.9 kN. The untreated samples (Figure 18) reach an average force of 2.6 kN, Figure 19.

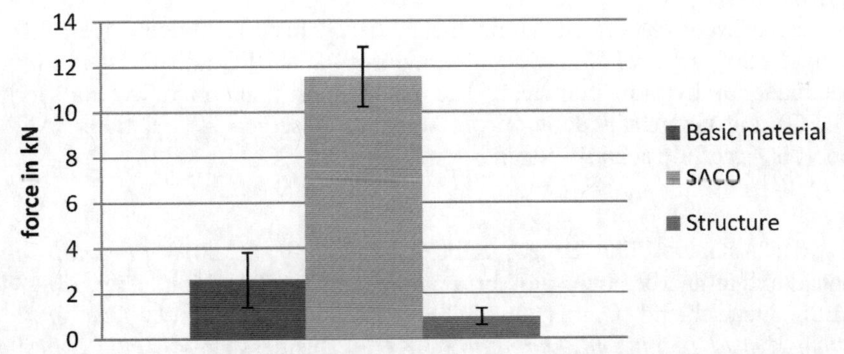


**Figure 19.** Average of the maximum resulting forces of the tensile tests

In comparison, the SACO-processed samples reach the highest values, but for a series production of this kind of sandwich parts a route without additional steps such as sandblasting was clearly preferred. Despite of the increase of the surface area to 140 %, the maximum force of the structured samples reached lower forces in comparison to the basic aluminium foam. One reason for the poorer results of the structured samples could be the entrapped air in the "pockets" between the ridges which prevent good bonding of the components. This assumption will be reviewed by microscopy in the next step. To provide better bonding, finer structures should be tested, taking into account the achievable reproducibility of such structures in the foaming process. Furthermore an additional layer of matrix material e.g. granulate is suggested to secure that no ambient air gets entrapped.

## 6. Summary and outlook

Until now, the manufacturing of metal-plastic hybrid components has still been very labour and cost intensive due to the separate production of metal and plastic components. For this reason, integrated processes are developed to enable the production of affordable hybrid components.

In the Federal Cluster of Excellence »Merge – Technologies for multifunctional lightweight structures« the focus lies on process combinations for manufacturing of sheet metal based and tube based hybrid components. The developed concept for the production of sheet metal-based hybrid components is a combination of deep drawing and injection moulding. Furthermore, the melt is used for media-based forming. For the production of tube-based hybrid components, the combination of hydroforming and injection moulding is following a similar approach.

The adhesion between the two components is very important in order to use the high potential for such a lightweight construction. In industry, the bond strength is mostly realized by a chemical bonding agent. The aim of the current work is the replacement of chemical bonding agents by surface structuring of the metal component. Especially the structuring by laser showed enormous potential in preliminary investigations.

Within the DFG–AiF Cluster 'Process chains for complex parts made of hybrid laminates' tempered forming strategies for single or even readily joined laminate com-pounds are elaborated. A suitable design of the interface between metallic and polymeric component dominates the research activities in this field. The production of the composite and the simultaneously occurring forming is determined as a complex hybrid process. To meet the demands of mass production, the different process routes are compared and evaluated in terms of forming limits as well as regarding the possible output quantity. For profound understanding and systematic improvement of the layer adhesion, the adhesive mechanisms are considered at different levels of scale and targeted through customized semi-finished properties and process parameters.

In addition using conventional materials, further research priorities are the development and efficient production of composites of aluminium foam and FRP layers with a high degree of lightweight construction with high functional integration.

## Acknowledgments

The presented results were elaborated within the Federal Cluster of Excellence EXC 1075 "Merge Technologies for Multifunctional Lightweight Structures" and DFG-AiF Cluster 'Process chains for complex parts made of hybrid laminates'. The projects are supported by the German Research Foundation (DFG) and German Federation of Industrial Research Associations (AiF) respectively. Financial support is gratefully acknowledged.

## Literature

[1] Abibe A B, Amancio-Filho S, dos Santos J, Hage Jr E: 2013 *Materials and Design 46* Mechanical and failure behavior of hybrid polymer-metal staked joints, pp. 338–347

[2] LANXESS Deutschland GmbH 06.08.2007 *Anwendungsbeispiel: Das erste Frontend in Aluminium-Hybrid-Technik mit Durethan® BKV 30* Business Unit SCP TI 2006-047 DE

[3] http://www.plasticsnews.com/article/20131015/NEWS/131019961, status 20.05.2014

[4] Neugebauer R, Lies C, Hohlfeld J, Hipke T 2007 *Production Engineering Volume 1* Adhesion in sandwiches with aluminium foam core pp 271-278.

[5] Hipke T, Lange G, Poss R *2007 Taschenbuch für Aluminiumschäume. 1st edn.* Aluminium-Verlag Marketing & Kommunikation GmbH pp 24–25

[6] Nestler D, Jung H, Trautmann M, Wielage B, Wagner G, Drossel W-G, Drebenstedt C, Hipke T 2015 *Materials Science Forum Volumes 825–826* New sandwich structures consisting of aluminium foam and thermoplastic hybrid laminate top layers pp 797–805

[7] Drebenstedt C, Lies C 2014 *Ingenieurspiegel 3* Herstellung und Einsatzpotenzial hybrider Leichtbaustrukturen mit Aluminiumschaumkern. pp 60–61

[8] Habenicht G 1973 *Kleben 6th edition*, Springer, Berlin Heidelberg New York

**Institute of Physics**
Dirac House, Temple Back
Bristol BS1 6BE UK

CURRAN ASSOCIATES INC.
proceedings.com

ISSN: 1757-8981
ISBN 978-1-5108-2233-7
9 781510 822337